THE PLANTS OF MOUNT KINABALU

2. ORCHIDS

THE PLANTS OF MOUNT KINABALU

2. ORCHIDS

JEFFREY J. WOOD

REED S. BEAMAN

JOHN H. BEAMAN

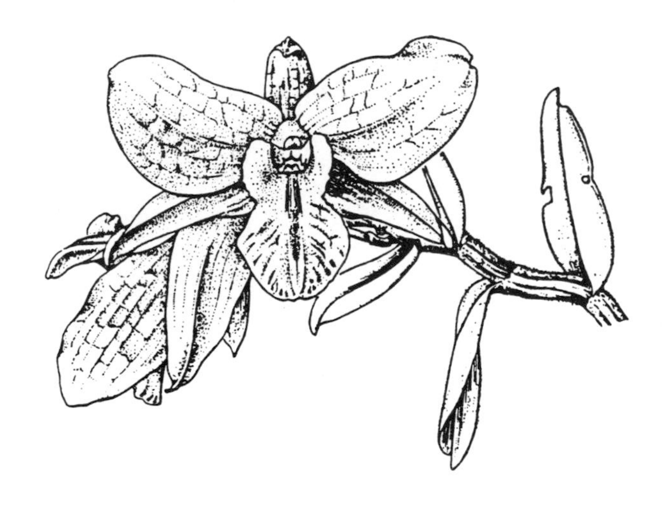

Royal Botanic Gardens, Kew

Published with the support of Sabah Parks, the TOBU Department Store, Ikebukuro, Tokyo, Japan, the Foundation for the Preservation and Study of Wild Orchids, Zürich, Switzerland, and the American Orchid Society.

First published 1993

ISBN 0 947643 46 X

Addresses of authors:

Jeffrey J. Wood, The Herbarium, Royal Botanic Gardens, Kew, Richmond, Surrey, TW9 3AB, England
Reed S. Beaman, 820 NE 5th Ave., Gainesville, Florida 32601, U.S.A.
John H. Beaman, Department of Botany & Plant Pathology, Michigan State University, East Lansing, Michigan 48824, U.S.A.

General Editor of Series: J. M. Lock

Cover Design by Media Resources, RBG, Kew. The terrestrial orchid on the cover is *Nephelaphyllum pulchrum*; the epiphyte is *Renanthera bella*. Both are adapted from a poster by Chan Chew Lun. The line drawing of the flower on the title page is *Dendrobium spectatissimum* by Chan Chew Lun. The book was typeset in New Baskerville® with Microsoft Word™ for Windows™ by R. S. Beaman and J. H. Beaman.

Printed & Bound in Great Britain by Whitstable Litho, Whitstable, Kent

In memory of

Oakes Ames (1874–1950)

and

Cedric Errol Carr (1892–1936)

CONTENTS

LIST OF FIGURES AND TABLES

Fig. 1. Oakes Ames. Painted by Blanche Ames, his wife, in 1942. Courtesy of the Orchid Herbarium of Oakes Ames, Harvard University.

Fig. 2. Cedric Errol Carr. Photograph by Professor R. E. Holttum. Courtesy of Mr. H. M. Burkill.

INTRODUCTION

The orchids are considered by many botanists to constitute the largest family of flowering plants, with an estimated 15,000 to 30,000 species. Their only rival in size is the aster or daisy family, Asteraceae (Compositae), which may have about the same number of species. In spite of this great diversity, the Orchidaceae are not a family upon which human survival depends. Few orchids are used as food plants, none provides shelter, their medicinal sources may be only of psychological value and only one, the vanilla plant, has economic importance outside the enormous enthusiast and horticultural interest. Nevertheless, orchids have many other remarkable features that render them among the most fascinating of plant families.

It is not possible to say that a particular group of plants is the most advanced, but the orchids surely represent one apex of flowering plant evolution. Not a single species is woody, generally considered to be a primitive feature. Orchids occur throughout the world from the hottest lowland tropics to the arctic tundra, although they are far more abundant in the tropics. In pollination they have some of the most specialised mechanisms found in any plants, deceiving and even trapping insects that transfer their pollen. In most of them the pollen is united into compact bundles, the pollinia, a character that occurs in only one other completely unrelated family, the milkweeds (Asclepiadaceae). Many of them produce vast quantities of dust-like seeds, even numbering into the millions, in a single fruit. The seeds lack endosperm; a fungal symbiont is therefore required for germination. All species have a close relationship with mycorrhizal fungi, at least in their seedling stages, and depend on the fungi to absorb nutrients. A high percentage of the orchid family, the most evolutionarily advanced, are epiphytes, using other plants for support although not obtaining nourishment directly from the host as would a parasite. A relatively limited number are saprophytes lacking chlorophyll. Two species in Australia even live and flower underground.

CONSERVATION

The extraordinary beauty and diversity of orchid flowers is a prime factor in their interest to people. The enormous variation in size, shape, colour and odour of the flowers gives them a fascination few other plants can match. In 19th century Europe many of the tropical species attained unrivaled popularity that made them status symbols of the nobility and the wealthy. In our own time a new surge of interest throughout the world has developed, with all levels of society participating. The demand, however, has become so great and remote regions where orchids flourish so accessible that many species are now in grave danger of extinction in the wild.

Mount Kinabalu is one of the formerly remote areas that is now experiencing great tourist interest with over of 200,000 visitors a year. Many people come to Kinabalu to see the orchids, and some, unfortunately, have come to steal them (Pain 1989). As a result it is no longer possible to find some of the particularly precious species, because they have been stripped from their former habitats. Lamb and Chan (1978) note that in the 1960s *Phalaenopsis amabilis,* once very common below Kundasang, was stripped and sold at roadside stalls and is now rarely seen. Even worse, however, has been the recent commercial stripping of *Paphiopedilum* species and the burning of much of their habitat.

All orchids have some legal protection in their native habitats, because they are listed in the Convention on International Trade in Endangered Species (CITES). This convention, now ratified by over 100 countries, prohibits virtually all international transport of the most endangered species (Appendix 1 species) and requires permits for all others (Appendix 2 species). The rules have been difficult to enforce, however, because of the problems in discovering and identifying illegally transported plants. In Kinabalu Park no plants can be collected without a permit.

As noted above, illegal collecting has sometimes occurred, but many of the Kinabalu orchids are even more endangered because of destruction of their habitats, a result from the increasing human population around the mountain, mining activities, demands for recreational areas (e.g. the golf course on the Pinosuk Plateau), and particularly the establishment of permanent agricultural plots for temperate vegetables. Beaman and Beaman (1990) listed 18 most important locations on Kinabalu on the basis of numbers of species that have been recorded from those localities. The natural vegetation of half of these areas is now destroyed. In view of the rapid and apparently inexorable destruction of forests on the lower slopes of Mount Kinabalu in the past 25 years, much of which formerly had Park status, it is questionable as to whether much, if any, of the original vegetation below 2000 metres will remain for long. The extreme fragility and endangered status of the whole Kinabalu flora are emphasised in that nearly 40 percent of the species are known from just one locality and about 25 percent have been collected only once.

The present study is offered in the hope that documentation of the great biological diversity of Kinabalu orchids will aid in their protection. We realise that making this information publicly available may open the way to further exploitation. Nevertheless, we are of the opinion that provision of the information may better serve the interests of conservation than would the secrecy and deception that has sometimes been practiced in the past. For a few species that are in high horticultural demand and cannot withstand further depredation, exact locality data have been withheld.

Historical Considerations

This treatment of the orchids of Mount Kinabalu continues the series that was initiated with the publication of the pteridophytes (Parris et al. 1992) of that area. The Orchidaceae are by a wide margin the largest angiosperm family on the mountain, and, like the pteridophytes, are appropriate to publish as an independent part of the total inventory of the flora. The general plan of the project was described by Beaman and Regalado (1989). It has been possible to extend the scope of the work for the Orchidaceae because of a parallel project by J. J. Wood and P. Cribb to prepare a checklist of the orchids of Borneo. Thus we have added a key to the genera, generic descriptions, citation of all basionyms (even though their type is not necessarily from Mount Kinabalu), and information on the general distribution of species. Also included are descriptions of ten new taxa; seven new combinations are made.

The Orchids of Mount Kinabalu by Oakes Ames and Charles Schweinfurth (1920) was the first major account of a single plant family on the mountain. As the subtitle of that work indicates, it was based chiefly on the collections of the Reverend Joseph Clemens, who, with his wife Mary Strong Clemens, spent about six weeks there from October 28 to December 12, 1915. In the preface Ames notes that "Chaplain Clemens, to keep his interest in the expedition constantly stimulated, decided to give his undivided attention to a single group of plants. He chose the Orchidaceae. As a result of his efforts, one hundred and fifty-five orchid species in a condition suitable for identification were secured. Of this number one hundred species proved to be new."

Ames noted that Joseph Clemens's collections "were gathered chiefly on Marei Parei Spur, near Kiau, at and around Lobong Cave, and in the neighborhood of Kamborangah. A few species were collected at Pakka and on the Gurulau Spur." The specimen labels and species descriptions frequently do not indicate elevations of the collection localities, but Ames summarizes this as follow: Kiau 3080 feet, Lobong 4790 feet, Gurulau Spur 5000–5500 feet, Marei Parei Spur 5000–7000 feet, Kamborangah 7040 feet, Pakka 9790 feet.

The enumeration of Kinabalu orchids by Ames and Schweinfurth also included a small collection made in 1916 by George Haslam and a complete listing of all previously reported species. Among these were 24 species in 12 genera collected by G. D. Haviland in 1892, identified by H. N. Ridley and published in Stapf's (1894) monumental paper *On the flora of Mount Kinabalu.* In 1910 Lilian Gibbs collected 41 species representing 22 genera. These were identified by Rolfe and reported in Gibbs (1914) *A contribution to the flora and plant formations of Mount Kinabalu and the highlands of British North Borneo.*

The Ames and Schweinfurth enumeration listed 52 genera and 222 species. The genera *Apostasia* and *Neuwiedia* were excluded on the grounds that they seemed to constitute a distinct family. Ames indicated that it was highly probable that intensive explorations throughout the year would bring to light many additions and confirm his belief "that Mount Kinabalu will prove to be one of the richest mountains in the world in the diversity and interest of its orchid flora."

Ames and Schweinfurth's account was an outstanding accomplishment for its time, with a surprisingly modern aspect. Synonymy was provided for the species, new taxa were described at length, specimens were fully and precisely cited, and type specimens were consistently indicated. Among the 122 new taxa described there, 74 are currently recognised, 12 have been transferred to other genera or reduced to varieties, and 19 have become synonyms.

In addition to the Ames and Schweinfurth contribution to the orchid flora of Mount Kinabalu, one other particularly important publication on this family for the area is that of C. E. Carr (1935) entitled *Two collections of orchids from British North Borneo, Part I.* Unfortunately, subsequent parts were not published, because Carr met an untimely end from blackwater fever while collecting in New Guinea in 1936 (van Steenis-Kruseman 1950). Carr enumerated 137 species, belonging to 40 genera; 39 species were described as new and 41 more were new to the flora of British North Borneo.

In an editorial note at the beginning of the Carr paper it is indicated "that in six months Mr. Carr collected flowering specimens of approximately 700 species." This statement must be incorrect, however, as we can account for a total of only 671 collections (not species) obtained by Carr. Most likely the intent was to say that Carr collected approximately 700 flowering specimens.

Carr had a keen ability to find orchids in the field and was very thorough in working up the descriptions of new species. In contrast to some other collectors, particularly the Clemenses, he also appears to have been very careful in recording locality and elevation data. Among the 39 species described as new in his paper only two are no longer recognised. One of his taxa has been transferred to a different genus and another lowered to varietal level.

In most instances Carr unfortunately did not designate type specimens for the taxa described, although it appears that he may have intended for one of his own collections to serve as the type. In his paper, types are explicitly designated for only three species, namely *Coelogyne craticulaelabris, Dendrochilum angustilobum,* and *D. lacteum.* In the case of *Dendrochilum pterogyne* Carr listed two of his collections, then discussed how the second cited differed from the type, which presumably was the first one listed. It seems likely that in all cases he intended the first cited specimen to be the type. Carr sometimes indicated on a specimen that it was a type. In such cases we have considered that particular specimen in SING to be the holotype. When more than one specimen was involved in the description of a species and there is no clear designation of the type, we have regarded the several specimens as syntypes.

The first non-technical account discussing the orchid family on Mount Kinabalu was published by Lamb and Chan (1978) as a chapter in the important book, *Kinabalu: Summit of Borneo.* A second edition of this book is now nearly ready for publication, and we have been able to examine the revised chapter on orchids by Lamb and Chan (in press).

Our own enumeration lists 686 species of Orchidaceae from Mount Kinabalu. These belong to 121 genera. An additional eight infraspecific taxa are included. Seventeen species are added in the Appendix. Thus the total number of orchid taxa now recognised from Kinabalu is 711. The Enumeration includes 596 fully named taxa; 100 more are incompletely identified. In some cases these are indicated as 'sp. 1', 'sp. 2', etc. In other instances they are designated as 'aff.' some other species. Seven taxa are questionably identified with 'cf.' separating the generic name and specific epithet.

New Taxa and New Generic Records

Ten new taxa described in the Enumeration are the following: *Appendicula fractiflexa* J. J. Wood, *Dendrobium beamanianum* J. J. Wood & A. Lamb, *D. hamaticalcar* J. J. Wood & Dauncey, *Liparis aurantiorbiculata* J. J. Wood & A. Lamb, *Pantlingia lamrii* J. J. Wood & C. L. Chan, *Phaius pauciflorus* (Blume) Blume subsp. *sabahensis* J. J. Wood & A. Lamb, *Porpax borneensis* J. J. Wood & A. Lamb, *Robiquetia crockerensis* J. J. Wood & A. Lamb, *R. pinosukensis* J. J. Wood & A. Lamb and *Trichoglottis collenetteae* J. J. Wood, C. L. Chan & A. Lamb. It should be noted that for the new taxa all specimens from which they are known are cited, not just those from Mount Kinabalu.

The following seven new combinations are made in the Enumeration: *Dilochia rigida* (Ridl.) J. J. Wood, *Platanthera borneensis* (Ridl.) J. J. Wood, *Robiquetia transversisaccata* (Ames & C. Schweinf.) J. J. Wood, *Trichotosia brevipedunculata* (Ames & C. Schweinf.) J. J. Wood, *T. mollicaulis* (Ames & C. Schweinf.) J. J. Wood, *T. pilosissima* (Rolfe) J. J. Wood and *Vanda hastifera* Rchb. f. var. *gibbsiae* (Rolfe) P. J. Cribb.

Eight new taxa included here (*Acanthephippium lilacinum, Bulbophyllum pugilanthum, Ceratochilus jiewhoeii, Dendrobium maraiparense, D. piranha, Ornithochilus difformis* var. *kinabaluensis, Phaius baconii, P. reflexipetalus*) will be formally described in *Orchids of Borneo,* Vol. 1. All of these except *Bulbophyllum pugilanthum* and *Ceratochilus jiewhoeii* are endemic to Mount Kinabalu.

Seven of the genera recognised here are reported for the first time from Borneo. These are: *Bogoria, Ceratochilus, Chamaeanthus, Microtatorchis, Pantlingia, Porpax* and *Porrorhachis. Ceratochilus, Pantlingia* and *Porpax* are each represented by a distinct new species in Borneo, one of which, *Pantlingia lamrii,* so far is known only from Mount Kinabalu. Inasmuch as 146 orchid genera are now known from all of Borneo (Wood, in prep.), the 121 genera represented on Mount Kinabalu constitute a high percentage of the total.

Future Research

In spite of the intensive work that has been devoted to the Orchidaceae of Mount Kinabalu at Kew and elsewhere over the past several years, there is still a residual accumulation of specimens that has not yet been determined. Some of these lack flowers and may be difficult or impossible to identify, but others are simply in need of further study. We considered excluding the undetermined specimens from the present enumeration, but have opted for their inclusion, partly to demonstrate where work on this family is still needed. Thus we list 373 collections of unidentified material in 35 genera. The majority of these (216) are Carr collections in SING; 100 are Clemens collections in BM, of which 43 are *Eria.* Clearly the last word has not yet been said on the orchids of Mount Kinabalu.

A number of genera are represented in the Kinabalu flora in which little work has been done with the Bornean species. Therefore, the delimitation of species is still very uncertain. The poorly known genera include *Calanthe* (particularly section *Styloglossum*), *Ceratostylis, Cleisostoma, Eria, Flickingeria,*

Liparis, Malaxis, Malleola, Oberonia, Phreatia, Poaephyllum, Taeniophyllum, Thrixspermum, Trichoglottis, Trichotosia and *Tropidia*.

METHODS

The overall concept of the botanical inventory of Mount Kinabalu as a taxonomic database was outlined by Beaman and Regalado (1989). An integrated system of computer programs used for data editing and printing the orchid enumeration (as well as those for other parts of the project) was written in the dBASE IV programming language by Reed Beaman. A management program, KINABALU, allows for accessing any aspect of the database through a menu system. Seven principal relational data files were employed. Two of these files contain data on specimens including types. Taxonomic, nomenclatural, geographical, and bibliographic information is linked from other files. Various procedures permit creating and editing a database. Menus facilitate inputting and editing specimen and taxon data, globally replacing various expressions such as changing an author's name or abbreviation, indexing and querying the database, computing a summary of elevation ranges for taxa, numbering taxa, making an index to numbered collections, and printing enumerations of all taxa in the database or selected families or genera.

One objective of the botanical inventory of Mount Kinabalu has been to examine critically all specimens upon which the Enumeration is based. For the orchids these include nearly 5000 specimens, of which 1900 are at K, 1689 at BM, 738 at SING, and 430 at AMES. The orchids represent a special situation in that two very important collection sets, those at AMES and SING, have not been readily accessible to us. The AMES specimens were competently studied by Ames and Schweinfurth, and many of those at SING (and some Clemens collections in BM) were critically determined by Carr. We have therefore decided in most cases to accept the determinations given in their publications, and cite those specimens even though we have not seen all of them. When necessary, we have updated the nomenclature Ames and Schweinfurth and Carr used. In the case of some records for *Bulbophyllum*, we have relied on data for specimens in L provided by J. J. Vermeulen.

Our former collaborator, J. C. Regalado, Jr., listed almost all Kinabalu orchid specimens in SING, thus providing an important record of collections in that herbarium, mostly material gathered by Carr. Some specimens collected by the Clemenses are in SING, but the majority cited by Carr (1935) were not found there. As the introductory editorial note in his paper indicates, however, he studied the Clemens specimens at the courtesy of the Keeper of the BM. We therefore have recorded Clemens specimens from SING only if they were actually listed by Regalado. We have found almost all the others Carr cited in BM.

Through a set of microfiche photographs of types in the AMES herbarium (Anonymous 1989) we have verified the presence there of all but one type cited by Ames and Schweinfurth. The one type not seen in the microfiches (*Clemens 113A, Bulbophyllum caudatisepalum*) is represented by a photograph at K. All types recorded in AMES seen in the microfiches are designated with 'mf' after the herbarium symbol where types are cited.

We have been reluctant to include in the Enumeration any taxa that could not be documented by specimen records. However, there has also been the desire to make the list as complete as possible. Thus we have included the following 12 taxa for which there are only sight records as indicated in the Enumeration: *Adenoncos* sp. 1, *Bulbophyllum turgidum, Coelogyne foerstermannii, C. planiscapa var. planiscapa, Cymbidium atropurpureum, Dendrobium sanguinolentum, Galeola nudifolia, Goodyera reticulata, Grammatophyllum speciosum, Renanthera matutina, Stereosandra javanica, Thrixspermum pensile.* Four additional taxa are included on the basis of photographs or drawings only: *Bogoria raciborskii, Chamaeanthus brachystachys, Trichoglottis lanceolaria, Vanilla pilifera.*

The location of the type specimens upon which Ridley's species described in Stapf (1894) were based is problematical. In that work 20 new species of orchids were described, all based on Haviland collections. We recognise 11 of these taxa under the name originally published for them. Six others have been transferred to different genera, one (*Eria major*) is under a name attributed by Stapf to Ridley, and two are in synonymy. No types (holotypes or isotypes) have been found for five names, i.e. *Bulbophyllum breviflorum, Calanthe ovalifolia, Eria angustifolia, Nephelaphyllum latilabre* (=*N. pulchrum*) and *Platanthera borneensis.*

Ridley was employed at the Singapore Botanic Gardens when the Stapf paper was published, but we are aware of only six of the relevant types there. Vermeulen (1991) considered *Haviland 1099* in SING to be the holotype of *Bulbophyllum montense.* The specimens there of *Coelogyne papillosa (Haviland 1098), Dendrochilum stachyodes (Haviland 1097)* and *Eria grandis (Haviland 1157)* appear to be holotypes. We are unsure about the status of the other two, i.e. *Haviland 1100 (Bulbophyllum coriaceum)* and *Haviland s.n. (Dendrochilum corrugatum).* Duplicate specimens for all these names are at K, except for *Dendrochilum corrugatum,* but it is not clear that the K specimens were actually used by Ridley in drawing up the descriptions. Ames (in Ames & Schweinfurth 1920: xiii) noted that the type of *Dendrochilum conopseum* is preserved in the Sarawak Museum (those collections now in SAR). A photograph of that specimen is in K. We attempted without success to confirm its presence at this time in SAR, and to learn if other Ridley types might also be there. Ridley drawings of the following species are in K: *Appendicula congesta, Bulbophyllum altispex (= B. mutabile var. mutabile), B. breviflorum, B. coriaceum, B. montense, B. montigenum (= B. flavescens)* and *Dendrochilum grandiflorum.*

For purposes of brevity, and especially to facilitate effective queries of the database, we have used standardised locality data in the Enumeration. A list of the standardised locality names, with their geographical coordinates, is provided by Beaman et al. (in press). It may be noted that some locality names are rather different from those on specimens and in the literature, because we have attempted to use spellings established in accordance with the modern Dusun language. Thus, the Dahobang River becomes the Tahubang, the Pinokok River becomes the Tinekuk, the Columbon (or Colombon) is the Kilembun, Bukit Hampuan is Hempuen Hill, and Kamborangah (and numerous other spellings) is standardised as Kemburongoh. For the orchids on Mount Kinabalu we have recorded 172 standard localities.

In the specimen citations, elevation data on labels given in feet have been converted to metres, and all elevations are rounded to the nearest 100 m. The stated elevations of 6000–13,500 ft for Clemens specimens from 'Upper Kinabalu' have been ignored. On field labels (see below under Collections: Consideration of the Clemens Collections) for these collections, a specific locality such as 'Lumu' or 'Kamboranga' is often indicated. When we have been able to associate specific localities with the numbers accompanying 'Upper Kinabalu' specimens, we have used the specific localities even though specimens in different herbaria may lack those data.

Elevation data are summarised for all taxa for which these data were available on specimen labels. The elevation range indicated for taxa is based on the lowest and highest elevations recorded (whether in feet or metres) for specimens of each taxon and rounded to the nearest 100 m. In some taxa certain specimens have no elevation data while others do. It may be apparent from the locality data that a particular taxon must occur at lower or higher elevations than indicated by the elevation printed. However, we have resisted the temptation to provide elevation ranges for taxa when the specimens do not provide this information, with the result that the elevations stated are sometimes misleading or incorrect.

In order to produce an index of determined specimens, it is necessary to have the taxa numbered. In the treatment of pteridophytes for the Kinabalu inventory, 30 families were recognised and numbered in alphabetical order with the fern allies first, followed by the ferns. Four gymnosperm families occur in the Kinabalu flora, and they are numbered 31 through 34. The monocotyledons comprise 28 families and start with the Araceae as family 35; in alphabetical order the Orchidaceae are family 55, the number used in the present numbering system. Families are recognized according to the list of Brummitt (1992).

Taxonomic treatments that the Enumeration follows for particular genera are indicated at the beginning of those genera, when such are available. Type specimens are cited only for taxa described from Mount Kinabalu. The synonymy includes names based on types from Mount Kinabalu, with the addition of basionyms for all recognised taxa. Names of authors of taxa are abbreviated in accordance with the standardised list of author abbreviations (Brummitt & Powell 1992).

VEGETATION

A vegetation map of Mount Kinabalu Park has been published recently by Kitayama (1991) in which 21 vegetation map units are recognised. He indicated that diagnostic canopy tree species could be used to distinguish vegetation zones. The species were mutually exclusive in distribution and were correlated with altitude. The upper boundary of lowland rainforest, where the majority of emergent trees (mostly Dipterocarpaceae) disappear from the canopy, is at 1200 m. The upper limit of lower montane forest is at between 2000 and 2350 m, and that of upper montane forest at between 2800 and 3000 m, the latter particularly marked by the upper limit of *Lithocarpus havilandii*. Above this level is a lower-subalpine coniferous forest dominated by *Dacrycarpus kinabaluensis* and two or three angiosperm species. The upper limit of this forest is at about 3400 m and corresponds to the closed forest line. A fragmented upper-subalpine forest extends above this level to the tree line at about 3700 m. Above this level is a zone of alpine rock-desert with scattered communities of alpine scrub. Kitayama suggests that the tree line may coincide with the lowest elevation where nocturnal ground frost is frequent. Not surprisingly, great variations were found in dominance type, species composition, and forest structure within each zone. These were attributed to altitudinal temperature effects, soil nutrient status in relation to topography (particularly ridge and valley differences), and slope aspect.

Kitayama considered the lowland forest to be six-layered with emergent trees characteristic and undergrowth sparse. Lower montane forest is five-layered and does not have emergents. Upper montane forest has a dense herb layer. The lower subalpine forest is sparser in undergrowth and lower in height.

Serpentine vegetation was noted by Kitayama to be strikingly different from that of surrounding forests on non-ultramafic substrates. He indicated that there are at least three altitudinally recognisable subdivisions in the woody serpentine vegetation. These are vegetation with *Tristania elliptica* dominance, that with *Leptospermum javanicum-Tristania* dominance, and that with *Leptospermum recurvum-Dacrydium gibbsiae* dominance. In addition, there are graminoid ultramafic communities at Marai Parai and on the summit of Mt. Tembuyuken. Kitayama did not comment on casuarina-dominated forests (i.e. those with *Gymnostoma sumatranum* or *Ceuthostoma terminale*), which we have found to be particularly distinctive markers of lower-elevation ultramafic vegetation.

In designating the habitat of orchids included in this treatment we have used elevation as a primary basis for recognition of forest types. 'Hill forest' is a much used term for the upper part of the dipterocarp-dominated lowland forest, and we have applied this to habitats between the levels of about 600 and 1200 m. Below 500–600 m is what Kitayama refers to as 'substituted vegetation' that we have designated as lowlands. Our designation of lower montane forest is in the elevation range of about 1200–2200 m, and corresponds essentially to that indicated by Kitayama. We have used the designation of upper montane forest for all forest vegetation above the lower montane forest. Additional considerations on the vegetation of Mount Kinabalu are to be found especially in Stapf (1894), Gibbs (1914) and other references cited by Kitayama. It should be noted that relatively few specimens have any indication of vegetation type beyond such vague terms as 'jungle'.

COLLECTIONS

The present study is based on over ten times the number of specimens available to Ames and Schweinfurth, i.e. about 3860 collections including nearly 5000 specimens. These have been obtained by 93 collectors or collecting teams and are deposited in 14 herbaria. Most of the specimens examined, however, are in BM, K and SING, particularly BM and K, which have provided the primary resource for this inventory of the Kinabalu Orchidaceae.

Collectors or collecting teams from whom we have recorded 20 or more specimens are listed below with the approximate date of their time in the field and number of collections made and taxa obtained.

C. Bailes & P. J. Cribb, April 1983: 105 collections, 83 taxa
J. H. Beaman, July 1983–August 1984: 293 collections, 190 taxa
C. E. Carr, February–August 1933: 671 collections, 372 taxa
C. L. Chan, 1982–1989: 22 collections, 21 taxa
W. L. Chew & E. J. H. Corner, January–May 1964: 53 collections, 44 taxa
J. Clemens, October–December 1915: 342 collections, 72 taxa
J. & M. S. Clemens, August 1931–December 1933: 1512 collections, 352 taxa
S. Collenette (née Darnton), 1955–1966, March 1981: 126 collections, 103 taxa
P. J. Cribb, November 1989, October 1991: 20 collections, 20 taxa
S. Darnton, February–March 1954: 76 collections, 68 taxa
L. S. Gibbs, February 1910: 45 collections, 42 taxa
G. D. Haviland, March–April 1892: 26 collections, 25 taxa
G. Haslam, July–August 1916: 50 collections, 50 taxa
A. Lamb, 1977–1991: 185 collections, 158 taxa
H. Lohok, 1991: 20 collections, 19 taxa
J. J. Vermeulen (sometimes with C. L. Chan, L. Duistermaat, or A. Lamb): 40 collections, 34 taxa
J. J. Wood, October 1985, May–June 1988: 38 collections, 36 taxa

Professional orchidologists who have collected on Mount Kinabalu include P. J. Cribb, J. J. Vermeulen, and J. J. Wood. Although the amount of time these botanists spent there was limited, they have made important discoveries. Particular attention must be called to the work of Carr, who, though not a professional orchidologist, provided some of the most significant contributions to our knowledge of the Kinabalu orchid flora. It is noteworthy that while the Clemenses made some 1854 collections recorded in this study, the number of taxa represented by their collections is 424 (only 352 in 1931–33), whereas we have records of 671 Carr collections representing 372 taxa. It may be noted that Professor R. E. Holttum, then Director of the Singapore Botanic Gardens, was instrumental in encouraging Carr to study orchids and probably helped arrange his expedition to Mount Kinabalu.

Also of great importance to our knowledge of the orchids of Kinabalu has been the work of Anthony Lamb. His familiarity in the field with many of the species is unrivalled, and he has generously collaborated with other scientists. Also noteworthy are the collections of Sheila Collenette (née Darnton). Her specimens (those from 1954–58 in BM, after that time in K) are particularly useful because of the detailed habitat and plant description data she recorded.

Among the 210 orchid taxa (144 fully named taxa) that have been collected only once on Mount Kinabalu, 57 were collected by Carr, 36 by Lamb, 34 by the Clemenses, 25 by Beaman, 9 by Bailes and Cribb, 9 by Collenette, 6 by Gibbs, 5 by Haviland, 5 by Lohok, 4 by Haslam, and 1–3 by 12 other collectors.

In the Enumeration, specimens obtained by the Royal Society expeditions are not cited by collector but only by the prefix RSNB (Royal Society North

Borneo). Collections of the first Royal Society expedition (1961) were by Chew, Corner and Stainton, and include numbers lower than 2765. Those of the second expedition (1964) are by Chew and Corner, and begin at 4031 (for orchids). A few RSNB collections, beginning with 6007, were obtained by Chai and Ilias in 1964. (See Index to Numbered Collections.)

Consideration of the Clemens Collections. Joseph and Mary Strong Clemens spent about two years on Mount Kinabalu in 1931–33 in addition to the six weeks they were there in 1915. In July 1932 they went to Bogor until mid-December, where preliminary determinations were made of their collections by Van Steenis and others. They collected more specimens from Mount Kinabalu in virtually all groups of plants than have any other collectors; their total collections probably amount to over 9,000 numbers.

Anyone concerned with collection data for plants from Mount Kinabalu must necessarily confront the eccentricities in numbering and labelling of Clemens specimens. We therefore record some observations resulting from the opportunity to examine large numbers of the Clemenses' specimens, particularly those in BM, where the so-called study set, the largest set but not containing every number, was deposited as a result of BM sponsorship of their 1931-33 field work. We have seen 1415 orchid collections in the BM coming from this expedition. Over 500 of these were unidentified or incompletely identified until they were examined for this study.

Orchid specimens collected by Joseph Clemens in 1915 were numbered more or less chronologically and present few problems. The labelling was done after the specimens were received by Oakes Ames. The lowest number we have seen with a specific date is *27*, collected 1 November, and the highest specifically dated number is *401*, collected 9 December. The highest number seen is *408*. Some specimens were not numbered; when that is the case the locality was given only as Kinabalu. Species other than orchids were collected by Mary Strong Clemens, with numbers beginning about *9773* and ending about *11201*. The non-orchid specimens were processed by E. D. Merrill in Manila.

Specimens from the 1931–33 expedition are far more numerous and widely distributed than those from 1915, and involve most of the confusing problems about the Clemenses' collecting procedures. When they obtained only one specimen (a unicate) of a particular plant at a locality, they ordinarily did not number it, but called it a 'single' or 'supplement'. This numbering philosophy is illustrated though not explained by the comment on the label for *Clemens 40220, Coelogyne rupicola* (BM), where it is stated under special notes "May be Suppl. Fear to leave unnumbered since Carr showed me large no. of varieties." In the Enumeration are 21 instances where the herbarium symbol 'BM' is repeated for *s.n.* collections. This indicates that two to five unnumbered specimens were collected from a particular locality at different times.

The numbering of their specimens must have been done at some time after the material was collected, even six or more months after the collection was made in certain instances. They usually typed the collection data (except for number) on a printed field label form headed 'FLORA OF THE MALAY ISLANDS: Clemens Expedition for the British Museum'. These labels appear to be of the design by Merrill (1916) and probably were used under his influence. Some of these, the ones used early on, had numbers stamped with a numbering machine, but these numbers were always crossed out and apparently have no relevance to the collections. Carbon copies were made of some field labels, and are to be found in various herbaria.

The supply of printed field labels was apparently exhausted in 1933. After they returned from Bogor in December, 1932, they often made labels with a rubber stamp, especially with the localities Penibukan, Gurulau Spur, and Masilau [sic] River. Additional information was typed on these stamped labels, but the number was added with pencil or pen. Some specimens bear an additional small label with observations that seem to have been written in the field or immediately on return to camp. Some of these scrap labels were written entirely by hand, others were typed.

Secondary labels were typed from data on the field labels, but the two forms are not always in perfect agreement. The secondary labels are headed 'PLANTS OF MOUNT KINABALU: B. N. Borneo: J. & M. S. Clemens'. Most specimens distributed to herbaria other than BM have only the printed secondary labels, which carry one of four basic data forms, i.e. Dallas, 3000 feet; Tenompok, 5000 feet; Upper Kinabalu, 6000–13500 feet, and a dual-purpose label with both Tenompok, 5000 feet, and Penibukan, 4000–5000 feet. When the last form was used, either Tenompok or Penibukan was struck out. Apparently they had an excess supply of some printed secondary labels and these, particularly the 'Dallas' and 'Upper Kinabalu' forms, were converted to field labels with the rubber stamp.

The numbers used began with 26000 and were written on the field label with pencil. The lowest numbered orchid we have seen is *26019*, collected on 3 August 1931. Numbers were sometimes changed one or more times with the old number crossed out and a different number added. Numbers from 30000 to 31000 seem to have been used erratically, apparently in considerable part for a residue of unnumbered specimens that had accumulated over several months. The same is true for numbers from 35000 to 36000 and 51000 to 51741 (the highest orchid number seen).

Although there is a degree of chronological order to the numbering of Clemens specimens, the numbering is by no means completely chronological. For example, *Clemens 27929, Calanthe kinabaluensis* was collected 25 January 1932, whereas *Clemens 28036, Calanthe pulchra* was collected 20 January 1932. Some of their collections of mixed species may be explained by their apparent attempts to identify the specimens before numbering. The large number of mixed-species orchid collections may have resulted largely from this sorting and numbering process, and makes one sceptical about the reliability of the Clemenses' label data. The collections can be particularly problematical when they occur with the same number but different data or mixed gatherings in different herbaria. The Index to Numbered Collections, therefore, should be used with caution with Clemens specimens. Mixed collections or use of a single number for more than one species can be seen in the Index in cases when more than one determination is given for a particular number.

The same kinds of problems pertain to Mrs. Clemens's New Guinea collections. Conn (1990) remarked: "However, the specimens were not numbered at the time of gathering. Hence, the numbering of her collections has created various problems. Some of the numbers are mixed collections and others are not in chronological order."

One of the most frustrating aspects of working with Clemens label data concerns instances when two or three different label forms with the same number appear on the same herbarium sheet. This problem is well exemplified by *Clemens 33174, Calanthe* aff. *gibbsiae* (BM). Only one plant is mounted on this sheet, and the field label gives the locality as 'Penibukan Ridge' and the habitat as 'Dahobang R. forest', altitude 3500 ft. Special notes indicate 'Via Gambod', one of their Dusun assistants. The date is written 'May 11/33'. The secondary label has 'Dallas, 3000 feet' printed, but this is struck out, and 'Marai Parai 8000 ft.' typed on. The date is given as 'May 11–19/33'. Since they made two other collections of this taxon, one unnumbered, the other *31965*, nearby at the Kinateki River Head in February and March, 1933, we believe that in this exceptional case the secondary label most likely carries the correct data for number *33174*.

Fairly numerous cases occur in which numbers were used more than once for entirely different species, but we are unsure if this was accidental or intentional. It is possible that different gatherings they considered to be the same species were sometimes intentionally given the same number. For example, one sheet of number *32608* of *Dendrobium alabense* in BM has two field labels, one with the elevation 4500 ft., 16 June 1933, the other with the elevation 5000–6000 ft., 10 August 1933. Both labels give the locality as Colombon (Kilembun) river, but the former has the additional information 'Nr. river'.

A particularly problematical example is illustrated by number *26874*. Three specimens from BM are different species with different dates and localities. These are *Liparis parviflora*, Lumu-lumu, 15 Nov. 1931, *Liparis viridiflora*, Lumu-lumu, 13 Nov. 1931, and *Ceratochilus jiewhoeii*, Tenompok, 27 Nov. 1931. Number *26874* is further confused by having been assigned to a specimen in K of *Liparis condylobulbon* from Upper Kinabalu, collected 13 Nov. 1931.

In numbering the Clemenses would often indicate that one number equalled another, for example *27853 = 26991*. This apparently means that they regarded the collection numbered *27853* to be the same taxon as that numbered *26991*. In determining the collection number to be assigned to a specimen, we have always taken the first number indicated.

They did not use all numbers between 26000 and 51000, but skipped the numbers between 36000 and 40000 almost entirely, and likewise those between 41000 and 50000. Numbers in the 30000s were used especially in 1933, but the number *30469*, for example, was used for a specimen dated 13 February 1932 and for another specimen of the same species (*Trichotosia* aff. *ferox*) dated 15 December 1932. Numbers in the 30000s apparently were first used in June, 1932, those in the 31000s in January, 1933, those in the 40000s in August, 1933, and those in the 50000s in November, 1933.

The Clemenses collected more specimens from Tenompok than from any other locality. This designation may have been used in a very broad sense, however, as for example number *29749, Bulbophyllum ionophyllum*. The K specimen of this collection has only a printed secondary label with 'Tenompok, 5000 ft' as the locality. A specimen bearing this number in BM, however, has a field label with the data '2 mi E Lodge, Tenompok, 4000 ft.' This locality would be about a mile east of present Park Headquarters. It is most unlikely that anyone now would designate the latter location as Tenompok.

We have seen many examples of minor discrepancies in locality data on different Clemens specimens with the same number. In questionable cases the field label data should generally be considered most reliable, and these labels are likely to be found in BM. There may be no satisfactory solution, however, when two Clemens labels with different locality data occur with the same specimen, which happens not infrequently.

It has sometimes been suggested that a high percentage of the Clemens specimens from Kinabalu were obtained by native helpers rather than by themselves. In the 'Special notes' field of some labels such expressions as 'Dusan coll.', 'Umpoh coll.', 'Via Gambod', 'Via Boys' etc., are indicated. These, however, are in the distinct minority, so we believe it likely that specimens they did not actually obtain themselves are duly attributed. A few specimens indicate 'Gill Coll'. Walter Gill was the Clemenses' nephew, who visited with them for a while in 1931 and constructed their 'Jungle Lodge' at Tenompok.

On four days the Clemenses collected in the orchid garden established by Carr at their Tenompok Lodge. We have seen 60 specimens representing 44 taxa dated November 9, November 17, November 23, and December 24, 1933, that have the locality as Tenompok, but typed on the label is an expression such as 'Jungle Lodge orchid garden – Carr.' Carr had already returned to Singapore when these specimens were collected, and the plants may originally have come from any of the localities he visited on Kinabalu. We have used the standardised locality 'Tenompok Orchid Garden' for such collections.

Collection notebooks for the Clemenses' Borneo specimens are unknown, and may have been destroyed with the herbarium in Manila (PNH) near the end of World War II. Four notebooks Mrs. Clemens made in Australia in the years 1943–44 are at MICH, and probably are representative of the way their notes were kept.

Although the preceding discussion of the Clemenses' collecting techniques may sound harshly critical, we neverthless have great respect for the work they accomplished. Without their specimens our knowledge of the rich Kinabalu flora would be far more limited. A discussion of their life and contributions to botany is beyond the scope of this work, but it may be noted that on leaving Borneo in January, 1934, they went to Manila, where their late 1932 and 1933 collections from Mount Kinabalu were processed. In August, 1935, they went to New Guinea. Joseph Clemens died there of food poisoning on 21 January 1936 at the age of 73 (van Steenis-Kruseman 1950; Turrill 1936). Mary Strong Clemens stoically continued with her plant collecting until she was evacuated to Australia at the beginning of the war on 26 December 1941 (Conn 1990). The rest of her life was spent in Brisbane, from where she continued to collect (Carter 1982). She died 13 April 1968 at the age of 95 (Langdon 1981; van Steenis 1969).

GEOGRAPHICAL DISTRIBUTION

General Distribution

The orchids of Mount Kinabalu have generally Southeast Asian distribution patterns. Fewer species are endemic than might have been expected. Among the completely determined taxa (the only ones for which we have attempted a geographical analysis) we record only 77 taxa as endemic; an additional 100 species are known only from Sabah; 283 taxa (48 percent of the Kinabalu orchid flora) are endemic to Borneo. Owing to the limited amount of collecting that has been done in most of Borneo, the designation of taxa as endemic is tentative. Truly endemic species are most likely to be the ones from high elevations. Those so-called endemics occurring in hill forest and lower montane forest are likely to be discovered elsewhere.

When we calculate on the basis of fully identified species, about 13 percent of the taxa are endemic. However, if this were calculated on the basis of all species, the figure would be closer to 11 percent. The percentage of endemism might be increased, however, if the taxonomy of all the species not yet determined had been worked out. Twenty-four of the endemics are known only from the type collection.

Several taxa previously thought to be endemic, e.g. *Epigeneium tricallosum, Phaius subtrilobus* and *Robiquetia transversisaccata,* have been discovered recently further south in the Crocker Range. We would expect that several more Kinabalu 'endemics' will be found in the Crocker Range as botanical exploration there increases. Detailed investigation of Mt. Alab (1932 m) and Mt. Trus Madi (2462 m), Malaysia's second highest mountain (excluding the Kinabalu satellite Mt. Tembuyuken [2580 m]), would be well rewarded.

Neoclemensia is notable in that it is the only orchid genus endemic to the mountain. The dull appearance and erratic saprophytic lifestyle of this taxon may account for its being overlooked elsewhere.

Forty-eight of the endemic taxa are known from no lower than 1200 m elevation; 32 from at and above 1500 m, 16 from 1800 m and above, and 6 from no lower than 2200 m.

A rather high proportion, 144 taxa (24 percent), of the total named Kinabalu orchid flora is known from a single collection. Among these little-known species, 24 are thought to be endemic to Mount Kinabalu, 33 are recorded only from Sabah, 64 are restricted to Borneo, and 80 have a wider distribution. The rare orchids of Kinabalu seem to illustrate a phenomenon prevalent in orchids in general (Case 1987), that certain species have wide geographical distributions, but are exceedingly rare where they occur. Only 38 Kinabalu orchid taxa are known also from Sulawesi. Wallace's line would therefore appear to be a significant phytogeographical boundary for even the small-seeded orchids (although orchids are undercollected in Sulawesi). A brief analysis of the occurrence in other geographical areas of the Kinabalu orchid species is provided in Table 1. It should be noted that numbers for some of the areas are lower than actually occur because only summary ranges are available for some species. Thus, certain areas in which the species occurs may have been omitted.

Table 1. Number of taxa occurring on Mount Kinabalu that also occur in other geographical areas.

Brunei	41	Java	168
Kalimantan	165	Sulawesi	38
Sarawak	232	Philippines	91
China	20	New Guinea	48
Japan	8	Australia	13
India	39	New Zealand	1
Burma	40	Pacific islands	13
Peninsular Malaysia	206	Africa	3
Thailand	123	Madagascar	2
Sumatra	179	Pantropical	1

Distribution on Mount Kinabalu

The greatest diversity of Kinabalu orchid species occurs at around 1500 m elevation (Fig. 3). Only 59 taxa are recorded from lowland habitats, whereas 319 occur in hill forest, 431 in lower montane forest, and 78 in upper montane forest. The curve for genera is skewed toward lower elevations. This indicates that the Kinabalu orchids belong largely to lower-elevation tropical genera.

Although much of the original lowland vegetation was destroyed before extensive collecting could be done there, it is likely that the majority of the orchid species, like the pteridophytes (Parris et al. 1992), are adapted to mid-elevation climatic conditions. The data indicating fewer species at the lower elevations probably are not just an artifact of inadequate collections from those areas.

More orchid taxa, 206 (Table 2), have been recorded from Tenompok than from any other single locality, and this agrees with the overall importance of Tenompok as a source of collections (Beaman & Beaman 1990). The figures for Tenompok may be inflated, however, because a high proportion of the collections (377 out of 514) were made by the Clemenses. They apparently used the locality designation of Tenompok in a very broad sense, and it is likely that many of their collections said to be from Tenompok were actually obtained a considerable distance from that locality (see the discussion above in the section on Consideration of the Clemens Orchid Collections). A number of species they reported from Tenompok otherwise have been found only at lower or higher elevations. Unfortunately, much of the natural vegetation at Tenompok has been destroyed subsequent to the time of the Clemens and Carr collections, and many of the species they collected may never be found there again.

In addition to Tenompok, other localities from which large numbers of species have been obtained (Table 2) include Penibukan (Peniguppan),

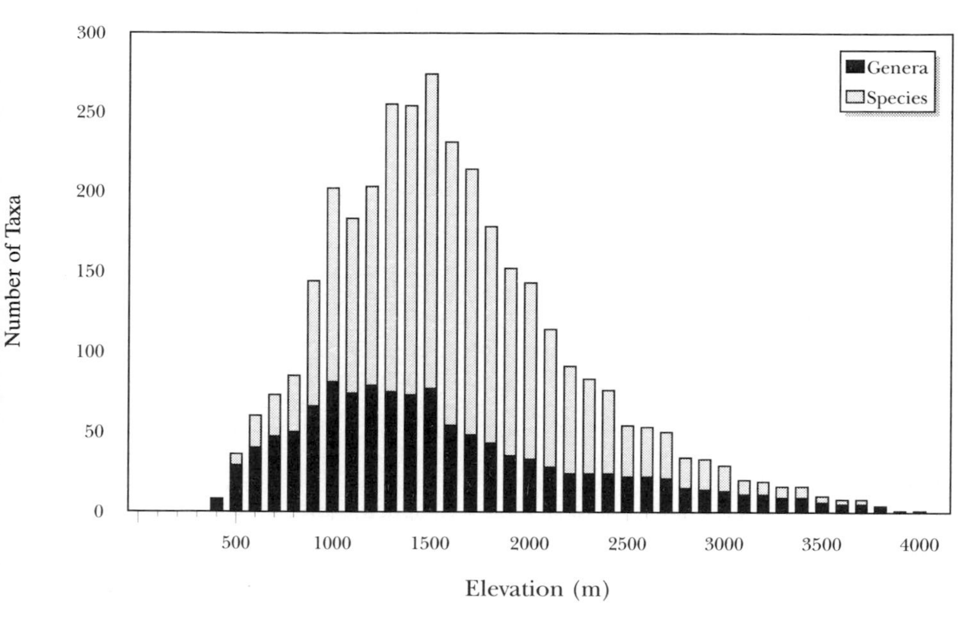

Fig. 3. Elevational distribution of orchid genera and species on Mount Kinabalu.

Table 2. Standard localities with more than 40 taxa.

Tenompok	206	Lohan River	60
Penibukan	166	Kaung	59
Marai Parai Spur	143	Lumu-lumu	56
Bundu Tuhan	112	Park Headquarters	53
Dallas	107	Lubang	47
Kiau	98	Summit Trail	45
Gurulau Spur	86	Pinosuk Plateau	44
Kemburongoh	61	Mesilau River	43
Hempuen Hill	61	Golf Course Site	43

Marai Parai, Bundu Tuhan and Dallas. At places that include the word 'Mesilau' in the locality, 226 taxa have been recorded. The East and West Mesilau Rivers where collections have been gathered are for the most part on the Pinosuk Plateau. This would give some credence to Corner's (1964) assertion that the Pinosuk Plateau was the richest area on the mountain. Unfortunately, most of the natural vegetation of the Pinosuk Plateau has been completely destroyed in the past ten years. Among developments there have been a dairy farm, golf course, expansion of permanent agricultural plots, particularly for cabbage, and subdivision into expensive house lots. Two photographs (Plate 5A, B) contrast the condition of the Pinosuk Plateau in 1964 and 1984. Unfortunately, the destruction there now is even worse than when the 1984 photograph was taken.

Many interesting species have been found on Hempuen Hill and Kulung Hill and along the Lohan River at the southeast base of the mountain. When these three localities are combined, we find 113 taxa recorded from there. Kiau also has been the source of many taxa (98), chiefly Clemens collections. We wonder, however, if Kiau like Tenompok may not have been used in a very broad sense by the Clemenses. Bundu Tuhan, not far from Tenompok and at a slightly lower elevation, has been the the source of 112 taxa, chiefly obtained by Carr, the Clemenses and Collenette. Bundu Tuhan is now under intensive cultivation with temperate vegetable gardens. Dallas, an area of hill forest already of mostly secondary vegetation at the time the Clemenses and Carr collected there, also has (or had) a rich orchid flora, with 107 species recorded. The Gurulau Spur, on the southwest side and an important locality for the Clemenses, Carr and Gibbs, is also well represented with 86 taxa. All localities from which more than 40 species are recorded are listed in Table 2.

It is difficult and rather imprecise to use individual standardised localities alone to express species richness and diversity for any particular area on Mount Kinabalu. Therefore, we have attempted to combine neighbouring localities for comparative purposes. Using this approach, we recognise eight locality groups for the Orchidaceae, as listed in Table 3. In addition, four other non-combined localities, i.e. Bundu Tuhan, Dallas, Gurulau Spur and Kaung, are included. With the grouped localities, we find a greater number of orchid taxa (301) from the Central West Side than from any other major locality on Kinabalu. This area has mostly ultramafic substrates. The large number of species emphasizes the importance of ultramafics in orchid diversity on Mount Kinabalu (see section below on Orchids of Ultramafic Substrates).

A similarity matrix based on the Jaccard community coefficient (Mueller-Dumbois & Ellenberg 1974) was constructed to compare localities listed in Table 3 and is shown in Table 4. Index of similarity coefficients were calculated using the formula

$$IS_J = \frac{c}{a + b + c}$$

where c is the number of taxa in common between two localities and a and b are the numbers of taxa unique to each locality compared. A summation of coefficients is given for each locality. This value is an indicator of how similar any one locality is to all the rest.

The highest coefficients shown in Table 4 occur between 1) the Central West Side and Tenompok-Park Headquarters, 2) the Central West Side and the Pinosuk Plateau, and 3) the Pinosuk Plateau and Tenompok-Park Headquarters. These are all areas of lower montane forest. High coefficients also occur between the Gurulau Spur and the Summit Trail, an

Table 3. Number of orchid species recorded from 12 major areas on Mount Kinabalu.

	Locality grouping	Code	No. of species
1.	Central West Side, centred on Marai Parai	CW	301
2.	Tenompok-Park Headquarters	TP	241
3.	Pinosuk Plateau	PP	225
4.	Southwest Side, centred on Kiau	SW	161
5.	Summit Trail	ST	140
6.	Hempuen Hill-Lohan River	HL	113
7.	Bundu Tuhan	BT	111
8.	Dallas	DA	107
9.	Gurulau Spur	GS	86
10.	Kaung	KA	67
11.	Northwest Side, centred on the Penataran River	NW	66
12.	Eastern Shoulder	ES	59

Table 4. Matrix of Jaccard similarity coefficients using number of orchid taxa at 12 major locality groups on Mount Kinabalu. For full names of the localities see Table 2.

BT	–										
CW	0.132	–									
DA	0.112	0.133	–								
ES	0.037	0.132	0.025	–							
GS	0.088	0.213	0.103	0.142	–						
HL	0.047	0.081	0.084	0.055	0.053	–					
KA	0.085	0.054	0.168	0.024	0.070	0.023	–				
NW	0.060	0.112	0.109	0.078	0.063	0.078	0.073	–			
PP	0.109	0.349	0.068	0.127	0.178	0.063	0.035	0.102	–		
SW	0.157	0.239	0.235	0.058	0.138	0.070	0.118	0.129	0.170	–	
ST	0.073	0.271	0.083	0.137	0.263	0.045	0.035	0.051	0.250	0.119	–
TP	0.197	0.365	0.180	0.091	0.160	0.079	0.062	0.100	0.294	0.237	0.221
Σ	1.097	2.081	1.300	0.906	1.471	0.678	0.747	0.955	1.745	1.549	1.986
	BT	CW	DA	ES	GS	HL	KA	NW	PP	SW	ST

intuitively logical correlation because these are parallel ridges with similar geology and elevation ranges. A high coefficient likewise occurs between the Central West Side and the Summit Trail, again reflecting similarity in geology (several ultramafic outcrops at relatively high elevations).

Conversely, the lowest coefficients are between the ultramafic Hempuen Hill-Lohan River and other localities. Hempuen Hill-Lohan River is the only low-elevation ultramafic locality in this comparison. The low coefficients between Hempuen Hill-Lohan River and the other localities point up the biological uniqueness of this area. When the coefficients between Hempuen Hill-Lohan River and all other localities are summed, the value is 0.678, compared with a value of 2.081 for the Central West Side with all other localities. The summing technique provides a rigorous means for expressing the uniqueness of Hempuen Hill-Lohan River. The sums of the coefficients of each locality with all others are shown in Table 4. Kaung, the only lowland locality, also has low similarity coefficients with all other localities.

ECOLOGICAL ASSOCIATIONS

Orchids of the Lowlands

Most of the low-elevation orchid species around Mount Kinabalu have been collected at Kaung, on the Kadamaian River on the west side of the mountain. Except for the Darnton (i.e. Sheila Collenette) collections from there in 1954, hardly any specimens have been obtained from that area since the Clemens expedition in 1931–33 and Carr's work in 1933. Kaung formerly was an important stop on the approach to Kinabalu, where a government rest house was used by various expeditions. The dominant vegetation there has been of a secondary nature since the time of the first botanical expeditions. Few of the low-elevation collections have any habitat data, with the notable exception of those by Collenette, and we therefore have limited understanding of the nature of the habitat of these species. Many of the epiphytes probably were found on trees along the Kadamaian River, where the Collenette specimens were collected. Among species found in that kind of habitat are the following: *Grammatophyllum kinabaluense, Poaephyllum pauciflorum, Thecostele alata, Thelasis capitata, T. micrantha,* and *Vanilla sumatrana.*

Orchids of Hill Forest

Among the important hill forest localities visited by Carr and the Clemenses are Bundu Tuhan, Dallas and Kiau on the south and west sides of the mountain. The late Professor R. E. Holttum has told us that Dallas, particularly important for the 1931 collections of the Clemenses, was an area of disturbed secondary vegetation at the time of his trip to the mountain with the Clemenses in 1931. Dallas is situated on the side of a ridge separated from the west side of Mount Kinabalu by the deep valley of the Kadamaian River. In the Dallas area are many sharp ridges and deep valleys, some of the latter of which still have a vestige of relatively undisturbed vegetation or old secondary forest. It is likely that most of the Clemens and Carr collections from Dallas came from forest remnants in valleys near there. On rare occasions a Clemens collection from Dallas includes brief habitat data, as for example two specimens of *Thelasis variabilis, 26309* (BM), 'Kinabalu Foothill jungle', and *s.n.* (BM, 26 September 1931), 'Kinabalu foothills gorge'. Among the Dallas taxa may be listed: *Aerides odorata, Apostasia odorata, Appendicula pendula, Bulbophyllum odoratum, Dendrochilum gibbsiae, Eria ornata, Goodyera rubicunda, Liparis viridiflora, Neuwiedia veratrifolia, Peristylus grandis, Pholidota imbricata* and *Vanilla kinabaluensis.*

The 1915 Clemens expedition obtained a large number of specimens with Kiau indicated as the locality. This is also an area with a long history of disturbance, certainly predating the Clemens expedition. Many of the orchids attributed by Joseph Clemens to Kiau are unlikely to have come from that immediate vicinity. However, there are valleys in the Kiau area retaining remnants of primary or old secondary vegetation that would harbour orchids, and old secondary forest probably was much more abundant in 1915. It is also probable that Clemens used the locality designation of Kiau in a broad sense with the likelihood that anything collected during a day trip from Kiau was attributed to that locality. It seems possible that some of the 1915 specimens said to be from Kiau could have come from the Tinekuk and Tahubang Rivers and perhaps even from Penibukan, locality names not used by Clemens in 1915.

Among hill forest species from Kiau are the epiphytes *Appendicula foliosa, Bulbophyllum caudatisepalum, B. ceratostylis, B. gibbosum, B. odoratum, Ceratostylis subulata, Coelogyne longibulbosa, C. venusta, Dendrochilum imbricatum, Eria carnosissima, Grammatophyllum kinabaluense, Liparis viridiflora, Malleola kinabaluensis* and *Vanda hastifera* var. *gibbsiae.* Terrestrial species recorded from Kiau include: *Ania borneensis, Anoectochilus longicalcaratus, Aphyllorchis pallida, Calanthe pulchra, C. triplicata, Goodyera rostellata, Lepidogyne longifolia, Malaxis variabilis, Phaius tankervilleae, Tainia ovalifolia* and *Vrydagzynea grandis.*

An interesting species of *Liparis* collected only once on Kinabalu, by Brentnall near Bundu Tuhan, is the Javan *L. rhodochila.* It has attractive flowers with pale yellow-green sepals and petals, a creamy-white column and brilliant scarlet-red recurved lip. The species is illustrated by Comber (1990: 137).

Hill forest areas that have been sources of collections by the Beaman team and by Anthony Lamb are Hempuen Hill and the Lohan River at the southeastern base of the mountain. This low-elevation area on ultramafic substrate has been the source of many unusual species, both terrestrial and epiphytic. Part of the area is very steep hillside and old landslide. In this area the forest is of exceptionally low stature for an area of this low elevation, and the canopy, from the air, has a yellowish green aspect in contrast to nearby areas on less extreme or non-ultramafic substrates. Particularly interesting and rare species from this area are the following: *Apostasia nuda, A. wallichii, Appendicula cristata, Chamaeanthus brachystachys, Chroniochilus minimus, C. virescens, Cleisostoma discolor, Dendrobium pachyanthum* (Lamb, pers. comm.), *Liparis mucronata, Macodes lowii, Malaxis metallica, Micropera callosa, Neuwiedia borneensis, Phaius reflexipetalus, Schoenorchis endertii, Spathoglottis gracilis* and *Tuberolabium rhopalorrhachis.* The area was degazetted from Kinabalu Park in 1984. After an announcement by the Chief Minister in 1990 that it would be regazetted as park, much of the area was intentionally burned, either spitefully or to prevent the regazettement.

ORCHIDS OF LOWER MONTANE FOREST

As illustrated in Figure 3, lower montane forest at 1500 m is the richest area for orchids on Mount Kinabalu. Great numbers of species have been collected at Tenompok, the Pinosuk Plateau, Penibukan Ridge, and Park Headquarters at about this elevation. Common lower montane epiphytes are *Ascidiera longifolia, Bulbophyllum lobbii, B. microglossum, Ceratostylis radiata, Coelogyne cuprea* var. *planiscapa, C. hirtella, C. kinabaluensis, C. monilirachis, C. moultonii, C. rhabdobulbon, Dendrobium cymbulipes, Eria latiuscula, E. leiophylla, F. magnicallosa, Liparis lobongensis, L. pandurata, Pholidota carnea* var. *carnea, Podochilus tenuis, Trichotosia aurea* and *T. brevipedunculata.*

One of the most noteworthy Kinabalu orchids of lower montane forest is, as its name would suggest, the beautiful white-flowered *Dendrobium spectatissimum.* This is a member of section *Formosae,* the so-called nigrohirsute dendrobiums, which have distinctive black hairs on the leaf sheaths. *Dendrobium spectatissimum* is one of a number of species having black hairs distributed over the entire leaf surface. The flowers may measure up to 10 cm across.

Many of the species in section *Formosae* are of great horticultural value, particularly those from Thailand, Borneo and the Philippines. Eleven members of the section are recorded from Borneo, all endemic, but only *D. parthenium, D. singkawangense* and *D. spectatissimum* occur on Kinabalu. Most species have conspicuous white flowers with a narrow extinctoriform, i.e. candle-snuffer-shaped, mentum and are most likely moth-pollinated.

Dendrobium spectatissimum is usually an epiphyte on *Leptospermum recurvum* and is often found growing in exposed positions. Collenette (pers. comm.) notes that for the first week or ten days after the flower has opened the petals are curled tightly back, completely enclosing the top edges of the lateral sepals. The petals then uncurl until they form right angles to the direction of the lip. During the last two or three days of the flower's life the petals move forward and fall parallel to the lip. The flowers remain open for up to six weeks.

Perhaps the most unusual lower montane locality (with elevation the primary criterion for calling the area lower montane) is Marai Parai. This remarkable site at the junction of the ridge that connects the main massif of Mount Kinabalu with its western satellite, Mt. Nungkek, seems to be unique on the mountain. It is an area of extreme ultramafic geology. The soil is exceedingly wet and boggy from seepage of water from the upper slopes of Kinabalu, with the plants frequently bathed in rain, fog and mist. The area is a mixture of vegetation types with some so-called 'grassland' dominated by sedges including *Costularia pilisepala* (from which the area takes its name; Beaman et al. in press) and low woody scrub dominated by *Leptospermum.* Various hummocks and pockets contribute to the diversity of the area and abundant *Sphagnum* adds to its remarkable nature.

Our database includes 143 orchid taxa recorded from Marai Parai and Marai Parai Spur (Table 2). The extraordinary conditions of the vegetation and flora at Marai Parai have been recognised since the time of the earliest botanical explorations of Kinabalu by St. John, Burbidge, and Haviland. Some of the especially interesting orchid species of Marai Parai, many of which have their type locality there, include: *Appendicula congesta, A.*

linearifolia, A. longirostrata, Bromheadia divaricata, Bulbophyllum disjunctum, Cleisocentron merrillianum, Coelogyne compressicaulis, C. kinabaluensis, Dendrobium maraiparense, D. piranha, Dendrochilum angustipetalum, D. lancilabium, Dilochia cantleyi, Epigeneium kinabaluense, Kuhlhasseltia javanica, Platanthera kinabaluensis, P. stapfii and *Trichotosia brevipedunculata.*

One of the most extraordinary orchids, endemic to Marai Parai, is *Dendrobium piranha,* which has rather sinister looking olive-brown to salmon or flesh-coloured flowers, which have acquired it the nickname of 'jaws'. This is in reference to the unusual excavated lip that gives the flowers the appearance of a piranha fish with its jaws open. Although related to *D. olivaceum,* also recorded from Kinabalu, *D. piranha* is instantly distinguished by its resupinate flowers with distinctive lip having a smaller, fleshier mid-lobe with a more lacerate callus. *Dendrobium piranha* and *D. olivaceum* are the largest flowered members of section *Distichophyllum,* which has its centre of speciation in Borneo.

Also from Marai Parai, but not endemic there, is *Dendrobium maraiparense.* This species also belongs to section *Distichophyllum* but has much smaller flowers reminiscent of the widespread *D. uniflorum.* It is a stiffly erect plant with white flowers, which turn to orange-apricot when ageing. The sepals and petals are strongly reflexed, a feature *D. maraiparense* shares with several members of the section.

One of the most interesting species at Marai Parai is *Cleisocentron merrillianum,* which exhibits leaf polymorphy as well as having unusually coloured flowers. An illustration (plate 97) by Blanche Ames appears in Ames and Schweinfurth's (1920) account. This shows a specimen with ligulate, acute leaves. Ames' original description gives the leaves as ". . . ± 8 mm. wide, distichous, linear-ligulate to linear-lanceolate, acute, conduplicate at the base, very thickly coriaceous, apparently very fleshy in the living plant."

The holotype of *Cleisocentron merrillianum* (*Clemens* s.n., AMES) consists of flowering stems displaying two leaf types, the lower terete and acute, the upper broader, ligulate and acute. The isotype at K consists only of two detached leaves, one ligulate, one terete, and some loose flowers. Ames clearly overlooked or ignored these odd leaf types when drawing up his description of *Sarcanthus merrillianus.* Among specimens we examined, *Carr SFN 27893* from Bundu Tuhan, *Collenette 597* from Mesilau Cave, *Lamb SAN 89677* from Pinosuk Plateau and *Sadau SAN 49690* from Mesilau Trail all have only narrow, terete leaves up to 18 cm long. *Lamb SAN 89678* from Marai Parai has dimorphic foliage up to 13 cm long, exactly matching the type. Another specimen, *Collenette 756* from the Eastern Shoulder, has ligulate leaves up to 13 × 1.2 cm, but with an unequally bilobed apex. The flowers are identical in all cases. Christenson (1992) comments that the "young shoots produce strap-shaped leaves very similar to *Cleisocentron trichromum,* but which grade in older shoots to terete leaves."

Cleisocentron merrillianum can grow up to 1 m long, as noted in the collection data by Collenette, although it is usually much shorter. The flowers are variously described as 'aquamarine blue', 'translucent Cambridge blue', or 'translucent blue-lavender', with deep indigo-blue markings on the lip. The intensity of blue seems to vary. *Cleisocentron merrillianum* has been recorded as an epiphyte on *Agathis* and *Podocarpus.*

Orchids Along the Summit Trail

Many visitors to Kinabalu will have encountered their first orchids growing in situ while climbing the well-trodden summit trail. A good initial impression of each vegetation zone and the representative orchid species can be gained this way. Among species most likely to be seen along the summit trail, between the Power Station and Paka-paka Cave, are the epiphytes *Bulbophyllum catenarium, B. montense, B. pugilanthum, Chelonistele kinabaluensis, C. sulphurea* var. *sulphurea* and var. *crassifolia, Coelogyne plicatissima, C. radioferens, Dendrochilum grandiflorum, D. kamborangense, Nabaluia clemensii* and *Thrixspermum triangulare.* Species likely to be encountered in ridge forest and scrub along the summit trail include *Eria grandis, E. pseudocymbiformis* var. *hirsuta, E. robusta, Entomophobia kinabaluensis* and *Liparis kamborangensis.*

On tree trunks just below the first hut above the Power Station can be found the delicate epiphyte *Bulbophyllum catenarium* (section *Monilibulbus*), which is well characterised by the peculiar shape of the lip. This has a distinctly swollen and coarsely verrucose apex and strongly and very abruptly recurved side-lobes. The sepals and petals are bright yellow, occasionally with somewhat orange veins, contrasting with the dark red or purple lip. The specific epithet is derived from the Latin *catenarius*, pertaining to a chain, and refers to the chain-like arrangement of the pseudobulbs.

The section *Monilibulbus* of *Bulbophyllum*, which ranges from China eastward to the Philippines and Sulawesi, is represented by about 32 species in Borneo. They are characteristic of lower and upper montane forest, where they form dense mats on trunks and branches. The pseudobulbs, which are prostrate and often flattened, are produced close together necklace fashion, covering and sometimes partly enveloping the rhizome. The often elegant little flowers are solitary and usually appear one or only a few at a time. The habit of many species of the section seems to depend on environmental factors, such as slope, availability of moisture, light and shade. For example, *B. anguliferum*, may be sturdy and compact in habit or elongated and larger-leaved, depending on habitat. *Bulbophyllum montense*, a species of open, often exposed *Dacrydium-Leptospermum* forest, grows on bark or in moss cushions and ascends to an elevation of 3400 m.

At lower elevations, usually between 1500 and 1800 m, can be found the strikingly different *Bulbophyllum pugilanthum*, which has greenish white or pale translucent orange-yellow flowers shaped like a boxing glove. The lip is pale yellow. Plants hang suspended from small horizontal branches and mossy saplings 2 to 3 m above ground level or low down from the trunks of larger trees. The long slender roots run along the length of and usually cover the rhizome and pseudobulbs below. The dark green acute fleshy leaves are semi-terete, appearing shallowly V-shaped in cross-section. The inflorescence of both *B. pugilanthum* and the smaller *B. ceratostylis* is pseudapical, i.e. arising from just below the apex of the pseudobulb, its base actually sunken into and partly adnate to it. *Aphanobulbon*, the section to which *B. pugilanthum* belongs, is prolific in Borneo and includes species of diverse habit ranging from robust and erect with many-flowered racemes to wiry and pendulous with solitary flowers. *Bulbophyllum flavescens* is the most widespread and common species of section *Aphanobulbon* on Mount Kinabalu.

A characteristic species of *Coelogyne* along the summit trail at around 2100 m is *C. plicatissima.* This belongs to a distinctive group of species which J. J. Smith (1931) referred to the section *Hologyne* and Carr (1935) to the section *Rigidiformes* (see below under Necklace Orchids and Their Allies). They are mostly plants of mossy montane forest and many, including *C. plicatissima,* have unusual brownish olive to brownish salmon-coloured flowers, often contrasting with a cream-coloured column. Other species have lime-green or olive-green flowers, whereas those of *C. clemensii* var. *clemensii,* also frequent at lower elevations, are creamy white and fragrant. The leaves of *C. plicatissima* are strongly corrugated (plicate) and the fleshy lip is strongly convolute for about half its length. It is the commonest species of *Coelogyne* in the scrub and moss forest on and below the summit of Mt. Trus Madi. There seems to be great variability among the taxa related to *C. plicatissima* and the interrelationship and delimitation of species within the group throughout Borneo needs further study.

One of the very few monopodial orchids to ascend as high as 3400 m along the summit trail is *Thrixspermum triangulare. Thrixspermum* is a large and taxonomically little-understood genus with around 23 species in Borneo, eight of which have been recorded from Mount Kinabalu. Vegetative habit, particularly inflorescence type and floral bract shape, are useful in grouping species. The flowers, although often pretty, are, like *Flickingeria,* very short-lived and become virtually impossible to determine after pressing and drying. Fresh flowers preserved in alcohol are essential for identification. *Thrixspermum triangulare* has fleshy leaves and inflorescences of varying length with many complanate, cymbiform, obtuse, distichous floral bracts. The flowers, which appear in succession usually one at a time, may be 3 cm in diameter and have broad rounded sepals and a distinctive fleshy, triangular callus on the lip. These are creamy white with crimson streaks on the side-lobes of the lip and a crimson central patch on the mid-lobe.

HIGH-ELEVATION GENERA

Two genera particularly prominent at high elevations are *Dendrochilum,* which includes epiphytic, lithophytic and terrestrial species, and the terrestrial genus *Platanthera. Dendrochilum* is particularly well represented in Borneo, with some 73 species having been recorded. Of these, 33 occur on Mount Kinabalu, along with several others as yet undetermined. Most Bornean species, including those on Mount Kinabalu, belong to section *Platyclinis.* This is distinguished by the erect, ovoid or fusiform pseudobulbs borne close together on the rhizome and the synanthous inflorescences appearing with the young leaves. The flowering portion curves gracefully and has a more or less drooping tip. Two of the commonest and most attractive endemic species of section *Platyclinis* to be seen along the summit trail are *D. grandiflorum* and *D. kamborangense.*

Dendrochilum grandiflorum has attractive flesh-pink or pinkish brown flowers, often with a dark olive column. The lip is divided into 3 distinctive sharp tooth-like lobes and has two flange-like calli near the base. *Dendrochilum kamborangense* usually has longer inflorescences of fragrant, greenish and lemon- or saffron-yellow flowers. The lip is brown and has rounded, fimbriate side-lobes and a larger elliptic, apiculate mid-lobe. Among the high-elevation species, most of which are endemic and probably of recent and local origin, are *D. acuiferum, D. alpinum, D. dewindtianum, D. grandiflorum, D. haslamii, D. kamborangense, D. lancilabium, D. pterogyne, D. scriptum, D. stachyodes* and *D. transversum.* Several of these species, particularly the white-flowered *D. stachyodes,* are conspicuous at higher elevations along the main summit trail.

Dendrochilum alpinum is one of the largest flowered species of the genus in Borneo, each flower measuring about 1 cm in diameter. It is found growing on the sides of granite rocks sheltered by stunted trees, particularly around Panar Laban and Sayat-sayat. The sepals and petals are yellow suffused with salmon-pink or entirely salmon-pink and the ovate lip has a brownish salmon median line, keels and apex. The deep salmon-pink column has distinctive brown lateral arms (stelidia). It was first collected by Carr in June 1933 and again by Sato of the Universiti Kebangsaan Malaysia in 1981.

Dendrochilum dewindtianum is one of the most attractive species in Borneo, with dense spikes of yellow-green resupinate flowers of varying size, the lip having bright green keels. Sepal and petal width varies; in some specimens, e.g. *Clemens 50774* from Gurulau Spur, they are broad, others, e.g. *Meijer SAN 20367* from above Kemburongoh, are quite narrow. The degree of toothing on the side-lobes of the lip also seems to vary, being more prominent in specimens having narrow sepals and petals. Figure 30 shows two extremes of variation. A Blanche Ames illustration (plate 82) of this species was published, as *D. perspicabile,* in Ames and Schweinfurth's 1920 account. A colour photograph of the plant, identified only as a *Dendrochilum,* was recently published by Cubitt and Payne (1990: 182), and is included in our plates.

Platanthera is best known as a temperate genus containing the sweetly scented white-flowered European butterfly orchids. This terrestrial genus contains about 85 species, seven of which occur in Borneo, and all are represented on Kinabalu. All, except *P. saprophytica,* have green or yellowish green flowers. *Platanthera angustata* is the most widespread, occurring in places as far apart as Thailand and New Ireland in the Bismarck Archipelago. Four high-elevation species, namely *P. angustata, P. borneensis, P. crassinervia,* and *P. gibbsiae,* are apparently very rare and little collected. Only one high-elevation *Platanthera, P. kinabaluensis,* is common. Three species are restricted to areas above 2700 m. Of these, *P. crassinervia* and *P. stapfii* seem distinct, but *P. gibbsiae* may prove to be conspecific with *P. kinabaluensis* after study of a wider range of material than is currently at hand. *Platanthera kinabaluensis* has one of the the widest altitudinal ranges (1500–3400 m) among the Bornean orchid species.

Dendrobium, although the second largest genus of orchids in Borneo, does not usually occur above an elevation of 1800 m. Only six of the 61 species (named and unnamed) recorded from Mount Kinabalu are found at higher elevations. Four of these, namely *D. alabense, D. lamelluliferum, D. tridentatum* and *D. ventripes,* belong to section *Rhopalanthe.* This section, which includes the well-known Pigeon orchid, *D. crumenatum,* is especially diverse in Borneo, with 26 named species recognised and several others undescribed. The other two species, *D. maraiparense* and *D. piranha,* are members of section *Distichophyllum,* which has 19 named species recorded from Borneo. *Dendrobium piranha* apparently holds the altitude record for *Dendrobium* on Kinabalu; a Clemens collection is recorded from 1800–2400 m on Marai Parai Spur.

Most Bornean species of *Dendrobium* (Wood 1990), including those from Mount Kinabalu, occur in hill and lower montane forest. The low number of high-elevation species is in striking contrast to New Guinea which, of course, has a considerably larger land area above 2400 m. Their paucity in Borneo may also be explained by the notable absence of sections such as *Calyptrochilus* (except *D. erosum (Blume) Lindl.*), *Latouria, Oxyglossum* and *Pedilonum* (except *D. secundum*), which form an important and colourful element in the montane flora of New Guinea.

In Borneo the upper montane element is only partially filled by the closely related, very much smaller genus *Epigeneium.* Although not exclusively montane, it does contain species that are a frequent constituent of upper montane moss forest and ridge vegetation. Three species are found in the upper montane zone on Mount Kinabalu, namely *E. kinabaluense* (with two-leaved pseudobulbs and rather broad leaves), *E. longirepens* (with one-leaved pseudobulbs) and *E. tricallosum* (with two-leaved pseudobulbs and narrower and usually longer linear-lanceolate to ligulate leaves). *Epigeneium kinabaluense* extends to 3400 m on Mount Kinabalu, and is also found elsewhere in Sabah including among the low scrub on the summit of Mt. Trus Madi, as well as on Mt. Mulu in Sarawak. *Epigeneium longirepens* has so far only been recorded from Mount Kinabalu, where it occurs at lower elevations than *E. kinabaluense,* particularly at around 1500 m. *Epigeneium tricallosum* is commonest around 1500 m but has also been found at 1800 m along the summit ridge of Mt. Alab in the Crocker Range. It often occurs as a scrambling terrestrial, in some places associated with *Chelonistele amplissima* and *Dilochia rigida.* Specimens can often be seen in flower in the mountain garden at Park Headquarters.

The vegetative habit of *Epigeneium kinabaluense* and particularly of *E. tricallosum* may vary, probably due to environmental factors such as degree of exposure and depth of shade. Some specimens of *E. tricallosum* become much-branched and have quite short leaves 1.5–3 cm long (*Madani SAN 76482* from Mesilau Trail). Others have leaves ranging from 3–7 cm long on the same plant (*Carr SFN 26955* from Tenompok). Plants growing in shade on Mt. Alab have leaves up to 10 cm long.

Orchids of the Summit Area

Common orchids of high elevations, particularly in the area through which the main trail to the summit passes, often occur as lithophytes, colonising in the granitic crevices, or at the bases of scrubby *Leptospermum* bushes or trees. These include *Bulbophyllum coriaceum, Coelogyne papillosa, Dendrochilum alatum, D. alpinum, D. grandiflorum, D. pterogyne, D. stachyodes, Eria grandis* and *Platanthera kinabaluensis.* All of these species are endemic to Mount Kinabalu.

Bulbophyllum coriaceum, belonging to section *Aphanobulbon,* is relatively common between 2700 and 3500 m. It has thick, fleshy leaves and racemes of pretty yellow flowers with a darker yellow or pale orange lip. A photograph of this species, incorrectly labelled as '*Coelogyne*' appears on page 98 of *Borneo* by Mackinnon (1975).

Coelogyne papillosa is the largest flowered and most conspicuous orchid of the summit area. It seems to favour open sites among *Dacrydium* and

Leptospermum, but is by no means restricted to them. The beautiful snow-white flowers are at their best from January to June. The lip has yellow-brown nerves on the side-lobes and a large central brown spot and yellow-brown keels on the mid-lobe. The leaves are strongly plicate and similar to those of *C. plicatissima*. This is also figured by Mackinnon, incorrectly labelled as '*Bulbophyllum*'.

Dendrochilum pterogyne was described by Carr from material collected around Paka-paka Cave where it is an abundant epiphyte, but it also occurs between 3400 and 3700 m in the summit area. The small ovoid pseudobulbs are often red and minutely wrinkled. The flowers are salmon-pink with a brown, oval, entire lip edged with salmon-pink and bearing a distinctive horseshoe-shaped callus formation. The fruit-capsule is provided with three low keels, hence the specific name.

Perhaps the most elegant species between 2100 and 3500 m is the creamy-white-flowered *Dendrochilum stachyodes*. This forms extensive colonies along rock crevices, usually in the open but sometimes under *Leptospermum*, and is obviously able to withstand adverse climatic conditions. Ridley (in Stapf, 1894) described the inflorescences as "reminding one of an ear of wheat, whence the specific name." *Clemens 33177* (BM) differs from other specimens in having longer leaves and arms on the column (Fig. 36, L–N).

The mass flowering of *D. stachyodes* among the bleak granitic rock faces is one of the delights of Mount Kinabalu. Such displays have, however, suffered in recent years from periodic droughts resulting from the the El Niño/Southern Oscillation phenomenon, that of 1983 being particularly severe (Beaman et al. 1986). Specific observations on drought effects on the vegetation at higher elevations on Kinabalu have been published by Lowry et al. (1973) and Smith (1979).

Eria (excluding *Trichotosia*) is a large and problematical genus and is represented by 83 named taxa in Borneo and numerous others that are little known or possibly undescribed. Conspecificity, however, may be expected to reduce this number. Thirty-six named taxa are recorded from Mount Kinabalu, but only a few attain high elevations – a similar situation to *Dendrobium*. One terrestrial species, *E. grandis*, however, is confined to areas above 2100 m, where it often forms continuous stands. Kitayama (1991: 42) noted that it is a dominant species in the upper-subalpine forest. It is a tall robust plant with long sword-shaped leaves and dense racemes of purple-pink or lilac flowers covered in white hairs on the outside. Lamb (pers. comm.) indicates that it occurs at a higher elevation on Mount Kinabalu than any other orchid.

Orchids of Ultramafic Substrates

Ultramafic substrates are significant in the occurrence of many Kinabalu orchids; 282 taxa have been found (on Mount Kinabalu) entirely or partly on ultramafic outcrops; 216 taxa that occur at least part of the time on ultramafics are epiphytic. Possibly the types of vegetation in ultramafic areas are particularly satisfactory for epiphytic species because of a light open canopy that may encourage the diversity of epiphytes, but the low, scrubby vegetation on the ultramafics may also allow collectors to readily find the epiphytes in those habitats. Some of the orchids on Mount Kinabalu that appear to be restricted to ultramafic substrates occur elsewhere in Borneo in other habitats, particularly kerangas or heath-forest vegetation, which also has an open aspect. Among the ultramafic species on Kinabalu that occur on other soil types elsewhere are *Ceratostylis ampullacea* and *Chroniochilus virescens.*

An analysis of the elevational distribution of species on ultramafic and non-ultramafic substrates is provided in Figure 4. Figure 5 shows the correlation of endemic species with elevation and ultramafic substrates. In these figures it should be noted that data are included for all species that have an occurrence at some time on the ultramafics. It is not intended to imply that they always occur on ultramafic substrates at the elevations indicated. Figure 4 shows that ultramafic and non-ultramafic species occur in nearly equal numbers at lower elevations. At elevations above 1500 m, however, ultramafic (or facultatively ultramafic) species substantially outnumber the non-ultramafic species. In the case of the endemics (Fig. 5), ultramafic species outnumber non-ultramafic species at all elevations.

Particularly noteworthy among species generally found on the ultramafics are *Appendicula congesta, Arachnis longisepala, Bulbophyllum nubinatum, Ceratostylis ampullacea, Chamaeanthus brachystachys, Chroniochilus minimus, C. virescens, Cymbidium elongatum, Dendrobium patentilobum, Dendrochilum kamborangense, D. lancilabium, Epigeneium kinabaluense, Eria cymbidifolia* var. *pandanifolia, E. major, Neuwiedia borneensis* and *Phaius pauciflorus subsp. sabahensis.*

One of the finest orchids found on the ultramafics is the Scorpion orchid, *Arachnis longisepala,* which was first described from eastern Sabah in 1981 (Wood & Bell 1992). It is related to *A. grandisepala* J. J. Wood, another species from the Crocker Range, also described in 1981. Both species have distinctive long dorsal sepals unique in the genus. *Arachnis longisepala* is distinguished by the longer inflorescences, which often attain 120 cm and bear 8 to 16 smaller but rather more elegant flowers. These have pale lemon-yellow lateral sepals and petals irregularly blotched with shiny maroon-purple above. The narrow strap-like dorsal sepal, which measures up to 7.4 cm long, is entirely shiny maroon-purple above. The lip has a bucket-shaped hypochile, whose lobes clasp the column, and a narrowly triangular-ovate epichile with an upcurved aristulate apex. *Arachnis longisepala* is restricted to hill forest where it grows on tall trees, sometimes 30 metres up in the canopy. It has also been reported growing in shade on tree trunks not far above ground level. The pollinating agent has not been identified, although C. L. Chan (pers. comm.) has observed day-flying cetonid beetles visiting the flowers.

ORCHIDS OF DISTURBED HABITATS

Probably the commonest orchid of disturbed habitats, a terrestrial of roadsides and ditch banks at medium elevations, is the very widespread and variable *Arundina graminifolia* with conspicuous cream and lilac flowers resembling a South American *Sobralia.* The plants produce tall cane-like stems up to 2 m high covered with many grass-like leaves, hence the common name 'Bamboo orchid'. Colonies quickly establish themselves on landslides and roadside cuttings away from forest shade. The plants are used for landscaping in the Mountain Garden at Park Headquarters. Three forms can be seen on Kinabalu. The showiest is the lowland form, which has

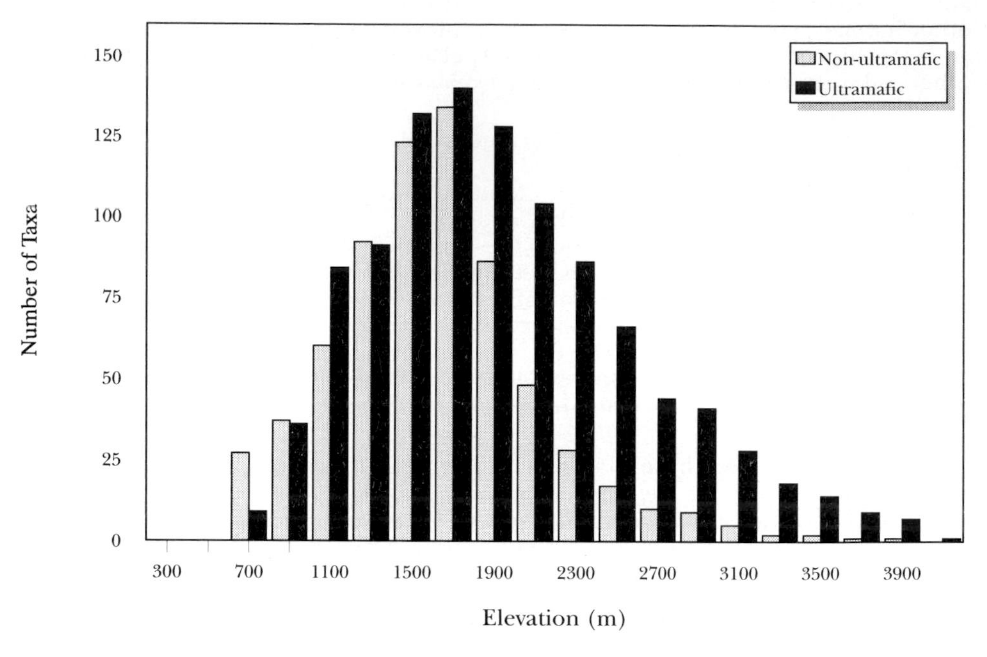

Fig. 4. Elevational distribution of orchid taxa on ultramafic and non-ultramafic substrates on Mount Kinabalu.

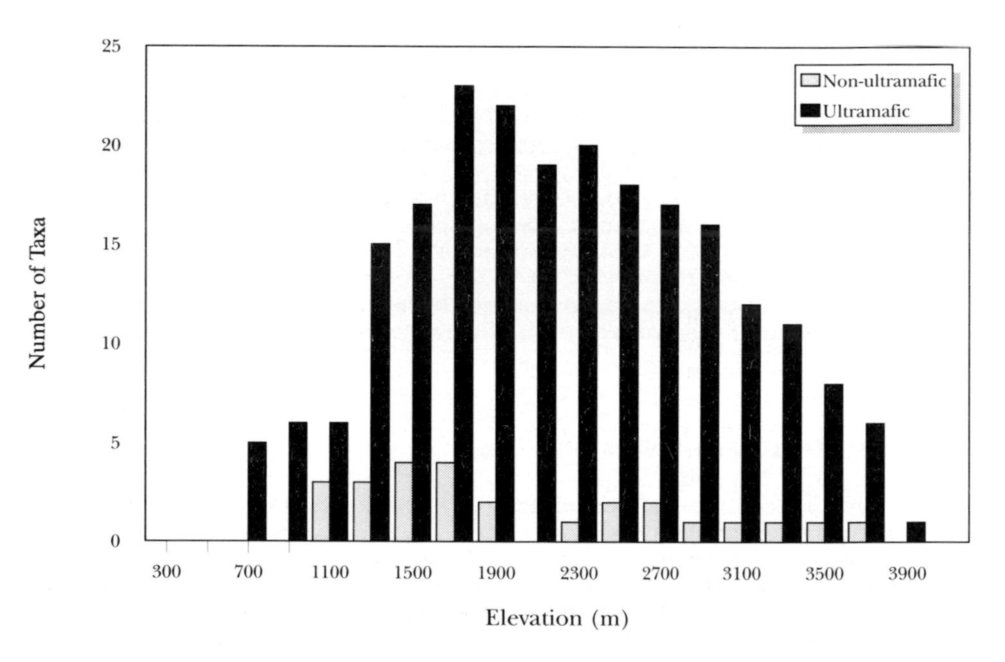

Fig. 5. Elevational distribution of endemic orchid taxa on ultramafic and non-ultramafic substrates on Mount Kinabalu.

pinkish sepals and petals and a large purple lip with the white base enclosing the column. The lip has a deep yellow patch on the mid-lobe. The cane-like stems have plain green leaves. This form is found rarely except on the lowermost slopes. The commonest form occurs between 1000 and 2000 m. It has purple-flushed leaves and smaller flowers with white sepals and petals and a less showy deep mauve and white lip with some yellow on the mid-lobe. Visitors will notice this plant, which often forms large clumps and is common along roadside banks around and within Park Headquarters. A pure white peloric form can also be found scattered among normal plants.

The yellow-flowered *Spathoglottis microchilina* is also common along ditch banks, frequently pushing up through tall ferns such as *Dicranopteris* and *Diplopterygium*. It can be seen around Park Headquarters. Originally described from Sumatra, this species is often cleistogamous and self-pollinating and is closely related to and easily confused with *S. aurea*. *Spathoglottis microchilina* has a very narrow, pointed, almost thread-like lip only 1 mm wide, whereas the lip of *S. aurea* may be as much as 4 mm wide and generally a little more expanded and obtuse at the tip. The flowers of *S. aurea* from Peninsular Malaysia, Sumatra and Java are generally larger and of a darker yellow hue. We have seen no specimens of it from Mount Kinabalu. Material previously determined as *S. aurea* is included under *S. microchilina* in the Enumeration. Holttum (1964) admitted that *S. microchilina* needed "more field study," and it seems possible that we are dealing here with one rather variable species.

Another open-habitat species found around Park Headquarters is the pink-flowered *Spiranthes sinensis*. It is widespread throughout Asia and the Pacific and closely related to the Lady's Tresses orchids of Europe and North America. Other terrestrials of open habitats, although not very often seen, are *Eulophia graminea*, a fire-resistant species found on sand banks along the Lohan River, *E. spectabilis, Oeceoclades pulchra, Peristylus kinabaluensis,* the common and widespread pink-flowered *Spathoglottis plicata,* and *Zeuxine strateumatica,* which may sometimes become a weed of padi at lower elevations in parts of Malaysia and Indonesia.

LIFE-FORMS AND HABITATS

The epiphytic life-form is predominant among Kinabalu orchids, with more than twice the number of epiphytes as terrestrials. Thus 515 taxa are recorded as epiphytic and 193 as terrestrial. Fourteen of the terrestrials are saprophytic (or mycotrophic parasites). Thirty-four taxa are recorded as lithophytic, 32 as both epiphytic and terrestrial, 24 as both epiphytic and lithophytic, 6 as both terrestrial and lithophytic (admittedly this is a subjective distinction), and 2 as epiphytic, terrestrial, and lithophytic. An analysis of the distribution of life-forms correlated with elevation is provided in Figure 6. This figure shows that as elevation increases, the proportion of terrestrials and lithophytes to epiphytes increases. Above 3000 m the epiphytes are outnumbered by the other two life-forms. The greatest number of epiphytes and terrestrials occurs between 1200 and 1500 m, but the greatest number of lithophytes is between 1500 and 1800 m.

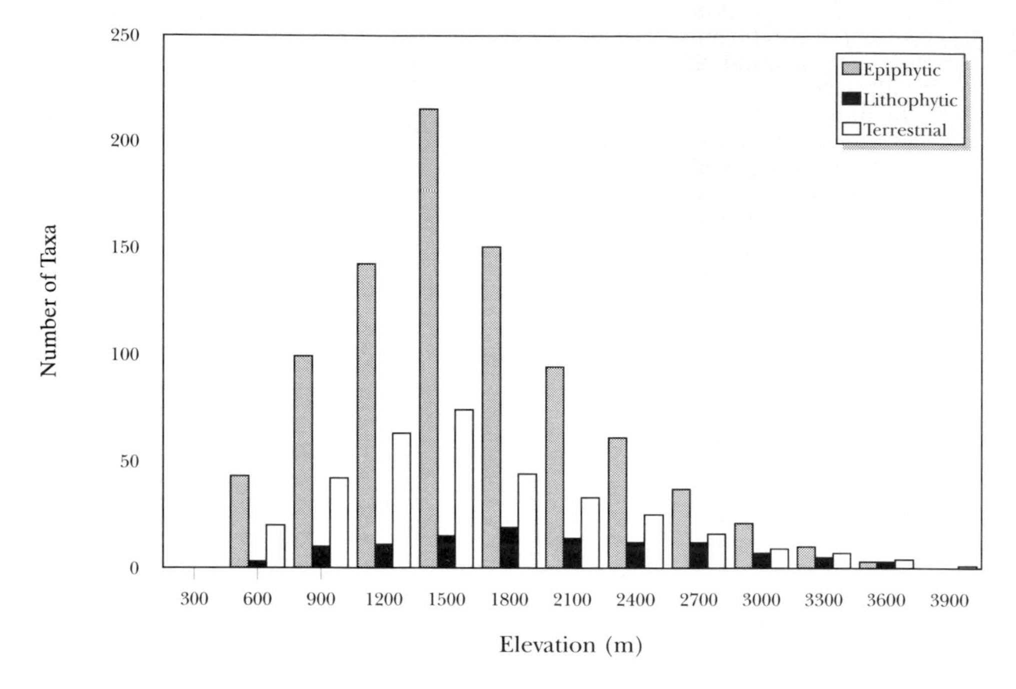

Fig. 6. Elevational distribution of orchids on Mount Kinabalu according to life-form.

TERRESTRIALS

Ania borneensis is the only representative of this genus, formerly included in *Tainia*, to occur on Mount Kinabalu from where it has only been recorded twice. Turner (1992: 47) recognises both genera, the main characters of *Ania* being the swollen, ovoid to ellipsoid pseudobulbs usually consisting of several internodes, lateral inflorescences, a usually distinctly spurred lip, and articulated petioles.

Both *Apostasia* and *Neuwiedia* are characteristic of damp, shady places in lowland and hill forest, although *Apostasia odorata* ascends to 1500 m on Mount Kinabalu. Nodular storage tubers on the roots, and stilt roots at the base of the stem, are found in *Apostasia*. Two species of *Apostasia* and three of *Neuwiedia* have been recorded from the ultramafic Hempuen Hill.

The calanthes of Kinabalu can be divided into two groups: those with fat pseudobulbs and deciduous leaves (subgenus *Preptanthe*, section *Preptanthe*, e.g. *Calanthe vestita*) and those without fat pseudobulbs and having persistent leaves (subgenus *Calanthe*). Within the latter group are species with persistent floral bracts (section *Calanthe*, e.g. *C. kinabaluensis* and *C. triplicata*) and those with deciduous floral bracts (section *Styloglossum*, e.g. *C. pulchra* and *C. speciosa*).

Both *Calanthe transiens* (section *Calanthe*) and *C. truncicola* (section *Styloglossum*) were formerly known only from Sumatra. *Calanthe transiens* occurs mostly around 2500–2700 m and was originally described from a similar altitude on Mt. Kerinci. It is related to the widespread and variable *C. triplicata,* but is distinguished by the longer floral bracts, less divided lip mid-lobe, shorter spur and puberulous column. The flowers, like *C. triplicata,* are white with an orange-yellow callus on the lip. The deep-orange flowered *C. truncicola* has only been recorded from ultramafic substrates on Kinabalu. These plants differ from the type, which was described from Bukit Djarat in Sumatra, in a few minor details, particularly the fewer-flowered inflorescence and slightly larger, less retuse lip mid-lobe.

Chrysoglossum reticulatum was originally described from Mount Kinabalu. The only specimens we have seen of it are the Carr syntypes. Lamb (pers. comm.) informs us that it has been seen on the Kiau View Trail near Park Headquarters. The sepals and petals are pale yellow, greenish or whitish, with many transverse pale purple spots. The white lip has attractive mauve-purple markings and three yellow or white keels.

The 'helmet-orchid' genus *Corybas* is represented by three species on Mount Kinabalu, one of which is endemic. These exquisite little plants consist of a single cordate leaf, often with coloured veins, above which sits the relatively large solitary flower. This has a hood- or helmet-shaped dorsal sepal that, together with the usually two-spurred lip, forms a tube with an expanded mouth. Many species, including *C. kinabaluensis* and *C. pictus,* have long, thread-like lateral sepals and petals.

Corybas species are exacting in their habitat requirements, occurring in moss carpets, on the boles of large trees or on mossy, well-drained banks. With some experience, it is possible to predict where *Corybas* will be found, although seemingly ideal habitats are sometimes devoid of plants. All three Kinabalu species have been recorded from ultramafic substrates but are not necessarily restricted to them.

The species of the genus *Cryptostylis,* although few in number, are not always easy to identify and further study over the entire range is desirable. Three species, *C. acutata, C. arachnites* and *C. clemensii* (*Chlorosa clemensii*) occur on the mountain, again mostly on ultramafics. *Cryptostylis clemensii* is distinguished by its petals which have a minutely tridentate apex. *Chlorosa clemensii* was described from Mount Kinabalu by Ames and Schweinfurth based on a peloric specimen in which the flowers do not open fully and the lip is shorter than normal. It is figured by Blanche Ames (plate 80) in Ames and Schweinfurth's 1920 account.

All but one of the ten taxa of *Cymbidium* known from Borneo are found on Mount Kinabalu. About half of these are epiphytic, but two, *C. borneense* and *C. ensifolium* subsp. *haematodes,* are terrestrial, and three others, *C. dayanum, C. elongatum* and *C. lancifolium,* may be either terrestrial or epiphytic. *Cymbidium dayanum* grows on mossy banks and stumps. *Cymbidium elongatum* is one of the most unusual species in the genus. It has a monopodial rather than sympodial habit, with indeterminately growing stems. These tend to lean on and scramble over the surrounding vegetation as they elongate. This atypical habit is reminiscent of climbing species of *Dipodium* (e.g. *D. scandens*). Material of the species at Kew was for a long time filed among the undetermined species of *Dipodium*. In Sarawak it has been found growing as an epiphyte.

Two of the four species of *Dilochia* recorded from Mount Kinabalu occur on ultramafic substrates. *Dilochia cantleyi* is quite an attractive species when seen growing in quantity. The tall, stiffly erect stems bear broad stiff leaves and terminate in a rather compact panicle of white or pale lemon-yellow flowers, which are subtended by cream cup-shaped bracts. The lip is yellow and suffused with brownish orange. *Dilochia rigida* is a much more slender plant with narrow leaves and was originally described as a *Bromheadia* by Ridley. It has a shorter, unbranched inflorescence bearing fewer cream-coloured flowers. The sepals are often flushed with pale green and purple on the reverse and the lip has a lemon-yellow patch on the mid-lobe, with purple markings between the three keels. *Dilochia rigida* is one of the most frequently seen species along the Summit Trail between the Power Station and Paka-paka Cave. It is a common constituent of ridge vegetation and flowers at regular intervals throughout the year, particularly during rainy periods. When not in bloom it may be confused with *Bromheadia crassiflora*. The *Bromheadia*, however, has rather fleshier, unequally bilobed leaves, often with wrinkled sheaths, and small, closely spaced, regularly alternate floral bracts resembling certain species of *Thrixspermum*.

Clemens 51302 collected from Mesilau proved to be *Goodyera bifida* and represents a new record for Borneo. It is found throughout Java above an elevation of 700 m and is figured by Comber (1990: 31).

Hetaeria is represented on the mountain by two species. It is one of only two genera in the *Goodyera* subtribe, the other being *Macodes*, to have non-resupinate flowers with the lip borne uppermost. It can be distinguished from *Macodes* by the green leaves usually lacking coloured nerves and by having a straight lip and column. *Hetaeria biloba*, originally described from Peninsular Malaysia, has pretty flowers with striking orbicular lip-epichile lobules which are strongly decurved in the natural position. The Sumatran *H. grandiflora* is conspecific.

Blume's genus *Dicerostylis* has been considered distinct from *Hylophila* because of two projecting stigma horns found in *Dicerostylis lanceolata* Blume. A very similar plant from Malaya was once considered the same as *Hylophila lanceolata* by Ridley and by Holttum. The latter, however, does not have stigma horns and was later named *H. cheangii* by Holttum. The presence of stigma horns is clearly not constant, and in other respects *Dicerostylis* is identical with *Hylophila*. Populations from Kinabalu have hirsute pink and red flowers with a rather broad column, whereas populations from elsewhere in the Crocker Range are less hairy, have a narrower column and are white flowered.

One of the most striking and certainly the tallest member of the *Goodyera* subtribe found on Mount Kinabalu is *Lepidogyne longifolia*. The whole plant attains a height of 1 metre with many sword-shaped leaves up to 40 × 3.5 cm grouped at the base. The short-haired inflorescence may be up to 40 cm long and is densely covered with many reddish brown flowers subtended by narrow, hairy bracts up to 3.5 cm long. The swollen base of the lip contains a transverse row of about six small calli.

Liparis is represented by 40 or more species in Borneo, some epiphytic and others terrestrial. Three of the most striking terrestrial species on Mount Kinabalu are *L. atrosanguinea, L. grandis* and *L. kamborangensis. Liparis atrosanguinea,* recorded mostly from ultramafic substrates, has large leaves up to 25 × 10.5 cm and beautiful well-spaced crimson-purple flowers with a broadly ovate lip up to 1.7 × 1.3 cm, the margin of which is finely toothed. Collenette in field data for her collection *A102* notes that the newly opened flowers have "a greenish-yellow lip turning dark orange-brown". J. and M. S. Clemens described the flowers of *40928* as "purple, lemon banner", where banner presumably refers to the lip. *Liparis grandis* is a robust species related to the newly described *L. aurantiorbiculata,* but has pale green to creamy flowers with a long slender pedicel-with-ovary. *Liparis kamborangensis* is altogether different, with a creeping stem producing flowering shoots at intervals, each bearing a solitary cordate leaf and a raceme of purple flowers, sometimes tipped green. The lip has a pectinate margin. *Liparis kamborangensis* occurs in damp mossy places, sometimes on rock faces.

Malaxis is a cosmopolitan genus, at present estimated to contain around 300 species. Ridley (1888) attempted a monograph over a century ago, in which he recognised less than 70 species, 40 of which were Asian. Since that time many new species have been described, particularly from Indonesia and New Guinea, by workers such as Schlechter and J. J. Smith. A modern revision is desperately needed, but would prove a very difficult task since our knowledge of the conspecificity and distribution of many species remains incomplete. Many species, including several from Borneo, remain to be described. Of the 33 named Bornean species 12 have been recorded from Mount Kinabalu, while many others remain unidentified.

Malaxis are commonly found throughout Borneo, particularly from sea-level to about 1600 m. *Malaxis graciliscapa, M. kinabaluensis* and *M. metallica* ascend to above 2000 metres on Mount Kinabalu. Nearly all species thrive in damp, shady places, growing in deep leaf litter on the forest floor. At least three sections are represented on the mountain. The most widespread species, *M. latifolia,* which has a particularly dense-flowered inflorescence, belongs to section *Gastroglottis.* Species in this section have a lip lacking auricles. These structures, found in the majority of species, stretch backward on both sides of the column. *Malaxis latifolia* has a 3-lobed lip apex, and a transverse callus in front of the column. Sections *Commelinodes* and *Malaxis* have auriculate lips. In section *Commelinodes* the lip has several teeth along the front edge, often appearing pectinate. The thin flowering shoots ascend from a creeping horizontal rhizome which roots at the nodes. Examples include *M. commelinifolia, M. graciliscapa, M. kinabaluensis, M. perakensis* and *M. variabilis.* Section *Malaxis* are mostly erect plants, with the front edge of the lip simple or more or less 3-lobed, the apical part often emarginate or bilobulate. Examples include *M. calophylla, M. multiflora* and *M. punctata.* Several species of *Malaxis* have colourful foliage and are noted in the section on Foliage Characteristics.

Mischobulbum scapigerum is the only representative in Borneo of this small but attractive genus. It is not a common plant and has only been found between 800 and 900 m in the hill-forest zone on Mount Kinabalu. The cordate-ovate leaves are dull dark green with an uneven or rather undulate surface, the mature leaves becoming puckered with rows of darker green concavities. The faintly sweet-scented flowers are variable in colour; the sepals and petals, ranging from pale greenish yellow to light brown, are veined with purple-red. The lip is white with fine purple spots and purple crests, an orange-yellow tip and an orange-yellow to purple base. It was figured, as *Nephelaphyllum scapigerum*, in *Curtis's Botanical Magazine*, plate 5390 (1863).

Myrmechis kinabaluensis is again the sole representative in Borneo of this little-known and easily overlooked genus. It is related to *Zeuxine* but is distinguished by the long-clawed lip, stigmas on short processes and 1–3-flowered inflorescences. The flowers of *M. kinabaluensis* are pure white. Apparently it is an extremely rare plant.

The slipper-orchid genus *Paphiopedilum* is represented by five species on Kinabalu. Four of these are terrestrial and one (*P. lowii*) epiphytic. Two of the terrestrials, *P. hookerae* var. *volonteanum* and *P. rothschildianum,* are confined to ultramafic substrates, whereas *P. dayanum* occurs on steep ridges, sometimes on ultramafics, and *P. javanicum* var. *virens* occurs among boulders on steep banks above rivers in lower montane forest. *Paphiopedilum dayanum* has been observed to have two different leaf forms at different elevations, one bright green, the other grey-green (Fowlie & Lamb 1983; Fowlie 1984).

Paphiopedilum rothschildianum is probably one of the most striking and best known of Kinabalu's endemic orchids. It was introduced into cultivation by M. Jean Linden in May 1887 and named in honour of Baron Ferdinand Rothschild, an eminent Victorian orchid grower. Reichenbach based his original description on a flower sent to him by the firm of Sander & Sons of St. Albans, who said it had originated from New Guinea. This appears to have been a deliberate attempt by Sander to mislead his competitors. Whitehead (1893: 116) reported seeing it on 19 February 1887 among piles of loose rock on the tops of hills.

Paphiopedilum rothschildianum must be one of the rarest species of *Paphiopedilum* in nature. Despite various searches over the past hundred years, it has been located from only two sites in hill forest on the lower slopes of Kinabalu. These are the very places that are today under most pressure from unauthorized logging, mining, shifting agriculture and forest fires. Attempts at artificial establishment of the species at two localities on Mount Kinabalu are currently under way (Grell et al. 1988). Excellent photographs taken in situ are provided by Fowlie and Lamb (1983).

The genus *Peristylus* is widespread throughout Asia, New Guinea and the Pacific islands but its species are rarely noticed by most people. The genus has, however, been the subject of much taxonomic discussion and most of the species have been moved around among different genera (including *Habenaria*) by various authors, or by the same author at different times. Despite difficulties in the development of a clearcut set of characters, it is possible, with experience, to distinguish a *Peristylus* from a *Habenaria.* The small flowers usually have an erect ovary placed close to the rachis, the two stigmas are not extended on stigmatophores in front of the column and the spur is very short. This is what Senghas (1973: 251) recognised as the '*Peristylus* Aspekt'. Seidenfaden (1977) grouped many of the species into four tentative sections based on lip characters.

Peristylus candidus is a graceful pure-white flowered species of open grassy places. According to Carr (1935) it was common in the neighbourhood of

Bundu Tuhan, an area subsequently converted to temperate vegetable gardens. The inflorescence is produced on a long, slender peduncle at the base of which are two oblong-elliptic greyish green leaves. The lip is rather broad and shortly trilobed. *Peristylus grandis* is a very different species from similar elevations having large leaves up to 30 × 8 cm and a stately densely flowered inflorescence borne on a robust stem over 1 m tall. The flowers are pale apple-green subtended by long, narrow greyish green or greenish white bracts. *Peristylus gracilis* is another graceful species with a cluster of leaves at the centre of the stem and a lax raceme of green flowers. The lip has long thread-like side-lobes which tend to curl and spiral.

One of the largest terrestrial orchids on Kinabalu is the widely distributed *Phaius tankervilleae*. Many fine specimens have been planted around Poring Hot Springs and in the orchid garden there. Several smaller species occur on ultramafic substrates, including the endemic *P. baconii* and *P. reflexipetalus*. *Phaius baconii* usually has only two leaves at the stem apex and a 3–6-flowered inflorescence borne from the middle to lower nodes of the stem. The flowers are about 5 cm in diameter and are white with purple stripes and a yellow disc on the lip. *Phaius reflexipetalus* has 3 or 4 leaves and an inflorescence with up to a dozen flowers with strongly reflexed petals. The sepals and petals are pale greenish yellow, stained pink or brown and the lip is yellow with a cream margin and reddish brown streaked disc. *Phaius subtrilobus*, which can be seen around Park Headquarters, has an inflorescence bearing 4–10 large flowers up to 7.5 cm across. The sepals and petals are white on the reverse and purple-maroon on the inside. The white lip has a pair of dark yellow, spotted purple fleshy keels that converge near the apex into a narrowly lanceolate papillose callus. The side-lobes are rudimentary, appearing semi-obcordate, and each bears two purple patches. The very short spur, measuring only 2.5–3.5 mm long, was noted by Ames and Schweinfurth (1920) to be "in every case pierced by an insect."

Vrydagzynea commemorates the Dutch pharmacologist Theodore Daniel Vrydag Zynen, a contemporary of the botanist Blume. The genus is widely distributed from northern India to the Pacific islands, with 15 species recorded from Borneo. The inflorescence is always rather short and densely many-flowered. The flowers seldom open widely and have a lip with a spur which projects between the lateral sepals and contains two stalked glands. *Vrydagzynea bractescens* is a small, graceful species of hill forest, originally described from Siberut Island, off the coast of Sumatra. The small flowers are white with green-tipped sepals and are subtended by conspicuous elliptic, acuminate bracts. It also occurs on Mt. Mulu in Sarawak. *Vrydagzynea grandis* is a much larger, more robust species up to 30 cm high with obliquely elliptic leaves measuring up to 10 × 5 cm. The flowers are white, except for the sepals, which are green tipped with white, and are densely packed in a 3–8 cm long inflorescence.

EPIPHYTES

Twig epiphytes. One of the most interesting and certainly most attractive twig epiphytes is *Ceratochilus jiewhoeii*, which has so far only been found on Mount Kinabalu and Mt. Alab. It probably will turn up elsewhere since, when not in flower, it can easily be confused with *Microsaccus griffithii*. The leaves of *C. jiewhoeii* are generally larger proportioned and more acute, and the flowers, which measure 1 cm in diameter, are pure white, sometimes with a small yellow spot in the centre of the lip. The generic name is derived from the Greek *keras* or *kerato*, horn, and *cheilos*, lip, and alludes to the rather obscure horn-like swellings at the mouth of the lip of the Javan *C. biglandulosus* Blume. These swellings, which are more flange-like than horn-like, probably represent rudimentary side-lobes and are developed as such in *C. jiewhoeii*. The flowers of the Javan species turn a vivid scarlet before falling. *Ceratochilus jiewhoeii* is generally found near the tips of thin branches and twigs of small saplings in the shade.

Two monopodial genera, *Microtatorchis* and *Taeniophyllum*, are highly specialised as twig epiphytes, often growing high up in the canopy and consequently not easily observed. They are most often seen when small branches and twigs have blown down after a storm or in logged forest. Some species of *Microtatorchis*, whose centre of speciation lies in New Guinea, may have small leaves (e.g. *M. javanica*) whereas others are 'leafless'. All species of *Taeniophyllum* are 'leafless'. The leaves in these 'leafless' epiphytes are reduced to tiny brown scales and photosynthesis is carried out by numerous green roots. These may be terete and spaghetti-like or flattened and tapeworm-like. The 'leafless' species of *Microtatorchis* resemble *Taeniophyllum*, but can be distinguished by having two instead of four pollinia and by the persistent leafy floral bracts. Many species of *Taeniophyllum* have an apical tooth or bristle on the lip. Both genera are very little collected, and were mostly found by Carr, who seemed to have had an eye for the inconspicuous. In fact, *Microtatorchis* is only known from Carr collections, and all six of the named species of *Taeniophyllum* likewise.

Some of the *Bulbophyllum* species, particularly those belonging to sections *Micromonanthe* and *Monilibulbus*, and small members of section *Aphanobulbon*, are also tiny twig epiphytes. These include *B. acutum* (section *Micromonanathe*), *B. ceratostylis* (section *Aphanobulbon*), *B. chanii* (section *Monilibulbus*), *B. longhutense* (section *Monilibulbus*) and *B. mutabile* var. *obesum* (section *Aphanobulbon*). Other orchids sometimes found as twig epiphytes are *Dendrobium alabense* and *D. cymbulipes* (section *Rhopalanthe*), *Oberonia* spp. and *Thrixspermum* spp.

Trunk epiphytes. Many species of *Bulbophyllum* occur as trunk epiphytes, occurring in humid, shady environments with little air movement. Among these is *B. sopoetanense*, a pretty species originally described from Mt. Sopoetan in Sulawesi, which often occurs on mossy trunks and fallen logs. A colour plate is provided by Sato (1991: 26). Other species occurring as trunk epiphytes include *B. dearei* (section *Sestochilus*), *B. heldiorum* (section *Globiceps*), *B. kestron* (section *Monilibulbus*), *B. lobbii*, *B. membranifolium* and *B. microglossum* (section *Sestochilus*), *B. minutulum*, *B. montense*, *B. nubinatum*, *B. pelicanopsis*, *B. puntjakense* and *B. schefferi* (all section *Monilibulbus*). These are figured and described in detail in Vermeulen (1991).

A remarkable trunk epiphyte that has been collected only twice, on opposite sides of the mountain, at Mamut and Penibukan, is *Eria pellipes*. This belongs to section *Strongyleria* which is characterised by one- to four-leaved stems, thick, terete leaves and synanthous inflorescences bearing one, two or rarely three flowers. *Eria pellipes* is closely related to a second species, *E. pannea*, so far only recorded on Kinabalu by Carr, who found it near Bundu Tuhan. This differs in having non-pseudobulbous stems with 2 to 4 leaves borne at intervals and 2–3-flowered inflorescences. *Eria pellipes* always has crowded 1-leaved pseudobulbs and solitary flowers. Both species are illustrated by Seidenfaden (1982: 46, 47), and *Eria pannea* was figured in *Curtis's Botanical Magazine*, n.s., plate 570 (1970). Neither species looks like a typical *Eria* from a foliage aspect.

Various species of *Podochilus* are frequent trunk epiphytes, which, with their diminutive habit and distichous foliage, often resemble leafy liverworts or certain mosses rather than orchids. *Podochilus lucescens* is a widespread species that has somewhat twisted, shiny leaves and usually terminal inflorescences bearing rows of minute alternate bracts. The minute flowers are white with a purple spot on the petals and at the base of the lip. It is often to be found growing among mosses low down on the trunks of saplings and full-grown trees, including the buttresses of lowland and hill forest species. *Podochilus serpyllifolius* is a smaller-leaved species with a moss-like habit, which often clothes the basal part of tree trunks. *Podochilus tenuis* is a very common species with narrow, closely appressed leaves and branching stems that, unlike most others, are held out away from the substrate.

Appendicula congesta is another common trunk epiphyte, frequently found in forest on ultramafic substrate. The tiny flowers are held in a compact terminal head, each subtended by translucent pale green to whitish bracts. The sepals are white, petals purple and lip white flushed yellow. It is figured (Plate 93) by Blanche Ames, as *Chilopogon kinabaluensis*, in Ames and Schweinfurth's (1920) account.

Pendent epiphytes. Although the majority of epiphytes are erect or porrect, a number hang from branches. The newly described *Appendicula fractiflexa*, known from only three collections, is a good example. Several species of *Bulbophyllum* have pendulous rhizomes, most of which belong to the polymorphic section *Aphanobulbon*. These include *B. ceratostylis, B. lygeron, B. mutabile* var. *mutabile, B. pocillum, B. pugilanthum* and *B. rhizomatosum.*

Bulbophyllum pugilanthum produces slender rhizomes up to 24 cm long with long, slender roots several of which are produced from the base of the pseudobulbs. The fleshy, semi-terete, grooved leaves measure 6–8 cm long and resemble those found in several species of *Dendrobium* section *Rhizobium* from New Guinea and Australia, such as *D. teretifolium* R. Br. Up to eight boxing glove-shaped flowers are produced one at a time in succession from the side of and partly adnate to the pseudobulb just below its apex where they emerge from several scarious sheaths.

Other species of *Bulbophyllum* with a pendulous habit include *B. salaccense* (section *Globiceps*) and *B. undecifilum* (section *Epicrianthes*). *Bulbophyllum undecifilum* is a typical member of section *Epicrianthes*, which can easily be recognised by the pendulous rhizome, one-flowered inflorescence and petals bearing a varying number of often hairy, mobile appendages. These differ in texture from the petal itself. Javan specimens of *B. undecifilum* have 11 appendages on each petal, but those from Mount Kinabalu have 16 or 17. The sides of the lip are often covered by a dense patch of large vescicles.

A distinctive *Coelogyne* with pendulous branching stems up to 1 m long is *C. craticulaelabris* (section *Rigidiformes*), which occurs between 900 and 2400 m on Kinabalu. The pseudobulbs become well spaced in large specimens and bear plicate elliptic leaves of variable length. The short inflorescences may have up to six or rarely ten small flowers subtended by translucent, very pale green bracts. The sepals are greenish cream or pale yellow suffused salmon and the petals, lip and column are usually white. The flowers do not open widely and are sweetly scented. *Coelogyne craticulaelabris* also occurs as an erect epiphyte along the summit ridges of Mt. Mulu in Sarawak and has been found as a terrestrial among sedges in scrubby forest on Mt. Monkobo in Sabah. It has recently been collected on Retak Hill in Brunei.

Very few species of *Dendrobium* in Borneo are exclusively pendent in habit. Some species in sections *Calcarifera* (*D. cumulatum, D. sanguinolentum*) and *Rhopalanthe* (*D. cymbulipes*) may become pendent as the stems elongate. *Dendrobium oblongum* (section *Oxystophyllum*) has thin stems covered in imbricate, laterally flattened and rather thick leaves and very soon becomes pendent as the plant matures.

The attractive *Dendrobium cumulatum* is often pendulous in habit. A plant collected from Mt. Nungkek (*Sands 3976*) has pale pink flowers with unusually dark pink veins on the sepals and petals, and yellow column wings. This species has otherwise only been recorded from mainland Asia.

The most spectacular pendent epiphytic orchid on Mount Kinabalu is undoubtedly *Paraphalaenopsis labukensis,* which only occurs in hill forest. Plants grow suspended from small trees 3–5 m above the ground or 12–20 m up on the trunks and branches of larger trees. The short stem produces 3–5 extraordinarily fleshy, terete, whip-like leaves which may measure up to 165 cm long. These are from 6–9 mm thick and slightly constricted 1.3–2.5 cm from the acute apex. The inflorescence bears 5 or 6 beautiful cinnamon-scented flowers, each measuring up to 6.2 cm wide and having sepals and petals that are purple-cinnamon speckled with yellow and edged with greenish yellow. The lip is yellow with many purple-red and orange spots and bars.

Among other monopodial taxa, *Robiquetia crockerensis, R. pinosukensis* and *R. transversisaccata* are sometimes pendulous. A most unusual truly pendent monopodial orchid is *Thrixspermum pensile.* The whole plant hangs straight down, with the 'upper' surface of the leaves facing down.

CLIMBERS

Most people do not think of orchids as climbers, yet one of the most familiar and commercially important genera, *Vanilla,* is a vigorous climbing terrestrial; some species climb many metres up through forest trees. The vanilla of commerce is derived from the fermented 'pods' of the Central American species, *V. fragrans* (Salisb.) Ames (syn.: *V. planifolia* Jacks.). The Malaysian species have large fruits, which are said to be sweet and edible, but they have little if any of the fragrance of *V. fragrans.*

The climbing monopodial habit in *Vanilla* is comparable to that found in the vandaceous orchids (subtribe Aeridinae). It can be distinguished though by the very regular occurrence of an adventitious root at each node, opposite the fleshy leaf. The roots act mainly as supporting organs and never grow very long. Those nearest the ground, however, grow downward and become branched, serving to absorb nutrients from forest-floor humus, similar to species of *Vanda.* Vanillas often grow to a large size, but rarely flower freely; identification is therefore often difficult. *Vanilla kinabaluensis* is a very robust species of hill forest. It climbs to 20 metres or more and has flaccid leaves up to 32 × 11 cm. The pale yellow flower has a lip with a central bundle of very thin fan-shaped plates with toothed edges, all pointing inward. At the apex of the lip is a dense group of branched dark red papillae. The species has also been seen growing in hill dipterocarp forest at 750 m on Mt. Trus Madi. *Vanilla pilifera*, originally described from Johor in Peninsular Malaysia, has a conspicuous tuft of fine hairs directed toward the base of the lip.

Climbing species are commonly found in several vandaceous genera, notably *Arachnis, Thrixspermum* and *Trichoglottis.*

The curious genus *Claderia* produces a creeping rhizome, which often climbs up the bases of tree trunks. The inflorescence bears a succession of many green flowers, one or two opening at a time.

Dipodium, with a centre of speciation in Australia, exhibits both the monopodial and sympodial habits. The two Kinabalu species are monopodial climbers with long spiralling stems covered in sword-shaped plicate leaves arranged in two ranks and overlapping at the base. *Dipodium pictum* has a lip about 2 cm long, much longer than the column. The lip is held in a horizontal position and has the mid-lobe folded; the sides are strongly deflexed. The side-lobes of the lip are inconspicuous and narrowly triangular. The flowers of *D. scandens* are smaller with a lip 1.4–1.7 cm long, often incurved toward the column apex. The side-lobes are distinct, ligulate, obtuse and measure 2–3 mm long. Flower resupination seems always to occur in *D. pictum,* but flowers of *D. scandens* are often non-resupinate. Resupination, however, is unlikely to be a reliable means for separating the two species.

SAPROPHYTES

Several interesting saprophytes occur in Borneo, but most are rarely seen because of their often inconspicuous colouring and ephemeral life style. Fourteen saprophytic species have been recorded from Mount Kinabalu: *Aphyllorchis montana, A. pallida, Cyrtosia javanica, Cystorchis aphylla, Didymoplexiella kinabaluensis, Epipogium roseum, Eulophia zollingeri, Galeola nudifolia, Gastrodia grandilabris, Lecanorchis multiflora, Neoclemensia spathulata, Platanthera saprophytica, Stereosandra javanica* and *Tropidia saprophytica.* All except *Epipogium roseum, Cyrtosia javanica* and *Tropidia saprophytica* have resupinate flowers, and all are usually found as scattered single plants. Several are widespread; *Epipogium roseum* even occurs in Africa, whereas *Didymoplexiella kinabaluensis* and *Neoclemensia spathulata* have only been recorded from Mount Kinabalu. The latter two may easily have been overlooked elsewhere.

The genus *Cyrtosia* is related to *Galeola,* an extraordinary genus of climbing saprophytes with dehiscent fruits containing large winged seeds. *Cyrtosia javanica* arises from a large underground rhizome that produces non-climbing erect fleshy stems bearing inflorescences which elongate slowly. The flowers are non-resupinate, pale yellow or greenish yellow, and papillose-mealy on the exterior of the sepals. The fleshy, indehiscent fruits are reddish in colour and contain wingless seeds.

Didymoplexiella can be distinguished from the closely related *Didymoplexis,* a genus so far unrecorded from Mount Kinabalu, by the long decurved stelidia on the column and the distinct column-foot. *Didymoplexiella kinabaluensis* has a dense many-flowered inflorescence up to 13 cm long. The dorsal sepal and petals are adnate into a cuneate 3-lobed blade, which is itself adnate at the base to the lateral sepals. The sepals and petals are dull brownish purple and are sparsely warty on the exterior. The white flushed purple 3-lobed lip is shortly clawed at the base, the claw provided with a conical bilobed callus. The pedicels elongate to many times their original length after pollination. This presumably ensures that the dehiscing seed capsules are held well above the ground and facilitates wind dispersal.

Gastrodia grandilabris, with its campanulate flowers, looks superficially like *Neoclemensia spathulata*, discussed below. The dull brownish purple sepals and petals are adnate, forming a tube with white margins. The 3-keeled lip is unusually large for the genus and is coloured white, suffused with pale dull brownish purple. The claw of the lip bears 2 transversely oblong blue-green calli at its apex.

Carr (1935) described the monotypic genus *Neoclemensia* from dried material. He distinguished it from *Gastrodia* principally by the fact that the petals are almost entirely free and much shorter than the sepals. The bright orange petals are linear with a fimbriate spathulate apex. In all other aspects the floral morphology resembles *Gastrodia*, and fresh material should be studied to ascertain whether or not it can be maintained as a separate genus.

Tropidia, together with *Corymborkis*, form the tribe Tropidieae, which exhibit a primitive growth habit and root tuberoids resembling the subfamily Apostasioideae, as well as advanced features such as the production of lateral inflorescences and sectile pollen. All species of *Tropidia*, except *T. saprophytica*, are rather tall plants with often tough stems bearing several large plicate leaves. The flowers are usually non-resupinate, borne in simple, unbranched inflorescences and have a lip which is shortly spurred or saccate at the base and narrowed to a slender, often recurved tip. The Tropidieae are restricted to shady habitats, mostly in primary and old secondary forest in lowland and montane zones.

Tropidia saprophytica is the only saprophyte in the Tropidieae. Its general appearance is quite unlike that of other members of the subtribe. The floral morphology is typical of the genus, however, and closely resembles *T. curculigoides* and *T. graminea* Blume. All three species have a saccate cymbiform lip with a recurved tip and a ramentaceous ovary. The wiry branching stems of *T. saprophytica*, which bear brown tubular sheaths, resemble those of *T. angulosa* Blume, although they are much narrower and smaller.

FOLIAGE CHARACTERISTICS

Leaf colouration

A large number of terrestrial genera have attractive leaves with either zones of variegation or coloured veins. These are often popularly referred to as 'jewel orchids'. The majority of epiphytic species have, by contrast, rather dull foliage.

Leaf sheaths. *Appendicula foliosa* has densely dark brown punctate leaf sheaths. These spots also occur on the undersurface of the leaf toward the base and sometimes on the floral bracts. These were incorrectly thought by Ames and Schweinfurth to be "probably fungous growth." *Robiquetia pinosukensis* usually has purple-stained and speckled leaf sheaths and those of *Trichoglottis collenetteae* are spotted and streaked with black.

Leaf blades – species with coloured veins. *Anoectochilus longicalcaratus* has dark olive-green leaves, usually purple or rose-lilac beneath with golden or pink veins. The leaves of *Corybas kinabaluensis* are finely veined with white or red and those of *C. pictus* with white or pink. Several species of Bornean *Goodyera* have leaves with coloured veins, including *G. kinabaluensis* with red or pink veins. It is also distinguished by the hairy flowers. *Goodyera reticulata* has grey-green leaves veined with white or yellowish. Its flowers are glabrous. *Goodyera ustulata* has silver or pink leaf veins and hairy flowers. *Macodes lowii* has impressive velvety blackish green leaves with irridescent red and pink veins, whereas *M. petola* has white or golden veins.

Leaf blades – species with zones of variegation, spotting or brightly coloured leaves. The epiphytes *Bulbophyllum heldiorum* and *B. trifolium* (section *Globiceps*) both have dark green leaves suffused with red or purple. It is among the terrestrials, however, that we find the most interestingly coloured leaves.

Cystorchis variegata has yellowish olive-green leaves, which are irregularly blotched with deeper green and yellow.

Kuhlhasseltia javanica is a mid-elevation forest species of very wet mossy habitats. It can usually be recognised even when not in flower, since the leaves are quite distinctive. These are most often dark velvety green with white or pale pink undulate margins on the upper surface and pale purple beneath. In dried condition the leaf margins are distinctive in being somewhat undulate or crisped. The genus commemorates H. Kuhl and J. C. van Hasselt, both 19th-century Dutch naturalists.

Some of the numerous Bornean species of *Malaxis* are very attractive, whereas many others are of more modest appeal. The leaves of *M. calophylla* are bronze with spotted green crisped edges. *Malaxis lowii* has leaves that are dull brownish purple with a broad green median zone above and pale green with purple veins beneath. Lamb, in a sketch, describes the leaves as golden brown with a greenish white median zone. *Malaxis metallica*, so far only recorded from hill forest on ultramafic substrate, is, to quote Reichenbach in the *Gardeners' Chronicle* of 1879, "quite a gem!" This species was imported into Britain from 'Borneo' by a Mr. Bull in the 1870s and later figured in *Curtis's Botanical Magazine* as plate 6668 (1883). The leaves were described by Reichenbach as being "of a light rose colour underneath, and blackish purple above, with quite an exquisite metallic lustre, and when dried after having been boiled in hot water they are green." Lamb in notes accompanying a sketch of *SAN 91557* describes them as "dark purple to amethyst when in sunlight." The pseudobulbs are dark blackish purple, whereas the peduncle and flowers are paler amethyst-purple with dark green column-wings. It has also been found on ultramafic soils on Mt. Tawai (Tavai) in the Sandakan District where, in one location, it was growing in moss on rocks in a riverbed above high-water level. *Malaxis punctata* is distinctive in that the leaf blades and sheaths are spotted purple above and beneath.

Malleola witteana has leaves that are green above, but a beautiful violet beneath.

Nephelaphyllum flabellatum has ovate-cordate leaves, which are pale olive to grey-green with darker purplish brown blotches and fine dark green veins. The leaves of *N. pulchrum* are marbled with lighter and darker green and flushed with purple. Carr (1935) described those of *N. verruculosum* as "brown reticulate dark brown with bright green nerves above, purple beneath."

Many of the slipper orchids also have attractive foliage. Three of the five that occur on Mount Kinabalu have tessellated markings on the upper surface of the leaf. *Paphiopedilum dayanum* has dark and light yellow- or bluish green tessellation and *P. hookerae* var. *volonteanum* is tessellated dark and light green above, but spotted purple beneath. The leaves of *P. javanicum* var. *virens* are pale green, veined and lightly mottled darker green.

Plocoglottis acuminata has plicate leaves, which are often covered with yellow blotches.

Tainia purpureifolia is a curious little plant of damp mossy places in lower montane forest. The elliptic to ovate leaves are dark green above, purple beneath, or entirely blackish purple with pinkish margins and paler purple midvein. The right margin of the leaf is strongly crenulate, while the left is not or only slightly crenulate. The inflorescence bears 1 or 2 greenish yellow flowers flushed with purple.

Vrydagzynea argentistriata has bright green leaves with darker green reticulation and a silver main vein and margins above and dark grey-green beneath.

A few orchids have colourful immature leaves. Notable among these are certain species of *Chelonistele*, e.g. *C. amplissima* and *C. sulphurea* var. *sulphurea*, whose immature leaves are olive-brown to bronze-coloured. Members of *Coelogyne* section *Longifoliae*, e.g. *C. cuprea* var. *planiscapa* and *C. planiscapa* var. *grandis* very often have salmon-pink or flesh-coloured immature leaves.

DISTINCTIVE LEAF FORMS

Terete and semi-terete leaves. Several genera contain species having leaves that are circular or nearly circular in cross-section. These include *Bulbophyllum* (*B. pugilanthum, B. teres*), leaves on mature shoots of *Cleisocentron merrillianum, Cordiglottis* (represented by an unidentifiable sterile collection), *Dendrobium* section *Rhopalanthe* (e.g. *D. setifolium*), *D.* section *Strongyle* (*D. uncatum*), *Eria* (*E. pannea, E. pellipes*), *Schoenorchis* (*S. juncifolia*), and *Thelasis* (*T. carnosa*). Two genera, *Luisia* and *Paraphalaenopsis*, have exclusively terete leaves.

Dendrobium uncatum is readily distinguished by zig-zag stems, which emerge upright but then hang down as they lengthen. The slightly curved, sharply acute terete leaves are placed 7 mm apart and held almost at right angles to the stem. It is figured by Comber (1990: 238).

Laterally compressed, often fleshy leaves. Three sections of *Dendrobium* on Kinabalu have exclusively laterally compressed leaves: *Aporum (D. grande, D. indivisum* var. *pallidum, D. kiauense* and *D. patentilobum); Oxystophyllum (D. concinnum, D. oblongum); Strongyle (D. kentrophyllum)*.

Other examples of genera with laterally compressed leaves are *Ceratochilus, Microsaccus* and *Oberonia.*

Leaves with distinctive outlines. Notable among these is the epiphytic *Bulbophyllum ionophyllum* (section *Aphanobulbon*), which has rather thick, elliptic leaves that are very abruptly constricted toward the base. The terrestrial *Liparis kamborangensis* has distinctly cordate leaves, and *Nervilia punctata* has shiny, rather orbicular leaves with 5–7 lobes, of which only the apical one is acute. These resemble certain species of European *Cyclamen,* particularly *C. repandum.*

INDUMENTUM

The majority of orchids have smooth, glabrous foliage, but there are exceptions. Three genera on Mount Kinabalu have leaves that are, to a greater or lesser extent, hairy, namely *Pilophyllum, Trichotosia* and members of *Dendrobium* sections *Conostalix* and *Formosae.*

Pilophyllum is a monotypic terrestrial genus related to *Chrysoglossum.* It is distinguished by having leaves that are densely covered with yellowish brown hairs. The leaves are oblong to broadly elliptic with an acuminate apex and measure from 16 to 24 by 7.5 to 12 cm. The blade and petiole are so densely covered with hairs that they feel softly furry to the touch and have a golden velvety sheen. The inflorescence and pedicel-with-ovary are also densely pubescent. The flowers are non-resupinate. The sepals and petals are orange-yellow with dark purple or rich crimson markings at the apex. The lip is white with pale mauve speckling and a dark yellow area near the base.

The generic name *Trichotosia* (formerly treated as a section of *Eria*) is derived from the Greek word *trichotos* meaning hairy and is indeed an apt name. The genus is represented by 25 named species in Borneo and several that are poorly known or undescribed. The majority of species have the leaves, stems and inflorescences densely covered by long and usually reddish brown hairs, e.g. *T. ferox* and *T. pilosissima* – perhaps the hairiest species of the genus! Some species, e.g. *T. dasyphylla* (Parish & Rchb. f.) Kraenzl. from mainland Asia, even have densely hairy roots. Others are only distinctly densely hairy on the inflorescences and along the edges of the leaf sheaths, sometimes with hairs on the leaf-sheath surfaces and young leaves. Examples are *T. aurea,* a narrow-leaved species with short, dense inflorescences of small yellow flowers and *T. brevipedunculata,* a more robust, broader leaved species with deeply concave floral bracts, and larger flowers in short, dense inflorescences.

Ames and Schweinfurth (1920) commented that *Trichotosia brevipedunculata* has a wide range of variability in vegetative parts and flowers: "*Clemens 393* shows shorter leaves which are very oblique and bears evidences of four inflorescences, while *Clemens 291* shows inflorescences up to 4 cm. long." The young leaf sheaths and blades have densely appressed cinnamon-coloured pilose hairs, which more or less disappear on older parts, although persisting at the nodes. Wood found this species in 1985 along the Summit Trail growing as a terrestrial and pushing up through a tangle of *Freycinetia* and ferns at the base of a tree. It occurs elsewhere in the Crocker Range and on Mt. Trus Madi. *Trichotosia microphylla* is a small species with creeping stems covered with a white indumentum and bearing solitary greenish yellow flowers. *Trichotosia mollicaulis* is another robust species with very densely rusty brown appressed-pilose or velutinous leaf sheaths. The undersurface of the mature leaves is similar to the leaf sheaths, but the upper surface is glabrous. The immature leaves are appressed-pilose above and beneath. The inflorescences, unlike *T. brevipedunculata,* are laxly flowered with pilose reflexed bracts.

Members of *Dendrobium* sections *Conostalix* and *Formosae* (often incorrectly referred to as '*Nigrohirsutae*') both have blackish hairs on the leaf sheaths. Section *Conostalix* always has thin wiry stems that are never fleshy and swollen at the internodes. The stems of section *Formosae* are mostly cylindric with fleshy internodes, or sometimes fusiform with the central internodes noticeably more swollen. Section *Conostalix* is represented on Mount Kinabalu by *D. beamanianum* and *D. orbiculare* (*D. fuscopilosum*), having densely black-hirsute and brownish-hirsute leaf sheaths, respectively. Section *Formosae* is represented by *D. parthenium, D. singkawangense* (originally described from low elevation at Singkawang in West Kalimantan) and the exquisite *D. spectatissimum.*

NECKLACE ORCHIDS AND THEIR ALLIES

It is remarkable that in the relatively small area encompassed by Mount Kinabalu are found all five of the currently recognised subfamilies of the Orchidaceae. Furthermore, representative genera of a high percentage of the tribes and subtribes are also present. Among the major units of classification within the family, however, the subtribe Coelogyninae is particularly noteworthy for its diversity in this area. This subtribe is a member of the tribe Coelogyneae in the subfamily Epidendroideae (see the section below on Classification). The subtribe is poorly represented in New Guinea, where only a few species of *Coelogyne* and *Dendrochilum* have been recorded. By contrast, in Borneo they form an important and conspicuous element in the orchid flora, particularly at higher elevations. Six genera occur in Borneo, two of which, namely *Entomophobia* and *Nabaluia,* are endemic.

Coelogyne is the most conspicuous genus, particularly the beautiful 'necklace orchids' belonging to section *Tomentosae,* most of which have densely tomentose inflorescences. De Vogel (1992) enumerates 24 species of section *Tomentosae,* 12 of which occur on Mount Kinabalu. One species (*C. radioferens*) is always proteranthous, two species (*C. dayana* and *C. hirtella*) are proteranthous to synanthous, a further three are heteranthous (*C. latiloba, C. rochussenii* and *C. swaniana*), five are synanthous (*C. moultonii, C. reflexa, C. rhabdobulbon, C. rupicola* and *C. venusta*) and one is synanthous or rarely proteranthous (*C. longibulbosa*).

In species with proteranthous to synanthous inflorescences, the inflorescence-bearing shoot produces at the base a pseudobulb with leaves. The leaves from the undeveloped pseudobulb at the base are initially present but not mature, either concealed among the scales of the young shoot or partly emerged. The leaves mature as fruit is set and the pseudobulb elongates and enlarges. The base of the decayed inflorescence is often persistent at the apex of the old pseudobulbs.

The vegetative growth of plants with heteranthous inflorescences is by means of separate vegetative shoots. The inflorescence-bearing shoot never produces a pseudobulb with leaves at its base.

Coelogyne radioferens is a frequent species along the summit trail. It has 6–15-flowered pendent inflorescences up to 28 cm long, and is easily recognised by the pale to deep ochre-yellow sepals and petals and white or cream lip with a brown area on the exterior of the side-lobes. The interior of the hypochile is brown with white lines and the epichile (mid-lobe) is brown with a white margin and bears a central mass and radiating rows of white to yellow papillae. Forms with bright lime-green as well as more typical brownish olive-green floral bracts occur on Mount Kinabalu. Some fine forms have also been found on Mt. Trus Madi. The species is usually found on tree trunks and major branches, but also occurs on mossy rocks.

Coelogyne hirtella has an erect inflorescence with up to 10 wide-opening, often scented flowers. These are white, sometimes flushed green, with an ochre or brownish patch on the exterior of the side-lobes at the front. The epichile is white with a yellow centre with 4–6 brown to almost black swollen keels. The leaves including the petioles are said to be hirsute, but this is not very apparent. It is found as an epiphyte often low down on tree trunks and on major branches, or sometimes among rocks.

Coelogyne latiloba was described by de Vogel (1992) from a collection made by Sheila Collenette on Marai Parai Spur. This endemic species has a short 2–3-flowered inflorescence and a lip with a broad hypochile about 1.8 cm wide with large rounded side-lobes and a much smaller epichile. It is known only from the type.

Coelogyne moultonii is a robust species with closely placed narrowly cylindric pseudobulbs up to 28 cm long and leaves as large as 52 × 9.5 cm. The limply pendent inflorescences may have as many as 50 small white flowers enclosed and almost or entirely overtopped by the broad, creamy yellow to ochre-brown boat-shaped bracts. The lip has various yellowish to brown markings mostly on the exterior of the side-lobes and over the epichile keels. The specific epithet honours J. C. Moulton, who was Curator of the Sarawak Museum from 1905 to 1915, Director of the Raffles Museum and Library from 1919 to 1923, and later Chief Secretary of Government in Sarawak. This species generally grows as an epiphyte on tree trunks or as a lithophyte among mossy rocks. Wood found a large colony growing with *C. hirtella* among large ultramafic boulders on a steep slope between *Rhododendron nervulosum* bushes.

Coelogyne reflexa is a recently described species with a distinctive, reflexed, ligulate lip epichile. The flowers are cream, greenish yellow or creamy green with a light yellow blotch on the epichile.

A conspicuous species planted around Park Headquarters is *C. rhabdobulbon*, which has been incorrectly referred to *C. pulverula* (conspecific with *C. dayana*). It is a large clump-forming plant producing many pendent inflorescences up to 63 cm long and is a lovely sight when in full bloom. The specific epithet refers to the slender cylindric, stick-like pseudobulbs that may be as much as 28 cm long. The young foliage has been described as 'brownish liver coloured'. The white flowers are sometimes flushed with pale violet or pink and have yellow keels on the lip. Collectors have reported them to have a 'rich chocolate smell' or sometimes to be unpleasantly scented.

The endemic *Coelogyne rupicola* can be found on Gurulau and Marai Parai Spurs. It has a semi-erect, curving inflorescence with between 5 and 19 wide-opening, sweetly scented flowers. The leaves are plicate. The white flowers are sometimes suffused with salmon and have a golden-yellow patch and golden keels on the lip. *Coelogyne rupicola* is often found as an epiphyte at the base of mossy shrubs, sometimes on rocks or as a terrestrial. It sometimes occurs in very damp localities and is equally at home in the shade or in exposed sites.

Coelogyne longibulbosa is another species with very slender pseudobulbs up to 25 cm long, often occurring low down on tree trunks. The floral bracts only enclose the base of the flowers which, in common with most, are white with yellow to brown on the lip.

Another group of *Coelogyne* widespread on Mount Kinabalu is section *Longifoliae*. These species often have a flexuous, zig-zag rachis bearing salmon-pink flowers. The flowers are produced one or a few together in succession and have very narrow linear petals. Some species have a strongly laterally flattened, sometimes winged peduncle, namely *C. compressicaulis, C. cuprea* var. *planiscapa* and *C. planiscapa* var. *grandis.* Ames and Schweinfurth (1920) commented that collections of *C. compressicaulis* made by Clemens show "striking variability in habit, some of them having clusters of imbricating bracts on the rachis."

Coelogyne cuprea var. *planiscapa* and *C. planiscapa* var. *grandis* both have flattened peduncles and very similar salmon-pink or flesh-coloured flowers and are virtually impossible to identify without reference to the leaves. In *C. cuprea* var. *planiscapa* the leaves range from (6–) 12–15 cm long and are always ovate, obovate or oblong-ovate, with an obtuse and mucronate apex. The margin is undulate and the lower surface is frequently stained purple-red. The leaves of *C. planiscapa* var. *grandis* are acute and measure 30–46 cm long. *Coelogyne cuprea* var. *cuprea* has a terete peduncle and is recorded from Sumatra and Mt. Silam in Sabah.

Two other members of section *Longifoliae* are *C. kinabaluensis* and *C. tenompokensis*. *Coelogyne kinabaluensis* is easily distinguished by its very narrowly cylindric pseudobulbs 20–28 cm long, recalling *C. longibulbosa.* It has a terete peduncle and is most closely related to *C. cuprea* var. *cuprea.* *Coelogyne tenompokensis* is a graceful species with flattened, often curved pseudobulbs, narrow, acute leaves, flattened peduncle and a lip with 3 simple keels and an oblong-elliptic, undulate, obtuse mid-lobe. It is a frequent species between 1200 and 2100 m. A closely related, but probably undescribed plant has been collected in three different localities by Beaman

(*7482*) and the Clemenses (*35193* and *40820*). It is a more robust plant and has an elliptic, acuminate lip mid-lobe resembling that of *C. rochussenii.*

The group of species associated with *C. clemensii* and its varieties, *C. exalata, C. plicatissima, C. rigidiformis* and *C. subintegra* J. J. Sm., belonging to Carr's section *Rigidiformes,* show considerable variation and species delimitation is still a problem. These species, except for *C. clemensii,* have green to brownish olive flowers. *Coelogyne clemensii* var. *clemensii* is quite distinct, having short heteranthous inflorescences bearing sweetly scented, creamy white flowers; *C. clemensii* var. *angustifolia* has narrower leaves and bright yellow-green or olive-green sepals and petals and appears more closely allied to the other green-flowered species than to *C. clemensii.* Section *Rigidiformes* is a characteristic element of the montane orchid flora of Borneo.

Coelogyne septemcostata is a member of section *Speciosae,* which contains several large-flowered species such as *C. lawrenceana* Rolfe from Vietnam and *C. speciosa* (Blume) Lindl. from Sumatra, Java and Borneo. *Coelogyne septemcostata* has small pseudobulbs each with a large solitary leaf. The flowers appear one at a time in succession from a pendent peduncle and rachis. The sepals and petals are greenish cream, contrasting with the side-lobes of the lip, which have a rusty orange-brown hue. The mid-lobe is greenish cream or white, often with a chocolate-brown margin. There are 7 keels on the lip, each covered with long rusty orange-brown hairs. This species generally grows on trees overhanging rivers and streams.

The other important genus in the Coelogyninae, *Dendrochilum,* is dealt with above in the section, High-Elevation Genera. There is, however, an interesting species which is quite unlike any other from Borneo. *Dendrochilum planiscapum* was described by Carr from Tenompok and is peculiar in having a laterally flattened peduncle and non-resupinate greenish yellow flowers with a coiled, ligulate brown lip. Lamb has also found this species on Mt. Alab. A related species from Sumatra, *D. mirabile* J. J. Wood, also has a flattened peduncle and similar flowers. The leaves of the Sumatran species, however, are shorter and broader, the inflorescence denser, a basal projection is present on the pedicel and the column is much longer and lacks the characteristic arms (stelidia).

Ames (in Ames & Schweinfurth 1920) when describing *Pholidota kinabaluensis,* noted that the lip is at its base "adherent to the sides of the column nearly up to the stigmatic surface," but he obviously did not feel that this was sufficient reason for the creation of a new genus. De Vogel (1984) pointed out that several morphological characters are present that together obstruct the entrance of flowers by pollinators. The flower remains virtually closed during anthesis, and is enclosed by the newly detached bract. The throat of the lip is narrow and the lip cannot be pulled backward because of the adnation to the column. Furthermore, a transverse, swollen callus on the lip blocks access to the stigma. De Vogel concluded that these features were unique and proposed the name *Entomophobia* for this endemic, monotypic genus. The name is well chosen since it is derived from the Greek words for insect and fear.

The flowers of *Entomophobia kinabaluensis* are white or cream, sometimes with a pink spot on the lip. They open together and are characteristically all turned to one side to form a secund raceme.

Ames (in Ames & Schweinfurth 1920) commemorated Southeast Asia's highest mountain, Kinabalu (cf. Beaman et al. in press), in the name *Nabaluia*, a genus related to *Pholidota*. It differs in having stellately arranged hairs on the inside of the sepals and petals, long slender side-lobes in front of the hypochile with a horseshoe-shaped callus in between. Two species, *N. angustifolia* and *N. clemensii* occur on Mount Kinabalu. A third, *N. exaltata* de Vogel, is curiously absent, although Wood has found it in scrubby forest below the summit of Mt. Trus Madi.

Nabaluia angustifolia has linear leaves up to 2 cm wide and flowers with a 4-lobed hypochile, whereas *N. clemensii* has ovate-oblong to obovate-lanceolate leaves up to 3.5 cm wide and a 2-lobed hypochile. Both species have greenish cream through lemon-green to yellow flowers and a white to pinkish lip with various brown patches and spots. Ames originally identified *Clemens 268* (*N. angustifolia*) as *N. clemensii* despite the fact that "vegetatively the plants are quite unlike the type . . ." and ". . . although the free ends of the callus are elongated." *Nabaluia clemensii* was illustrated by Blanche Ames (plate 87) in Ames and Schweinfurth (1920). A good colour plate is provided by Sato (1991: 44).

Pholidota is represented by 10 species in Borneo (de Vogel 1988), 6 of which occur on Kinabalu. *Pholidota clemensii* is a pretty species with pink floral bracts contrasting with resupinate white flowers arranged in a dense erect inflorescence. Normally epiphytic, it is also recorded as a terrestrial in kerangas forest elsewhere in Borneo. Wood has found large colonies of this species growing epiphytically and as a terrestrial in lower montane forest with *Agathis* at 1560 m on Mt. Trus Madi. *Pholidota gibbosa* has 2-leaved pseudobulbs and elegant pendulous inflorescences of cream to pale salmon non-resupinate flowers arranged in 2 rows on a zig-zag rachis. The flowers all appear resupinate because of the pendulous habit of the inflorescence. *Pholidota pectinata* is a slender, graceful species known on Kinabalu only from the type collection. It has narrowly linear leaves and a wiry curving inflorescence bearing many small closely arranged non-resupinate creamy white to tan flowers. De Vogel (1988) comments that plants are very variable in size, especially the length and width of the leaves. Populations from Sabah all have a long scape, the inflorescence when stretched out exceeding the leaves. Populations from Kalimantan and Sarawak have a smaller scape.

POLLINATION

Little work has been done on the pollination of Bornean orchids, but a few general observations can be made. Agents of pollination are insects for the most part, primarily bees and wasps, butterflies and moths, and flies. One species of *Dendrobium* is known to be pollinated by birds. Interesting observations on pollination of orchids in various parts of the world are provided by van der Pijl and Dodson (1966).

BEES AND WASPS

These insects have an excellent sense of smell and taste and their vision extends toward the blue end of the spectrum into the ultraviolet. Orchids that are bee- or wasp-pollinated are therefore brightly coloured, but never red, and pleasantly scented. The lip is developed into a prominent landing platform with nectar guides in the form of lines or ridges often invisible to the human eye. Examples are *Arundina graminifolia* (bee), *Bromheadia* spp. (bee), *Cymbidium finlaysonianum* (bee), *Dendrobium crumenatum* (bee), *Grammatophyllum speciosum* (bee and wasp), *Phalaenopsis amabilis* (bee), *Spiranthes sinensis* (bee), *Vanda* spp. (bee) and *Vanilla* spp. (bee).

Some orchids deceive the male bee into believing the lip is the body of a female, and pollination is effected through 'pseudocopulation'. The European genus *Ophrys* is a classic example. The lip of *Cryptostylis*, only checked in Australian species, is thought to serve a similar function in attracting male wasps.

BUTTERFLIES AND MOTHS

Nectar is the food of these insects, which feed using a long hollow proboscis or tongue. The orchid flowers correspondingly have copious nectar contained in a spur or sac of varying length. Butterflies are active during the day and require a good landing place, so the orchid flower often has a horizontal lip, vivid colouration including red, and is sometimes pleasantly scented. Most moths fly at night and are attracted to white or green flowers with a strong nocturnal scent. Examples are *Dendrobium cumulatum* (probably butterfly), *D. parthenium* and *D. spectatissimum* (moth), *Habenaria* spp. and *Platanthera* spp. (moth) and *Renanthera bella* (probably butterfly). Deception is probably also involved to a certain degree, particularly in those species with a saccate spur or mentum.

FLIES

The behaviour of flies is more erratic than that of bees, butterflies or moths and the flowers have to guide them in a more devious way to effect pollination. Attraction is often by smell, either sweet, or more often, foetid and resembling rotting meat (e.g. *Bulbophyllum foetidolens*). The flowers are often dull coloured, usually purplish and may have a variety of fine hairs or appendages that move in the wind, e.g. *Bulbophyllum* sections *Cirrhopetalum* and *Epicrianthes.* Many have balanced versatile lips (*Bulbophyllum*), which move suddenly and throw the fly exactly onto the column. Other examples include *Corybas* spp., *Gastrodia* spp., *Liparis* spp., *Paphiopedilum rothschildianum* and *Plocoglottis* spp. *Corybas* have a fungus-like odour and are reputed to be pollinated by fungus flies.

Atwood (1985) demonstrated the significance of the strange geniculate, bifid staminode of *Paphiopedilum rothschildianum* in its pollination by syrphid flies. He suggested that the glandular hairs on the staminode mimic an aphid colony, the normal brood site of the syrphid larvae. The staminode attracts females of the fly *Dideopsis aegrota* to deposit their eggs on its surface. On alighting upon the staminode to lay their eggs, the flies sometimes fall into the lip. Their only exit is through a gap between the base of the lip and the column, and the flies pass beneath the stigma and pollinia, picking up a pollinium. A visit to a second flower followed by a similar scenario with the pollinium rubbed off onto the stigma will thereby effect pollination.

Male fruit flies are attracted by fragrant fruity odours (ketones) present as secretions on the flower parts of several orchids including *Bulbophyllum* section *Sestochilus*, namely *B. cornutum*, *B. macranthum* and *B. vinaceum*, *Dendrobium anosmum* (despite the epithet, smells of raspberries), and probably *Bromheadia finlaysoniana*. Other compounds, now being studied, seem to be linked to the development of the reproductive organs of female fruit flies.

The pollination mechanism of *Plocoglottis lowii* was discussed in detail by Burkill (1913). This species uses an explosive mechanism to effect pollination. The base of the lip acts as a strong spring, against the tension of which the lip is held back by the right-hand lateral sepal. The lip is pushed into this position as the flower opens by the expanding petals. When the lip is in position, the right-hand lateral sepal moves across a little and holds it, after which the petals expand fully. When an insect touches the right-hand sepal, the lip is released and springs sharply up into contact with the face of the column. The insect is thereby brought into contact with the column and the pollinia are removed as it struggles to get free. Holttum (1964) noted that no insect has actually been observed to do this.

Birds

Orchid flowers that are attractive to birds are bright pink, red, yellow or orange, are tubular in shape, contain plenty of nectar and are unscented. The many beautiful montane species of *Dendrobium* sections *Calyptrochilus*, *Oxyglossum* and *Pedilonum* from New Guinea have long-lasting flowers that are bird-pollinated. The only orchid occurring on Mount Kinabalu known to be bird-pollinated is the deep pink flowered *D. secundum* (section *Pedilonum*).

CLASSIFICATION

In the Enumeration of Taxa we have used an alphabetical arrangement of orchid genera and species to facilitate computer processing of the data and to make it easier for those unfamiliar with orchid classification to find the names of taxa. Such an arrangement, however, gives no indication of evolutionary relationships among the genera. The most recent fully published classification of the orchids was provided by Dressler (1981, 1990a, b, c). In addition to the floral and vegetative characters used by previous workers, Dressler has included much additional information from anatomy, cytology and micromorphology to improve and substantiate his classification. He divides the orchids into five subfamilies: Apostasioideae (the most primitive), Cypripedioideae, Spiranthoideae, Orchidoideae and Epidendroideae (the most advanced). In 1990 he reassigned the primitive tribe Neottieae, formerly included within subfamily Orchidoideae, to subfamily Epidendroideae.

Apostasia and *Neuwiedia* comprise the subfamily Apostasioideae and are usually considered to be the most primitive orchids and have been placed into a separate family, the Apostasiaceae, by some workers (e.g. Rasmussen 1985). Unlike the majority of orchids, which only have one fertile stamen, *Apostasia* has two and *Neuwiedia* three. The pollen grains of *Apostasia* and *Neuwiedia* are powdery and granular and never aggregated into pollinia, which is otherwise universal (except in some slipper orchids). Nevertheless, both have a column, which is the most characteristic structure found in the orchid flower, and today the two genera are generally accepted as forming a distinct subfamily within the Orchidaceae.

The Cypripedioideae (slipper orchids) have also, at times, been treated as a separate family, Cypripediaceae (Rasmussen 1985). The flowers have a short column with two lateral ventral fertile anthers and a large, usually shield-shaped, sterile anther or staminode. The stigma is ventral and stalked. They, too, have a column with fused stamens and style and are best included in the Orchidaceae.

The remaining orchid subfamilies have a column with a single anther and pollen aggregated into separate pollinia.

The Spiranthoideae include the attractive 'jewel orchids' typical of the forest floor flora. The subfamily is characterised by the dorsal erect anther, which is subequal to the rostellum, the mealy pollinia attached to a viscidium at the apex of the rostellum, and usually a creeping, fleshy rhizome.

The Orchidoideae include terrestrials such as *Corybas* and *Habenaria*. They have root-stem tuberoids, commonly called tubers, sectile pollinia and an anther firmly attached by its base to the column.

The most advanced and largest subfamily is the Epidendroideae, which is characterised by hard discrete pollinia in an apically attached anther. It includes almost all the epiphytic orchids, and is the most diverse subfamily in number of species. Its genera exhibit both sympodial and monopodial growth forms. Within the subfamily the vandoid orchids, represented by subtribe Aeridinae on Mount Kinabalu, are particularly numerous. They are distinguished by having lateral inflorescences, an anther with reduced partitions, superposed pollinia attached to a stipes, and viscidium.

Readers will notice that a number of species are referred to in the Enumeration by the modifiers 'aff.' or 'cf.' These identifications are tentative because we have had access to only limited material of many of the critical taxa and in some cases to only a single specimen or a fragment. The variation within a species is often difficult to ascertain. Therefore, we have used these modifiers to indicate probable affinities. The current trend in some taxonomic circles to use a very narrowly defined species concept is not, in our opinion, useful when applied to the poorly known Bornean orchids for which limited material is available.

The following summary lists the 121 currently known orchid genera of Mount Kinabalu according to the subfamilies, tribes, subtribes and groups into which they have been classified. The keys presented below are organised largely along the lines of this classification.

Classification of Kinabalu Orchids
(Adapted from Dressler, 1981, 1990a, b, c)

Subfamily Apostasioideae
Apostasia, Neuwiedia

Subfamily Cypripedioideae
Paphiopedilum

Subfamily Spiranthoideae

Tribe Tropidieae
Corymborkis, Tropidia

Tribe Cranichideae

Subtribe Goodyerinae
Anoectochilus, Cystorchis, Erythrodes, Goodyera, Hetaeria, Hylophila, Kuhlhasseltia, Lepidogyne, Macodes, Myrmechis, Pristiglottis, Vrydagzynea, Zeuxine

Subtribe Spiranthinae
Spiranthes

Subtribe Cryptostylidinae
Cryptostylis

Subfamily Orchidoideae

Tribe Diurideae

Subtribe Acianthinae
Corybas, Pantlingia

Tribe Orchideae

Subtribe Orchidinae
Platanthera

Subtribe Habenariinae
Habenaria, Peristylus

Subfamily Epidendroideae

Tribe Gastrodieae

Subtribe Gastrodiinae
Didymoplexiella, Gastrodia, Neoclemensia

Subtribe Epipogiinae
Epipogium, Stereosandra

Tribe Neottieae

Subtribe Limodorinae
Aphyllorchis

Tribe Nervilieae
Nervilia

Tribe Vanilleae

Subtribe Galeolinae
Cyrtosia, Galeola

Subtribe Vanillinae
Vanilla

Subtribe Lecanorchidinae
Lecanorchis

Tribe Malaxideae
Liparis, Malaxis, Oberonia

Tribe Cymbidieae

Subtribe Bromheadiinae
Bromheadia

Subtribe Eulophiinae
Dipodium, Eulophia, Geodorum, Oeceoclades

Subtribe Thecostelinae
Thecopus, Thecostele

Subtribe Cyrtopodiinae
Chrysoglossum, Claderia, Cymbidium, Grammatophyllum, Pilophyllum

Subtribe Acriopsidinae
Acriopsis

Tribe Arethuseae

Subtribe Bletiinae
Acanthephippium, Ania, Calanthe, Mischobulbum, Nephelaphyllum, Phaius, Plocoglottis, Spathoglottis, Tainia

Subtribe Arundinae
Arundina, Dilochia

Tribe Glomereae

Subtribe Glomerinae
Agrostophyllum

Subtribe Polystachyinae
Polystachya

Tribe Coelogyneae

Subtribe Coelogyninae
Chelonistele, Coelogyne, Dendrochilum, Entomophobia, Nabaluia, Pholidota

Tribe Podochileae

Subtribe Eriinae
Ascidieria, Ceratostylis, Eria, Porpax, Trichotosia

Subtribe Podochilinae
Appendicula, Poaephyllum, Podochilus

Subtribe Thelasiinae
Octarrhena, Phreatia, Thelasis

Tribe Dendrobieae

Subtribe Dendrobiinae
Dendrobium, Epigeneium, Flickingeria

Subtribe Bulbophyllinae
Bulbophyllum

Tribe Vandeac

Subtribe Aeridinae

Group 1
Adenoncos, Microsaccus, Taeniophyllum

Group 2
Abdominea, Arachnis, Bogoria, Ceratochilus, Cleisocentron, Cleisostoma, Cordiglottis, Dimorphorchis, Gastrochilus p.p., Micropera, Ornithochilus, Pomatocalpa, Renanthera, Schoenorchis, Thrixspermum, Trichoglottis

Group 3
Aerides, Macropodanthus, Paraphalaenopsis, Phalaenopsis, Pteroceras, Robiquetia, Vanda

Group 4
Gastrochilus p.p., Luisia

Group 5
Chamaeanthus, Chroniochilus, Grosourdya, Malleola, Microtatorchis, Pennilabium, Porrorhachis, Tuberolabium

KEYS

KEY TO SUBFAMILIES, TRIBES AND SUBTRIBES ON MOUNT KINABALU

1. Flowers with 2 or 3 fertile anthers. Pollen sticky but not forming pollinia

  2. Leaves plicate. Perianth segments more or less similar, lip not deeply saccate. Column with 3 or 2 anthers. Staminode, if present, not shield-like. Ovary with 3 chambers .... **Subfamily Apostasioideae**

  2. Leaves more or less coriaceous, conduplicate. Perianth segments very unequal, lateral sepals united, lip deeply saccate. Column with 2 lateral anthers and a large shield-like median staminode .... **Subfamily Cypripedioideae**

1. Flowers with a single fertile anther. Pollen aggregated into pollinia

  3. Anther remaining erect or bending back, never short and operculate at the apex of the column. Pollinia 2 or 4, soft, mealy, sectile or not. Leaves usually spirally arranged, convolute, not articulated at base

    4. Rostellum equalling the anther, usually elongate. Viscidium at the apex of the anther and attached to the apex of the pollinia or caudicles. Root-stem tuberoids (tubers) absent .... **Subfamily Spiranthoideae**

      5. Habit 'shrubby'. Nodular root tuberoids sometimes present (in *Tropidia*). Stems tough and rigid, elongate. Leaves plicate, or absent (*Tropidia saprophytica* only) .... **Tribe Tropidieae**

      5. Habit never 'shrubby'. Nodular root tuberoids absent. Stems fleshy, rarely elongate. Leaves convolute, conduplicate, or rarely absent (in saprophytic species of *Cystorchis*) .... **Tribe Cranichideae**

        6. Roots usually scattered along rhizome. Pollinia sectile. Lip often with emergent glands within a sac or spur .... **Subtribe Goodyerinae**

        6. Roots usually clustered. Pollinia not sectile. Lip sometimes saccate, never spurred, emergent glands absent

          7. Flowers resupinate. Lip usually saccate and containing 2 appendages .... **Subtribe Spiranthinae**

          7. Flowers non-resupinate. Lip simple .... **Subtribe Cryptostylidinae**

    4. Rostellum usually shorter than the anther. Anther usually projecting beyond the rostellum. Viscidium, if present, usually at the base or in the middle of pollinia. Plants usually with root-stem tuberoids (tubers) .... **Subfamily Orchidoideae**

8. Viscidium single or absent. Pollinia sectile or not, usually lacking caudicles. Column with a restriction below the anther. Lip with two spurs, or spurs absent .... **Tribe Diurideae (Subtribe Acianthinae)**

8. Viscidium usually double. Pollinia sectile, usually with prominent caudicles. Anther firmly attached to the column. Lip with a single spur, or absent .... **Tribe Orchideae**

9. Viscidia not borne on long rostellar stalks. Stigma entire, concave .... **Subtribe Orchidinae**

9. Viscidia borne on long rostellar stalks. Stigma entire or two-lobed, convex .... **Subtribe Habenariinae**

3. Anther eventually bending downward over column apex to become operculate, or operculate at column apex but not bending downward. Pollinia 2, 4, or 8, usually hard, waxy, sometimes mealy or sectile. Leaves distichous, usually articulated at base .... **Subfamily Epidendroideae**

10. Anther eventually bending downward over column apex to become operculate on column apex. Pollinia 2, 4, 6, or 8, usually flattened or clavate, with or without caudicles and viscidia, usually lacking a stipes

11. Pollinia mealy or sectile, 2 or 4, never superposed, lacking caudicles. Leaves, if present, non-articulated

12. Pollinia sectile

13. Leaves absent. Plants saprophytic

14. Sepals fused to varying degrees. Spur always absent .... **Tribe Gastrodieae (Subtribe Gastrodiinae)**

14. Sepals free. Spur present or absent

15. Lip without a pair of globular basal calli .... **Tribe Neottieae (Subtribe Limodorinae)**

15. Lip with a pair of globular basal calli .... **Tribe Gastrodieae (Subtribe Epipogiinae)**

13. Leaves solitary, subcircular or reniform, plicate, appearing after flowers mature. Plants autotrophic......... **Tribe Nervilieae**

12. Pollinia soft and mealy, non-sectile. Plants often liana-like climbers .... **Tribe Vanilleae**

16. Plants saprophytic

17. Flowers without a prominent calyculus or abscission layer at base of perianth .... **Subtribe Galeolinae**

17. Flowers with a prominent calyculus and an abscission layer at base of perianth .... **Subtribe Lecanorchidinae**

16. Plants non-saprophytic climbers .... **Subtribe Vanillinae**

11. Pollinia either mealy or hard, variously shaped, sometimes superposed; if soft, usually 8 and with caudicles. Leaves articulated

18. Pollinia variously shaped, never superposed, 2 to 8. Plants with or without pseudobulbs. Inflorescence terminal or lateral

19. Pollinia 2 or 4, quite naked, without caudicles, rarely with viscidia or stipes

20. Flowers lacking a column-foot. Leaves plicate or conduplicate .... **Tribe Malaxideae**

20. Flowers with a distinct column-foot. Leaves always conduplicate .... **Tribe Dendrobieae**

21. Lip usually immobile, not hinged at base. Mentum often spur-like .... **Subtribe Dendrobiinae**

21. Lip mobile, hinged at base. Mentum saccate .... **Subtribe Bulbophyllinae**

19. Pollinia 2 to 8, with distinct caudicles

22. Pollinia usually 8 and rather soft. Plants usually cormous. Leaves usually plicate. Inflorescence lateral or terminal .... **Tribe Arethuseae**

23. Inflorescence lateral. Leaves convolute or plicate .... **Subtribe Bletiinae**

23. Inflorescence terminal. Leaves conduplicate .... **Subtribe Arundinae**

22. Pollinia 2 to 8, usually rather hard. Plants with pseudobulbs, corms or slender stems. Leaves usually conduplicate. Inflorescence lateral or terminal..................... **Tribe Podochileae**

24. Pollinia always 8. Inflorescence usually lateral

25. Column-foot usually prominent. Stem slender or forming pseudobulbs (flattened in *Porpax*), usually of several internodes. Flowers variously coloured. Leaves rarely flattened and fleshy .... **Subtribe Eriinae**

25. Column-foot absent, or short, but never prominent. Stem with or without pseudobulbs. Flowers white or greenish yellow. Leaves sometimes laterally flattened and fleshy .... **Subtribe Thelasiinae**

24. Pollinia 4, 6, or rarely 8. Inflorescence terminal, lateral, or both .... **Subtribe Podochilinae**

18. Pollinia 4, or 2, superposed, or ovoid, 4, or 2. Plants usually with pseudobulbs of one internode. Inflorescences terminal .... **Tribe Coelogyneae (Subtribe Coelogyninae)**

10. Anther usually operculate at column apex, but not bending downward during development. Pollinia 2 or 4, usually dorsiventrally flattened, with reduced caudicles; stipes or stipites usually present

26. Habit usually sympodial

27. Pollinia 4. Plants with pseudobulbs of several internodes, on elongate stems. Inflorescence usually terminal .... **Tribe Glomereae**

28. Lip often basally saccate or spurred, without pseudopollen. Inflorescence terminal, often a dense globose head .... **Subtribe Glomerinae**

28. Lip neither saccate nor spurred, commonly with pseudopollen. Inflorescence terminal or lateral, simple or branched ....**Subtribe Polystachyinae**

27. Pollinia 2, often cleft, or 4. Plants usually cormous, or with pseudobulbs of several internodes. Inflorescence lateral

29. Pollinia laterally flattened .... **Tribe Cymbidieae (Subtribe Acriopsidinae)**

29. Pollinia thick and dorsiventrally flattened....... **Tribe Cymbidieae (Subtribes Bromheadiinae, Eulophiinae, Thecostelinae and Cyrtopodiinae)**

30. Inflorescence terminal. Stems always narrow and reed-like. Flowers always resupinate .... **Subtribe Bromheadiinae**

30. Inflorescence usually lateral, rarely terminal. Stems cormous, pseudobulbous, rarely slender and without pseudobulbs. Flowers resupinate or non-resupinate

31. Lip joined at its base with an outgrowth from the sigmoid or arcuate column and with the column-foot to form a tube at right angles to the base of the column ....**Subtribe Thecostelinae**

31. Lip otherwise

32. Plants usually cormous, or stems pseudobulbous, of one or a few internodes. Lip spurred, saccate, or not spurred (*Dipodium*). Leaves usually plicate ....**Subtribe Eulophiinae**

32. Plants not cormous, pseudobulbs usually present, rarely absent (*Claderia, Cymbidium elongatum*). Lip rarely saccate, not spurred. Leaves convolute or plicate .... **Subtribe Cyrtopodiinae**

26. Habit always monopodial .... **Tribe Vandeae(Subtribe Aeridinae)**

Key to Genera (excluding Saprophytes and Tribe Vandeae, Subtribe Aeridinae)

1. Flowers with 2 or 3 fertile anthers

 2. Perianth segments similar, the lip never deeply saccate

  3. Anthers 2, with or without a staminode. Inflorescence usually branched, curved and spreading, never erect .... **Apostasia**

  3. Anthers 3, staminode absent. Inflorescence simple, erect...**Neuwiedia**

 2. Perianth segments very unequal, the lip deeply saccate, slipper- or pouch-shaped. Anthers 2, lateral. Staminode median, large and shield-shaped .... **Paphiopedilum**

1. Flowers with a single fertile anther

 4. Plants terrestrial or climbing

  5. Plants terrestrial

   6. Rostellum elongate, equalling the anther. Root-stem tuberoids (tubers) absent

    7. Stems tough and rigid. Leaves plicate

     8. Lip widest at apex. Column long. Inflorescence often branched.. **Corymborkis**

     8. Lip widest at base. Column short. Inflorescence simple .... **Tropidia**

    7. Stems weak and fleshy, often brittle, never tough and rigid. Leaves convolute or conduplicate

     9. Pollinia sectile. Roots scattered along rhizome

      10. Anther ± erect. Rostellum erect or suberect

       11. Flowers resupinate

        12. Spur or saccate base of lip containing neither glands nor hairs (hairs may occur near mid-lobe only)

         13. Lip with spur which projects between lateral sepals .... **Erythrodes**

         13. Lip saccate, entirely enclosed by lateral sepals .... **Hylophila**

        12. Lip hairy within or having papillae or glands on either side near base or in spur or sac

14. Lip hairy within .... **Goodyera**

14. Lip otherwise

15. Apex of lip not abruptly widened into a distinct spathulate or transverse, bilobed blade

16. Saccate base of lip with a transverse row of small calli. Plant robust, up to 100 cm tall......................... **Lepidogyne**

16. Saccate base of lip or spur containing stalked or sessile glands. Plants much smaller

17. Hypochile swollen at base into twin lateral sacs each containing a sessile gland. Epichile with fleshy involute margins, forming a tube .... **Cystorchis**

17. Hypochile otherwise, containing 2 stalked glands. Epichile otherwise .... **Vrydagzynea**

15. Apex of lip abruptly widened into a distinct spathulate or transverse, bilobed blade

18. Claw of lip with a toothed or pectinate flange on either side .... **Anoectochilus**

18. Claw of lip otherwise

19. Dorsal sepal and petals connivent, forming a hood

20. Column without appendages .... **Kuhlhasseltia**

20. Column with 2 narrow wings which project into the base of the lip.................................. **Pristiglottis**

19. Dorsal sepal and petals free

21. Lip with a long claw. Stigmas on short processes. Inflorescence 1–2-flowered.................... **Myrmechis**

21. Lip with a short claw. Stigmas sessile. Inflorescence several- flowered .... **Zeuxine**

11. Flowers non-resupinate

22. Lip and column twisted to one side...................................... **Macodes**

22. Lip and column straight.. **Hetaeria**

9. Pollinia not sectile. Roots in a close fascicle

23. Flowers non-resupinate, small, arranged spirally in a dense inflorescence..**Spiranthes**

23. Flowers resupinate, large, arranged in all directions in a lax inflorescence..**Cryptostylis**

6. Rostellum usually shorter than the anther. Root-stem tuberoids (tubers) present or absent

24. Root-stem tuberoids (tubers) present

25. Lip with 2 spurs, or spurs absent

26. Lip 2-spurred, tubular below. Flowers helmet-shaped ....**Corybas**

26. Lip spurless

27. Inflorescence produced with the leaf. Lip orbicular. Column with a tooth-like process below..................**Pantlingia**

27. Inflorescence produced before the leaf. Lip 3-lobed, the base embracing the column. Column without a tooth-like process ....**Nervilia**

25. Lip with 1 spur

28. Stigmas not freely extending in front of column, sometimes connate or adpressed to lip hypochile

29. Lip simple, strap-shaped. Spur cylindric, rather long, not swollen at apex. Stigmas joined to form a concave structure, free from hypochile..**Platanthera**

29. Lip 3-lobed. Spur short, usually globular, saccate or fusiform. Stigmas convex, cushion-like, connate with or adpressed to hypochile ....**Peristylus**

28. Stigmas each on a stigmatophore extending from the column, free from hypochile ....**Habenaria**

24. Root-stem tuberoids (tubers) absent

30. Pollinia 2 or 4, naked, i.e. without caudicles, viscidia and stipes usually absent

31. Column-foot absent

32. Column long. Flowers usually resupinate. Lip without 2 large basal auricles, apex rarely pectinate ... **Liparis** (in part)

32. Column short. Flowers non-resupinate. Lip with 2 large basal auricles, apex often pectinate .... **Malaxis**

31. Column-foot prominent

33. Rhizomatous part of shoot (sometimes also the non-rhizomatous part) carrying one-noded pseudobulbs .... **Epigeneium** (in part, sometimes *E. kinabaluense*)

33. Non-rhizomatous part of shoot consisting of several internodes, wholly or partly fleshy, with or without pseudobulbs .... **Dendrobium** (in part, some species in sections *Conostalix* and *Distichophyllum*)

30. Pollinia 2 to 8, with caudicles (sometimes reduced)

34. Inflorescences numerous or not, borne along a slender, leafy stem, lateral or terminal

35. Inflorescences lateral

36. Pollinia 2. Inflorescences not numerous. Flowers c. 4 cm across .... **Cymbidium** (in part, *C. elongatum*)

36. Pollinia 6 or 8. Inflorescences numerous. Flowers much smaller

37. Pollinia 6. Leaf sheaths and flowers glabrous .... **Appendicula** (in part, *A. longirostrata*)

37. Pollinia 8. Leaf sheaths and flowers usually covered in reddish brown hispid hairs .... **Trichotosia** (in part, *T. brevipedunculata, T. rubiginosa*)

35. Inflorescences terminal

38. Pollinia 8

39. Flowers large, up to 8 cm across (sometimes peloric). Petals much broader than sepals. Inflorescence usually unbranched. Floral bracts small, acute, persistent ....**Arundina**

39. Flowers much smaller. Petals similar to sepals. Inflorescence branching. Floral bracts conspicuous, concave, deciduous .... **Dilochia**

38. Pollinia 2 .... **Bromheadia** (in part, *B. finlaysoniana*)

34. Inflorescences never numerous, usually solitary, never borne along a slender, leafy stem, usually lateral, sometimes axillary or terminal

40. Pollinia 2

41. Plant glabrous. Flowers resupinate .... **Chrysoglossum**

41. Plant densely covered in yellowish-brown hairs. Flowers non-resupinate .... **Pilophyllum**

40. Pollinia 4 or 8

42. Pollinia 4

43. Inflorescence arcuate, strongly decurved .... **Geodorum**

43. Inflorescence otherwise

44. Lip not spurred

45. Lip convex, adnate to sides and apex of column-foot to form a sac, usually with an elastic hinge that springs when touched. Pseudobulbs often elongated into a leafy stem ....**Plocoglottis**

45. Lip never convex, free or fused at base to base of column, without an elastic hinge. Pseudobulbs short, often enclosed in sheathing leaf-bases ...... **Cymbidium** (in part, *C. borneense, C. ensifolium* subsp. *haematodes, C. lancifolium*)

44. Lip spurred

46. Lip entire, or 3-lobed (mid-lobe not bilobulate) .... **Eulophia** (in part, *E. graminea, E. spectabilis*)

46. Lip '4-lobed', mid-lobe bilobulate........... **Oeceoclades**

42. Pollinia 8

47. Lip spurred, or gibbous and partially adnate to and embracing column to form a tube

48. Pseudobulbs 1-leaved. Plants remaining green when damaged. Lip spurred

49. Inflorescence lateral ....**Ania**

49. Inflorescence terminal ....**Nephelaphyllum**

48. Pseudobulbs 2 to several-leaved. Plants turning bluish-black when damaged. Lip spurred or gibbous

50. Column margins fused with the base of the lip over nearly their entire length. Lip spurred .... **Calanthe**

50. Column margins fused with lip only at or near the base. Lip shortly spurred or gibbous .... **Phaius**

47. Lip not spurred

51. Pseudobulbs 1-leaved

52. Leaf base cordate in mature plants, petiole absent .... **Mischobulbum**

52. Leaf base ± decurrent along a petiole.................................**Tainia**

51. Pseudobulbs 2- to several-leaved

53. Flowers urn-shaped, sepals fleshy, fused to form a swollen tube, free at the apices. Lip movably hinged to a column-foot, not clawed or callose .... **Acanthephippium**

53. Flowers with free, usually spreading sepals. Lip not movably hinged, mid-lobe very narrowly clawed, with 2 ovoid, often pubescent basal calli. Column-foot absent .... **Spathoglottis**

5. Plants climbing

54. Leaves fleshy, never plicate. Stems fleshy. Pollinia soft and mealy, as monads ....**Vanilla**

54. Leaves plicate. Stem never fleshy, often rather tough, sometimes brittle. Pollinia 2, cleft

55. Habit monopodial. Stems not distant on a creeping rhizome. Leaves distichous, imbricate, ensiform. Flowers pale yellowish with pink to crimson blotches .... **Dipodium**

55. Habit sympodial. Stems placed distantly on a creeping rhizome. Leaves elliptic, neither distichous or imbricate. Flowers green .... **Claderia**

4. Plants epiphytic or lithophytic

56. Pollinia 2 or 4, naked, i.e. without caudicles

57. Column-foot absent. Leaves equitant, distichous, bilaterally flattened .... **Oberonia**

57. Column-foot prominent. Leaves dorsiventral, or occasionally bilaterally flattened (in *Dendrobium* sections *Aporum*, *Oxystophyllum* and *Strongyle* only)

58. Lip usually immobile, not hinged at base. Mentum often spur-like

59. Rhizomatous part of shoot (sometimes also the non-rhizomatous part) bearing one-noded, 1- or 2-leaved pseudobulbs. Flowers long-lasting........**Epigeneium** (in part)

59. Non-rhizomatous part of shoot (when present) consisting of several internodes, with or without 1- to several-noded pseudobulbs. Flowers ephemeral or long-lasting

60. Stems superposed, the non-rhizomatous part of the shoot consisting of several quite long thin internodes, the uppermost pseudobulbous and 1-leaved. Flowers always ephemeral..**Flickingeria**

60. Stems not superposed; either 1) rhizomatous, 2) erect and many-noded, 3) erect and 1-noded or several-noded from a many-noded rhizome, or 4) rhizome absent, new stems of many nodes arising from base of old ones. Leaves 1 to many. Flowers long-lived or ephemeral
.... **Dendrobium** (in part)

58. Lip movably hinged to column-foot. Mentum saccate
....**Bulbophyllum**

56. Pollinia 2 to 8, with distinct, though sometimes reduced, caudicles

61. Stems slender, leafy, without pseudobulbs

62. Pollinia 2 or 4

63. Pollinia 2. Leaves sometimes laterally flattened
.... **Bromheadia** (in part)

63. Pollinia 4. Leaves never laterally flattened............**Podochilus**

62. Pollinia 6 or 8

64. Pollinia 6..**Appendicula**

64. Pollinia 8

65. Inflorescence terminal, usually globose, surrounded by bracts. Flowers white or yellow................**Agrostophyllum**

65. Inflorescence lateral, terminal or subterminal, never of globose heads. Flowers variously coloured

66. Column-foot absent

67. Leaves laterally compressed or terete, distichous. Flowers yellowish green .... **Octarrhena**

67. Leaves dorsiventral, linear to linear-elliptic or strap-shaped

68. Inflorescence and sepals white-tomentose. Flowers arranged in whorls, non-resupinate
....**Ascidieria**

68. Inflorescence and sepals glabrous. Flowers not arranged in whorls, resupinate
.......**Thelasis** (in part, *T. carinata, T. micrantha*)

66. Column with a short or long foot

69. Leaf sheaths covered with reddish brown, or rarely white, hispid hairs. Leaves never fleshy and subterete .... **Trichotosia**

69. Leaf sheaths glabrous. Leaves sometimes fleshy and subterete

70. Stems one-leaved .... **Ceratostylis**

70. Stems few- to many-leaved

71. Stems elongate, leafy throughout entire length

72. Inflorescence axillary, few-flowered, glabrous

73. Floral bracts large and coloured .... **Eria** (section *Cylindrolobus*)

73. Floral bracts minute, green or brownish .... **Poaephyllum**

72. Inflorescence terminal or subterminal, usually densely many-flowered, densely hirsute. Floral bracts small .... **Eria** (section *Mycaranthes*)

71. Stems short, entirely enclosed by imbricate leaf sheaths. Inflorescence a densely flowered raceme with small bracts .... **Phreatia** (in part, *P. amesii, P. densiflora, P. monticola, P. secunda*)

61. Stems pseudobulbous, pseudobulbs sometimes small and entirely enclosed by imbricate leaf sheaths

74. Pollinia 2

75. Lip joined at its base with an outgrowth from the column and with column-foot to form a tube at right angles to base of column .... **Thecostele**

75. Lip otherwise

76. Plants very large, with pseudobulbs up to 3 m or more long. Flowers up to 10 cm across. Sepals and petals up to 2.6 cm wide, with large irregular blotches. Stipes U-shaped .... **Grammatophyllum**

76. Plants much smaller. Flowers up to 5.7 cm across. Sepals and petals narrow, without blotching. Stipes absent .... **Cymbidium** (all epiphytic species except *C. lancifolium*)

74. Pollinia 4 or 8

77. Pollinia 4

78. Inflorescence terminal

79. Flowers with a distinct column-foot, always non-resupinate .... **Polystachya**

79. Flowers without a column-foot, resupinate, or, more rarely, non-resupinate

80. Basal half of the narrow, saccate lip adnate to basal half of column. Apical half of lip separated by a transverse, high, fleshy callus .... **Entomophobia**

80. Lip otherwise

81. Lip hypochile with long, slender lateral front lobes .... **Nabaluia**

81. Lip hypochile without such lobes

82. Column usually with lateral arms (stelidia), although normally absent in *D. stachyodes* .... **Dendrochilum**

82. Column without lateral arms

83. Lip hypochile saccate, distinctly separate from epichile. Lip rarely 3-lobed............... **Pholidota**

83. Lip hypochile, although often concave, not sharply distinct from epichile. Lip almost always 3-lobed

84. Side-lobes of lip (when present) narrow, borne from front part of hypochile at right angles to the epichile. Hypochile narrow, saccate .... **Chelonistele**

84. Side-lobes of lip broad, widening gradually from base of lip. Hypochile ± concave, broader and rarely saccate .... **Coelogyne** (in part)

78. Inflorescence lateral

85. Lip joined at its base with an outgrowth from the column and with column-foot to form a tube at right angles to base of column... **Thecopus**

85. Lip otherwise

86. Lateral sepals united into a synsepalum. Stipes long, linear .... **Acriopsis**

86. Lateral sepals free. Stipes absent .... **Cymbidium** (in part, *C. lancifolium* only)

77. Pollinia 8

87. Sepals connate to varying degrees, forming a tube. Pseudobulbs flattened .... **Porpax**

87. Sepals free

88. Column with a prominent foot. Rachis usually hirsute or woolly. Pseudobulbs rarely flattened .... **Eria** (in part)

88. Column absent or short. Rachis glabrous. Pseudobulbs sometimes flattened

89. Column with a short foot. Anther cap horizontal on top of column, not beaked .... **Phreatia** (in part, *P. listrophora, P. sulcata*)

89. Column-foot absent. Anther cap vertical behind column, beaked .... **Thelasis** (in part, *T. capitata, T. carnosa, T. variabilis*)

Key to Genera of the Subtribe Aeridinae

1. Pollinia 4

2. Pollinia more or less equal, globular, free from each other ..... (**Group 1**)

3. Plants with normal leaves borne on a distinct stem. Roots lacking chlorophyll

4. Leaves dorsiventral. Inflorescence 1–4-flowered. Flowers green .... **Adenoncos**

4. Leaves bilaterally flattened/compressed. Inflorescence 2-flowered. Flowers white .... **Microsaccus**

3. Plants without leaves, or leaves reduced to minute brown scales. Stem minute. Roots terete or flattened, with chlorophyll ..... **Taeniophyllum**

2. Pollinia appearing as 2 pollen masses, each completely divided into either rather unequal, or more or less equal, semiglobular free halves .... (**Group 2**)

5. Flowers without a distinct column-foot

6. Leaves bilaterally flattened, distichous, with sheathing bases, resembling those of *Microsaccus.* Mid-lobe of lip expanded into a broadly oblong-elliptic, emarginate blade .... **Ceratochilus**

6. Leaves otherwise

7. Lip adnate to column, not movable

8. Spur without a longitudinal internal septum

9. Hypochile of lip globose-saccate, the side-lobes reduced to low, often fleshy edges of the sac, mid-lobe fan-shaped
....**Gastrochilus** (*G. patinatus* only – not recorded from Kinabalu)

9. Hypochile otherwise

10. Calli and outgrowths absent from back-wall of spur

11. Mid-lobe of lip distinctly pectinate-fringed. Stipes linear, about 4 times as long as diameter of pollinia
.... **Ornithochilus**

11. Mid-lobe of lip otherwise. Stipes about twice as long as diameter of pollinia

12. Spur separated from apical portion of lip by a fleshy transverse wall .... **Abdominea**

12. Spur without a transverse wall

13. Flowers vivid orange-red, large and showy. Lip much shorter than dorsal sepal.............. **Renanthera**

13. Flowers white to pink, bluish or mauve, small. Lip as long as or longer than dorsal sepal............ **Schoenorchis**

10. Back-wall of spur ornamented with calli and/or outgrowths

14. Spur with a usually hairy ligulate tongue at the entrance
.... **Trichoglottis**

14. Spur with a tongue or valvate callus, often forked at the tip, projecting diagonally from deep inside .... **Pomatocalpa**

8. Spur with a distinct longitudinal median septum. Rostellum projection well developed, turned obliquely sideward and upward
.... **Micropera**

7. Lip not adnate to column, movable. Sepals and petals narrow, usually rather spathulate. Spur short and conical................................ **Arachnis**

5. Flowers with a distinct, though sometimes short column-foot

15. Flowers large, showy, dimorphic, the basal two always strongly scented, differently coloured from those above............................. **Dimorphorchis**

15. Flowers much smaller, not dimorphic

16. Lip without a distinct spur or sac, but hypochile often somewhat concave. Flowers ephemeral

17. Leaves terete .... **Cordiglottis**

17. Leaves dorsiventral .... **Thrixspermum**

16. Lip with a distinct spur or sac. Flowers long-lasting

18. Spur without a median longitudinal septum

19. Stems very short. Inflorescence borne below the leaves. Flowers small, greenish yellow and white, marked with crimson on lip. Lip deeply saccate .... **Bogoria**

19. Stems long, usually pendent. Inflorescences axillary. Flowers translucent lavender-blue. Lip distinctly spurred .... **Cleisocentron**

18. Spur with a median longitudinal septum .... **Cleisostoma**

1. Pollinia 2

20. Pollinia sulcate or porate

21. Pollinia sulcate, i.e. more or less, but not completely cleft or split .... (**Group 3**)

22. Column-foot absent or very indistinct

23. Flowers small, crowded on to a usually densely many-flowered raceme or panicle. Spur sometimes with calli or scales inside. Stipes linear, spathulate, uncinnate, rarely hamate. Viscidium usually small .... **Robiquetia**

23. Flowers often large and showy, usually only a few, well spaced on a raceme. Spur lacking internal ornamentation. Stipes short and broad. Viscidium broadly orbicular .... **Vanda**

22. Column-foot distinct, though sometimes short

24. Leaves very long and terete .... **Paraphalaenopsis**

24. Leaves dorsiventral, much shorter

25. Spur or sac, if present, developed from the hypochile. Epichile dorsiventral

26. Spur absent, or rudimentary. Lip with at least one forward-pointing forked appendage. Flowers few, sometimes large and showy, distichous .... **Phalaenopsis**

26. Spur well developed. Forked appendages absent. Flowers many, facing in every direction, developing simultaneously .... **Aerides**

25. Spur or sac borne centrally on lip. Epichile reduced, fleshy

27. Rostellum projection conspicuously elongated. Lip bent upward so as to make a right angle with the column-foot, distinctly unguiculate. Flowers long-lasting, developing simultaneously .... **Macropodanthus**

27. Rostellum projection inconspicuous. Lip continuing the line of the column-foot, subsessile. Flowers ephemeral, usually developing successively .... **Pteroceras**

21. Pollinia porate .... (**Group 4**)

28. Leaves dorsiventral. Hypochile globose-saccate ...... **Gastrochilus**

28. Leaves terete. Lip neither saccate or spurred .... **Luisia**

20. Pollinia entire .... (**Group 5**)

29. Column with a distinct foot. Lip movable

30. Lip without a spur, side-lobes sometimes fimbriate .... **Chamaeanthus**

30. Lip spurred or saccate

31. Spur-like conical portion of lip more or less solid. Peduncle short, glabrous .... **Chroniochilus**

31. Sac or spur thin-walled, without interior fleshiness. Peduncle longer, often prickly-hairy .... **Grosourdya**

29. Column without a foot. Lip not movable

32. Lip with a bristle or tooth inside near the apex. Floral bracts conspicuous, triangular, leafy .... **Microtatorchis**

32. Lip without an apical bristle or tooth. Floral bracts not leafy

33. Side-lobes of lip very large, often fringed .... **Pennilabium**

33. Side-lobes of lip small, never fringed

34. Lip spurred. Lateral sepals not adpressed to lip

35. Rachis slender, never thickened and sulcate, or clavate. Column hammer-shaped. Stipes linear-spathulate, much broadened at apex .... **Malleola**

35. Rachis fleshy, sulcate, or clavate. Column short and stout. Stipes linear, much reduced ...... **Tuberolabium**

34. Lip not truly spurred, but with a spur-like tubular cavity. Lateral sepals adpressed to lip .... **Porrorhachis**

KEY TO SAPROPHYTIC GENERA (LEAFLESS TERRESTRIALS LACKING CHLOROPHYLL)

1. Flowers with sepals and petals fused (connate) to a varying degree, often appearing campanulate and always resupinate

    2. Petals fimbriate, bright orange .... **Neoclemensia**

    2. Petals otherwise

        3. Petals fused for about half their length with dorsal sepal. Column with long decurved arms (stelidia), without a foot. Stigma near top of column .... **Didymoplexiella**

        3. Petals fused equally to dorsal and lateral sepals. Column without long, decurved stelidia. Stigma at base of column .... **Gastrodia**

1. Flowers with free, spreading or connivent sepals and petals, not appearing campanulate; resupinate or non-resupinate

    4. Flowers always resupinate, lip lowermost

        5. Stem simple, slender or fleshy. Sepals and petals not encircled by a shallow denticulate calyculus (cup)

            6. Lip divided into a distinct hypochile and epichile. Hypochile with or without twin lateral sacs

                7. Hypochile swollen at base into twin lateral sacs, each containing a globular sessile gland. Epichile with fleshy involute margins, tube-like. Flowers pink to reddish, tipped with white .... **Cystorchis** (*C. aphylla* only)

                7. Hypochile without such sacs. Epichile 3-lobed. Flowers greenish white or creamy white and purple .... **Aphyllorchis**

            6. Lip not divided into a distinct hypochile and epichile

                8. Flowers large, reddish brown. Lip 3-lobed, saccate .... **Eulophia** (*E. zollingeri* only)

                8. Flowers small, white, or white flushed with purple at apex. Lip entire, with or without a spur

                    9. Lip spurred, strap-shaped, margin not undulate .... **Platanthera** (*P. saprophytica* only)

                    9. Lip not spurred, narrowly elliptic, margin undulate .... **Stereosandra**

        5. Stem branching, tough and wiry. Sepals and petals surrounded by a shallow denticulate calyculus (cup) .... **Lecanorchis**

    4. Flowers non-resupinate or resupinate

10. Stem simple. Flowers non-resupinate. Lip with a short spur .... **Epipogium**

10. Stem branching. Flowers resupinate or non-resupinate. Spur absent

11. Stems long and climbing. Flowers resupinate. Sepals brownish mealy on reverse. Fruits dry, dehiscent .... **Galeola**

11. Stems short, never climbing. Flowers non-resupinate. Sepals brownish mealy or blackish ramentaceous on reverse. Fruits succulent and indehiscent or dry and dehiscent

12. Plant robust, with several thick, fleshy stems borne from each rhizome. Sepals obtuse, concave, brownish mealy on reverse. Fruits succulent, indehiscent..**Cyrtosia**

12. Plant slender, with a single narrow, wiry stem borne from each rhizome. Sepals acute, reflexed, blackish ramentaceous on reverse. Fruits dry, dehiscent....... **Tropidia** (*T. saprophytica* only)

ENUMERATION OF TAXA

55. ORCHIDACEAE

Dressler, R. L. (1981). The Orchids: Natural History and Classification. Harvard Univ. Press. Dressler, R. L. (1990a). The Orchids: Natural History and Classification. Ed. 2. Harvard Univ. Press. Dressler, R. L. (1990b). The Spiranthinae: grade or subfamily? Lindleyana 5: 110–116. Dressler, R. L. (1990c). The major clades of the Orchidaceae-Epidendroideae. Lindleyana 5: 117–125.

55.1. **ABDOMINEA** J. J. Sm.

Bull. Jard. Bot. Buit., ser. 2, 14: 52 (1914).

Tiny short-stemmed monopodial epiphytes. *Leaves* obovate, 3–5 cm long. *Inflorescences* many-flowered racemes with persistent bracts. *Flowers* very small, about 1.2 mm across; *sepals* and *petals* similar, cinnamon-orange with black spots; *lip* sac-shaped with a pointed apex, epichile pure white with a red-brown spot, spur or sac translucent green; *column* short and stout; *anther-cap* purple; *rostellar projection* longer and broader than the rest of the column, from a narrow base widening into a cordate, acuminate blade; *stipes* linear-clavate, acute, more than 3 times the diameter of the pollinia; *viscidium* very inconspicuous if at all present; *pollinia* 4, appearing as 2 pollen masses.

A monotypic genus distributed from Thailand to the Philippines.

55.1.1. Abdominea minimiflora (Hook. f.) J. J. Sm., Bull. Jard. Bot. Buit., ser. 2, 25: 98 (1917).

Saccolabium minimiflorum Hook. f., Fl. Brit. Ind. 6: 59 (1890).

Epiphyte. Lowlands and hill forest, sometimes on ultramafic substrate. Elevation: 500–900 m.

General distribution: Sabah, Peninsular Malaysia, Thailand, Java, Philippines.

Collections. MINITINDUK: 900 m, *Carr SFN 27829* (SING); PENATARAN RIVER: 500 m, *Beaman 8856* (K); TENOMPOK: *Clemens s.n.* (BM, BM), *26739* (BM).

55.2. **ACANTHEPHIPPIUM** Blume

Bijdr., 353 (1825).

Terrestrial herbs. *Pseudobulbs* few-noded, rather long, fleshy, subcylindrical, covered by sheaths when young, 2- to 3-leaved at apex. *Leaves* suberect, large, plicate. *Inflorescence* lateral, shorter than leaves, fleshy, few-flowered. *Flowers* erect, rather large, urn-shaped; *sepals* fleshy, connate to form a swollen tube, free at apices; *lateral sepals* forming an obtuse mentum with column-foot;

petals narrower, enclosed together with lip in sepal tube; *lip* 3-lobed, movably hinged to column-foot; *column* long, foot curved; *pollinia* 8, unequal.

About 15 species distributed from tropical Asia to the Pacific islands.

55.2.1. Acanthephippium javanicum Blume, Bijdr., 354, t. 47 (1825).

Terrestrial. Lowlands, hill forest, lower montane forest in wet situations. Elevation: 300–1200 m.

General distribution: Sabah, Peninsular Malaysia, Sumatra, Java.

Collections. BUNGOL/KAUNG: 300–600 m, *Clemens 26020* (BM, K); KAUNG: 400 m, *Carr SFN 27449* (SING); KIAU: *Clemens 356* (AMES); PENIBUKAN: 1200 m, *Clemens 32137* (BM).

55.2.2. Acanthephippium lilacinum J. J. Wood & C. L. Chan, ined. Plate 7A.

Terrestrial. Lowlands. Elevation: 300 m.

General distribution: Sabah.

Collection. RANAU: 300 m, *Collenette 33* (BM).

55.3. ACRIOPSIS Reinw. ex Blume

Bijdr., 376 (1825).

Minderhoud, M. E. & de Vogel, E. F. (1986). A taxonomic revision of the genus *Acriopsis* Reinwardt ex Blume (Acriopsidinae, Orchidaceae). Orch. Monogr. 1: 1–16.

Epiphytic herbs. *Rhizome* creeping, branched. *Roots* slender, branched, with tufts of erect, acuminate branches at base of pseudobulb. *Pseudobulbs* crowded, ovoid, covered at the base by thin, silvery bracts. *Leaves* apical, (1–)2–3(–4), oblong or linear, the midrib sunken above, prominent below, petiolate. *Inflorescence* a raceme or panicle, usually many-flowered, arising from the base of the pseudobulb on a short rooting rhizome; peduncle relatively long, terete; floral bracts persistent. *Flowers* small, more or less twisted but not resupinate, widely open; *sepals* lanceolate, concave at the apex; *petals* spreading, oblong to obovate, about as long as the sepals; *lip* trilobed, pandurate or entire, the disc with 2 keels; *column* more or less straight or slightly S-shaped, with 2 long parallel, porrect or decurved, elongate stelidia; *pollinia* 4.

Six species distributed from India (Sikkim), Burma, Thailand and Indochina eastward through Malaysia and Indonesia to the Philippines, New Guinea, the Solomon Islands and Australia.

55.3.1. Acriopsis indica Wight, Icon. Pl. Ind. Orient. 5: 20, t. 1748-1 (1852).

Epiphyte with many ascending, branched catch-roots. Hill forest, lower montane forest. Elevation: 900–1500 m.

General distribution: Sabah, Kalimantan, India?, Burma, Peninsular Malaysia, Thailand, Indochina, Java, Timor, Philippines.

Collections. BUNDU TUHAN: 1100 m, *Lamb AL 1144/89* (K); DALLAS: 900 m, *Clemens 26304* (K); KINUNUT VALLEY HEAD: 1200 m, *Carr SFN 27154* (SING); MOUNT KINABALU: *Haslam s.n.* (AMES, BM, K); TENOMPOK: 1500 m, *Clemens 28136* (BM, K), 1500 m, *28224* (BM, K), 1500 m, *29816* (BM, K), 1500 m, *29902* (BM, K).

55.3.2. Acriopsis javanica Reinw. ex Blume, Bijdr., 377 (1825).

a. var. **javanica**

Epiphyte. Hill forest; interface between old secondary and primary forest; lower montane forest. Elevation: 400–1200 m.

General distribution: Sabah, Brunei, Kalimantan, Sarawak, Peninsular Malaysia, Thailand, Philippines, New Guinea, Australia, Solomon Islands.

Collections. DALLAS: 900 m, *Clemens 27531* (BM, K); DALLAS/TENOMPOK: 1200 m, *Clemens 29121* (BM, K); KAUNG: 400 m, *Darnton 372* (BM); MELANGKAP KAPA: 700–1000 m, *Beaman 8812a* (K); PINOSUK: 900 m, *Carr SFN 27191* (SING); TAHUBANG RIVER: 1100 m, *Clemens 33068* (BM); TENOMPOK: *Clemens s.n.* (BM).

55.3.3. Acriopsis ridleyi Hook. f., Fl. Brit. Ind. 6: 79 (1891).

Epiphyte. Low-stature hill forest on ultramafic substrate. Elevation: 800 m.

General distribution: Sabah, Kalimantan, Sarawak, Peninsular Malaysia, Singapore.

Collection. LOHAN RIVER: 800 m, *Beaman 10800* (K).

55.4. ADENONCOS Blume

Bijdr., 381 (1825).

Small monopodial epiphytes. *Stems* 10–30 cm long. *Leaves* short, narrow, to 7 × 1 cm, very fleshy, acute, arranged in 2 rows. *Inflorescences* short, 1–5 flowered. *Flowers* small, green or yellowish, lasting a long time; *sepals* and *petals* free; *dorsal sepal* 1.5–5 mm; *petals* narrower than sepals; *lip* entire or lobed, concave, somewhat saccate, with a papillose basal keel; *column* short and erect; *stipes* linear or clavate, 2 times the diameter of the pollinia; *viscidium* narrowly elliptic; *pollinia* 4, ± equal.

About 16 species, distributed from Thailand and Indochina to Indonesia and New Guinea. They most often occur growing upright on the topmost branches of large forest trees.

55.4.1. Adenoncos sp. 1

Epiphyte. Hill forest on ultramafic substrate. No specimens seen. Sight record of A. Lamb (pers. comm.) on Kulung Hill.

55.5. AERIDES Lour.

Fl. Cochinch. 2: 525 (1790).

Seidenfaden, G. (1973). Contributions to the orchid flora of Thailand, V. Bot. Tidssk. 68: 68–80. Christenson, E. A. (1987). The taxonomy of *Aerides* and related genera. Proc. 12th World Orchid Conference, 35–40.

Medium-sized, rather coarse monopodial epiphytes. *Stems* short to elongate, up to 1 m long. *Leaves* oblong-ligulate to linear or terete, often very fleshy, apex usually distinctly bilobed, to 60 cm long. *Inflorescences* variable, simple, usually densely many-flowered, often pendent. *Flowers* showy, to 3 cm or more across, white and rose-violet; *sepals* and *petals* broad, spreading; *lateral sepals* decurrent on column-foot; *lip* 3-lobed, immobile, with a basal, often forward-curving spur, side-lobes decurrent on the column, mid-lobe often erose; *column* short, often broadened at the apex, foot short; *rostellar projection* usually long and pointed; *stipes* usually long and slender; *pollinia* 2, sulcate.

Some 20 species distributed from Sri Lanka, India and the Himalayan region to Thailand and Indochina south to Malaysia and Indonesia, north to the Philippines.

55.5.1. Aerides odorata Lour., Fl. Cochinch. 2: 525 (1790). Plate 7B.

Epiphyte. Hill forest, sometimes on ultramafic substrate, and lower montane forest. Elevation: 800–1500 m.

General distribution: Sabah, widespread in mainland & SE Asia east to the Philippines.

Collections. BUNDU TUHAN: 1200 m, *Carr SFN 27218* (SING); DALLAS: 900 m, *Clemens s.n.* (BM), 900 m, *26776* (BM, K), 900 m, *26780* (BM, K), 900 m, *26791* (K); KIAU: 900 m, *Clemens 43* (AMES), 900 m, *69* (AMES), 900 m, *81* (AMES), 900 m, *183* (AMES), *307* (AMES); LOHAN RIVER: 800 m, *Beaman 8370* (K); MOUNT KINABALU: *Clemens s.n.* (AMES); PENIBUKAN: 900 m, *Clemens 50098* (BM, K); TENOMPOK: 1500 m, *Clemens 29369* (BM, K).

55.6. AGROSTOPHYLLUM Blume

Bijdr., 368 (1825).

Epiphytic herbs. *Stems* clustered, without pseudobulbs, erect or pendent, bilaterally flattened, of many internodes, leafy. *Leaves* distichous, narrow, rather thin, usually twisted at base to lie in one plane, with black or brown edged imbricate sheaths. *Inflorescences* terminal, usually globose heads on an elongate axis, or in a panicle, surrounded by bracts. *Flowers* small, resupinate, numerous, white or yellow, often self-pollinating; *sepals* and *petals* similar, free, petals narrower; *lateral sepals* forming a mentum which contains the saccate lip base; *lip* entire or 3-lobed, saccate base divided from the blade by a transverse partition; *column* short or rather long, foot rudimentary; *pollinia* 8, attached to a solitary viscidium.

Between 40 and 50 species distributed in the Old World tropics from the Seychelles and tropical Asia east to the Pacific islands, with the centre of distribution in New Guinea.

55.6.1. Agrostophyllum arundinaceum Ridl., Sarawak Mus. J. 1: 36 (1912).

Epiphyte. Lower montane forest in open area on ultramafic substrate. Elevation: 1700 m.

General distribution: Sabah, Sarawak.

Collection. MARAI PARAI SPUR: 1700 m, *Bailes & Cribb 832* (K).

55.6.2. Agrostophyllum bicuspidatum J. J. Sm., Icon. Bogor. 2: 55 (1903).

Epiphyte. Lowlands. Elevation: 400–500 m.

General distribution: Sabah, Brunei, Kalimantan, Sarawak, Burma, Peninsular Malaysia, Thailand, Sumatra, Java, Mentawai, Krakatau, distribution further east uncertain.

Collections. KAUNG: 500 m, *Carr SFN 27990* (SING), 400 m, *Darnton 354* (BM).

55.6.3. Agrostophyllum cyathiforme J. J. Sm., Orch. Java, 291, f. 223 (1905).

Epiphyte. Lower montane forest. Elevation: 1200 m.

General distribution: Sabah, Peninsular Malaysia, Sumatra, Java.

Collection. PENIBUKAN: 1200 m, *Clemens 40632* (BM).

55.6.4. Agrostophyllum aff. **cyathiforme** J. J. Sm., Orch. Java, 291, f. 223 (1905).

Epiphyte. Lower montane forest, probably on ultramafic substrate. Elevation: 1500 m.

Collection. KINATEKI RIVER: 1500 m, *Clemens 50398* (BM, K).

55.6.5. Agrostophyllum globigerum Ames & C. Schweinf., Orch. 6: 138 (1920). Type: MARAI PARAI SPUR, *Clemens 241* (holotype AMES mf).

Epiphyte. Lower montane forest on ultramafic substrate. Elevation: 1500–2100 m.

Endemic to Mount Kinabalu.

Additional collections. MARAI PARAI: 1500 m, *Clemens 32351* (BM), 1500 m, *Collenette A 34* (BM); MARAI PARAI SPUR: 2100 m, *Gibbs 4034* (BM, K).

55.6.6. Agrostophyllum aff. **globigerum** Ames & C. Schweinf., Orch. 6: 138 (1920).

Epiphyte. Lower montane forest. Elevation: 1600 m.

Collection. PINOSUK PLATEAU: 1600 m, *Bailes & Cribb 530* (K).

55.6.7. Agrostophyllum glumaceum Hook. f., Fl. Brit. Ind. 5: 821 (1890).

Epiphyte. Hill forest. Elevation: 900 m.

General distribution: Sabah, Brunei, Kalimantan, Sarawak, Peninsular Malaysia, Sumatra.

Collections. DALLAS: 900 m, *Clemens 26645* (BM), 900 m, *27481* (BM, K), 900 m, *27745* (K).

55.6.8. Agrostophyllum javanicum Blume, Bijdr., 369, t. 53 (1825). Plate 7C.

Epiphyte. Lower montane forest. Elevation: 1400–2000 m.

General distribution: Sabah, Sumatra, Java.

Collections. PARK HEADQUARTERS: 1700–2000 m, *Vermeulen & Chan 409* (K); TENOMPOK: 1500 m, *Clemens 29455* (BM, K); TINEKUK RIVER: 1400 m, *Clemens 40919* (BM); WEST MESILAU RIVER: 1700 m, *Brentnall 159* (K).

55.6.9. Agrostophyllum majus Hook. f., Fl. Brit. Ind. 5: 824 (1890).

Epiphyte. Hill forest, lower montane forest, sometimes on ultramafic substrate; growing on stilt roots. Elevation: 900–1500 m.

General distribution: Sabah, Kalimantan, Sarawak, Peninsular Malaysia, Singapore, Sumatra, New Guinea, Solomon Islands, Vanuatu.

Collections. DALLAS: 900 m, *Clemens 26777* (BM); MARAI PARAI: 1500 m, *Clemens 32358* (BM); PENIBUKAN: 1100 m, *Clemens 40956* (BM, K), 1500 m, *50201* (BM); TENOMPOK: *Clemens 50174* (BM); TENOMPOK ORCHID GARDEN: *Clemens 50174* (K).

55.6.10. Agrostophyllum saccatum Ridl., J. Linn. Soc. Bot. 31: 286 (1895).

Epiphyte. Lowlands, hill forest, lower montane forest on ultramafic substrate. Elevation: 400 m.

General distribution: Sabah, Sarawak.

Collections. KAUNG: 400 m, *Darnton 352* (BM); KIAU: *Clemens 338* (AMES, BM); MARAI PARAI SPUR: *Clemens 276* (AMES).

55.6.11. Agrostophyllum sp. 1, sect. Agrostophyllum

Epiphyte. Lower montane forest. Elevation: 1500–1800 m.

Collections. MESILAU RIVER: 1800 m, *Clemens 51588* (BM); TENOMPOK: 1500 m, *Carr SFN 27350* (K, SING), 1500 m, *Clemens 30144* (K).

55.6.12. Agrostophyllum sp. 2, sect. Appendiculopsis

Epiphyte. Lower montane forest. Elevation: 900–1500 m.

Collections. DALLAS: 900 m, *Clemens 27324* (BM); TENOMPOK: 1500 m, *Clemens 26774* (BM, K).

55.6.13. Agrostophyllum indet.

Collections. BUNDU TUHAN: 1200 m, *Carr SFN 27426* (SING); TAHUBANG RIVER: 1000 m, *Carr SFN 26394* (SING); TENOMPOK: 1500 m, *Carr 3317* (SING), 1500 m, *Clemens s.n.* (SING).

55.7. ANIA Lindl.

Gen. Sp. Orch., 129 (1831).

Turner, H. (1992). A revision of the orchid genera *Ania* Lindley, *Hancockia* Rolfe, *Mischobulbum* Schltr. and *Tainia* Blume. Orchid Monogr. 6: 43–100 (1992).

Terrestrial herbs. *Shoots* arising from base of last pseudobulb; sterile shoots with 1 terminal leaf. *Pseudobulb* with 1 or several internodes, usually erect, conical, rarely ovoid to ellipsoid. *Leaf* petiolate, elliptic to slightly obovate, acute to acuminate. *Inflorescence* lateral, erect, arising from base of pseudobulb of previous shoot. *Flowers* resupinate, glabrous (except in *A. ponggolensis* A. Lamb); *sepals* and *petals* free, elliptic to obovate; *lateral sepals* slightly decurrent along column-foot; *lip* entire or 3-lobed, usually distinctly spurred, disc with 3–7 keels; *column* alate; *pollinia* 8, in 4 pairs.

Eight species distributed in India, Burma, China, Hong Kong, Vietnam, Thailand and Peninsular Malaysia eastward to the Philippines and New Guinea.

55.7.1. Ania borneensis (Rolfe) Senghas, Schltr. Orch. ed. 3, 1: 863 (1984).

Ascotainia borneensis Rolfe in Gibbs, J. Linn. Soc. Bot. 42: 154 (1914). Type: KIAU, 900 m, *Gibbs 3958* (holotype BM!; isotype K!).

Tainia rolfei P. F. Hunt, Kew Bull. 26: 182 (1971).

Terrestrial. Secondary hill forest. Elevation: 900–1200 m.

General distribution: Sabah, Sumatra, Java, Maluku, New Guinea.

Additional collections. DALLAS: 900 m, *Clemens 30125* (K); KIAU: 1200 m, *Lamb AL 275/84* (K).

55.8. ANOECTOCHILUS Blume

Bijdr., 411 (1825).

Terrestrial herbs. *Leaves* green, or with colourful silvery or red nerves, stalked. *Inflorescence* with a rather short peduncle, few-flowered. *Flowers* resupinate, showy, white, flushed pink; *dorsal sepal* and *petals* connivent, forming a hood; *lip* in contact with base of column, either with a projecting spur or a saccate base enclosed by the sepals, inside which are 2 large, sessile glands; middle part of lip narrowed into a channelled claw, the edges of which are involute, with a toothed or fringed flange on either side, sometimes with distinct side-lobes at base of claw, apex of claw widened into a transverse bilobed blade; *column* 2-winged in front, either small, or prolonged downwards as free parallel plates into spur; *anther* acute, short or elongate; *stigmas* 2; *pollinia* 2.

When considered in the broad sense to include *Odontochilus* Blume, about 40 species can be recognised distributed from Sri Lanka and India to Japan, south to Malaysia and Indonesia, and eastward to the Pacific islands.

55.8.1. Anoectochilus integrilabris Carr, Gardens' Bull. 8: 186 (1935). Type: MINITINDUK GORGE, 800 m, *Carr 3162, SFN 26634* (holotype SING!; isotype AMES mf).

Terrestrial. Young secondary hill forest. Elevation: 800 m. Known only from the type.

Endemic to Mount Kinabalu.

55.8.2. Anoectochilus longicalcaratus J. J. Sm., Bull. Jard. Bot. Buit., ser. 3, 5: 18 (1922).

Terrestrial. Lower montane forest; in leaf litter in shade; sometimes on ultramafic substrate. Elevation: 900–1800 m. Specimens in AMES were determined as *A. reinwardtii* Blume.

General distribution: Sabah, Sumatra.

Collections. KIAU: 900 m, *Clemens 164* (AMES); KIAU VIEW TRAIL: 1500 m, *Lamb s.n.* (K); LUGAS HILL: 1300 m, *Beaman 10535* (K); LUMU-LUMU?: 1500 m, *Clemens s.n.* (BM); MARAI PARAI SPUR: *Clemens 263* (AMES), *400* (AMES); MOUNT KINABALU: *Thrower s.n.* (K); MT. NUNGKEK: 1100 m, *Darnton 452* (BM); PARK HEADQUARTERS: 1500–1800 m, *Lamb LKC 3156* (K); PENIBUKAN: 1200 m, *Carr 3084, SFN 26533* (BM, K, SING), 1200–1500 m, *Clemens s.n.* (BM), 1200 m, *30517* (BM), 1400 m, *40513* (K).

55.9. **APHYLLORCHIS** Blume

Bijdr., t. 77 (1825).

Leafless saprophytic herbs. *Rhizome* short, rather thin, erect, with thick spreading roots. *Stem* erect, slender. *Inflorescence* terminal, many-flowered. *Flowers* resupinate; *sepals* and *petals* about equal, free; *lip* divided into a short narrow basal hypochile and a 3-lobed apical epichile; *column* long; *anther* erect, dorsal; *pollinia* 2, powdery.

Around 30 species have been proposed, although the true figure is probably much lower. Distributed from Sri Lanka, India and the western Himalayas to China, through Indochina, Malaysia, Indonesia, Taiwan and the Philippines, eastward to New Guinea and Australia.

55.9.1. **Aphyllorchis montana** Rchb. f., Linnaea 41: 57 (1876). Plate 8A.

Saprophyte. Hill forest, lower montane forest; in leaf litter in dense shade. Elevation: 800–1500 m.

General distribution: Sabah, Sarawak, Kalimantan, widespread from Himalayas, India, Sri Lanka, China, Taiwan, Ryukyu Islands, Peninsular Malaysia & Thailand to the Philippines.

Collections. EASTERN SHOULDER: 800 m, *RSNB 214* (K, SING); PARK HEADQUARTERS: *Lamb s.n.* (K), 1500 m, *Wood 622* (K).

55.9.2. **Aphyllorchis pallida** Blume, Bijdr., t. 77 (1825).

Saprophyte. Hill forest, lower montane forest; in leaf litter in dense shade. Elevation: 1100–1600 m.

General distribution: Sabah, Kalimantan, Sarawak, Peninsular Malaysia, Thailand, Sumatra, Java, Philippines.

Collections. DALLAS/TENOMPOK: 1200 m, *Clemens 27567* (BM, K, SING); GURULAU SPUR: 1400 m, *Carr 3140, SFN 26590* (K, SING); KIAU: *Clemens 252* (AMES), *323* (AMES, BM, K, SING); LUGAS HILL: 1300 m, *Beaman 8480* (K), 1300 m, *9525* (K); MAMUT COPPER MINE: 1500 m, *Collenette 1044* (K); MARAI PARAI: 1500 m, *Clemens 33106* (BM); MT. NUNGKEK: 1500 m, *Clemens 32787* (BM); PENIBUKAN: 1100 m, *Clemens 40333* (BM, K); TAHUBANG RIVER: 1200 m, *Clemens 30473* (BM); TENOMPOK: 1500 m, *Carr 3654* (K); WEST MESILAU RIVER: 1600 m, *Collenette 637* (K).

55.10. **APOSTASIA** Blume

Bijdr., 423 (1825).

De Vogel, E. F. (1969). Monograph of the tribe Apostasieae (Orchidaceae). Blumea 17: 313–350.

Erect, rhizomatous terrestrials with nodular storage roots. *Stems* thin, often branched, leafy throughout. *Leaves* usually linear, acute, not plicate, the lowermost smaller, often dying off. *Inflorescences* terminal, often branched, branches decurved or spreading, many flowered; *floral bracts* small. *Flowers* yellow or white; *sepals, petals* and *lip* free, almost equal, spreading, sometimes revolute; *column* straight or curved; *fertile stamens* 2, with short filaments adnate to the style to various degrees, *anthers* clasping the style; *staminode* present or absent, usually adnate to the style, free at the apex; *stigma* small; *pollen grains* as monads, not aggregated into pollinia; *ovary* narrowly cylindric, 3-angular in cross section.

Seven species distributed from the Himalayan region, India and Sri Lanka east to the Ryukyu Islands, New Guinea and Australia.

55.10.1. Apostasia nuda R. Br. in Wall., Pl. Asiat. Rar. 1: 76 (1830).

Terrestrial. Hill forest on ultramafic substrate; in leaf litter in dense shade. Elevation: 800–1000 m.

General distribution: Sabah, Brunei, Kalimantan, Sarawak, Burma, Peninsular Malaysia, Cambodia, Vietnam, Sumatra, Java, Bangka.

Collection. HEMPUEN HILL: 800–1000 m, *Beaman 7424* (K).

55.10.2. Apostasia odorata Blume, Bijdr., 423 (1825).

Terrestrial. Hill forest, lower montane forest, sometimes on ultramafic substrate; in leaf litter in shade. Elevation: 800–1500 m.

General distribution: Sabah, China (Yunnan), India, Peninsular Malaysia, Thailand, Vietnam, Sumatra, Java, Sulawesi, Bangka, Belitung.

Collections. EAST MESILAU RIVER: 1400 m, *RSNB 1351* (K); LIWAGU/MESILAU RIVERS: 1400 m, *RSNB 2764* (K); MAMUT RIVER: 1400 m, *RSNB 1670* (K); PENIBUKAN: 1100 m, *3090, SFN 27083* (K, SING), 1200–1500 m, *Clemens s.n.* (BM), 1200 m, *51576* (BM); SAYAP: 800–1000 m, *Beaman 9792* (K).

55.10.3. Apostasia wallichii R. Br. in Wall., Pl. Asiat. Rar. 1: 75, t. 84 (1830).

Terrestrial. Hill forest on ultramafic substrate. Elevation: 500–1200 m.

General distribution: Sabah, Sarawak, Kalimantan, widespread from India & Sri Lanka, east to New Guinea & Australia.

Collections. HEMPUEN HILL: 800–1000 m, *Beaman 7414* (K); MELANGKAP TOMIS: 500–900 m, *Beaman 8709* (K); PENIBUKAN: 1200 m, *Clemens 30484* (BM).

55.11. APPENDICULA Blume

Bijdr., 297 (1825).

Epiphytic, lithophytic, rarely terrestrial herbs. *Stems* erect or pendulous, simple or branched, pseudobulbs absent. *Leaves* distichous, flat, twisted at the base so that the blades all lie in one plane. *Inflorescence* terminal, lateral or both, short or long. *Flowers* small, resupinate, white or greenish. *Lateral sepals* connate at base to column-foot to form a mentum. *Lip* with a round or concave basal appendage, sometimes lengthened into small keels; *mid-lobe* often with a median keel or callus. *Column* with a foot. *Pollinia* 6, on a solitary forked caudicle or 2 separate ones.

About 60 species distributed from tropical Asia to the Pacific islands.

55.11.1. Appendicula anceps Blume, Bijdr., 299 (1825).

Epiphyte. Lowlands, hill forest, lower montane forest. Elevation: 400–1500 m.

General distribution: Sabah, Brunei, Kalimantan, Sarawak, Peninsular Malaysia, Thailand, Sumatra, Java, Sulawesi, Bangka, Natuna Islands, Riau Archipelago, Philippines.

Collections. DALLAS: 900 m, *Clemens 26302* (BM, K), 900 m, *26769* (BM, K), 900 m, *27456* (BM); KAUNG: 400 m, *Carr 3013, SFN 26268* (SING), *Clemens 27697* (BM), 400 m, *Darnton 302* (BM); MELANGKAP KAPA: 600–700 m, *Beaman 8604* (K); TENOMPOK: 1500 m, *Clemens 27257* (BM), 1500 m, *30171* (K).

55.11.2. Appendicula buxifolia Blume, Bijdr., 300 (1825).

Epiphyte. Lower montane forest. Elevation: 1500 m.

General distribution: Sabah, Sarawak, Sumatra, Java, Philippines.

Collections. TENOMPOK: *Clemens s.n.* (BM), 1500 m, *29954* (K).

55.11.3. Appendicula calcarata Ridl., J. Linn. Soc. Bot. 31: 302 (1896).

Epiphyte. Hill forest. Elevation: 800–900 m.

General distribution: Sabah, Kalimantan, Sarawak.

Collections. BUNDU TUHAN: 800 m, *Carr SFN 27882* (SING); DALLAS: 900 m, *Clemens 26520* (BM, K).

55.11.4. Appendicula congesta Ridl. in Stapf, Trans. Linn. Soc. Bot. 4: 239 (1894). Plate 8B. Type: TINEKUK RIVER, 1200 m, *Haviland 1302* (holotype? K!).

Chilopogon kinabaluensis Ames & C. Schweinf., Orch. 6: 141, pl. 93 (1920). Type: MARAI PARAI SPUR, *Clemens 230* (holotype AMES mf; isotypes BM!, K!).

Appendicula kinabaluensis (Ames & C. Schweinf.) J. J. Sm., Bull. Jard. Bot. Buit., ser. 3, 5: 65 (1922).

Epiphyte. Hill forest, lower montane forest, frequently on ultramafic substrate. Elevation: 900–2100 m.

General distribution: Sabah.

Additional collections. DALLAS: 900 m, *Clemens 27138* (BM), 900 m, *27324* (BM, K); KIAU: 900 m, *Clemens 61* (AMES), *348* (AMES); KILEMBUN RIVER HEAD: 1400 m, *Clemens 32433* (BM, K), 1800 m, *32523* (BM); KINATEKI RIVER: 1200–1500 m, *Clemens 31043* (BM); KINATEKI RIVER HEAD: 1200–2100 m, *Clemens 31732* (BM); MAHANDEI RIVER: 1100 m, *Carr 3038, SFN 26506* (BM, SING); MARAI PARAI: 1200 m, *Clemens 32771* (BM); MARAI PARAI SPUR: 1500 m, *Bailes & Cribb 823* (K), *Clemens 237A* (AMES); MESILAU RIVER: 1500 m, *RSNB 4093* (K); MOUNT KINABALU: *Clemens s.n.* (AMES); MT. NUNGKEK: 1200 m, *Clemens 32879* (BM); MURU-TURA RIDGE: 1500 m, *Clemens 34362* (BM); PENIBUKAN: 1200–1500 m, *Clemens s.n.* (BM, K), 1200 m, *30523* (BM), 1800 m, *Gibbs 4058* (BM); TENOMPOK: 1500 m, *Clemens s.n.* (BM), 1500 m, *27138* (K), 1500 m, *30143* (K); TENOMPOK ORCHID GARDEN: 1500 m, *Clemens 50352* (K); WEST MESILAU RIVER: 1600–1700 m, *Beaman 7459* (K), 1600 m, *7526* (K), 1600 m, *9035* (K).

55.11.5. Appendicula cristata Blume, Bijdr., 240, t. 40B (1825). Plate 8C.

Appendicula divaricata Ames & C. Schweinf., Orch. 6: 143 (1920), **syn. nov.** Type: KIAU, 900 m, *Clemens 137* (holotype AMES mf).

Epiphyte. Hill forest, lower montane forest, sometimes on ultramafic substrate. Elevation: 600–1500 m.

General distribution: Sabah, Brunei, Sumatra, Java, Sulawesi.

Additional collections. DALLAS: 900 m, *Clemens 27397* (BM, K); HEMPUEN HILL: 900 m, *Wood 838* (K); KIAU: *Clemens 360* (AMES), 600 m, *Darnton 596* (BM); LOHAN RIVER: 700–900 m, *Beaman 9256* (K); MAHANDEI RIVER: 1200 m, *Carr 3030, SFN 26310* (SING); MOUNT KINABALU: *Clemens s.n.* (AMES), *Haslam s.n.* (AMES); MT. NUNGKEK: 800 m, *Clemens 32905* (BM); PENIBUKAN: 1200 m, *Clemens 30708* (BM); TENOMPOK: 1500 m, *Clemens 28950* (BM, K).

55.11.6. Appendicula floribunda (Schltr.) Schltr., Feddes Repert. 1: 355 (1912).

Podochilus floribundus Schltr., Mém. Herb. Boissier 21: 58 (1900).

Epiphyte. Hill forest on ultramafic substrate. Elevation: 600 m.

General distribution: Sabah, Sumatra.

Collection. LOHAN RIVER: 600 m, *Lamb AL 504/85* (K).

55.11.7. Appendicula foliosa Ames & C. Schweinf., Orch. 6: 145 (1920). Type: KIAU, *Clemens 361* (holotype AMES mf; isotypes BM!, K!, SING!).

Epiphyte or lithophyte. Hill forest, lower montane forest, sometimes on ultramafic substrate. Elevation: 800–2400 m.

General distribution: Sabah, Sarawak.

Additional collections. KEMBURONGOH: 2400 m, *Clemens s.n.* (BM); KIAU: 900 m, *Clemens 63* (AMES); LUBANG: *Clemens 208* (AMES), *222* (AMES); MAHANDEI RIVER: 1100 m, *Carr 3027, SFN 26309* (SING); MARAI PARAI SPUR: *Clemens 237* (AMES); MESILAU CAVE TRAIL: 1700–1900 m, *Beaman 8003* (K); PENATARAN BASIN: 1100 m, *Clemens 34050* (BM); PENIBUKAN: 1200–1500 m, *Clemens 30584* (BM, K); SAYAP: 800–1000 m, *Beaman 9782* (K); TAHUBANG FALLS: 1200 m, *Clemens 30697* (BM, K); TENOMPOK: 1500 m, *Clemens s.n.* (BM, BM), 1500 m, *28736* (BM, K), 1500 m, *30168* (K), 1500 m, *30169* (K).

55.11.8. Appendicula fractiflexa J. J. Wood, **sp. nov.** Fig. 7.

Appendiculae ramosae Blume, species sumatrana et javanica, valde affinis, sed inflorescentiis eximiis pedunculis fractiflexis usque 3.5 cm longis bracteis acicularibus distinguitur.

Type: EAST MALAYSIA, SABAH, MOUNT KINABALU, EAST MESILAU/MENTEKI RIVERS, 1650 m, 31 May 1988, *Wood 827* (holotype K!, herbarium and spirit material).

Pendulous epiphytic herb. *Stems* to 70 cm long, much branched, slender, flexuous, branches 10–35 cm long, internodes 5–8 mm long. *Leaves* 1–2.1(–2.5) × 0.2–0.4(–0.7) cm, those on lower part of stem usually broader, those on flowering branches narrower, narrowly oblong or narrowly obovate, apex unequally retuse, with a mucro in the sinus, sheaths 5–8 mm long. *Inflorescences* borne from middle or upper part of stem, one flower open at a time; *peduncle* 1.5–3.5 cm long, strongly fractiflex; *sterile bracts* 4–5.5 mm long, acicular; *rachises* usually 2, emerging from several imbricate acicular bracts 3–4 mm long, each rachis 3–4 mm long, fractiflex; *floral bracts* 1–2 mm long, triangular-ovate, acuminate, concave. *Flowers* non-resupinate, pale lilac to purple, with a cream lip. *Pedicel-with-ovary* 1.8–2 mm long. *Sepals* with prominent and slightly raised median nerve. *Dorsal sepal* 2.5 × 1.5 mm, ovate, slightly aristate, concave. *Lateral sepals* 2.5 × 1.5 mm, fused to column-foot to form a rounded mentum 3 mm long, obliquely triangular-ovate, aristate. *Petals* 1.8 × 0.8 mm, oblong, apiculate. *Lip* 4.5 × 1.9 mm, entire, oblong, retuse, apex sharply decurved, margin below deflexed apex erect, disc provided with a minutely papillose, concave, bilobed basal appendage which is continued as 2 smooth fleshy keels which terminate c. 1 mm below the apex; a small obscure low subapical keel is also present. *Column* 0.8 mm long, foot 3 mm long, fleshy and sulcate above, apex incurved. *Anther-cap* 0.8 × 0.8 mm, ovate, cucullate.

Appendicula fractiflexa is closely related to *A. ramosa* Blume from Sumatra and Java, but is readily distinguished by the distinctive inflorescences which have fractiflex peduncles up to 3.5 cm long and acicular bracts. The floral morphology of both species is rather similar.

The specific epithet is derived from the Latin *fractiflexus*, zigzag, in reference to the peduncle and rachis.

Lower montane forest. Elevation: 1200–1650 m.

General Distribution: Sabah.

Additional collections: MALIAU BASIN (KINABATANGAN DISTRICT): 1400 m, *Lamb AL 1409/92* (K); PENIBUKAN: 1200–1500 m, *Clemens 31567* (BM).

55.11.9. Appendicula linearifolia Ames & C. Schweinf., Orch. 6: 147 (1920). Type: MARAI PARAI SPUR, *Clemens 286* (holotype AMES mf; isotypes BM!, K!, SING!).

Epiphyte. Lower montane forest on ultramafic substrate. Elevation: 1400–2100 m.

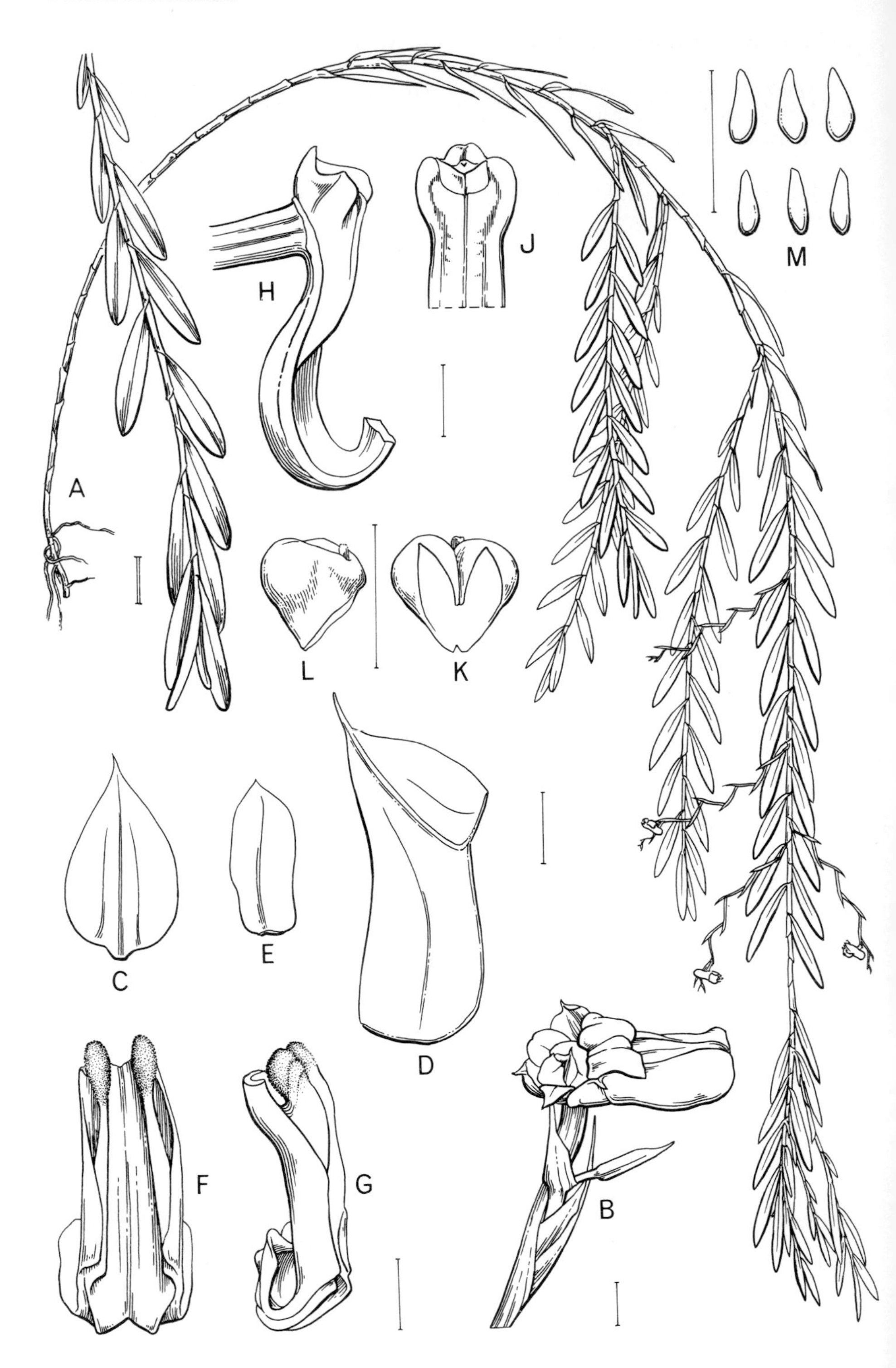
A
B
C
D
E
F
G
H
J
K
L
M

Endemic to Mount Kinabalu.

Additional collections. MARAI PARAI: 1500 m, *Clemens 32271* (BM), 1400 m, *32353* (BM), 1800 m, *32411* (BM), 2100 m, *33104* (BM); MARAI PARAI SPUR: *Clemens 404* (AMES), 2000 m, *Collenette A 57* (BM); MOUNT KINABALU: *Clemens s.n.* (AMES).

55.11.10. Appendicula longirostrata Ames & C. Schweinf., Orch. 6: 149 (1920). Fig. 8, Plate 8D. Type: MARAI PARAI SPUR, *Clemens 387* (holotype AMES mf; isotypes BM!, K!, SING!).

Epiphytic and terrestrial. Lower montane forest, sometimes on ultramafic substrate. Elevation: 1100–2700 m.

General distribution: Sabah, Brunei.

Additional collections. KINATEKI RIVER HEAD: 2100 m, *Clemens 31788* (BM); MESILAU RIVER: 2700 m, *Clemens 51434* (BM, K), *51599* (BM, K); PENIBUKAN: 1200 m, *Clemens 35168* (BM); PIG HILL: 1700–2000 m, *Sutton 11* (K); SUMMIT TRAIL: 1600 m, *Mikil SAN 33935* (K); TAHUBANG RIVER: 1200 m, *Clemens 30704* (BM), 1100 m, *40958* (BM, K); TAHUBANG RIVER HEAD: 2100 m, *Clemens 32977* (BM); TENOMPOK: 1500 m, *Carr SFN 27344* (SING), 1500 m, *Clemens 27241* (BM, K), 1500 m, *28516* (BM), 1500 m, *29843* (BM), 1500 m, *30170* (K), 1500 m, *30170A* (K); TENOMPOK ORCHID GARDEN: *Clemens 50359* (BM), 1500 m, *50367* (BM); TINEKUK FALLS: 1800 m, *Clemens 40906* (BM, K).

55.11.11. Appendicula lucida Ridl., J. Linn. Soc. Bot. 32: 392 (1896).

Epiphyte. Hill forest on ultramafic substrate; on dead tree trunks and live tree roots. Elevation: 500–1000 m.

General distribution: Sabah, Kalimantan, Sarawak, Peninsular Malaysia, Singapore, Riau Archipelago.

Collections. HEMPUEN HILL: 800–1000 m, *Beaman 7421* (K), 500–600 m, *Wood 604* (K); LOHAN RIVER: 800–1000 m, *Beaman 10010* (K).

55.11.12. Appendicula magnibracteata Ames & C. Schweinf., Orch. 6: 151 (1920). Type: MARAI PARAI SPUR, *Clemens 282* (holotype AMES mf).

Epiphyte. Hill forest, lower montane forest, sometimes on ultramafic substrate.

General distribution: Sabah.

Additional collections. KIAU: *Clemens 351* (AMES); MOUNT KINABALU: *Clemens s.n.* (AMES).

Fig. 7. Appendicula fractiflexa. A, habit; **B**, flower and part of rachis; **C**, dorsal sepal; **D**, lateral sepal; **E**, petal; **F**, lip (front view); **G**, lip (side view); **H**, column and pedicel-with-ovary (side view); **J**, column (front view); **K**, anther-cap (front view); **L**, anther-cap (side view); **M**, pollinia. All from *Wood 827.* Scale: single bar = 1 mm; double bar = 1 cm. Drawn by Eleanor Catherine.

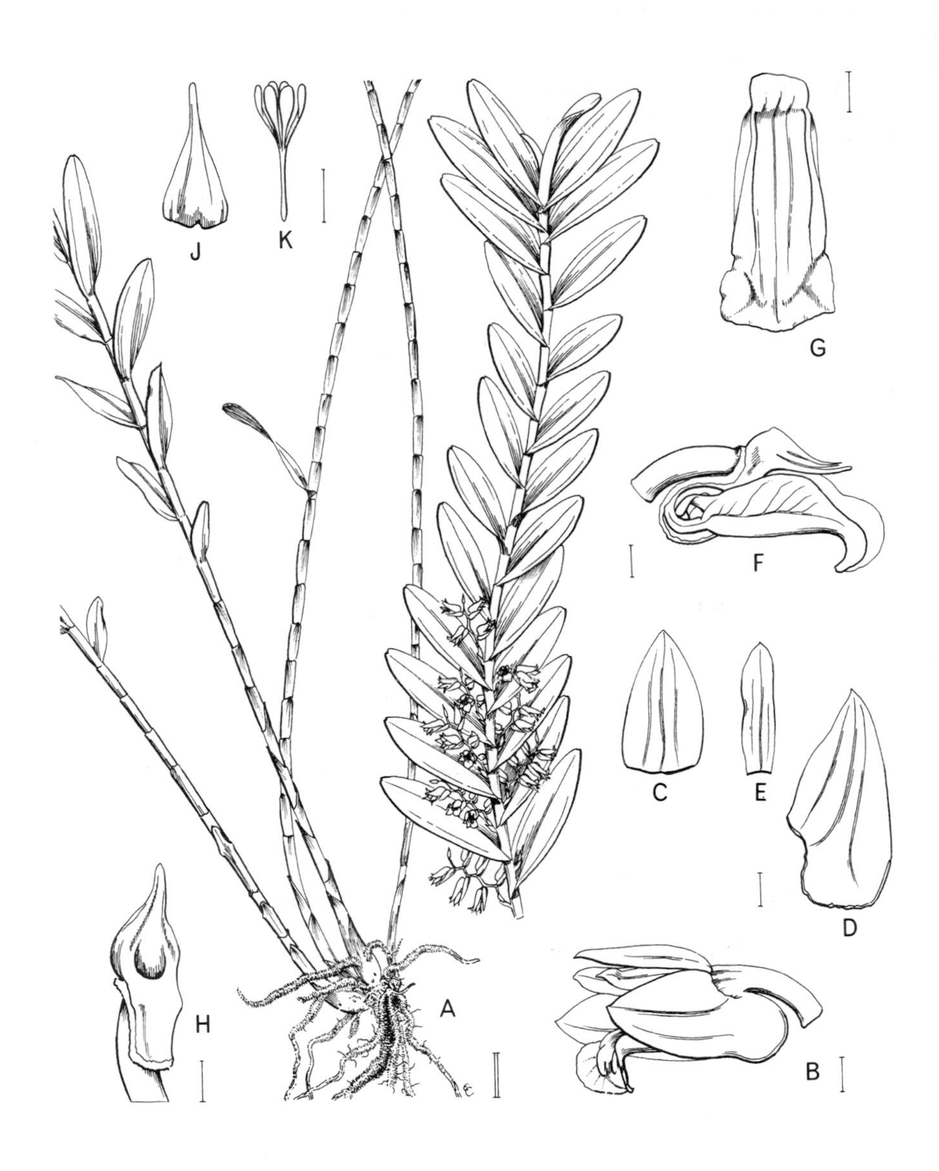

Fig. 8. Appendicula longirostrata. **A**, habit; **B**, flower (side view); **C**, dorsal sepal; **D**, lateral sepal; **E**, petal; **F**, pedicel-with-ovary, column and lip (side view); **G**, lip (front view); **H**, column (front view); **J**, anther-cap (back view); **K**, pollinia. **A** from *Beaman 10496a* and *Wood 792*; **B–K** from *Bailes & Cribb 619*. Scale: single bar = 1 mm; double bar = 1 cm. Drawn by Eleanor Catherine.

55.11.13. Appendicula minutiflora Ames & C. Schweinf., Orch. 6: 153 (1920). Type: KIAU, *Clemens 333* (holotype AMES mf).

Epiphyte. Lowlands. Elevation: 400–500 m.

General distribution: Sabah.

Additional collections. KAUNG: 400 m, *Carr SFN 27374* (K, SING), 400 m, *Darnton 291* (BM); TAKUTAN: 500 m, *Shea & Aban SAN 77158* (K).

55.11.14. Appendicula cf. **minutiflora** Ames & C. Schweinf., Orch. 6: 153 (1920).

Pendent epiphyte. Hill forest on ultramafic substrate. Elevation: 800–1000 m.

Collection. HEMPUEN HILL: 800–1000 m, *Beaman 7420* (K).

55.11.15. Appendicula pendula Blume, Bijdr., 298 (1825).

Epiphyte, lithophyte. Hill forest, lower montane forest, sometimes on ultramafic substrate. Elevation: 400–1800 m.

General distribution: Sabah, Kalimantan, Sarawak, Peninsular Malaysia, Thailand, Sumatra, Java, Natuna Islands.

Collections. DALLAS: 900 m, *Clemens s.n.* (BM), 900 m, *26312* (BM, K), 900 m, *26605* (BM, K), 900 m, *26772* (BM, K), 900 m, *27279* (BM, K), 900 m, *27456* (K); GURULAU SPUR: 1500 m, *Clemens 50448* (BM, K), 1500 m, *50540* (BM, K); KAUNG: 400 m, *Darnton 316* (BM); KIAU: *Clemens 358* (AMES); KILEMBUN BASIN: 1200 m, *Clemens 34488* (BM); LUBANG: 1500 m, *Clemens 124* (AMES), 1500 m, *129* (AMES), *138* (AMES); MARAI PARAI: 1800 m, *Clemens 32666* (BM); MOUNT KINABALU: 1000 m, *Haslam s.n.* (AMES); PENATARAN RIVER: 500 m, *Beaman 9310* (K); PENIBUKAN: 1200–1500 m, *Clemens s.n.* (BM), 1200 m, *30593* (BM), 1200 m, *30855* (BM), 1500 m, *31006* (BM), 1200–1500 m, *35169* (BM), 1200 m, *51712* (BM); TENOMPOK: 1500 m, *Clemens 27257* (K), 1500 m, *28817* (BM, K).

55.11.16. Appendicula recondita J. J. Sm., Bull. Jard. Bot. Buit., ser. 3, 11: 128 (1931).

Epiphyte. Lowlands. Elevation: 500–600 m.

General distribution: Sabah, Kalimantan.

Collections. KAUNG: 500 m, *Carr SFN 27924* (SING); PINAWANTAI: 600 m, *Shea & Aban SAN 76910* (K).

55.11.17. Appendicula rupicola (Ridl.) Rolfe in Gibbs, J. Linn. Soc. Bot. 42: 159 (1914).

Podochilus rupicolus Ridl., J. Straits Branch Roy. Asiat. Soc. 50: 142 (1908).

Epiphyte. Lowlands in secondary forest. Elevation: 300 m.

General distribution: Sabah, Sarawak.

Collection. KAUNG: 300 m, *Gibbs 4301* (BM, K).

55.11.18. Appendicula torta Blume, Bijdr., 303 (1825). Plate 9A.

Epiphyte. Hill forest, lower montane forest, mostly on ultramafic substrate. Elevation: 800–1500 m.

General distribution: Sabah, Kalimantan, Sarawak, Peninsular Malaysia, Sumatra, Java.

Collections. BUNDU TUHAN: 1400 m, *Darnton 195* (BM); DALLAS: 900 m, *Clemens 2929?* (K), 900 m, *27223* (BM, K); HEMPUEN HILL: 800–1000 m, *Beaman 7410* (K); MAMUT RIVER: 1400 m, *Collenette 1041* (K); PENATARAN BASIN: 1200 m, *Clemens 34052* (BM); PENIBUKAN: 1200 m, *Clemens s.n.* (BM, K); TENOMPOK: 1500 m, *Clemens s.n.* (BM), 1500 m, *28243* (BM), 1500 m, *29297* (BM).

55.11.19. Appendicula indet.

Collections. BUNDU TUHAN: 900 m, *Carr SFN 27777* (SING), 900 m, *SFN 27803* (SING), 800 m, *SFN 27989* (SING); GURULAU SPUR: 1500 m, *Clemens s.n.* (K); KUNDASANG: 600 m, *Carr SFN 26314A* (SING); LUBANG: 1400 m, *Carr 3193, SFN 26796* (SING); MAHANDEI RIVER: 1200 m, *Carr SFN 26314* (SING); MINITINDUK RIVER: 900 m, *Carr 3161* (SING); PALUAN RIVER: 500 m, *Carr SFN 27390* (SING); PENIBUKAN: 1200 m, *Carr 3046, SFN 26346* (SING), 1200 m, *Clemens 30784* (BM).

55.12. ARACHNIS Blume

Bijdr., 365 (1825).

Tan, K. W. (1975–1976). Taxonomy of *Arachnis, Armodorum, Esmeralda & Dimorphorchis*, Orchidaceae, Part I & Part II. Selbyana 1(1): 1–15; 1(4): 365–373.

Large robust monopodial terrestrials or epiphytes. *Stems* often scrambling, occasionally branching, sometimes several m long. *Leaves* strap-shaped, rigid, to 30 cm long (in Asiatic species), apex bilobed. *Inflorescences* rigid, very often long and branched, few- to many-flowered. *Flowers* often large and showy, to ca 7 cm across, fragrant, green or yellow with maroon blotches or bars; *sepals* and *petals* narrowly oblong to linear, spreading; *lip* much shorter, articulated to the short column-foot by a short strap, 3-lobed, mid-lobe with a raised central ridge or callus, basally saccate or with a short spur; *column* short and stout; *stipes* short, broad; *viscidium* broadly ovate; *pollinia* 4, appearing as 2 unequal masses.

Thirteen species and one natural hybrid widely distributed from India (Sikkim) in the west, China (Yunnan) in the north, Java and Bali in the south, eastward to the Solomon Islands. Borneo is the centre of speciation.

55.12.1. Arachnis calcarata Holttum, Sarawak Mus. J. 5: 172 (1949).

Epiphyte. Lower montane forest. Elevation: 1500 m.

General distribution: Sabah, Sarawak.

Collection. PARK HEADQUARTERS: 1500 m, *Lamb SAN 89685* (K).

55.12.2. Arachnis flosaeris (L.) Rchb. f., Bot. Centralbl. 28: 343 (1886).

Epidendrum flosaeris L., Sp. Pl., 952 (1753).

Epiphyte. Lowlands and hill forest? Elevation: 300–900 m. The *Clemens 26019* locality data are confused; Kaung is more likely correct.

General distribution: Sabah, Kalimantan, India?, Peninsular Malaysia, Thailand, Sumatra, Java, Bali, Philippines.

Collections. DALLAS: 900 m, *Clemens 26019* (K); KAUNG: 400 m, *Carr 3015, SFN 26276* (SING), 300 m, *Clemens 26019* (BM, K).

55.12.3. Arachnis longisepala (J. J. Wood) Shim & A. Lamb, Orchid Digest 46: 178 (1982).

Arachnis calcarata Holttum subsp. *longisepala* J. J. Wood, Orchid Rev. 89(1050): 113, f. 96 (1981).

Epiphyte. Hill forest on ultramafic substrate. Elevation: 800 m.

General distribution: Sabah.

Collection. LOHAN RIVER: 800 m, *Bailes & Cribb 654* (K).

55.13. ARUNDINA Blume

Bijdr., 401 (1825).

Terrestrial herbs. *Stems* slender, erect, leafy, often swollen at base, pseudobulbs absent. *Leaves* grass-like with imbricate sheaths. *Inflorescence* terminal, sometimes branched, producing a succession of flowers one or two at a time. *Flowers* resupinate, large and showy, purple to white; *sepals* and *petals* free; *sepals* narrow, dorsal erect, laterals close together behind lip; *petals* broader than sepals, spreading; *lip* trumpet-shaped, embracing column, subentire, apex emarginate, disc with 3 thin longitudinal keels; *column* without a foot; *pollinia* 8.

Between 2 and 5 species distributed from India and Sri Lanka eastward to Sulawesi, north to China; *A. graminifolia* is widely naturalised in the Pacific islands.

55.13.1. Arundina graminifolia (D. Don) Hochr., Bull. New York Bot. Gard. 6: 270 (1910).

Bletia graminifolia D. Don, Prodr. Fl. Nep., 29 (1825).

Terrestrial. Open grassy areas, roadsides, secondary vegetation, rocky banks. Elevation: 600–1500 m.

General distribution: Sabah, widespread in S & SE Asia, east to the Pacific islands.

Collections. DALLAS: 900 m, *Clemens 27445* (K); KIAU: *Clemens 186* (AMES), *265* (AMES), *277* (AMES); MINITINDUK/KINATEKI DIVIDE: 1100 m, *Carr 3171, SFN 26800* (SING); MOUNT KINABALU: *Clemens s.n.* (AMES, K), *Haslam s.n.* (AMES), 600 m, *Puasa 1540* (K); PENATARAN RIVER: *Clemens 34321* (BM); TENOMPOK: 1500 m, *Clemens 27985* (BM), 1500 m, *30148* (K), 1500 m, *30149* (K).

55.14. **ASCIDIERIA** Seidenf.

Nordic J. Bot. 4: 44 (1984).

Epiphytic, rarely terrestrial, herb. *Stems* up to 25 cm long, rather slender, covered with thin sheaths below, 2-leaved at apex. *Leaves* up to 25 × 1.2 cm, linear to linear-elliptic, acute to acuminate. *Inflorescences* erect, 1 or 2 emerging from near stem apex, densely pubescent, up to 16 cm long, with many flowers arranged in whorls about 5 mm apart, each whorl containing up to 10 flowers. *Flowers* non-resupinate, very small, 4 × 3 mm, white; *sepals* densely pubescent; *dorsal sepal* free, ovate-elliptic; *lateral sepals* united at base and attached directly to base of column, triangular-ovate, acute; *petals* smaller than sepals; *lip* cymbiform, deeply concave; *column* short, foot absent; *pollinia* 8.

A monotypic genus distributed in Thailand, Peninsular Malaysia, Sumatra and Borneo.

55.14.1. Ascidieria longifolia (Hook. f.) Seidenf., Nordic J. Bot. 4: 44 (1984). Fig. 9.

Eria longifolia Hook. f., Fl. Brit. Ind. 5: 790 (1890).

Epiphyte or lithophyte. Lower montane forest, sometimes on ultramafic substrate. Elevation: 1200–1800 m.

General distribution: Sabah, Brunei, Sarawak, Peninsular Malaysia, Thailand, Sumatra.

Collections. GOLF COURSE SITE: 1700–1800 m, *Beaman 10674* (K); GURULAU SPUR: 1800 m, *Clemens 50412* (BM); KILEMBUN RIVER: 1400 m, *Clemens 33966* (BM); KUNDASANG: 1400 m, *Kidman Cox 2517* (K); MOUNT KINABALU: *Haslam s.n.* (AMES); PENATARAN RIVER: 1700 m, *Clemens 34441* (BM, K); PENIBUKAN: 1200 m, *Clemens 30705* (BM); TENOMPOK: 1500 m, *Beaman 10518* (K, MSC), 1500 m, *Carr SFN 27049* (SING), 1200 m, *Clemens s.n.* (BM), 1500 m, *29823* (K), 1500 m, *29904* (K).

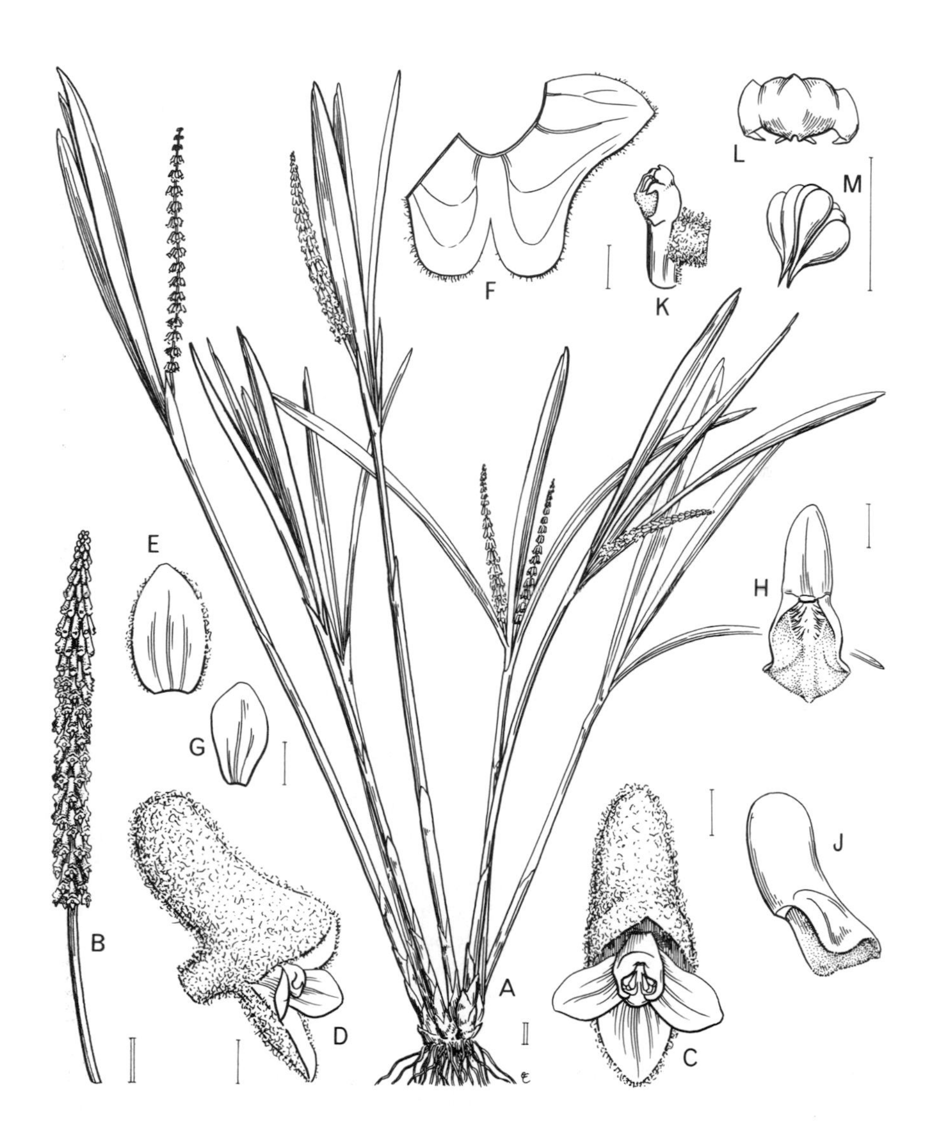

Fig. 9. Ascidieria longifolia. **A**, habit; **B**, inflorescence; **C**, flower (front view); **D**, flower (side view); **E**, dorsal sepal; **F**, lateral sepals; **G**, petal; **H**, lip and spur (front view); **J**, lip and spur (side view); **K**, column and part of pedicel-with-ovary (side view); **L**, anther-cap (back view); **M**, pollinia. **A**, from *Beaman 10674*; **B–M** from *Clements s.n.*, Crocker Range, Sabah. Scale: single bar = 1 mm; double bar = 1 cm. Drawn by Eleanor Catherine.

55.15. **BOGORIA** J. J. Sm.

Orch. Java, 566 (1905).

Small monopodially branching epiphytes. *Stems* very short with thick, flattened, greenish roots. *Leaves* flat, strap-shaped, apex unequally bilobed. *Inflorescences* borne from the stem below the leaves, simple; peduncle narrow and terete; rachis deeply sulcate, angular, with 3–4 flowers open simultaneously. *Flowers* small, greenish-yellow, lip white marked with crimson; *sepals* and *petals* free, spreading; *lip* 3-lobed, deeply saccate, immobile, side-lobes converging, mid-lobe thickened and merely forming the rim of the pouch; *column* with a foot; *anther-cap* large; *pollinia* 4, appearing as 2 unequal masses.

Four species distributed in Java, Borneo, Sumatra, Philippines and New Guinea.

55.15.1. Bogoria raciborskii J. J. Sm., Orch. Java, 566 (1905). Plate 9B.

Epiphyte. Hill forest? Elevation: 900 m. Recorded on the basis of a drawing, specimen not seen.

General distribution: Sabah, Java.

Collection. LIWAGU RIVER: 900 m, *Bacon in Lamb AL 68/83* (K, sketch).

55.16. **BROMHEADIA** Lindl.

Bot. Reg. 27: misc. 89, no. 184 (1841).

Terrestrial or epiphytic herbs. *Stems* slender, usually leafy, pseudobulbs absent. *Leaves* normal and oblong or laterally compressed. *Inflorescence* terminal, occasionally also lateral, producing a succession of flowers one or few at a time in two opposite ranks; bracts regularly alternate. *Flowers* resupinate, small to medium-sized; *sepals* and *petals* similar, spreading; *lip* straight, parallel with column, side-lobes erect and touching column, disc between them thickened, hirsute, mid-lobe ovate, recurved, with a median warty or papillose area; *column* long and slender, without a foot, winged above; *pollinia* 2, on a short, broad stipes.

Some 17 species distributed from Sri Lanka, Thailand and Indochina through Malaysia and Indonesia to the Philippines, New Guinea and Australia.

55.16.1. Bromheadia aff. **aporoides** Rchb. f., Otia Bot. Hamburg. 1: 44 (1878).

Epiphyte. Lower montane forest on ultramafic substrate. Elevation: 1400 m.

Collection. MAHANDEI RIVER HEAD: 1400 m, *Carr SFN 27245* (SING).

55.16.2. Bromheadia brevifolia Ridl., J. Linn. Soc. Bot. 32: 340 (1896).

Epiphyte. Lower montane forest. Elevation: 1100 m.

General distribution: Sabah, Sarawak, Peninsular Malaysia.

Collection. PENIBUKAN: 1100 m, *Clemens 40465* (BM).

55.16.3. Bromheadia crassiflora J. J. Sm., Bull. Jard. Bot. Buit., ser. 3, 11: 149 (1931).

Epiphyte. Lower montane forest. Elevation: 1500–1600 m.

General distribution: Sabah, Kalimantan.

Collections. LUMU-LUMU: 1600 m, *Carr SFN 27783* (K, SING); PENIBUKAN: 1500 m, *Clemens 31011* (BM); TENOMPOK: *Clemens 50490* (BM); TENOMPOK ORCHID GARDEN: 1500 m, *Clemens 50490* (K).

55.16.4. Bromheadia divaricata Ames & C. Schweinf., Orch. 6: 155 (1920). Plate 10A. Type: MARAI PARAI SPUR, *Clemens 389* (holotype AMES mf; isotypes BM!, K!, SING!).

Epiphyte. Lower montane forest, upper montane forest, on ultramafic substrate. Elevation: 2100–2700 m.

General distribution: Sabah.

Additional collections. KEMBURONGOH: 2200 m, *Carr SFN 27467* (SING), 2600 m, *Clemens 27177* (BM); KINATEKI RIVER HEAD: 2700 m, *Clemens 31930* (BM); MARAI PARAI SPUR: 2100 m, *Clemens 32887* (BM).

55.16.5. Bromheadia finlaysoniana (Lindl.) Miq., Fl. Ned. Ind. 3: 709 (1859).

Grammatophyllum finlaysonianum Lindl., Gen. Sp. Orch., 173 (1833).

Terrestrial. Hill forest, lower montane forest, in open damp areas. Elevation: 900–1500 m.

General distribution: Sabah, Brunei, Kalimantan, Sarawak, SE Asia.

Collections. DALLAS: 1000 m, *Carr SFN 27368* (SING), *Clemens s.n.* (BM, BM), 900 m, *26849* (BM), 900 m, *27445* (K); TENOMPOK: 1500 m, *Clemens s.n.* (BM).

55.16.6. Bromheadia scirpoidea Ridl., J. Bot. 38: 71 (1900).

Epiphyte. Lower montane forest. Elevation: 1400–1500 m.

General distribution: Sabah, Peninsular Malaysia.

Collections. GURULAU SPUR: 1500 m, *Clemens 50541* (K); PENIBUKAN: 1400 m, *Carr SFN 27578* (K, SING); TENOMPOK: 1500 m, *Clemens 28818* (K).

55.16.7. Bromheadia truncata Seidenf., Opera Bot. 72: 14, f. 5 (1983).

Epiphyte. Cultivated at Poring, of local origin.

General distribution: Sabah, Peninsular Malaysia, Singapore, Thailand, Sumatra.

Collection. PORING ORCHID GARDEN: *Lohok 28* (K).

55.16.8. Bromheadia indet.

Collections. GURULAU SPUR: 1400 m, *Carr 3176, SFN 27455A* (SING), 1400 m, *SFN 27455* (SING).

55.17. BULBOPHYLLUM Thouars

Hist. Orchid., tabl. esp. 3 sub u. (1822).

Vermeulen, J. J. (1991). Orchids of Borneo 2: *Bulbophyllum.* Royal Botanic Gardens, Kew & Toihaan Publishing Co., Kota Kinabalu.

Epiphytic, lithophytic or rarely terestrial herbs. *Rhizome* short to long, creeping or pendulous. *Pseudobulbs* clustered or remote, 1- to several-leaved. *Leaves* thin-textured to coriaceous. *Inflorescences* lateral, arising from base of pseudobulbs, or from nodes on the rhizome, racemose or pseudoumbellate, 1- to many-flowered. *Flowers* usually resupinate, minute to large. *Dorsal sepal* usually free. *Lateral sepals* connate at base to column-foot forming a saccate mentum. *Petals* usually free, shorter than sepals. *Lip* simple to 3-lobed, often sigmoid and recurved, versatile, fleshy, sometimes ciliate or pubescent. *Column* short, with apical aristate teeth or wings (stelidia) and a foot. *Pollinia* 2 or 4, naked, occasionally with a viscidium or viscidia, or a stipes.

A cosmopolitan genus containing around 1000 species, particularly well represented in Africa, SE Asia and New Guinea. The largest orchid genus in Borneo and on Mount Kinabalu.

55.17.1. Bulbophyllum acutum J. J. Sm., Orch. Java, 466 (1905).

Epiphyte. Hill forest, lower montane forest; on tree trunks and main branches; sometimes on ultramafic substrate. Elevation: 1700–2000 m.

General distribution: Sabah, China, India, Thailand, Sumatra, Java.

Collections. LOHAN RIVER: *Vermeulen 1297* (L); PARK HEADQUARTERS: 1700–2000 m, *Vermeulen & Lamb 365* (L).

55.17.2. Bulbophyllum alatum J. J. Verm., Orchids of Borneo 2: 167, f. 54 (1991). Type: PINOSUK PLATEAU, *Lamb AL 726/86* (holotype K!).

Epiphyte. Lower montane forest. Elevation: 1200–1800 m.

General distribution: Sabah.

Additional collections. GOLF COURSE SITE: 1700–1800 m, *Beaman 10669* (K); KINUNUT VALLEY HEAD: 1200 m, *Carr SFN 27348* (SING); LUMU-LUMU: 1800 m, *Carr SFN 27843* (L); MOUNT KINABALU: *Lamb AL 828/87* (K); PINOSUK PLATEAU: 1500–1700 m, *Vermeulen 479* (L); TENOMPOK: 1500 m, *Carr SFN 26864* (K, SING).

55.17.3. Bulbophyllum anguliferum Ames & C. Schweinf., Orch. 6: 164 (1920). Type: LUBANG, *Clemens 133* (holotype AMES mf).

Epiphyte. Lower montane forest, upper montane forest. Elevation: 1500–2300 m.

General distribution: Sabah, Sarawak.

Additional collections. KEMBURONGOH: 2000 m, *Carr SFN 27549* (L, SING), 2300 m, *Gunsalam 5* (K); MOUNT KINABALU: 2000 m, *Lamb AL 585/86* (K); TENOMPOK: 1600 m, *Carr SFN 27110* (L, SING); TENOMPOK/RANAU: 1500 m, *Carr SFN 27020* (SING).

55.17.4. Bulbophyllum antenniferum (Lindl.) Rchb. f., Walp. Ann. Bot. Syst. 6: 250 (1861).

Cirrhopetalum antenniferum Lindl., Bot. Reg. 29: sub t. 49 (1843).

Epiphyte. Lowlands, hill forest, lower montane forest. Elevation: 500–1500 m.

General distribution: Sabah, Java, Philippines, New Guinea, Solomon Islands.

Collections. DALLAS: 900 m, *Clemens 27491* (BM); KAUNG: 500 m, *Carr SFN 27292* (K, SING); TENOMPOK: 1500 m, *Clemens 30109* (K).

55.17.5. Bulbophyllum apheles J. J. Verm., Orchids of Borneo 2: 245, f. 85 (1991). Type: PINOSUK PLATEAU, 1500–1700 m, *Vermeulen 487* (holotype L n.v.; isotype K!).

Epiphyte. Lower montane forest. Elevation: 1500–1700 m.

General distribution: Sabah.

Additional collections. PINOSUK PLATEAU: 1700 m, *Collenette 2361* (K); TENOMPOK: 1500 m, *Carr SFN 27899* (SING).

55.17.6. Bulbophyllum apodum Hook. f., Fl. Brit. Ind. 5: 766 (1890).

Epiphyte. Hill forest, lower montane forest on ultramafic substrate. Elevation: 800–2000 m.

General distribution: Sabah, Kalimantan, Sarawak, Peninsular Malaysia, Thailand, Vietnam, Sumatra, Java, Sulawesi, Philippines.

Collections. MARAI PARAI: 1500 m, *Clemens 32207* (BM); MARAI PARAI SPUR: 2000 m, *Collenette A 133* (BM); MINITINDUK GORGE: 800 m, *Carr 3153, SFN 26605* (SING).

55.17.7. Bulbophyllum armeniacum J. J. Sm., Bull. Jard. Bot. Buit., ser. 2, 25: 70 (1917). Plate 10B.

Epiphyte. Lower montane forest. Elevation: 1200–1500 m.

General distribution: Sabah, Kalimantan, Sarawak, Peninsular Malaysia, Sumatra.

Collections. PENIBUKAN: 1200–1500 m, *Clemens 31252* (BM), 1400 m, *31975* (BM).

55.17.8. Bulbophyllum biflorum Teijsm. & Binn., Natuurk. Tijdschr. Ned.-Ind. 5: 488 (1853).

Epiphyte. Hill forest. Elevation: 1200 m.

General distribution: Sabah, Peninsular Malaysia, Thailand, Sumatra, Java, Bali, Lombok, Philippines.

Collection. BUNDU TUHAN: 1200 m, *Carr SFN 27231* (SING).

55.17.9. Bulbophyllum breviflorum Ridl. in Stapf, Trans. Linn. Soc. Bot. 4: 236 (1894). Type: MOUNT KINABALU, 1800 m, *Haviland s.n.* (holotype unlocated).

Epiphyte. Lower montane forest. Elevation: 1500–1800 m.

General distribution: Sabah, Sarawak.

Additional collections. EAST MESILAU RIVER: 1500–1600 m, *Bailes & Cribb 520* (K); KEMBURONGOH: 1800 m, *Carr SFN 27741* (K, SING).

55.17.10. Bulbophyllum calceolus J. J. Verm., Orchids of Borneo 2: 157, f. 51 (1991). Plate 10C. Type: PINOSUK PLATEAU, 1500–1700 m, *Vermeulen 489* (holotype L n.v.; isotype K!).

Epiphyte. Lower montane forest; on branches of shrubs and trees. Elevation: 1500–1700 m.

General distribution: Sabah.

Additional collection. PINOSUK PLATEAU: 1500 m, *Lamb AL 881/87* (K).

55.17.11. Bulbophyllum carinilabium J. J. Verm., Orchids of Borneo 2: 233, f. 81 (1991).

Epiphyte. Lower montane forest, mossy forest. Elevation: 1600 m.

General distribution: Sabah.

Collection. TENOMPOK: 1600 m, *Carr SFN 26987* (L, SING).

55.17.12. Bulbophyllum catenarium Ridl. in Stapf, Trans. Linn. Soc. Bot. 4: 235 (1894). Plate 11A. Type: MOUNT KINABALU, 2000 m, *Haviland 1164* (lectotype of Vermeulen (1991) K!).

Epiphyte. Lower montane forest, upper montane forest, mossy forest, sometimes on ultramafic substrate. Elevation: 1500–2300 m.

General distribution: Sabah, Kalimantan, Sarawak, Peninsular Malaysia.

Additional collections. EASTERN SHOULDER: 1500 m, *Collenette 753* (K); KEMBURONGOH: 2100 m, *Carr SFN 27461* (L, SING), 2300 m, *Gunsalam 4* (K), 1700 m, *Molesworth-Allen 3313* (K), 2100 m, *Price 223* (K), 2100 m, *Sinclair et al. 9057* (SING); MOUNT KINABALU: *Lamb SAN 88566* (K).

55.17.13. Bulbophyllum caudatisepalum Ames & C. Schweinf., Orch. 6: 166 (1920). Type: LUBANG, 1500 m, *Clemens 113A* (holotype AMES n.v.; K photo).

Bulbophyllum cuneifolium Ames & C. Schweinf., Orch. 6: 172 (1920). Type: KIAU, 900 m, *Clemens 195* (holotype AMES mf).

Bulbophyllum pergracile Ames & C. Schweinf., Orch. 6: 190 (1920). Type: KIAU, *Clemens 326* (holotype AMES mf).

Epiphyte. Hill forest, lower montane forest, often on ultramafic substrate. Elevation: 800–2400 m.

General distribution: Sabah, Kalimantan, Sarawak, Peninsular Malaysia.

Additional collections. BAMBANGAN RIVER: 1600 m, *RSNB 4941* (K); BUNDU TUHAN: 800 m, *Carr SFN 27919* (L, SING), 800 m, *SFN 27919a* (SING), 800 m, *SFN 27919b* (SING), 900 m, *Darnton 536* (BM); GOLF COURSE SITE: 1700 m, *Beaman 9048a* (K, MSC); KIBAMBANG LUBANG: 1400 m, *Clemens 32557* (BM); KIBAMBANG RIVER: 1200 m, *Clemens 34305* (BM, K); KINATEKI RIVER HEAD: 2100 m, *Clemens 31736* (BM), 2400 m, *31836* (BM, L); MARAI PARAI SPUR: 1700 m, *Clemens 34375* (BM); MINETUHAN: *Clemens 35202* (L); PENIBUKAN: 1200–1500 m, *Clemens s.n.* (K); TENOMPOK: 1500 m, *Clemens 27871* (BM, K), 1500 m, *28952* (BM), 1500 m, *30105* (K), *50247* (K); TENOMPOK ORCHID GARDEN: *Clemens 50247* (BM).

55.17.14. Bulbophyllum aff. **caudatisepalum** Ames & C. Schweinf., Orch. 6: 167 (1920).

Epiphyte. Lower montane forest. Elevation: 600 m.

Collection. SUMMIT TRAIL: 600 m, *Vermeulen & Duistermaat 539* (K).

55.17.15. Bulbophyllum ceratostylis J. J. Sm., Recueil Trav. Bot. Néerl. 1: 154 (1904).

Bulbophyllum eximium Ames & C. Schweinf., Orch. 6: 178 (1920). Type: KIAU, *Clemens 317* (holotype AMES mf; isotypes K!, SING!).

Epiphyte. Hill forest, lower montane forest; on small branches of trees and shrubs. Elevation: 900–1600 m.

General distribution: Sabah, Sumatra.

Additional collections. KUNDASANG: 1400 m, *Kidman Cox 918* (K); MAHANDEI RIVER: 1100 m, *Carr 3061, SFN 26493* (SING); MT. LENAU (BETWEEN TENOMPOK AND KEMBURONGOH): 1600 m, *Sinclair et al. 9011* (K); TAHUBANG RIVER: 1000 m, *Carr 3056, SFN 26373* (SING), 900 m, *Clemens 50146* (BM, K); TENOMPOK: 1500 m, *Carr SFN 26877* (K, L, SING), 1500 m, *Clemens s.n.* (BM), 1500 m, *29674* (BM, K).

55.17.16. Bulbophyllum chanii J. J. Verm. & A. Lamb, Orchids of Borneo 2: 177, f. 58 (1991).

Epiphyte. Lower montane forest. Elevation: 1400–1500 m.

General distribution: Sabah.

Collections. EAST MESILAU RIVER: *Lamb s.n.* (K); GURULAU SPUR: 1400 m, *Carr 3141, SFN 27925* (SING); TENOMPOK/RANAU: 1500 m, *Carr SFN 27020* (SING).

55.17.17. Bulbophyllum compressum Teijsm. & Binn., Natuurk. Tijdschr. Ned.-Ind. 24: 307 (1862).

Epiphyte. Lowland dipterocarp forest. Elevation: 500 m.

General distribution: Sabah, Sumatra, Java, Sulawesi.

Collections. MAMUT RIVER: 500 m, *Lamb AL 785/87* (K); PORING: 500 m, *Lamb SAN 89693* (K).

55.17.18. Bulbophyllum coniferum Ridl., J. Fed. Malay States Mus. 4: 67 (1909).

Bulbophyllum reflexum Ames & C. Schweinf., Orch. 6: 192 (1920). Type: KIAU, *Clemens 384* (holotype AMES mf).

Epiphyte. Lower montane forest. Elevation: 1100–2200 m.

General distribution: Sabah, Kalimantan, Sarawak, Peninsular Malaysia, Sumatra, Java.

Additional collections. BUNDU TUHAN: 1400 m, *Carr SFN 27152* (SING); KINATEKI RIVER: 1200 m, *Clemens 31042* (BM); KINATEKI RIVER HEAD: 2100 m, *Clemens 31735* (BM); LUMU-LUMU: 1600 m, *Carr SFN 27030* (K, SING); MESILAU CAVE: 1900–2200 m, *Beaman 9575* (K); PINOSUK PLATEAU: 1500 m, *Lamb AL 12/82* (K), 1500–1700 m, *Vermeulen 471* (L); SUMMIT TRAIL: 2000 m, *Vermeulen & Duistermaat 546* (L); TAHUBANG RIVER: 1100 m, *Carr 3058, SFN 27826* (SING); TENOMPOK: 1500 m, *Carr SFN 27971* (L, SING).

55.17.19. Bulbophyllum coriaceum Ridl. in Stapf, Trans. Linn. Soc. Bot. 4: 235 (1894). Fig. 10, Plate 11B. Type: MOUNT KINABALU, 3200 m, *Haviland 1100* (holotype? SING!; isotype? K!).

Bulbophyllum kinabaluense Rolfe in Gibbs, J. Linn. Soc. Bot. 42: 148 (1914). Type: LUBANG/PAKA-PAKA CAVE, 1800–2900 m, *Gibbs 4252* (holotype BM!).

Bulbophyllum venustum Ames & C. Schweinf., Orch. 6: 198 (1920). Type: PAKA-PAKA CAVE, *Clemens 113* (holotype AMES mf; isotypes BM!, K!).

Epiphyte or lithophyte. Lower montane forest, upper montane forest, *Leptospermum* scrub among rocks, sometimes on ultramafic substrate. Elevation: 1200–3500 m. Its relationship with *B. sopoetanense* needs investigation.

Endemic to Mount Kinabalu.

Additional collections. GURULAU SPUR: 2100–2700 m, *Clemens 50775* (BM); KEMBURONGOH: 2400 m, *Clemens 29125* (BM); KEMBURONGOH/LUMU-LUMU: 1800–2400 m, *Clemens 27150* (BM, K); KILEMBUN BASIN: 1500–1800 m, *Clemens 34442* (BM); LAYANG-LAYANG/PAKA-PAKA CAVE: 2700–3500 m, *Lamb SAN 91579* (K); LUMU-LUMU: 1800 m, *Carr SFN 27833A* (K, SING); MARAI PARAI: 2400–3000 m, *Clemens 33175* (BM); MESILAU CAVE: 1800 m, *Chew & Corner RNSB 4784* (K, SING); MINETUHAN SPUR: 1400 m, *Clemens 34194* (BM); PAKA-PAKA CAVE: 3100 m, *Carr SFN 27596* (SING), 3400 m, *Clemens 27863* (BM, K); PANAR LABAN: 3500 m, *Beaman 8303* (K); PENIBUKAN: 1200–1500 m, *Clemens 31256* (BM, K), 1200 m, *40602* (BM, K); PINOSUK PLATEAU: 1500–1700 m, *Vermeulen 472* (K); SUMMIT TRAIL: 2100 m, *Mikil SAN 46571* (K); TENOMPOK ORCHID GARDEN: 1500 m, *Clemens 50485* (BM).

55.17.20. Bulbophyllum aff. **coriaceum** Ridl. in Stapf, Trans. Linn. Soc. Bot. 4: 235 (1894).

Epiphyte. Lower montane forest. Elevation: 2000 m.

Collection. SUMMIT TRAIL: 2000 m, *Beaman 6766* (K).

55.17.21. Bulbophyllum cornutum (Blume) Rchb. f., Walp. Ann. Bot. Syst. 6: 247 (1861).

Ephippium cornutum Blume, Bijdr., 310 (1825).

Bulbophyllum concavum Ames & C. Schweinf., Orch. 6: 168 (1920). Type: KIAU, 900 m, *Clemens 94* (holotype AMES mf).

Epiphyte. Hill forest, lower montane forest. Elevation: 900–1700 m.

General distribution: Sabah, Java, Philippines.

Additional collections. KINUNUT VALLEY HEAD: 1200 m, *Carr SFN 27133* (SING); PINOSUK PLATEAU: 1500–1700 m, *Vermeulen 513* (K).

55.17.22. Bulbophyllum crepidiferum J. J. Sm., Bull. Jard. Bot. Buit., ser. 3, 2: 88 (1920).

Epiphyte. Lower montane forest. Elevation: 1700 m.

General distribution: Sabah, Sumatra.

H
C
J
K
L
A
B
E
G
D
F

Collection. PINOSUK PLATEAU: 1700 m, *Lamb AL 880/87* (K).

55.17.23. Bulbophyllum dearei Rchb. f., Flora 71: 156 (1888).

Epiphyte. Hill forest. Elevation: 1200 m.

General distribution: Sabah, Kalimantan?, Sarawak, Peninsular Malaysia, Philippines.

Collections. BUNDU TUHAN: 1200 m, *Carr SFN 27448* (SING); MOUNT KINABALU: *Haslam s.n.* (AMES).

55.17.24. Bulbophyllum deltoideum Ames & C. Schweinf., Orch. 6: 174 (1920). Type: LUBANG, 1500 m, *Clemens 115* (holotype AMES mf; isotype SING!).

Epiphyte. Lower montane forest, sometimes on ultramafic substrate. Elevation: 1200–2000 m.

General distribution: Sabah, Kalimantan.

Additional collections. MARAI PARAI: 1500 m, *Clemens 32207* (BM), 1800 m, *32445* (BM, K); MARAI PARAI SPUR: 2000 m, *Collenette A 133* (BM); MESILAU RIVER: 1500 m, *RSNB 4856* (K); PENATARAN BASIN: 1700 m, *Clemens 40133* (BM); PENIBUKAN: 1200–1500 m, *Clemens s.n.* (K), 1400 m, *31254* (BM, K); SUMMIT TRAIL: 1800 m, *Vermeulen & Duistermaat 538* (K).

55.17.25. Bulbophyllum aff. **deltoideum** Ames & C. Schweinf., Orch. 6: 174 (1920).

Epiphyte. Lower montane forest, upper montane forest, probably on ultramafic substrate. Elevation: 1200–2100 m.

Collections. KEMBURONGOH: 2100 m, *Price 228* (K); PENIBUKAN: 1200 m, *Clemens 51713* (BM, K); TAHUBANG RIVER: 1200 m, *Clemens 40340* (BM, K).

55.17.26. Bulbophyllum disjunctum Ames & C. Schweinf., Orch. 6: 176 (1920). Type: MARAI PARAI SPUR, *Clemens 254* (holotype AMES mf; isotypes K!, SING!).

Epiphytic or terrestrial. Lower montane forest, sometimes on ultramafic substrate. Elevation: 900–2000 m.

Fig. 10. Bulbophyllum coriaceum. A and **B**, habit; **C**, flower (oblique view); **D**, dorsal sepal; **E**, lateral sepal; **F**, petal; **G**, lip (front view, flattened); **H**, pedicel-with-ovary, column and lip (side view); **J**, column and anther-cap (front view); **K**, anther-cap (side view); **L**, pollinia. **A** from *Haviland 1100*; **B** from *Mikil SAN 46571*; **C–L** from *Cult. Kew EN 683-66* (collector unknown). **B** corresponds to *B. venustum* Ames & C. Schweinf. Scale: single bar = 1 mm; double bar = 1 cm. Drawn by Eleanor Catherine.

General distribution: Sabah, Kalimantan, Sarawak, Thailand.

Additional collections. GOLF COURSE SITE: 1700–1800 m, *Beaman 7238* (K); KIBAMBANG LUBANG: 1400 m, *Clemens 32554* (BM); LITTLE MAMUT RIVER: 1600 m, *Collenette 1002* (K); LUMU-LUMU: 1800 m, *Clemens 27361* (BM, L); MT. NUNGKEK: 900 m, *Clemens 32787* (BM); PARK HEADQUARTERS: 1500 m, *Aban SAN 56304* (K); PENIBUKAN: 1200 m, *Clemens 40226* (BM); SUMMIT TRAIL: 2000 m, *Beaman 6764* (K).

55.17.27. Bulbophyllum dryas Ridl., J. Fed. Malay States Mus. 6: 175 (1915).

Epiphyte. Upper montane forest. Elevation: 2000 m.

General distribution: Sabah, Peninsular Malaysia, Sumatra.

Collections. KEMBURONGOH: 2000 m, *Carr SFN 27559* (SING); PENATARAN BASIN: 2000 m, *Clemens 40130* (BM).

55.17.28. Bulbophyllum cf. **dryas** Ridl., J. Fed. Malay States Mus. 6: 175 (1915).

Epiphyte. Lower montane forest. Elevation: 1800 m.

Collection. PENIBUKAN: 1800 m, *Clemens 50291* (BM, K).

55.17.29. Bulbophyllum aff. **elachanthe** J. J. Verm., Orchids of Borneo 2: 49, f. 13 (1991).

Epiphyte. Lower montane forest.

Collection. PARK HEADQUARTERS: *Lamb AL 715/86* (K).

55.17.30. Bulbophyllum flammuliferum Ridl., J. Bot. 36: 211 (1898). Plate 12A.

Epiphyte. Hill forest on ultramafic substrate. Elevation: 600 m.

General distribution: Sabah, Peninsular Malaysia, Sumatra.

Collection. HEMPUEN HILL: 600 m, *Lamb & Surat AL 1302/91* (K).

55.17.31. Bulbophyllum flavescens (Blume) Lindl., Gen. Sp. Orch., 54 (1830).

Diphyes flavescens Blume, Bijdr., 313 (1825).

Bulbophyllum montigenum Ridl. in Stapf, Trans. Linn. Soc. Bot. 4: 235 (1894). Type: MOUNT KINABALU, 1800 m, *Haviland 1252* (holotype? K!).

Bulbophyllum lanceolatum Ames & C. Schweinf., Orch. 6: 180 (1920). Type: GURULAU SPUR, *Clemens 305* (holotype AMES mf).

Epiphyte. Lower montane forest, rarely in hill forest and upper montane forest, sometimes on ultramafic substrate. Elevation: 800–3800 m.

General distribution: Sabah, Kalimantan, Sarawak, Peninsular Malaysia, Sumatra, Java, Philippines.

Additional collections. BAMBANGAN RIVER: 3800 m, *Collenette 2376* (K); BUNDU TUHAN: 1200 m, *Carr SFN 27830* (SING), 1400 m, *Darnton 211* (BM); KEMBURONGOH: 2300 m, *Carr SFN 27502* (K, SING); KILEMBUN RIVER HEAD: *Clemens 35203* (BM, L); KINATEKI RIVER: 1400 m, *Carr 3131, SFN 26551* (SING); LUMU-LUMU: 1800 m, *Clemens 27364* (BM); MAHANDEI RIVER: 1200 m, *Carr 3033, SFN 26313* (L, SING); MARAI PARAI/KIAU TRAIL: 1000 m, *Bailes & Cribb 805* (K); MESILAU CAMP: 1500 m, *RSNB 6031* (K); MESILAU CAVE: 1600 m, *Bailes & Cribb 687* (K), 1900–2200 m, *Beaman 9569* (K); MESILAU RIVER: 2100 m, *Clemens 29255* (BM, K); MESILAU TRAIL: 1500–1800 m, *Chow & Leopold SAN 76434* (K); MOUNT KINABALU: 800 m, *Puasa 1552* (K); PENIBUKAN: 1200–1500 m, *Clemens 30592* (BM), 1200–1500 m, *31003A* (L), 1200 m, *40521* (BM, K), 1200 m, *40526* (BM, K), 1100 m, *50063* (BM, K), 1500 m, *50230* (BM, K); PINOSUK PLATEAU: 1500–1700 m, *Vermeulen 484* (K); TAHUBANG RIVER: 1000 m, *Carr 3061, SFN 26393* (K), *Clemens s.n.* (K), 1100 m, *40143* (BM), 900–1400 m, *40343* (BM, K), 1200 m, *50394* (BM, K); TENOMPOK: 1600 m, *Carr SFN 27082* (SING), 1500 m, *Clemens s.n.* (K), 1500 m, *28198* (BM, K), 1500 m, *28606* (AMES, BM, K), 1500 m, *30112* (K), 1500 m, *30114* (K); TENOMPOK RIDGE: 1400–1500 m, *Beaman 8193* (K).

55.17.32. **Bulbophyllum** aff. **flavescens** (Blume) Lindl., Gen. Sp. Orch., 54 (1830).

Epiphyte. Lower montane forest. Elevation: 1500–1600 m.

Collection. EAST MESILAU RIVER: 1500–1600 m, *Bailes & Cribb 515* (K).

55.17.33. **Bulbophyllum foetidolens** Carr, Gardens' Bull. 5: 135, t. 3, f. 2 (1930). Plate 12B.

Epiphyte, lithophyte. Lower montane forest; on lower parts of tree trunks, fallen logs, and mossy rocks. Elevation: 1500 m.

General distribution: Sabah, Kalimantan, Sarawak, Peninsular Malaysia.

Collection. LIWAGU RIVER TRAIL: 1500 m, *Lamb AL 21/82* (K).

55.17.34. **Bulbophyllum gibbosum** (Blume) Lindl., Gen. Sp. Orch., 54 (1830).

Diphyes gibbosa Blume, Bijdr., 312 (1825).

Bulbophyllum magnivaginatum Ames & C. Schweinf., Orch. 6: 186 (1920). Type: KIAU, 900 m, *Clemens 36* (holotype AMES mf).

Epiphyte. Hill forest. Elevation: 800–900 m.

General distribution: Sabah, Peninsular Malaysia, Sumatra, Java, Bali.

Additional collections. BUNDU TUHAN: 800 m, *Carr SFN 27819* (K); KIAU: 900 m, *Clemens 48* (AMES), *325* (AMES), *352* (AMES); MOUNT KINABALU: *Clemens s.n.* (AMES), *Haslam s.n.* (AMES).

55.17.35. Bulbophyllum gibbsiae Rolfe in Gibbs, J. Linn. Soc. Bot. 42: 149 (1914). Type: PENIBUKAN, 1800 m, *Gibbs 4059* (holotype BM!; isotype K!).

Bulbophyllum minutiflorum Ames & C. Schweinf., Orch. 6: 188 (1920). Type: MOUNT KINABALU, *Haslam s.n.* (holotype AMES mf).

Epiphyte, lithophyte. Lower montane forest, sometimes on ultramafic substrate. Elevation: 1100–2200 m.

General distribution: Sabah, Sarawak.

Additional collections. BUNDU TUHAN: 1400 m, *Darnton 210* (BM), 1100 m, *541* (BM); GURULAU SPUR: 1500 m, *Clemens 50707* (BM, K); MAHANDEI RIVER HEAD: 1400 m, *Carr 3071, SFN 26429* (K, SING); MESILAU CAVE: 1900–2200 m, *Beaman 9568* (K); PENIBUKAN: 1500 m, *Clemens 31003* (BM, K), 1200–1500 m, *31253* (BM); PINOSUK PLATEAU: 1500–1700 m, *Vermeulen 480* (K); TAHUBANG RIVER: 1200 m, *Clemens 40337* (K); TENOMPOK: 1600 m, *Carr SFN 26984* (K, SING), 1400 m, *SFN 27236* (K, SING), 1600 m, *SFN 27410* (K, SING), 1500 m, *Clemens 30146* (K); ULAR HILL TRAIL: *Collenette 2315* (K).

55.17.36. Bulbophyllum heldiorum J. J. Verm., Orchids of Borneo 2: 143, f. 46 (1991). Type: SUMMIT TRAIL, 2000–2500 m, *Vermeulen & Duistermaat 547* (holotype L n.v.; isotypes K!, SNP).

Epiphyte. Lower montane forest, upper montane forest, mossy forest, sometimes on ultramafic substrate. Elevation: 1800–2500 m.

Endemic to Mount Kinabalu.

Additional collections. LUMU-LUMU: 1800 m, *Carr SFN 27821* (SING); SUMMIT TRAIL: 2500 m, *Chan & Gunsalam 34/87* (K, L).

55.17.37. Bulbophyllum ionophyllum J. J. Verm., Orchids of Borneo 2: 65, f. 18 (1991).

Epiphyte. Lower montane forest. Elevation: 1200–1700 m.

General distribution: Sabah.

Collections. EAST AND WEST MESILAU RIVERS: 1500 m, *Collenette 586* (K); PINOSUK PLATEAU: 1700 m, *Vermeulen 477* (L); TENOMPOK: 1500 m, *Carr SFN 27407* (K, SING), 1500 m, *Clemens s.n.* (K), 1200 m, *29749* (BM, K).

55.17.38. Bulbophyllum kestron J. J. Verm. & A. Lamb, Malayan Orchid Rev. 22: 45 (1988).

Epiphyte. Lower montane forest. Elevation: 1400 m.

General distribution: Sabah.

Collection. KINUNUT VALLEY HEAD: 1400 m, *Carr SFN 27198* (SING).

55.17.39. Bulbophyllum lambii J. J. Verm., Orchids of Borneo 2: 185, f. 62 (1991). Type: PINOSUK PLATEAU, *Lamb AL 577/86* (holotype K!).

Epiphyte. Lower montane forest. Elevation: 1500–1900 m.

General distribution: Sabah.

Additional collections. MESILAU CAVE: 1900 m, *Collenette 917* (K); PINOSUK PLATEAU: 1500 m, *Bailes & Cribb s.n.* (K), 1500–1700 m, *Vermeulen 482* (L); WEST MESILAU RIVER: 1700 m, *Brentnall 134* (K).

55.17.40. Bulbophyllum latisepalum Ames & C. Schweinf., Orch. 6: 182 (1920). Type: MOUNT KINABALU, *Clemens s.n.* (holotype AMES mf).

Epiphyte. Lower montane forest. Elevation: 1400–1900 m.

General distribution: Sabah.

Additional collections. KOLOPIS RIVER HEAD: 1400 m, *Carr 3192, SFN 27849* (SING); MESILAU: *Lamb s.n.* (K); MESILAU CAVE: 1900 m, *Collenette 910* (K); MESILAU RIVER: *Lamb & Phillipps s.n.* (K); PINOSUK PLATEAU: *Lamb AL 829/87* (K), 1700 m, *Vermeulen & Lamb 357* (L); TENOMPOK: 1500 m, *Clemens 30117* (K).

55.17.41. Bulbophyllum laxiflorum (Blume) Lindl., Gen. Sp. Orch., 57 (1830).

Diphyes laxiflora Blume, Bijdr., 316 (1825).

Epiphyte. Hill forest, lower montane forest. Elevation: 900–1400 m.

General distribution: Sabah, Kalimantan, Sarawak, Burma, Peninsular Malaysia, Thailand, Laos, Sumatra, Java, Sulawesi, Philippines.

Collections. KIBAMBANG LUBANG: 1400 m, *Clemens 40230* (BM); MT. NUNGKEK: 900 m, *Clemens 32880* (BM).

55.17.42. Bulbophyllum lissoglossum J. J. Verm., Orchids of Borneo 2: 25, f. 3 (1991). Plate 14B. Type: PARK HEADQUARTERS, 1700–2000 m, *Vermeulen & Chan 390* (holotype L n.v.; isotype K!).

Epiphytic or terrestrial. Lower montane forest; on trunks and branches, fallen logs and terrestrial in moss cushions. Elevation: 1500–2000 m.

General distribution: Sabah, Sarawak.

Additional collections. LUMU-LUMU: 1600 m, *Carr SFN 27249* (SING); PARK HEADQUARTERS: 1700 m, *Lamb SAN 91580* (K); TENOMPOK: 1500 m, *Clemens 29820* (K).

55.17.43. Bulbophyllum lobbii Lindl., Bot. Reg. 33: sub t. 29 (1847).

Epiphyte. Hill forest, lower montane forest, infrequently on ultramafic substrate. Elevation: 800–1800 m.

General distribution: Sabah, Kalimantan, Sarawak, widespread in India & SE Asia east to the Philippines.

Collections. BAMBANGAN RIVER: 1700 m, *RSNB 1324* (K); DALLAS: 800 m, *Clemens 29303* (BM, K); EAST MESILAU RIVER: 1500–1600 m, *Bailes & Cribb 509* (K), 1400 m, *Sands 3904* (K); EAST AND WEST MESILAU RIVERS: 1600 m, *Collenette 538* (K); GOLF COURSE SITE: 1700 m, *Beaman 8557* (K); KEGIITAN AGAYE HEAD: 1200 m, *Carr 3165, SFN 26742* (SING); LOHAN RIVER: 800 m, *Beaman 8367* (K); MARAI PARAI/KIAU TRAIL: 1500 m, *Bailes & Cribb 803* (K); MESILAU CAMP: 1500 m, *RSNB 6008* (K), *Poore H 87* (K); MESILAU CAVE TRAIL: 1800 m, *Beaman 7479* (K); MESILAU RIVER: 1500 m, *RSNB 4046* (K); MINITINDUK: 900 m, *Carr SFN 26742a* (SING); PENATARAN BASIN: 1400 m, *Clemens 32588* (BM); TENOMPOK: 1500 m, *Clemens 27659* (BM); WEST MESILAU RIVER: 1600–1700 m, *Beaman 8700* (K), 1600 m, *8996* (K), 1700 m, *Brentnall 146* (K).

55.17.44. Bulbophyllum longhutense J. J. Sm., Bull. Jard. Bot. Buit., ser. 3, 11: 141 (1931).

Epiphyte. Lower montane mossy forest. Elevation: 1500–1700 m.

General distribution: Sabah, Kalimantan.

Collections. PARK HEADQUARTERS: 1700 m, *Vermeulen 497* (L); TENOMPOK: 1500 m, *Carr SFN 26863* (SING), 1500 m, *Clemens 30118* (K).

55.17.45. Bulbophyllum longimucronatum Ames & C. Schweinf., Orch. 6: 184 (1920). Type: KIAU, 900 m, *Clemens 56* (holotype AMES mf).

Epiphyte. Hill forest, lower montane forest. Elevation: 900–1900 m.

General distribution: Sabah.

Additional collections. BAMBANGAN LUBANG: 1100 m, *Clemens 34411* (BM); DALLAS: 900 m, *Clemens s.n.* (BM); SUMMIT TRAIL: 1900 m, *Darnton 514* (BM); TENOMPOK: 1500 m, *Clemens 26843* (BM, K); TENOMPOK ORCHID GARDEN: 1500 m, *Clemens 50172* (BM).

55.17.46. Bulbophyllum lygeron J. J. Verm., Orchids of Borneo 2: 69, f. 20 (1991). Type: PARK HEADQUARTERS, 1700–2000 m, *Vermeulen & Lamb 353* (holotype L n.v.).

Pendulous epiphyte. Lower montane forest. Elevation: 1200–2000 m.

General distribution: Sabah.

Additional collections. PENIBUKAN: 1200 m, *Clemens 32140* (BM); TINEKUK FALLS: 1400 m, *Clemens 40821* (BM).

55.17.47. Bulbophyllum macranthum Lindl., Bot. Reg. 30: t. 13 (1844).

Epiphyte. Hill forest, lower montane forest, sometimes on ultramafic substrate. Elevation: 600–1400 m.

General distribution: Sabah, Brunei, Kalimantan, Sarawak, Burma, Peninsular Malaysia, Thailand, Vietnam, Sumatra, Java, Philippines.

Collections. KUNDASANG: 1400 m, *Kidman Cox 2518* (K); LOHAN RIVER: 600–700 m, *Lamb AL 503/85* (K).

55.17.48. Bulbophyllum mandibulare Rchb. f., Gard. Chron. ser. 2, 17: 366 (1882). Plate 13A.

Epiphyte. Lowlands. Elevation: 300–1000 m.

General distribution: Sabah.

Collections. BUNDU TUHAN: 1000 m, *Brentnall 108* (K); KAUNG: 400 m, *Carr 3020, SFN 27921* (K, SING), 300 m, *Collenette 1* (BM), 400 m, *Darnton 394* (BM); KIAU: 900 m, *Clemens 40502* (BM); PORING ORCHID GARDEN: *Lohok 26* (K).

55.17.49. Bulbophyllum membranaceum Teijsm. & Binn., Ned. Kruidk. Arch. 3: 397 (1855).

Epiphyte. Lowlands. Elevation: 400 m.

General distribution: Sabah, Peninsular Malaysia, Thailand, Sumatra, Sulawesi, New Guinea?

Collection. KAUNG: 400 m, *Carr 3016, SFN 27375* (SING).

55.17.50. Bulbophyllum membranifolium Hook. f., Fl. Brit. Ind. 5: 756 (1890).

Epiphyte. Lower montane forest, sometimes on ultramafic substrate. Elevation: 1500–2100 m.

General distribution: Sabah, Kalimantan, Sarawak, Peninsular Malaysia, Sumatra, Philippines.

Collections. KINATEKI RIVER HEAD: 2100 m, *Clemens 31737* (BM, L); LUMU-LUMU: 1700 m, *Carr SFN 27577* (SING); MARAI PARAI: 1700 m, *Clemens 32311* (BM, L), 1800 m, *34335* (BM, L); MESILAU RIVER: *Clemens 28899* (BM, L); MURU-TURA RIDGE: 1700 m, *Clemens 31966* (BM); SUMMIT TRAIL: 1800 m, *Vermeulen & Duistermaat 542* (L); TENOMPOK: 1500 m, *Clemens 29750* (BM, K).

55.17.51. Bulbophyllum microglossum Ridl., J. Linn. Soc. Bot. 38: 325 (1908). Plate 13B.

Epiphyte. Lower montane forest. Elevation: 1200–2200 m.

General distribution: Sabah, Peninsular Malaysia.

Collections. MESILAU CAVE: 1900–2200 m, *Beaman 9572* (K); MT. NUNGKEK: 1200 m, *Clemens 32785* (BM), *32903* (BM); PENIBUKAN: 1200 m, *Clemens 30604* (BM), 1500 m, *50341* (BM); TENOMPOK: 1500 m, *Clemens 28553* (BM), 1500 m, *29633* (BM, K), 1500 m, *30106* (K); WEST MESILAU RIVER: 1600–1700 m, *Beaman 8701* (K).

55.17.52. Bulbophyllum minutulum Ridl., Fl. Malay Penins. 4: 62 (1924).

Epiphyte. Lower montane forest; on trunks and branches usually in shady places. Elevation: 1100–2000 m.

General distribution: Sabah, Kalimantan, Peninsular Malaysia.

Collections. KINATEKI RIVER: 1500 m, *Clemens 51731* (BM, K); MAHANDEI RIVER: 1100 m, *Carr 3099, SFN 27364* (L, SING); PARK HEADQUARTERS: 1700–2000 m, *Vermeulen & Lamb 359* (K).

55.17.53. Bulbophyllum montense Ridl. in Stapf, Trans. Linn. Soc. Bot. 4: 234 (1894). Type: MOUNT KINABALU, 3400 m, *Haviland 1099* (holotype fide Vermeulen SING n.v.; isotype K!).

Bulbophyllum vinculibulbum Ames & C. Schweinf., Orch. 6: 202 (1920). Type: LUBANG/PAKA-PAKA CAVE, *Topping 372* (holotype AMES mf).

Epiphyte. Upper montane forest, frequently on ultramafic substrate. Elevation: 2200–3400 m.

Endemic to Mount Kinabalu.

Additional collections. DACHANG: 2700 m, *Clemens 29233* (BM, K); GURULAU SPUR: 3200 m, *Clemens 50903* (BM, K); JANET'S HALT/SHEILA'S PLATEAU: 2800 m, *Collenette 553* (K); KADAMAIAN RIVER, JUST BELOW PAKA-PAKA CAVE: 3000 m, *Sinclair et al. 9172* (K, SING); KEMBURONGOH: 2200 m, *Carr SFN 27520* (SING); LIWAGU RIVER HEAD: 3200 m, *Carr SFN 27520c* (L, SING); MARAI PARAI: 2400–3000 m, *Clemens 33179* (BM, K); MOUNT KINABALU: 3000 m, *Smith & Everard 150* (K); PAKA-PAKA CAVE: 3000 m, *Carr SFN 27520a* (SING), 3000 m, *SFN 27520b* (SING), 3000 m, *Clemens 29127* (BM, K); SUMMIT TRAIL: 2700–3200 m, *Gunsalam 11* (K), 2300 m, *Vermeulen & Lamb 350* (L), 2300 m, *Vermeulen & Duistermaat 543* (L); UPPER KINABALU: *Clemens 30016* (L).

55.17.54. Bulbophyllum mutabile (Blume) Lindl., Gen. Sp. Orch., 48 (1830).

a. var. **mutabile**

Diphyes mutabilis Blume, Bijdr., 312 (1825).

Bulbophyllum altispex Ridl. in Stapf, Trans. Linn. Soc. Bot. 4: 236 (1894). Type: MOUNT KINABALU, 2400 m, *Haviland 1143* (holotype? K!).

Epiphyte. Lower montane forest, upper montane forest?, sometimes on ultramafic substrate. Elevation: 1100–2700 m.

General distribution: Sabah, Kalimantan, Sarawak, Peninsular Malaysia, Thailand, Sumatra, Java, Sulawesi, Philippines.

Additional collections. BUNDU TUHAN: 1200 m, *Clemens 40116* (BM, L); MAHANDEI RIVER: 1100 m, *Carr 3097, SFN 26495* (SING); MESILAU CAVE: 1900 m, *Collenette 918* (K); MESILAU RIVER: 2100–2700 m, *Clemens 51424* (BM); PENIBUKAN: 1200 m, *Clemens 32059* (BM); PINOSUK PLATEAU: *Chow & Leopold SAN 74535* (K); TAHUBANG RIVER HEAD: 2100 m, *Clemens 32978* (BM); TENOMPOK: 1500 m, *Carr SFN 26894* (L, SING), 1500 m, *SFN 27746* (L, SING), 1500 m, *Clemens 29000* (BM).

b. var. **obesum** J. J. Verm., Orchids of Borneo 2: 74, f. 21D (1991). Plate 13C. Type: PARK HEADQUARTERS, 1700 m, *Vermeulen 468* (holotype L n.v.).

Epiphyte. Lower montane forest. Elevation: 1400–2400 m.

General distribution: Sabah.

Collections. BUNDU TUHAN: 1400 m, *Darnton 208* (BM); LUMU-LUMU: 1600 m, *Carr SFN 27106* (SING); MINETUHAN SPUR: 2400 m, *Clemens 33862* (BM); PINOSUK PLATEAU: 1500–1700 m, *Vermeulen 485* (L); TENOMPOK: 1600 m, *Carr SFN 26976* (L, SING), 1700 m, *Vermeulen & Duistermaat 662* (K).

55.17.55. Bulbophyllum nubinatum J. J. Verm., Malayan Orch. Rev. 22: 46 (1988). Type: SUMMIT TRAIL, 3000 m, *Chan 63/87* (holotype L n.v.; isotype K!).

Epiphyte. Lower montane forest, upper montane *Dacrydium/ Leptospermum* forest, on ultramafic substrate. Elevation: 1800–3000 m.

Endemic to Mount Kinabalu.

Additional collections. MARAI PARAI: 1800 m, *Carr SFN 27935* (SING); TIBABAR RIVER: 2100 m, *Carr SFN 27576* (SING).

55.17.56. Bulbophyllum obtusum (Blume) Lindl., Gen. Sp. Orch., 56 (1830).

Diphyes obtusa Blume, Bijdr., 315 (1825).

Epiphyte. Lower montane forest. Elevation: 1200–1800 m.

General distribution: Sabah, Java.

Collections. GURULAU SPUR: 1500 m, *Clemens 50512* (BM, K); MESILAU CAVE TRAIL: 1800 m, *Beaman 7481* (K); PENIBUKAN: 1200 m, *Clemens 40635* (BM, K), 1500 m, *50206* (BM, K), 1700 m, *50365* (BM, K), 1200 m, *50369* (K), 1200 m, *51729* (BM).

55.17.57. Bulbophyllum odoratum (Blume) Lindl., Gen. Sp. Orch., 54 (1830). Plate 13D.

Diphyes odorata Blume, Bijdr., 312 (1825).

Bulbophyllum crassicaudatum Ames & C. Schweinf., Orch. 6: 170 (1920). Type: MOUNT KINABALU, *Clemens s.n.* (holotype AMES mf; isotype K!).

Epiphyte. Hill forest, lower montane forest, sometimes on ultramafic substrate. Elevation: 900–2400 m.

General distribution: Sabah, Kalimantan, Sarawak, Peninsular Malaysia, Sumatra, Java, Sulawesi, Lesser Sunda Islands, Maluku, Philippines.

Additional collections. DALLAS: 900 m, *Clemens 26050* (BM, K), 900 m, *27162* (BM, K), 900–1200 m, *27264* (BM, K); GURULAU SPUR: 2100–2400 m, *Clemens 50962* (BM, K); KIAU: 900 m, *Clemens 21* (AMES), 900 m, *97* (AMES), 900 m, *99* (AMES), 900 m, *154* (AMES), 900 m, *172* (AMES), *408* (AMES), 900 m, *50344* (BM, K); PARK HEADQUARTERS: 1700 m, *Lamb AL 738/87* (K); PENIBUKAN: 1200 m, *Clemens 31851* (BM); SOSOPODON: 1400 m, *Mikil SAN 34516* (K); TENOMPOK: 1500 m, *Clemens 27530* (BM); TENOMPOK ORCHID GARDEN: 1500 m, *Clemens 51514* (BM).

55.17.58. Bulbophyllum ovalifolium (Blume) Lindl., Gen. Sp. Orch., 49 (1830).

Diphyes ovalifolia Blume, Bijdr., 318 (1825).

Epiphyte. Lower montane forest. Elevation: 1200–2000 m.

General distribution: Sabah, Peninsular Malaysia, Thailand, Sumatra, Java, Sulawesi, Flores.

Collections. PARK HEADQUARTERS: *Lamb AL 665/86* (K); SUMMIT TRAIL: 2000 m, *Bailes & Cribb 779* (K); TAHUBANG RIVER: 1200 m, *Clemens 40348* (K); TENOMPOK RIDGE: 1400–1500 m, *Beaman 8192* (K).

55.17.59. Bulbophyllum pelicanopsis J. J. Verm., Malayan Orch. Rev. 22: 47 (1988). Plate 14A.

Epiphyte. Lower montane forest.

General distribution: Sabah.

Collections. MESILAU: *Lamb s.n.* (K); PIG HILL: *Lamb AL 742/87* (L).

55.17.60. Bulbophyllum placochilum J. J. Verm., Orchids of Borneo 2: 29, f. 5 (1991). Type: MOUNT KINABALU, 1500–1700 m, *Clemens 34065* (holotype L n.v.; isotype BM n.v.).

Epiphyte. Lower montane forest. Elevation: 1500–1700 m.

Endemic to Mount Kinabalu.

Additional collection. TENOMPOK: 1500 m, *Carr SFN 27938* (L, SING).

55.17.61. Bulbophyllum planibulbe (Ridl.) Ridl., Mat. Fl. Malay. Penins. 1: 79 (1907).

Cirrhopetalum planibulbe Ridl., Trans. Linn. Soc. Bot. 3: 364, t. 64 (1893).

Epiphyte. Hill forest on ultramafic substrate.

General distribution: Sabah, Kalimantan, Peninsular Malaysia, Thailand, Sumatra.

Collection. LOHAN RIVER: *Vermeulen 1308* (L).

55.17.62. Bulbophyllum pocillum J. J. Verm., Orchids of Borneo 2: 83, f. 24 (1991). Plate 14C.

Pendulous epiphyte. Lower montane forest, sometimes on ultramafic substrate. Elevation: 1200–2100 m.

General distribution: Sabah.

PLATE 1.

A. Mount Kinabalu from Mount Trus Madi. Photo M. J. S. Sands.

B. Mount Kinabalu from the southwest, near Dallas. Photo R. S. Beaman.

PLATE 2.

A. Mount Kinabalu from Park Headquarters. Photo A. Lamb.

B. Granite rock face with pinnacles above East Mesilau River basin; ridges in foreground are about 2400 m and clothed in upper montane mossy forest. Photo A. Lamb.

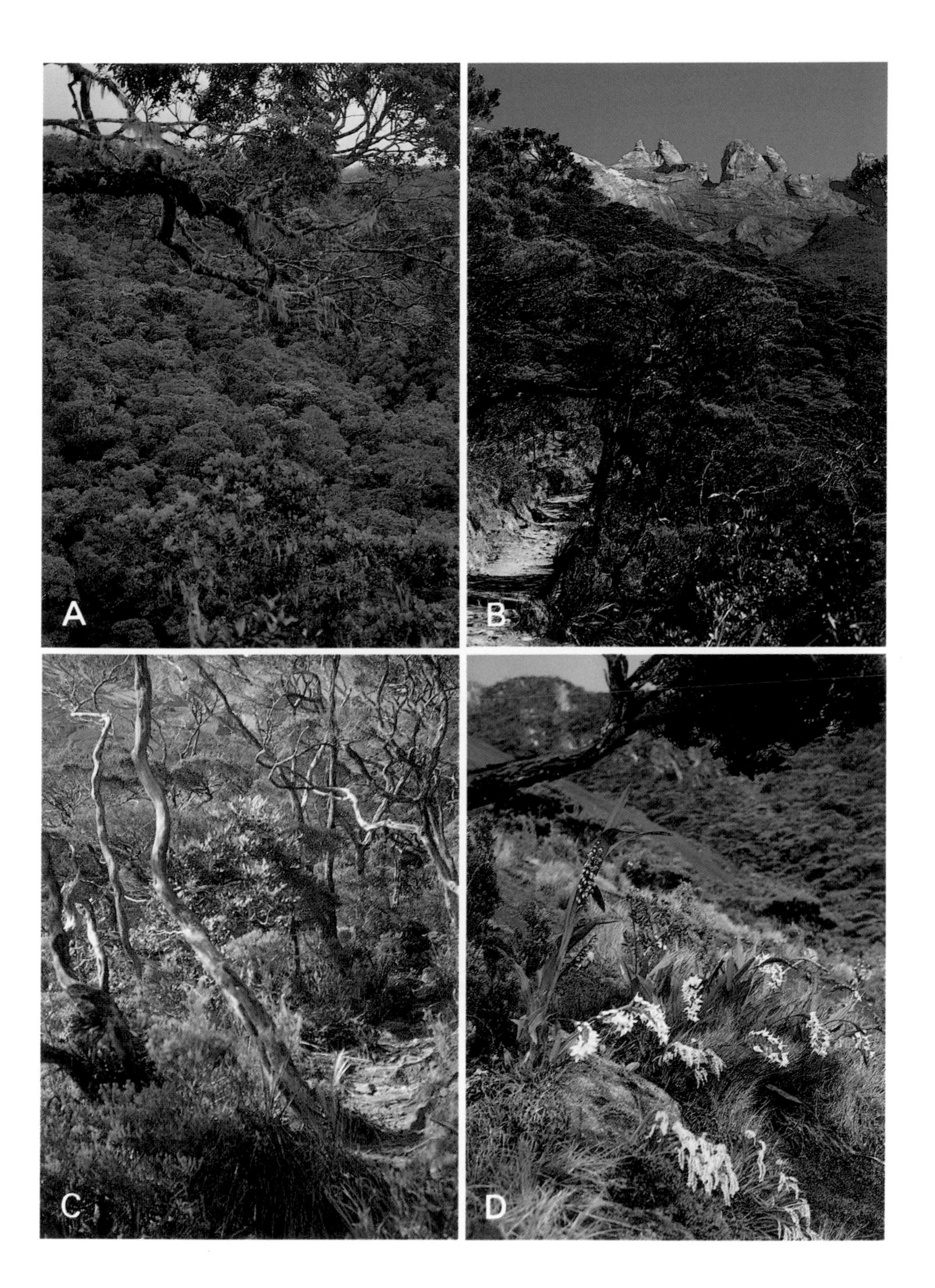

PLATE 3.

A. Liwagu River valley below Kemburongoh. Photo J. Dransfield.

B. *Dacrydium-Leptospermum* forest along Summit Trail above Layang-layang at about 2800 m. Photo A. Phillipps.

C. *Leptospermum* scrub on the Summit Trail at about 3000 m. Photo A. Lamb.

D. *Coelogyne papillosa* (left and middle distance), *Dendrochilum stachyodes* (foreground) and *Eria grandis* (extreme left) at Panar Laban. Photo A. Lamb.

PLATE 4.

A. Paka-paka Cave. Photo P. J. Cribb.

B. Vegetation at Marai Parai. Photo P. J. Cribb.

PLATE 5.

A. Eastern Shoulder with the Pinosuk Plateau below, in 1964. Photo W. Meijer.

B. Pinosuk Plateau in 1984, showing much of the primary forest destroyed. Photo J. H. Beaman.

PLATE 6.

A. Hempuen Hill on the left with extreme ultramafic geology. The vegetation of this area was completely destroyed by forest fires in September, 1990. Lohan River in centre, Mamut Copper Mine tailings at upper right. Photo J. H. Beaman.

B. Lohan River valley filled with Mamut Copper Mine tailings. Photo J. H. Beaman.

PLATE 7.

A. Acanthephippium lilacinum. Sabah, Crocker Range. *Lamb K51* (holotype). Photo RBG, Kew.

B. Aerides odorata. Kalimantan. Photo P. J. Cribb.

C. Agrostophyllum javanicum. Mount Kinabalu, *Brentnall 159.* Photo RBG, Kew.

PLATE 8.

A. Aphyllorchis montana. Sarawak, Mt. Mulu. Photo J. Dransfield.

B. Appendicula congesta. Mount Kinabalu. Photo P. J. Cribb.

C. Appendicula cristata. North Sumatra. Photo J. B. Comber.

D. Appendicula longirostrata. Sabah, Crocker Range. Photo R. B. G. Kew.

PLATE 9.

A. **Appendicula torta.** Sabah. Photo A. Lamb.

B. **Arachnis breviscapa.** Mount Kinabalu, Lohan River. Photo A. Lamb.

C. **Bogoria raciborskii.** Mount Kinabalu, Liwagu River. Photo P. J. Cribb.

PLATE 10.

A. Bromheadia divaricata. Mount Kinabalu, Summit Trail. Photo A. Lamb.

B. Bulbophyllum armeniacum. Sabah, Sipitang District. Photo J. B. Comber.

C. Bulbophyllum calceolus. Sabah, Sipitang District. Photo J. B. Comber.

PLATE 11.

A. Bulbophyllum catenarium. Sabah, Mt. Alab. Photo J. B. Comber.

B. Bulbophyllum coriaceum. Mount Kinabalu, Paka-paka Cave. Photo A. Lamb.

PLATE 12.

A. Bulbophyllum flammuliferum. Mount Kinabalu, Hempuen Hill. Photo A. Lamb.

B. Bulbophyllum foetidolens. Sabah, Mt. Lumaku. Photo J. B. Comber.

PLATE 13.

A. Bulbophyllum mandibulare. Cult. Poring Orchid Garden, *Lohok 26.* Photo P. J. Cribb.

B. Bulbophyllum microglossum. Mount Kinabalu, Mesilau Cave, *Beaman 9572.* Photo R. S. Beaman.

C. Bulbophyllum mutabile var. **obesum.** Sabah, Mt. Alab. Photo J. B. Comber.

D. Bulbophyllum odoratum. Java, Mt. Slamet. Photo J. B. Comber.

PLATE 14.

A. Bulbophyllum pelicanopsis. Sabah, Sipitang District. Photo J. B. Comber.

B. Bulbophyllum lissoglossum. Mount Kinabalu, Mesilau Cave. Photo S. Collenette

C. Bulbophyllum pocillum. Mount Kinabalu, Park Headquarters, *Wood 617.* Photo J. B. Comber.

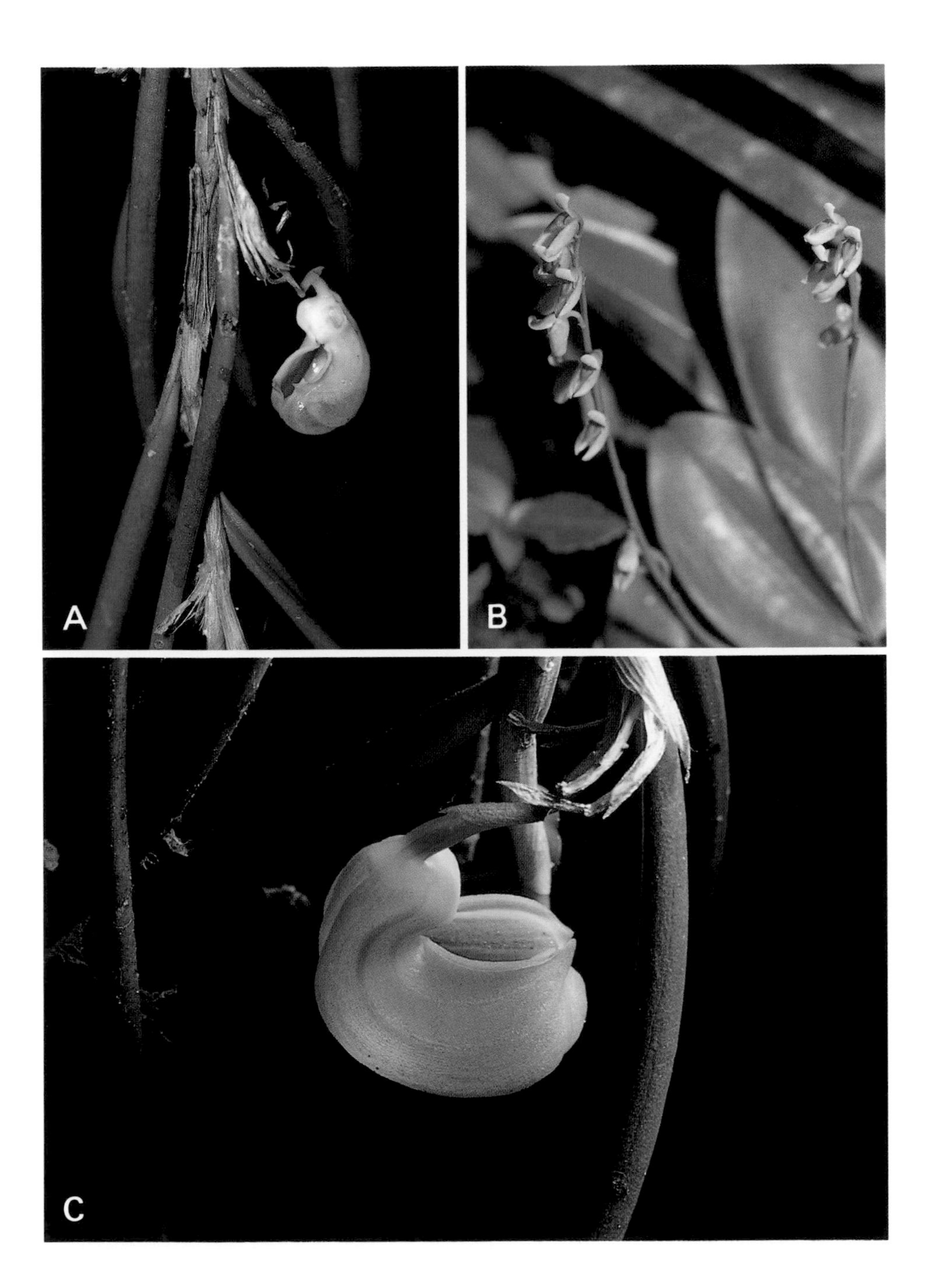

PLATE 15.

A. Bulbophyllum pugilanthum. Mount Kinabalu, Park Headquarters. Photo A. Lamb.

B. Bulbophyllum sopoetanense. Mount Kinabalu, Park Headquarters. Photo S. Collenette.

C. Bulbophyllum pugilanthum. Sabah, Sipitang District. Photo J. B. Comber.

PLATE 16.

A. Bulbophyllum uniflorum. Cult. Tenom Orchid Centre, Sabah. Photo S. Collenette.

B. Calanthe aff. **aureiflora.** Mount Kinabalu. Photo P. J. Cribb.

C. Calanthe kinabaluensis. Mount Kinabalu, Tenompok, *Beaman 9455*. Photo J. H. Beaman.

PLATE 17.

A. Calanthe kinabaluensis. Mount Kinabalu, Park Headquarters. Photo A. Lamb.

B. Calanthe pulchra. Cult. Poring Orchid Garden. Photo P. J. Cribb.

C. Calanthe pulchra. Peninsular Malaysia. Photo J. Dransfield.

PLATE 18.

A. Calanthe transiens. Mount Kinabalu, Layang-layang. Photo A. Lamb.

B. Calanthe truncicola. Mount Kinabalu. Photo A. Lamb.

C. Ceratochilus jiewhoeii. Mount Kinabalu, Pinosuk Plateau. Photo A. Lamb.

PLATE 19.

A. **Chelonistele amplissima.** Mount Kinabalu. Photo P. J. Cribb.

B. **Chelonistele lurida** var. **lurida.** Mount Kinabalu. Photo P. J. Cribb.

C. **Chelonistele sulphurea** var. **sulphurea.** Sabah, Crocker Range. Photo R. B. G. Kew.

PLATE 20.

A. **Chelonistele sp. 1.** Mount Kinabalu, Summit Trail. Photo P. J. Cribb.

B. **Chroniochilus virescens.** Mount Kinabalu, Lohan River. Photo A. Lamb.

C. **Chrysoglossum reticulatum.** Mount Kinabalu, Kiau View Trail. Photo A. Lamb.

PLATE 21.

A. **Claderia viridiflora.** Sarawak. Photo J. Dransfield.

B. **Cleisocentron merrillianum.** Mount Kinabalu, Pinosuk Plateau. Photo A. Lamb.

C. **Cleisocentron sp. 1.** Mount Kinabalu, Marai Parai Spur, cult. Kinabalu Mountain Garden, *Lamb AL 1211/90*. Photo A. Lamb.

PLATE 22.

A. Coelogyne clemensii var. **clemensii.** Mount Kinabalu, Kiau View Trail. Photo P. J. Cribb.

B. Coelogyne clemensii var. **clemensii.** Sabah, Tenom District. Photo J. B. Comber.

PLATE 23.

A. Coelogyne compressicaulis. Cult. RBG Edinburgh. Photo J. B. Comber.

B. Coelogyne craticulaelabris. Sabah, Mt. Alab. Photo J. B. Comber.

PLATE 24.

A. Coelogyne cumingii. Sabah, Nabawan, cult. R. B. G. Kew. Photo R. B. G. Kew.

B. Coelogyne section **Rigidiformes,** related to *C. exalata.* Mount Kinabalu, East Mesilau/Menteki Rivers. Photo J. B. Comber.

C. Coelogyne section **Rigidiformes,** related to *C. exalata.* Mount Kinabalu, East Mesilau/Menteki Rivers. Photo J. B. Comber.

PLATE 25.

A. Coelogyne hirtella. Sabah, Sipitang District. Photo J. B. Comber.

B. Coelogyne monilirachis. Mount Kinabalu, West Mesilau River, *Beaman 8997.* Photo R. S. Beaman.

C. Coelogyne moultonii. Mount Kinabalu, Golf Course Site, *Bailes & Cribb 703.* Photo R. B. G. Kew.

D. Coelogyne moultonii. Mount Kinabalu, Mesilau Cave Trail, *Wood 850.* Photo J. B. Comber.

PLATE 26.

A. Coelogyne pandurata. Mount Kinabalu, Lohan River, *Lamb K57*. Photo R. B. G. Kew.

B. Coelogyne papillosa. Mount Kinabalu, Panar Laban. Photo J. B. Comber.

C. Coelogyne papillosa. Mount Kinabalu, 3000 m. Photo P. J. Cribb.

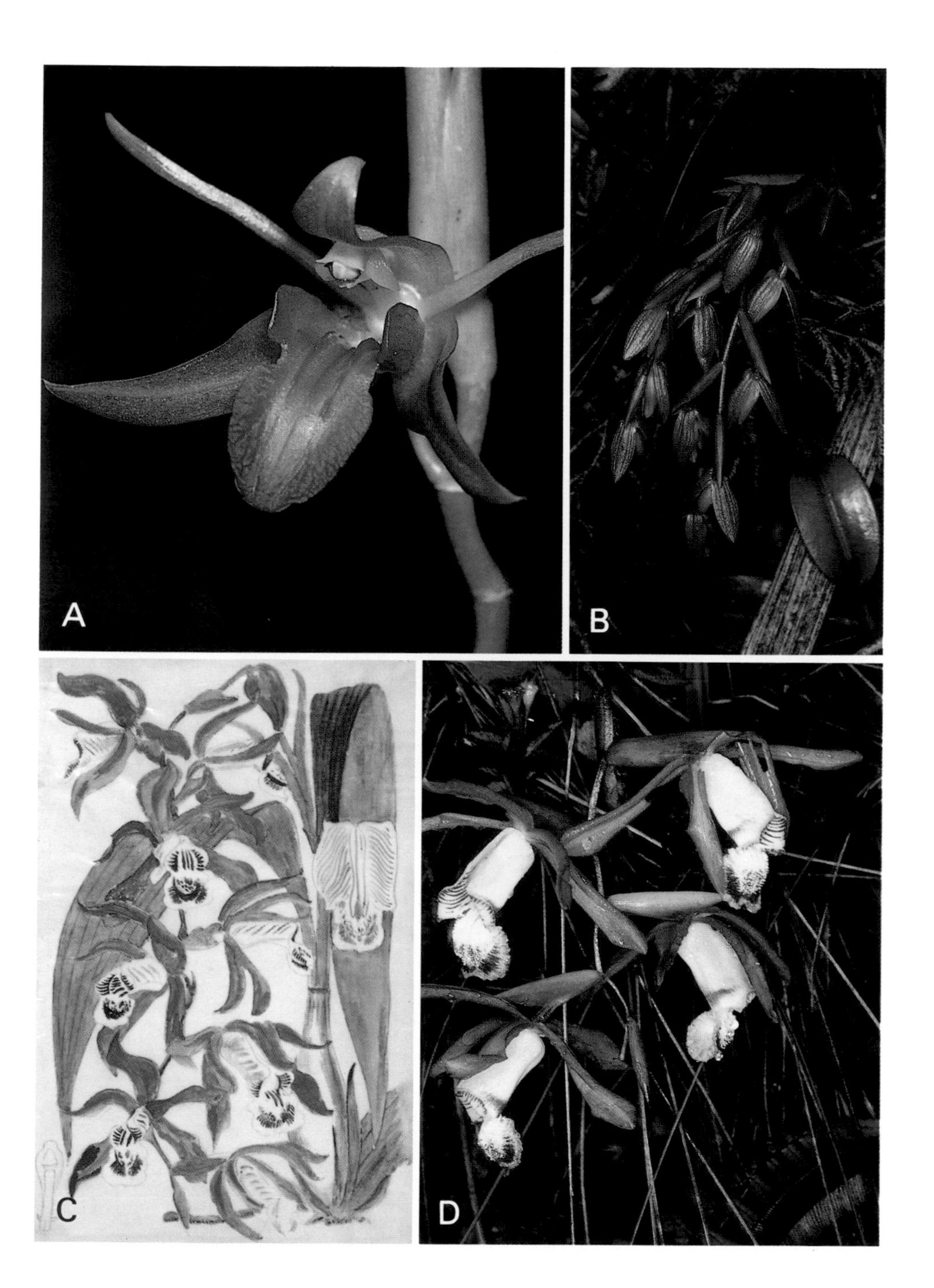

PLATE 27.

A. Coelogyne planiscapa var. **grandis.** Mount Kinabalu. Photo A. Lamb.

B. Coelogyne plicatissima. Mount Kinabalu, 3000 m. Photo P. J. Cribb.

C. Coelogyne radioferens. Painting by F. W. Burbidge, numbered 110, at Natural History Museum, London (BM). Photo R. B. G. Kew.

D. Coelogyne radioferens. Mount Kinabalu, Summit Trail. Photo P. J. Cribb.

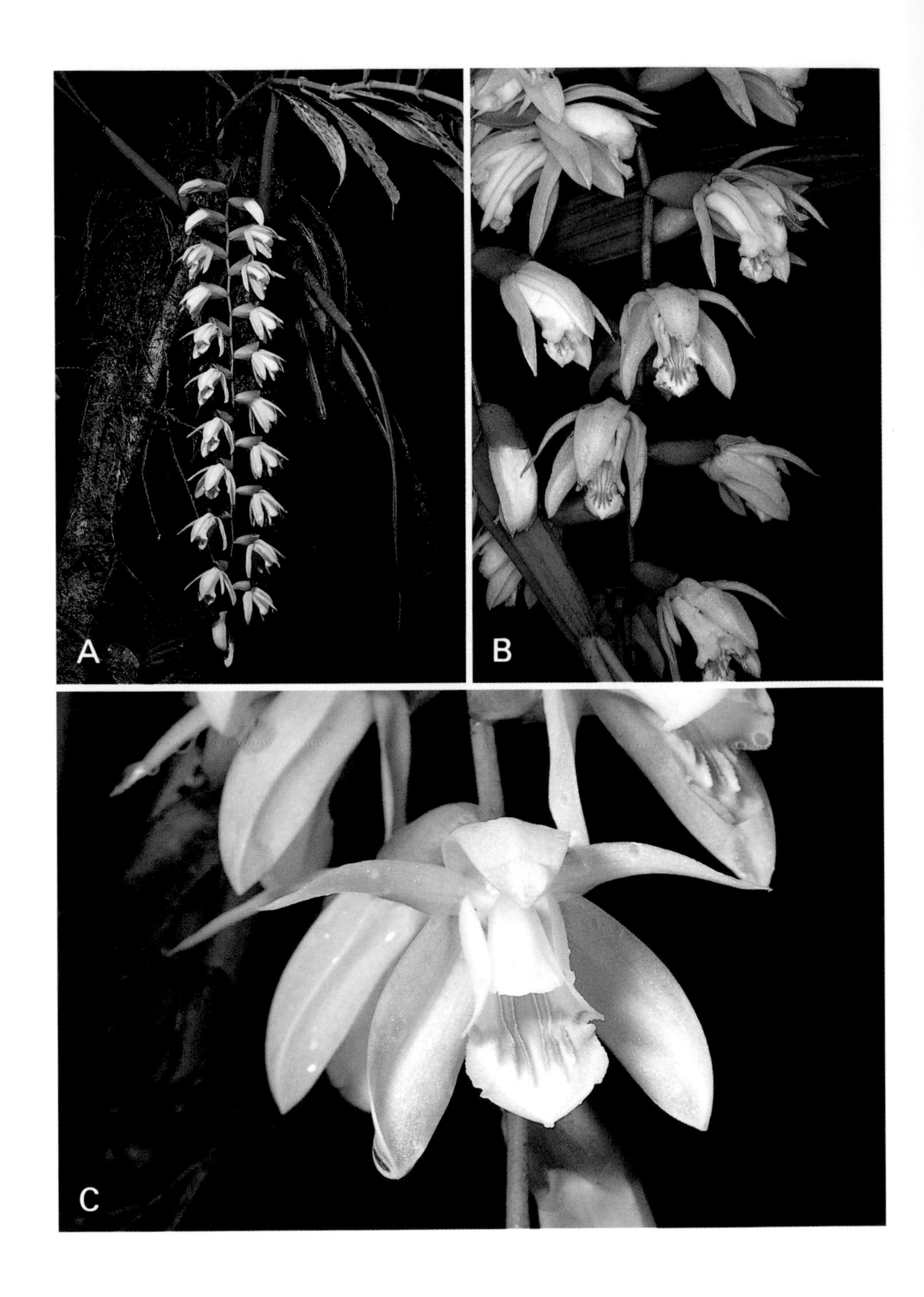

PLATE 28.

A. Coelogyne rhabdobulbon. Mount Kinabalu, 1500 m. Photo J. B. Comber.

B. Coelogyne rhabdobulbon. Mount Kinabalu. Photo P. J. Cribb.

C. Coelogyne rhabdobulbon. Mount Kinabalu, 1500 m. Photo J. B. Comber.

PLATE 29.

A. Coelogyne rochussenii. Cult. R. B. G. Kew. Photo P. J. Cribb.

B. Coelogyne rochussenii. Sarawak, cult. R. B. G. Kew. Photo R. B. G. Kew.

PLATE 30.

A. Coelogyne rupicola. Mount Kinabalu, Marai Parai Spur. Photo J. Dransfield.

B. Coelogyne septemcostata. Sarawak. Photo J. B. Comber.

C. Coelogyne tenompokensis. Mount Kinabalu, Golf Course Site, *Bailes & Cribb 729.* Photo R. B. G. Kew.

PLATE 31.

A. Coelogyne venusta. Mount Kinabalu. Photo P. J. Cribb.

B. Corybas pictus. Mount Kinabalu, Park Headquarters, *Wood 616.* Photo J. B. Comber.

C. Cryptostylis acutata. Mount Kinabalu, Hempuen Hill, *Wood 839.* Photo J. B. Comber.

D. Cryptostylis arachnites. Mount Kinabalu, Mesilau Cave Trail, *Wood 845.* Photo J. B. Comber.

PLATE 32.

A. **Cymbidium borneense.** Sabah, cult. R. B. G. Kew. Photo D. J. Du Puy.

B. **Cymbidium borneense.** Sabah, cult. R. B. G. Kew. Photo P. J. Cribb.

C. **Cymbidium bicolor** subsp. **pubescens.** Sabah, cult. R. B. G. Kew. Photo D. J. Du Puy.

Collections. KEMBURONGOH/LUMU-LUMU: 1800–2100 m, *Clemens 27807* (BM, K); KIAU VIEW TRAIL: 1700 m, *Gunsalam 9* (K); LUMU-LUMU: 1700 m, *Clemens 27856* (BM); MT. NUNGKEK: 1200 m, *Clemens 32878* (BM); MURU-TURA RIDGE: 1700 m, *Clemens 34104* (BM); NUMERUK RIDGE: 1400 m, *Clemens 32114* (BM); PARK HEADQUARTERS: 1500 m, *Wood 617* (K); PENIBUKAN: 1200 m, *Carr 3049, SFN 26349* (L), 1200 m, *Clemens 30600* (BM, L), 1200 m, *30852* (BM), 1200 m, *50107* (BM, K); PINOSUK PLATEAU: 1500–1700 m, *Vermeulen 475* (K, L).

55.17.63. Bulbophyllum pugilanthum J. J. Wood, ined. Fig. 11, Plate 15A, 15C. Type: PARK HEADQUARTERS, 1500 m, *Lamb AL 56/83* (holotype K!).

Pendulous epiphyte. Lower montane forest, upper montane forest, sometimes on ultramafic substrate; low on tree trunks and on small branches. Elevation: 1400–2400 m.

General distribution: Sabah.

Additional collections. KEMBURONGOH: 2400 m, *Clemens s.n.* (BM, BM); LUMU-LUMU: 1800 m, *Clemens s.n.* (BM, BM); MOUNT KINABALU: 1400 m, *Price 151* (K); NUMERUK RIDGE: 1400 m, *Clemens 40055* (BM); TENOMPOK: 1500 m, *Clemens s.n.* (BM, BM, BM); TENOMPOK/TOMIS: 1700 m, *Clemens 29440* (BM).

55.17.64. Bulbophyllum puntjakense J. J. Sm., Bull. Dép. Agric. Indes Néerl. 13: 46 (1907).

Epiphyte. Lower montane forest. Elevation: 1500–1700 m.

General distribution: Sabah, Java, Bali.

Collections. MOUNT KINABALU: *Phillipps SNP 2403* (K); PARK HEADQUARTERS: 1700 m, *Vermeulen 499* (L); TENOMPOK: 1500 m, *Carr SFN 26861* (L, SING), 1500 m, *Clemens 27277* (K); UPPER KINABALU: *Clemens 30087* (K).

55.17.65. Bulbophyllum purpurascens Teijsm. & Binn., Natuurk. Tijdschr. Ned.-Ind. 24: 308 (1862).

Epiphyte. Hill forest. Elevation: 900–1200 m.

General distribution: Sabah, Brunei, Kalimantan, Sarawak, Peninsular Malaysia, Thailand, Sumatra, Java, Bangka, Krakatau.

Collections. DALLAS: 900 m, *Clemens 26786* (K), 900 m, *26913* (K); KINUNUT VALLEY HEAD: 1200 m, *Carr SFN 27247* (SING).

55.17.66. Bulbophyllum rhizomatosum Ames & C. Schweinf., Orch. 6: 194 (1920). Type: LUBANG, 1500 m, *Clemens 106* (holotype AMES mf; isotype BM!).

Pendulous epiphyte. Lower montane forest. Elevation: 1100–1500 m.

General distribution: Sabah, Brunei, Peninsular Malaysia, Philippines.

Additional collections. PARK HEADQUARTERS: 1400–1500 m, *Lamb SAN 93471* (K); PENATARAN BASIN: 1500 m, *Clemens 40195* (BM); TAHUBANG RIVER: 1100 m, *Carr SFN 26510*

A
B
C
D
E
F
G
H
J
K
L
M
N
O

(AMES, SING), 1200 m, *Clemens 40338* (BM, K); TENOMPOK: 1500 m, *Carr SFN 26994* (AMES, SING), 1500 m, *Clemens 29000* (BM), 1500 m, *29440* (K).

55.17.67. Bulbophyllum rubiferum J. J. Sm., Bull. Jard. Bot. Buit., ser. 2, 26: 73 (1918).

Epiphyte. Lower montane mossy forest.

General distribution: Sabah, Java.

Collection. PARK HEADQUARTERS: *Lamb AL 592/86* (K).

55.17.68. Bulbophyllum salaccense Rchb. f., Bonplandia 5: 57 (1857).

Epiphyte. Hill forest, upper montane forest? Elevation: 1000–3000 m.

General distribution: Sabah, Sarawak, Peninsular Malaysia, Sumatra, Java.

Collections. KILEMBUN RIVER HEAD: 2700–3000 m, *Clemens 33982* (BM, L); TAHUBANG RIVER: 1000 m, *Carr 3060, SFN 26374* (L, SING).

55.17.69. Bulbophyllum scabrum J. J. Verm. & A. Lamb, Malayan Orch. Rev. 22: 46 (1988). Type: PINOSUK PLATEAU, 1500–1700 m, *Vermeulen 483* (holotype L n.v.; isotype K!).

Epiphyte. Lower montane forest. Elevation: 1500–1700 m.

Endemic to Mount Kinabalu.

Additional collection. KUNDASANG: *Lamb AL 846/87* (K).

55.17.70. Bulbophyllum schefferi (Kuntze) Schltr., Beih. Bot. Centralbl. 28, 2: 417 (1915).

Phyllorchis schefferi Kuntze, Revis. Gen. Pl. 2: 676 (1891).

Epiphyte. Lowlands. Elevation: 400 m.

General distribution: Sabah, Kalimantan, Sarawak, Sumatra, Java, Lombok, Philippines.

Collection. KAUNG: 400 m, *Carr SFN 27376* (L).

Fig. 11. Bulbophyllum pugilanthum. **A**, habit; **B**, portion of stem and leaf showing insertion of inflorescence; **C**, leaf (transverse section); **D**, flower (side view); **E**, flower (front view); **F**, dorsal sepal; **G**, lateral sepal; **H**, petal; **J**, lip (front view, flattened); **K**, lip (side view); **L**, pedicel-with-ovary, column and part of a lateral sepal (side view); **M**, anther-cap (back view); **N**, anther-cap (side view); **O**, pollinia. **A** from *Wood 669* and *712* (Sipitang District, Ulu Long Pa Sia); **B–D** and **H** from *Wood 669*; **E–G** and **J–O** from *Lamb AL 56/83*. Scale: single bar = 1 mm; double bar = 1 cm. Drawn by Eleanor Catherine.

55.17.71. Bulbophyllum sessile (Koenig) J. J. Sm., Orch. Java, 448 (1905).

Epidendrum sessile Koenig in Retz., Observ. Bot. 6: 60 (1791).

Epiphyte. Hill forest. Elevation: 1200 m.

General distribution: Sabah, widespread from Burma, Thailand, Indochina to Peninsular Malaysia and Indonesia east to New Guinea, the Solomon Islands & Fiji.

Collection. BUNDU TUHAN: 1200 m, *Carr SFN 27446* (SING).

55.17.72. Bulbophyllum sigmoideum Ames & C. Schweinf., Orch. 6: 196 (1920). Type: GURULAU SPUR, *Clemens 316* (holotype AMES mf).

Epiphyte. Hill forest, lower montane forest. Elevation: 900–1500 m.

General distribution: Sabah.

Additional collections. KIAU: 900 m, *Clemens 168* (AMES), *327* (AMES), *354* (AMES); MESILAU RIVER: 1500 m, *RSNB 4191* (K); PENIBUKAN: 1100 m, *Clemens 50057* (K), 1200 m, *50141* (BM, K).

55.17.73. Bulbophyllum sopoetanense Schltr., Feddes Repert. 10: 181 (1912). Plate 15B.

Epiphyte. Lower montane forest, mossy forest, on small trees or low on trunks of larger trees or on logs. Elevation: 1200–2000 m.

General distribution: Sabah, Kalimantan, Sarawak, Sulawesi.

Collections. BUNDU TUHAN: 1200 m, *Carr SFN 27453* (L, SING); GURULAU SPUR: 1800 m, *Clemens 50422* (BM, K); MURU-TURA RIDGE: 1500 m, *Clemens 34365* (BM); NUMERUK RIDGE: 1500 m, *Clemens 40056* (BM); PENIBUKAN: *Clemens 40602* (K), 1200 m, *40631* (BM); SUMMIT TRAIL: 2000 m, *Vermeulen & Duistermaat 541* (L); TAHUBANG RIVER HEAD: 1500 m, *Clemens 40678* (BM, K); TENOMPOK: 1500 m, *Carr SFN 27910* (SING).

55.17.74. Bulbophyllum aff. **sopoetanense** Schltr., Feddes Repert. 10: 181 (1912).

Epiphyte. Lower montane forest. Elevation: 2000 m.

Collection. SUMMIT TRAIL: 2000 m, *Beaman 6762* (K).

55.17.75. Bulbophyllum stipitatibulbum J. J. Sm., Bull. Jard. Bot. Buit., ser. 3, 11: 148 (1931).

Epiphyte. Lower montane forest. Elevation: 2100 m.

General distribution: Sabah, Kalimantan.

Collection. KINATEKI RIVER HEAD: 2100 m, *Clemens 31741* (BM).

55.17.76. Bulbophyllum subclausum J. J. Sm., Bull. Dép. Agric. Indes Néerl. 22: 35 (1909).

Epiphytic or terrestrial. Lower montane forest. Elevation: 1200–2100 m.

General distribution: Sabah, Sumatra.

Collections. BUNDU TUHAN: 1200 m, *Carr SFN 27441* (SING); LUMU-LUMU: 1800 m, *Carr SFN 27889* (SING); SUMMIT TRAIL: 1800–2100 m, *Wood 613* (K); TENOMPOK: 1500 m, *Beaman 10517* (K, MSC), 1500 m, *Carr SFN 27418* (K, SING), 1500 m, *Clemens 29791* (BM, K), 1500 m, *29846* (BM, K), 1500 m, *29944* (K); WEST MESILAU RIVER: 1500 m, *Collenette 587* (K).

55.17.77. Bulbophyllum tenompokense J. J. Sm., Feddes Repert. 36: 116 (1934). Type: TENOMPOK, 1500 m, *Clemens 29674* (holotype BO n.v.).

Epiphyte. Lower montane forest. Elevation: 1500 m. Known only from the type collection.

Endemic to Mount Kinabalu.

55.17.78. Bulbophyllum tenuifolium (Blume) Lindl., Gen. Sp. Orch., 50 (1830).

Diphyes tenuifolia Blume, Bijdr., 316 (1825).

Epiphyte. Lower montane forest. Elevation: 1500 m.

General distribution: Sabah, Kalimantan, Sarawak, Peninsular Malaysia, Thailand, Java.

Collections. TENOMPOK: 1500 m, *Carr SFN 27442* (K), 1500 m, *Clemens 30116* (K).

55.17.79. Bulbophyllum teres Carr, Gardens' Bull. 8: 115 (1935).

Epiphyte. Lower montane forest. Elevation: 1500–1800 m.

General distribution: Sabah, Sarawak.

Collections. KILEMBUN RIVER: 1500–1800 m, *Clemens 35204* (BM); MURU-TURA RIDGE: 1500–1800 m, *Clemens 34364* (BM); PINOSUK PLATEAU: 1500–1700 m, *Vermeulen 481* (K).

55.17.80. Bulbophyllum tortuosum (Blume) Lindl., Gen. Sp. Orch., 50 (1830).

Diphyes tortuosa Blume, Bijdr., 311 (1825).

Epiphyte. Hill forest on ultramafic substrate. Leaves broader than typical.

General distribution: Sabah, Bhutan, Peninsular Malaysia, Thailand, Laos, Vietnam, Sumatra, Java.

Collection. HEMPUEN HILL: *Lamb AL 1272/90* (L).

55.17.81. Bulbophyllum trifolium Ridl., J. Linn. Soc. Bot. 32: 278 (1896).

Epiphyte. Lower montane forest on ultramafic substrate. Elevation: 1400–1500 m.

General distribution: Sabah, Sarawak, Peninsular Malaysia.

Collections. MARAI PARAI: 1500 m, *Collenette A 78* (BM); PENIBUKAN: 1400 m, *Carr SFN 27429* (SING), 1500 m, *Clemens 30946* (BM); PIG HILL: *Lamb AL 737/87* (K).

55.17.82. Bulbophyllum turgidum J. J. Verm., Orchids of Borneo 2: 103, f. 33 (1991).

Epiphyte. Lower montane forest. Sight record of J. J. Vermeulen, 1987, near Park Headquarters; no specimens seen.

General distribution: Sabah.

55.17.83. Bulbophyllum undecifilum J. J. Sm., Bull. Jard. Bot. Buit., ser. 3, 9: 51, t. 4, v (1927).

Epiphyte. Lower montane forest; on tree trunks in shade. Elevation: 1200–1700 m.

General distribution: Sabah, Java.

Collections. BUNDU TUHAN: 1200 m, *Carr SFN 27835* (SING); PARK HEADQUARTERS: 1700 m, *Vermeulen 701* (K).

55.17.84. Bulbophyllum uniflorum (Blume) Hassk., Cat. Hort. Bot. Bog., 39 (1844). Plate 16A.

Ephippium uniflorum Blume, Bijdr., 309 (1825).

Terrestrial, lithophyte, or climbing on tree trunks. Lower montane forest. Elevation: 1600–1700 m.

General distribution: Sabah, Sarawak, Peninsular Malaysia, Sumatra, Java, Philippines.

Collections. KIAU VIEW TRAIL: *Saikeh SAN 82800* (K); MESILAU RIVER: 1700 m, *RSNB 7123* (K); MOUNT KINABALU: *Aban SAN 76693* (K); TENOMPOK: 1600 m, *Carr SFN 27968* (SING).

55.17.85. Bulbophyllum vaginatum (Lindl.) Rchb. f., Walp. Ann. Bot. Syst. 6: 261 (1861).

Cirrhopetalum vaginatum Lindl., Gen. Sp. Orch., 59 (1830).

Epiphyte. Hill forest. Elevation: 800 m.

General distribution: Sabah, Peninsular Malaysia, Singapore, Thailand, Sumatra, Java, Bangka, Maluku.

Collection. BUNDU TUHAN: 800 m, *Carr SFN 27807* (SING).

55.17.86. Bulbophyllum vinaceum Ames & C. Schweinf., Orch. 6: 200 (1920). Type: MARAI PARAI SPUR, *Clemens 240* (holotype AMES mf).

Epiphyte. Hill forest, lower montane forest on ultramafic substrate. Elevation: 600 m.

General distribution: Sabah, Kalimantan.

Additional collection. HEMPUEN HILL: 600 m, *Lamb & Surat AL 1297/91* (K).

55.17.87. Bulbophyllum sp. 1, sect. Cirrhopetalum

Epiphyte. Hill forest on ultramafic substrate. Elevation: 600 m.

Collection. HEMPUEN HILL: 600 m, *Lamb AL 1289/90* (K).

55.17.88. Bulbophyllum indet.

Collections. BUNDU TUHAN: 1200 m, *Carr SFN 27424* (SING), 800 m, *SFN 27813* (SING), 900 m, *SFN 27850* (SING), 900 m, *SFN 27858* (SING), 600 m, *SFN 27890* (SING), 800 m, *SFN 27936* (SING), 800 m, *SFN 27957* (SING); GURULAU SPUR: 1400 m, *Carr 3138, SFN 26594* (K), *Clemens 50954* (BM); HEMPUEN HILL: 1000 m, *Cribb 89/38* (K); KEMBURONGOH: 2200 m, *Carr SFN 27469* (SING), 2300 m, *SFN 27564* (SING), 2000 m, *SFN 27768* (SING), 2100 m, *SFN 27868* (SING), 2000 m, *SFN 27937* (SING), 2000 m, *SFN 27939* (SING), 2600 m, *Clemens 27151* (BM); KILEMBUN BASIN: 1400 m, *Clemens 35162* (BM); KINATEKI RIVER HEAD: 1800 m, *Clemens 33028* (BM); KINUNUT VALLEY HEAD: 1200 m, *Carr SFN 27220* (SING), 1200 m, *SFN 27248* (SING); LUBANG: 1400 m, *Carr 3191* (SING); LUMU-LUMU: 1600 m, *Carr SFN 27109* (SING), 1600 m, *SFN 27354* (SING), 1800 m, *SFN 27822* (SING), 1800 m, *SFN 27833* (SING), 1800 m, *SFN 27883* (SING), 1800 m, *SFN 27889* (K), 1700 m, *SFN 27902* (SING), 1600 m, *SFN 27906* (SING), 1800 m, *Clemens 27859* (BM); MAHANDEI RIVER: 1100 m, *Carr SFN 26458* (SING), 1100 m, *SFN 26532* (SING), 1100 m, *SFN 27987* (SING); MESILAU CAVE TRAIL: 1700–1900 m, *Beaman 9144* (K); PENIBUKAN: 1400 m, *Carr SFN 26456* (SING), 1200 m, *Clemens 30854* (BM); SEDIKEN RIVER: 1200–1500 m, *Clemens 34371* (BM); TENOMPOK: 1500 m, *Carr SFN 26856* (SING), 1600 m, *SFN 26914* (SING), 1500 m, *SFN 27194* (SING), 1600 m, *SFN 27237* (SING), 1500 m, *SFN 27403* (SING), 1500 m, *SFN 27909* (SING), 1500 m, *SFN 27926* (SING), 1600 m, *SFN 27956* (SING), 1500 m, *SFN 27999* (SING), 1500 m, *SFN 28005* (SING); TENOMPOK/RANAU: 1500 m, *Carr SFN 26996* (SING); TIBABAR RIVER: 1800 m, *Carr SFN 27548* (SING).

55.18. CALANTHE R. Br.

Bot. Reg. 7: sub t. 573 (1821).

Terrestrial, rarely epiphytic, herbs. *Roots* thick, long and hairy. *Stems* pseudobulbous, short or long, usually 2- to several-leaved, often clustered. *Pseudobulbs* small, ovoid, covered by leaf bases. *Leaves* persistent, rarely deciduous, narrowly elliptic or linear-elliptic, plicate, petiolate, up to 1 m long. *Inflorescence* axillary, terminal or arising from base of leafy pseudobulb, erect, long or short, few- to many-flowered, racemose; floral bracts persistent or deciduous. *Flowers* resupinate, small to rather large and showy, usually turning blue-black when damaged; *sepals* and *petals* similar, free, spreading, rarely connivent; *lip* entire, 3- or 4-lobed, adnate to column at base, spurred, mid-lobe often deeply bifid, disc usually with basal verrucose calli or lamellae; *column* short, rarely long, fleshy, truncate, without a foot; *pollinia* 8.

About 100 species from the tropical, subtropical and warm temperate regions from Africa to Asia and the Pacific islands, with a single species in the tropical Americas.

55.18.1. Calanthe aff. **aureiflora** J. J. Sm., Bull. Jard. Bot. Buit., ser. 3, 5: 67 (1922). Plate 16B.

Terrestrial. Cultivated at Poring, of local origin.

Collection. PORING ORCHID GARDEN: *Lohok 27* (K).

55.18.2. Calanthe crenulata J. J. Sm., Bot. Jahrb. Syst. 48: 97 (1912).

Terrestrial. Lower montane forest on ultramafic substrate. Elevation: 1200 m.

General distribution: Sabah, Kalimantan.

Collection. KINATEKI RIVER/MARAI PARAI: 1200 m, *Collenette A 134* (BM).

55.18.3. Calanthe flava (Blume) C. Morren, L'Hortic. Belge 2: 238, t. 46 (1834).

Amblyglottis flava Blume, Bijdr., 370, t. 64 (1825).

Calanthe parviflora Lindl., Paxt. Fl. Gard. 3: 37, t. 61 (1852–53).

Terrestrial. Lower montane forest. Elevation: 2100 m. Reported as *Calanthe parviflora* Lindl. by Ridley in Stapf (1894).

General distribution: Sabah, Sumatra, Java.

Collection. MOUNT KINABALU: 2100 m, *Haviland s.n.* (unlocated).

55.18.4. Calanthe aff. **flava** (Blume) C. Morren, L'Hortic. Belge 2: 238, t. 46 (1834).

Terrestrial. Lower montane forest. Elevation: 1900–2200 m.

Collection. MESILAU CAVE: 1900–2200 m, *Beaman 9548* (K).

55.18.5. Calanthe gibbsiae Rolfe in Gibbs, J. Linn. Soc. Bot. 42: 156 (1914).

Terrestrial. Hill forest, lower montane forest. Elevation: 800–1700 m.

General distribution: Sabah, Kalimantan.

Collections. BAMBANGAN RIVER: 900–1200 m, *Lamb SAN 91509* (K); DALLAS: 900 m, *Clemens 26724* (BM, K), 900 m, *26866* (K); MINITINDUK GORGE: 800 m, *Carr 3154, SFN 26726* (SING); SUMMIT TRAIL: 1700 m, *Cribb 89/34* (K).

55.18.6. Calanthe aff. **gibbsiae** Rolfe in Gibbs, J. Linn. Soc. Bot. 42: 156 (1914).

Terrestrial. Upper montane forest on ultramafic substrate. Elevation: 1800–2700 m.

Collections. KINATEKI RIVER HEAD: 2700 m, *Clemens s.n.* (BM), 2700 m, *31965* (BM); MARAI PARAI: 2400 m, *Clemens 33174* (BM); MESILAU CAVE: 1800 m, *RSNB 4725* (K).

55.18.7. Calanthe kinabaluensis Rolfe in Gibbs, J. Linn. Soc. Bot. 42: 156 (1914). Plate 16 C, 17 A. Type: LUBANG, 1500 m, *Gibbs 4108* (holotype K!).

Calanthe cuneata Ames & C. Schweinf., Orch. 6: 159 (1920), **syn. nov.** Type: MOUNT KINABALU, *Haslam s.n.* (holotype AMES mf).

Terrestrial. Lower montane forest, rarely hill forest. Elevation: 900–2100 m.

General distribution: Sabah.

Additional collections. BAMBANGAN RIVER: 1500 m, *RSNB 1299* (K); DALLAS: 900 m, *Clemens 26122* (BM, K); MESILAU BASIN: 2100 m, *Clemens 29253* (BM); MESILAU RIVER: 1500 m, *RSNB 7113* (K), 2100 m, *Clemens 51471* (BM); PENATARAN BASIN: 1200 m, *Clemens 34053* (BM); TAHUBANG RIVER: 1100 m, *Clemens 31681* (BM); TENOMPOK: 1500–1600 m, *Beaman 9455* (K), 1500 m, *Carr SFN 26918* (SING), 1500 m, *SFN 26918A* (SING), 1500 m, *Clemens 27979* (BM), 1500 m, *28324* (BM, K), 1500 m, *30119* (K); WEST MESILAU RIVER: 1600–1700 m, *Beaman 8622* (K), 1500 m, *Collenette 584* (K).

55.18.8. Calanthe cf. **lyroglossa** Rchb. f., Otia Bot. Hamburg. 1: 53 (1878).

Terrestrial. Lower montane forest, on large rock in middle of river. Elevation: 1300 m.

Collection. KINATEKI RIVER: 1300 m, *Bailes & Cribb 858* (K).

55.18.9. Calanthe ovalifolia Ridl. in Stapf, Trans. Linn. Soc. Bot. 4: 239 (1894). Type: TINEKUK RIVER, 900 m, *Haviland s.n.* (holotype unlocated).

Terrestrial. Hill forest. Elevation: 900 m. Known only from the type collection.

Endemic to Mount Kinabalu.

55.18.10. Calanthe pulchra (Blume) Lindl., Gen. Sp. Orch., 250 (1833). Plate 17B, 17C.

Amblyglottis pulchra Blume, Bijdr., 371 (1825).

Terrestrial. Hill forest, lower montane forest, sometimes on ultramafic substrate. Elevation: 900–2400 m.

General distribution: Sabah, Sarawak, Peninsular Malaysia, Thailand, Sumatra, Java, Philippines.

Collections. DALLAS: 900 m, *Clemens s.n.* (BM); GURULAU SPUR: 1500 m, *Clemens 50350* (BM, K); KIAU: 900 m, *Clemens 98* (AMES, BM, K), 900 m, *126* (AMES), 900 m, *155* (AMES), *302* (AMES), *402* (AMES); MARAI PARAI: 1500 m, *Clemens 32451* (BM); MARAI PARAI SPUR: 2400 m, *Clemens 381* (AMES); PARK HEADQUARTERS: 1500 m, *Lamb SAN 91581* (K); PENIBUKAN: 1200 m, *Clemens 40612* (BM, K), 1200 m, *50109* (BM); TENOMPOK: 1600 m, *Carr SFN 26937* (SING), 1500 m, *Clemens 28036* (BM); TENOMPOK RIDGE: 1400–1500 m, *Beaman 8191* (K); TINEKUK FALLS: 1700 m, *Clemens 40828* (BM, K); UPPER KINABALU: *Clemens 27170* (K).

55.18.11. Calanthe speciosa (Blume) Lindl., Gen. Sp. Orch., 250 (1833).

Amblyglottis speciosa Blume, Bijdr., 371 (1825).

Terrestrial. Hill forest, lower montane forest. Elevation: 900–1900 m.

General distribution: Sabah, Sarawak, Peninsular Malaysia, Sumatra, Java.

Collections. EAST MESILAU RIVER: 1600–1700 m, *Bailes & Cribb 744* (K); LOHAN/MAMUT COPPER MINE: 900 m, *Beaman 10623* (K, MSC); MESILAU CAVE TRAIL: 1700–1900 m, *Beaman 7994* (K, MSC).

55.18.12. Calanthe tenuis Ames & C. Schweinf., Orch. 6: 161 (1920). Type: MOUNT KINABALU, *Haslam s.n.* (holotype AMES mf).

Terrestrial. Known only from the type collection.

Endemic to Mount Kinabalu.

55.18.13. Calanthe transiens J. J. Sm., Bull. Jard. Bot. Buit., ser. 3, 5: 70 (1922). Plate 18A.

Terrestrial. Upper montane forest on ultramafic substrate. Elevation: 1700–2700 m.

General distribution: Sabah, Sumatra.

Collections. LAYANG-LAYANG: 2700 m, *Lamb AL 587/86* (K); MESILAU RIVER?: 1700 m, *Lamb 91528* (K, sketch); SUMMIT TRAIL: 2500 m, *Bailes & Cribb 768* (K).

55.18.14. Calanthe triplicata (Willemet) Ames, Philipp. J. Sci. 2C: 326 (1907).

Orchis triplicata Willemet, Ann. Bot. (Usteri) 18: 52 (1796).

Terrestrial. Lowland dipterocarp forest, hill forest, lower montane forest. Elevation: 600–1500 m.

General distribution: Sabah, Sarawak, Kalimantan, Madagascar, Mascarene Is., China, Japan, India, through Malaysia, Indonesia east to Australia & the Pacific islands.

Collections. BUNDU TUHAN: 1400 m, *Carr SFN 27055* (SING), 1400 m, *SFN 27055A* (SING), 1400 m, *SFN 27055B* (SING), 1400 m, *SFN 27055D* (SING); KIAU: 900 m, *Clemens 42* (AMES), 900 m, *53* (AMES), 900 m, *132* (AMES), *343* (AMES), *363* (AMES); KUNDASANG: 1300 m, *Collenette 596* (K); LOHAN RIVER: 700–900 m, *Beaman 9258* (K); LUBANG GORGE: *Clemens 301* (AMES); MAHANDEI RIVER: 1100 m, *Carr SFN 27055C* (SING); MESILAU RIVER: 1400 m, *RSNB 7093* (K); MOUNT KINABALU: *Haslam s.n.* (AMES); PORING HOT SPRINGS: 600 m, *Beaman 7593* (K); TENOMPOK: 1500 m, *Clemens 27979* (K).

55.18.15. Calanthe truncicola Schltr., Bot. Jahrb. Syst. 45, Beibl. 104: 26 (1911). Plate 18B.

Terrestrial. Lower montane forest on ultramafic substrate, among bamboo and scrub. Elevation: 1400–1500 m.

General distribution: Sabah, Sumatra.

Collections. LITTLE MAMUT RIVER: 1400 m, *Collenette 1013* (K); MARAI PARAI SPUR: 1500 m, *Lamb AL 44/83* (K).

55.18.16. Calanthe vestita Lindl., Gen. Sp. Orch., 250 (1833).

Terrestrial. Low-stature hill forest on ultramafic substrate. Elevation: 800–900 m.

General distribution: Sabah, Sarawak, Burma, Peninsular Malaysia, Thailand, Vietnam, Java, Sulawesi, Seram.

Collections. LIWAGU RIVER: *Bacon s.n.* (K); LOHAN RIVER: 800–900 m, *Beaman 9482* (K).

55.18.17. Calanthe indet.

Collections. BUNDU TUHAN: 1400 m, *Carr SFN 27866* (SING); KADAMAIAN RIVER: 2000 m, *Carr SFN 27745* (SING); KIBAMBANG LUBANG: 1400 m, *Clemens 32439* (BM); KINATEKI RIVER: 1400 m, *Carr SFN 26587* (SING); LUMU-LUMU: 1600 m, *Carr SFN 27027* (SING); TAHUBANG RIVER: 1200–1500 m, *Clemens 31109* (BM); TENOMPOK: 1600 m, *Carr SFN 27012* (SING), 1400 m, *SFN 27857* (SING), 1700 m, *Clemens s.n.* (BM), 1500 m, *28244* (BM).

55.19. CERATOCHILUS Blume

Bijdr., 358 (1825).

Very small monopodial epiphytes. *Stems* to 8 cm. *Leaves* very short, fleshy, laterally flattened. *Inflorescences* arising laterally from near the stem apex, 1-flowered. *Flowers* large in proportion to the rest of the plant, transparent white, sometimes fading to scarlet, 1–2.2 cm across; *sepals* and *petals* free, spreading; *dorsal sepal* 0.5–1.3 × 0.5 cm; *lip* immobile, spurred, with very small side-lobes which clasp the column, mid-lobe tiny (*C. biglandulosus*) or expanded into a broadly oblong-elliptic, emarginate blade (*C. jiewhoeii*);

column short, foot absent; *stipes* linear-clavate; *pollinia* 4, appearing as 2 unequal masses.

One species, *C. biglandulosus*, endemic to Java, another, *C. jiewhoeii*, endemic to Borneo. A third species, undescribed, has been collected on Mt. Alab in the Crocker Range (Lamb, pers. comm.).

55.19.1. Ceratochilus jiewhoeii J. J. Wood & Shim, ined. Fig. 12, Plate 18C. Type: PINOSUK PLATEAU, *Lamb AL 58/83* (holotype K!).

Epiphyte. Lower montane forest; near tips of thin branches of small saplings in shade. Elevation: 900–1800 m.

General distribution: Sabah.

Additional collections. BUNDU TUHAN: 900 m, *Darnton 520* (BM); GOLF COURSE SITE: 1700–1800 m, *Beaman 10673* (K); LUMU-LUMU (KIAU VIEW GAP): 1700 m, *Lamb AL 1516/92* (K). PENIBUKAN: 1200 m, *Clemens 50106* (BM, K); SEDIKEN RIVER/MARAI PARAI: 1500 m, *Clemens 35173* (BM); TENOMPOK: 1500 m, *Clemens 26874* (BM), 1500 m, *30150* (K).

55.20. CERATOSTYLIS Blume

Bijdr., 304 (1825).

Epiphytic herbs. *Roots* fibrous. *Stems* simple or branched, tufted, 1-leaved, sometimes terete and rush-like, with thin, brown, often reticulate basal sheaths, pseudobulbs absent. *Leaves* narrow, coriaceous, fleshy or subterete, rarely thin-textured. *Flowers* resupinate, small, rarely large and showy, solitary or a few in a small cluster of bracts. *Sepals* erect, connivent; *lateral sepals* forming a saccate or spur-like mentum with column-foot. *Petals* narrower than sepals. *Lip* adnate to column-foot by a long incumbent claw, entire, usually with longitudinal calli. *Column* short, dilated above, 2-lobed or with 2 spathulate erect arms, foot long. *Pollinia* 8, sessile.

About 100 species distributed from tropical Asia to the Pacific islands.

55.20.1. Ceratostylis ampullacea Kraenzl., Bot. Jahrb. Syst. 17: 487 (1893).

Epiphyte. Lower montane forest, upper montane forest; low mossy and xerophyllous scrub forest on extreme ultramafic substrate. Elevation: 1200–3200 m.

General distribution: Sabah, Peninsular Malaysia, Thailand, Sumatra.

Collections. MESILAU RIVER: 2100 m, *Clemens 51468* (BM, K), 2300 m, *Lamb LKC 3176* (K); PAKA-PAKA CAVE: 2700 m, *Carr SFN 27907* (K, SING), 3200 m, *Clemens 27868* (BM); PENIBUKAN: 1200–1500 m, *Clemens s.n.* (BM), 1200–1500 m, *31566* (BM), 1200 m, *35174* (BM), 1200 m, *40835* (BM); PIG HILL: 2000–2300 m, *Beaman 9892* (K); UPPER KINABALU: *Clemens 28994* (BM, K).

55.20.2. Ceratostylis crassilingua Ames & C. Schweinf., Orch. 6: 135 (1920). Type: MARAI PARAI SPUR, *Clemens 260* (holotype AMES mf; isotype K!).

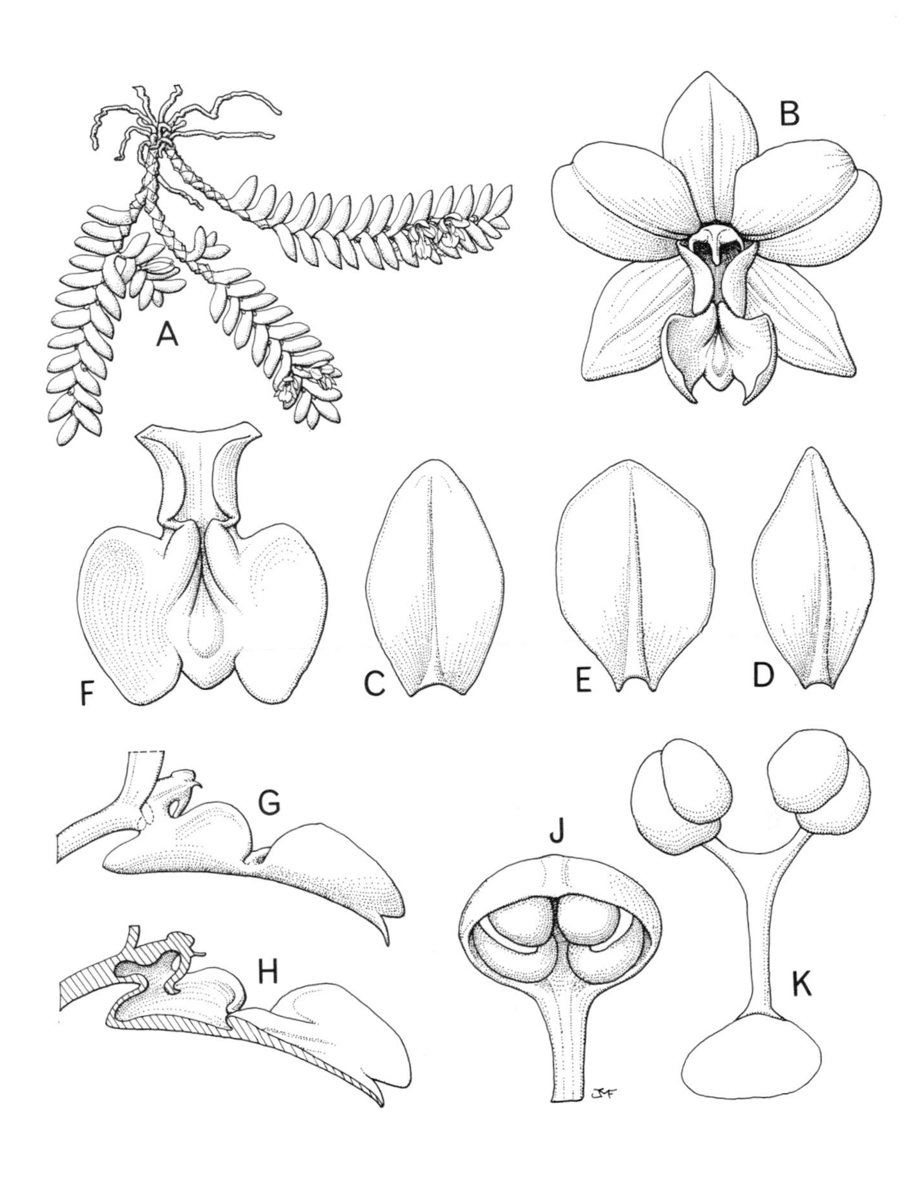

Fig. 12. Ceratochilus jiewhoeii. A, habit × 1; **B**, flower (front view) × 5; **C**, dorsal sepal × 6; **D**, lateral sepal × 6; **E**, petal × 6; **F**, lip (front view) × 6; **G**, column and lip (side view) × 6; **H**, column and lip (longitudinal section) × 6; **J**, anther-cap with pollinia (front view) × 30; **K**, pollinarium × 30. **A** from *Lamb AL 611/86*, Mt. Alab, Sabah. **B–K** from *Beaman 10673* and *Lamb AL 58/83*. Drawn by Mark Fothergill.

Epiphyte. Lower montane forest; upper montane forest, usually on ultramafic substrate, in open areas or in low xerophyllous scrub. Elevation: 1200–2300 m.

Endemic to Mount Kinabalu.

Additional collections. BUNDU TUHAN: 1400 m, *Darnton 198* (BM); MARAI PARAI: 1500 m, *Clemens 33159* (BM); MESILAU CAVE: 2000–2100 m, *Beaman 8153* (K); PENIBUKAN: 1200–1500 m, *Clemens s.n.* (BM), 1200 m, *30584* (BM), 1200 m, *31046* (BM); PIG HILL: 2000–2300 m, *Beaman 9887* (K).

55.20.3. Ceratostylis longisegmenta Ames & C. Schweinf., Orch. 6: 136 (1920). Type: KIAU, 900 m, *Clemens 93* (holotype AMES mf).

Epiphyte. Hill forest. Elevation: 900 m.

General distribution: Sabah, Kalimantan.

Additional collection. KIAU: 900 m, *Clemens 189* (AMES).

55.20.4. Ceratostylis radiata J. J. Sm., Orch. Java, 295 (1905).

Epiphyte. Lower montane forest. Elevation: 1100–1900 m.

General distribution: Sabah, Peninsular Malaysia, Thailand, Sumatra, Java.

Collections. EAST MESILAU/MENTEKI RIVERS: 1700 m, *Beaman 8760* (K); GOLF COURSE SITE: 1600 m, *Bailes & Cribb 734* (K), 1700–1800 m, *Beaman 7236* (K); KINATEKI RIVER: *Clemens 30603* (K); MARAI PARAI: 1200 m, *Clemens 32781* (BM); MESILAU CAVE TRAIL: 1700–1900 m, *Beaman 8012* (K); MESILAU RIVER: 1500 m, *RSNB 4100* (K); PENIBUKAN: 1200–1500 m, *Clemens 30603* (BM), 1200 m, *32170* (BM), 1100 m, *50060* (BM, K), 1500 m, *50231* (BM), 1500 m, *50323* (BM, K); PINOSUK PLATEAU: 1500 m, *Lamb AL 15/82* (K); TENOMPOK: 1500 m, *Clemens 27918* (BM, K); WEST MESILAU RIVER: 1600 m, *Bailes & Cribb 535* (K).

55.20.5. Ceratostylis subulata Blume, Bijdr., 306 (1825).

Epiphyte. Hill forest, lower montane forest. Elevation: 900–1700 m.

General distribution: Sabah, Nicobar Islands, Peninsular Malaysia, Thailand, Cambodia?, Sumatra, Java, Seram, Philippines, New Guinea, Solomon Islands, Vanuatu.

Collections. KIAU: 900 m, *Clemens 70* (AMES, K, SING), 900 m, *177* (AMES, K, SING), *320* (AMES); PENIBUKAN: 1200–1500 m, *Clemens 30522* (BM, K), 1200–1500 m, *30945* (BM, K); WEST MESILAU RIVER: 1600 m, *Beaman 7524* (K), 1700 m, *Brentnall 130* (K).

55.20.6. Ceratostylis sp. 1

Epiphyte. Hill forest on ultramafic substrate. Elevation: 800–1000 m.

Collection. HEMPUEN HILL: 800–1000 m, *Beaman 7409* (K).

55.20.7. Ceratostylis indet.

Collections. BUNDU TUHAN: 900 m, *Carr SFN 27913* (SING); KAUNG: 500 m, *Carr SFN 27959* (SING); MARAI PARAI/KIAU TRAIL: 1200–1500 m, *Bailes & Cribb 808* (K); PENIBUKAN: 900–1800 m, *Carr SFN 27560* (SING); TENOMPOK: 1500 m, *Carr SFN 27417* (SING); TENOMPOK/RANAU: 1500 m, *Carr SFN 27004* (SING).

55.21. CHAMAEANTHUS Schltr. in J. J. Sm.

Orch. Java, 552 (1905).

Small monopodial epiphytes. *Stems* very short or up to a few cm long. *Leaves* few, somewhat fleshy, to 7 × 1 cm. *Inflorescences* simple, with up to 20 flowers, rachis usually club-shaped. *Flowers* small, ephemeral, only a few open at a time, greenish-yellow, looking superficially like a *Bulbophyllum*; *sepals* and *petals* similar, often linear and acuminate, to 4–6 × 3 mm; *lateral sepals* adnate to the column-foot; *lip* mobile, 3-lobed, side-lobes ear-like, margins sometimes fimbriate; mid-lobe conical, fleshy; *column* short and stout with a 3 mm long foot, rostellar projection elongate; stipes strap-shaped, about 2 times the diameter of the pollinia, viscidium small, obovate; *pollinia* 2, entire, although some specimens have a very tiny notch.

About 3 species distributed from S. Thailand east to Java and New Guinea, north to the Philippines. Senghas (1988) has transferred 10 species with 4 unequal pollinia, all from New Guinea and the Pacific islands, to *Gunnarella* Senghas.

55.21.1. Chamaeanthus brachystachys Schltr. in J. J. Sm., Orch. Java, 552 (1905).

Epiphyte. Hill forest. Known only from a colour slide by A. Lamb.

General distribution: Sabah, Kalimantan, Thailand, Java.

Collection. HEMPUEN HILL: *Lamb s.n.* (K photo).

55.22. CHELONISTELE Pfitzer in Pfitzer & Kraenzl.

Pflanzenr. IV. 50. II. B. 7: 136 (1907).

De Vogel, E. F. (1986). Revisions in Coelogyninae (Orchidaceae) II. The genera *Bracisepalum, Chelonistele, Entomophobia, Geesinkorchis* & *Nabaluia*. Orch. Monogr. 1: 23–40.

Epiphytic or lithophytic herbs. *Rhizome* creeping, rather short. *Pseudobulbs* ovoid, conical or cylindrical, all turned to one side of rhizome, 1- or 2-leaved. *Leaves* elliptic to linear, usually tough and coriaceous, petiole deeply sulcate. *Inflorescence* proteranthous to synanthous, racemose, erect to spreading, few- to many-flowered; rachis fractiflex; floral bracts caducous. *Flowers resupinate*, small to medium-sized, tender to rather fleshy. *Sepals* free, concave, spreading. *Petals* narrower, often rolled backward. *Lip* 3-lobed, base narrow, saccate, with ± parallel sides; side-lobes narrow, arising above the base at right angles to the blade and almost in same plane; mid-lobe lingulate to broadly spathulate or almost orbicular, acute or deeply retuse; disc with 2, rarely 4, keels. *Column* long, with an entire or toothed apical wing. *Pollinia* 4, each attached to a small caudicle.

Eleven species were formerly known, all, except *C. sulphurea* var. *sulphurea*, endemic to Borneo. An unidentified species was photographed (Plate 20A) at 2250 m on Mount Kinabalu in 1991 by P. J. Cribb, but a specimen was not collected. De Vogel (pers. comm.) reports a further two, or possibly five, new species from Brunei.

55.22.1. Chelonistele amplissima (Ames & C. Schweinf.) Carr, Gardens' Bull. 8: 218 (1935). Plate 19A.

Coelogyne amplissima Ames & C. Schweinf., Orch. 6: 21 (1920). Type: KIAU, 900 m, *Clemens 80* (holotype AMES mf; isotype L).

Epiphyte. Lower montane forest, mossy forest, ridge forest, frequently on ultramafic substrate. Elevation: 900–2100 m.

General distribution: Sabah, Kalimantan, Sarawak.

Additional collections. GURULAU SPUR: 1700 m, *Clemens 50414* (BM, K), 1700 m, *50703* (BM, K), 1500–2100 m, *50740* (BM); KEMBURONGOH: 2100 m, *Carr 3755* (SING); LUMU-LUMU: 2100 m, *Clemens 27181* (BM); MAMUT COPPER MINE: 1400–1500 m, *Beaman 10350* (K), 1400–1500 m, *10355* (K); MARAI PARAI SPUR: *Clemens 231* (AMES), *233* (AMES); PENIBUKAN: 1700 m, *Clemens 50286* (BM, K); TENOMPOK: *Clemens 27659* (K), *27659A* (BM); TENOMPOK RIDGE: 1400–1500 m, *Beaman 8181* (K); UPPER KINABALU: *Clemens 27174* (K), *27362* (K).

55.22.2. Chelonistele kinabaluensis (Rolfe) de Vogel, Blumea 30: 203 (1984).

Sigmatochilus kinabaluensis Rolfe in Gibbs, J. Linn. Soc. Bot. 42: 155, pl. 3 (1914). Type: PAKA-PAKA CAVE, *Gibbs 4260* (holotype BM!).

Pholidota sigmatochilus J. J. Sm., Blumea 5: 299 (1943).

Epiphyte. Upper montane forest, often on ultramafic substrate, on branches of shrubs in thick moss cushions. Elevation: 1800–3500 m.

Endemic to Mount Kinabalu.

Additional collections. GURULAU SPUR: 2400–2700 m, *Clemens 50658* (BM, K), 1800 m, *51192* (BM, K); KEMBURONGOH: 2900 m, *Carr 3524, SFN 27533* (K), 2900 m, *Fuchs 21072* (L); KEMBURONGOH/PAKA PAKA CAVE: 2900 m, *Carr 3524, SFN 27533A* (BM); KEMBURONGOH/PAKA-PAKA CAVE: 2900 m, *Carr 3524, SFN 27533* (SING), 2900 m, *SFN 27533A* (SING); LUBANG: *Clemens 206* (AMES), *223* (AMES); MESILAU CAVE: 2000–2100 m, *Beaman 8154* (K); MOUNT KINABALU: *Haslam s.n.* (AMES, K); PAKA-PAKA CAVE: 3500 m, *Clemens 27153* (BM); PIG HILL: *Lamb AL 734/87* (K).

55.22.3. Chelonistele lurida (L. Linden & Cogn.) Pfitzer, Pflanzenr. IV. 50. II. B. 7: 138 (1907).

a. var. **lurida.** Plate 19B.

Chelonanthera lurida L. Linden & Cogn., Lindenia 11: 80 (1895).

Epiphyte. Lower montane forest, upper montane forest, on ultramafic substrate; on trunks and moss-covered roots. Elevation: 1400–2700 m.

General distribution: Sabah, Kalimantan, Sarawak.

Collections. GURULAU SPUR: 2400–2700 m, *Clemens 50649* (BM); KEMBURONGOH: 2200 m, *Carr 3518, SFN 27713* (K, L, SING), 2400 m, *Sinanggul SAN 38318* (L); KINATEKI RIVER HEAD: 2400 m, *Clemens s.n.* (BM); LUMU-LUMU: 2300 m, *Clemens s.n.* (BM); MARAI PARAI SPUR: 2100 m, *Gibbs 4095* (BM, K); MARAI PARAI/KIAU TRAIL: 1400–1500 m, *Bailes & Cribb 804* (K); PIG HILL: 1700–2000 m, *Sutton 12* (K); SUMMIT TRAIL: 2000 m, *Bailes & Cribb 783* (K), 2000 m, *Beaman 6765* (K), *Wood 614* (K).

55.22.4. Chelonistele sulphurea (Blume) Pfitzer, Pflanzenr. IV. 50. II. B. 7: 137 (1907).

a. var. **sulphurea.** Plate 19C.

Chelonanthera sulphurea Blume, Bijdr., 383 (1825).

Epiphyte. Hill forest, lower montane forest, frequently on ultramafic substrate; often low on tree trunks. Elevation: 800–2400 m.

General distribution: Sabah, Kalimantan, Sarawak, Peninsular Malaysia, Sumatra, Java, Philippines.

Collections. BUNDU TUHAN: 1200 m, *Carr 3343, SFN 27180* (K, SING), 1200 m, *3488, SFN 27443* (SING); HEMPUEN HILL: 800–1200 m, *Beaman 7716* (K); KEMBURONGOH: 2000 m, *Carr 3763* (SING); KIAU: *Clemens 375* (AMES); KILEMBUN RIVER: 1400 m, *Clemens 33700* (BM); KILEMBUN RIVER HEAD: 2100–2400 m, *Clemens 33937* (BM); LUMU-LUMU: 1600 m, *Carr 3292, SFN 27028* (K, SING), 1700 m, *3580, SFN 27792* (K, SING), 1600 m, *SFN 27028* (SING), 1700 m, *Clemens 29994* (BM); MARAI PARAI SPUR: *Clemens 239* (AMES); PENIBUKAN: 1200 m, *Clemens 30590* (BM), 1200 m, *30856* (BM), 1500 m, *31004* (BM), 1200–1500 m, *31111* (BM), 1200–1500 m, *31257* (BM); SUMMIT TRAIL: 1800 m, *Fuchs 21032* (K, L); TENOMPOK: 1500 m, *Clemens 29819* (K), 1500 m, *29824* (BM, K); TENOMPOK ORCHID GARDEN: *Clemens 50244* (BM); UPPER KINABALU: *Clemens 29994* (K); WEST MESILAU RIVER: 1600 m, *Bailes & Cribb 536* (K), 1600 m, *539* (K).

b. var. **crassifolia** (Carr) de Vogel, Blumea 30: 205 (1984).

Chelonistele crassifolia Carr, Gardens' Bull. 8: 218 (1935). Type: KEMBURONGOH, 1700–2400 m, *Carr 3565, SFN 28027* (holotype SING!).

Terrestrial. Lower montane forest, upper montane forest on ultramafic substrate; on rocks and boulders, often in rather exposed sites. Elevation: 1500–2400 m.

General distribution: Sabah.

Additional collections. EAST MESILAU RIVER: 1500–1600 m, *Bailes & Cribb 516* (K); KEMBURONGOH: 2400 m, *Clemens 27365* (AMES, BM); LUMU-LUMU: 2000 m, *Clemens 27152* (AMES, BM, K), 2100 m, *27169* (AMES, BM, K, SING); MESILAU CAVE: 1600 m, *Bailes & Cribb 696* (K); MESILAU TRAIL: 1900 m, *Collenette 635* (K); SUMMIT TRAIL: 1800–2000 m, *Chan & Gunsalam 40/87* (K).

55.22.5. Chelonistele sp. 1. Plate 20A.

Epiphyte. Upper montane forest. Elevation: 2300 m.

Collection. SUMMIT TRAIL: 2300 m, *Cribb s.n.* (K photo).

55.22.6. Chelonistele indet.

Collections. LUMU-LUMU: 1700 m, *Carr 3638* (SING); TENOMPOK ORCHID GARDEN: 1500 m, *Clemens 50257* (SING).

55.23. CHRONIOCHILUS J. J. Sm.

Bull. Jard. Bot. Buit., ser. 2, 26: 81 (1918).

Small monopodial epiphytes. *Stems* very short. *Leaves* few, flat. *Inflorescences* simple, few-flowered; rachis sometimes flattened. *Flowers* small; *sepals* and *petals* free, spreading; *lip* mobile, sessile, arrow-head shaped, conical, with large ear-like side-lobes, spur absent, the conical median part solid; *column* with a distinct foot; *rostellar projection* prominent; *stipes* linear-oblong; *viscidium* large, oval, at least half as long as the stipes; *pollinia* 2, entire.

Four species distributed from S Thailand through Malaysia to Indonesia.

55.23.1. Chroniochilus minimus (Blume) J. J. Sm., Bull. Jard. Bot. Buit., ser. 3, 8: 366 (1927).

Dendrocolla minima Blume, Bijdr., 290 (1825).

Epiphyte. Low-stature hill forest on ultramafic substrate. Elevation: 700 m.

General distribution: Sabah, Peninsular Malaysia, Java.

Collection. HEMPUEN HILL: 700 m, *Lamb AL 838/87* (K).

55.23.2. Chroniochilus virescens (Ridl.) Holttum, Kew Bull. 14: 273 (1960). Plate 20B.

Sarcochilus virescens Ridl., J. Straits Branch Roy. Asiat. Soc. 39: 85 (1903).

Epiphyte. Low-stature hill forest on ultramafic substrate. Elevation: 500–1000 m.

General distribution: Sabah, Peninsular Malaysia, Thailand.

Collections. LOHAN RIVER: 800–1000 m, *Beaman 10014* (K), 500 m, *Lamb AL 390/85* (K); PORING: *Lohok 5* (K).

55.24. **CHRYSOGLOSSUM** Blume

Bijdr., 337 (1825).

Terrestrial, rarely epiphytic, herbs. *Rhizome* creeping. *Pseudobulbs* articulated, one-leaved. *Leaves* convolute, linear-elliptic to narrowly elliptic, acuminate, petiolate, sometimes variegated. *Inflorescence* erect, racemose, many-flowered. *Flowers* resupinate; *sepals* and *petals* free, similar; *lateral sepals* inserted on column-foot; *lip* 3-lobed, mobile, fleshy, hypochile pleated at base, disc with 3 keels, epichile recurved, concave; *column* erect, with distinct foot having a saccate spur, and a small lobe in front; *pollinia* 2, waxy.

Four species distributed from Sri Lanka and India eastward to New Guinea and the Pacific islands.

55.24.1. Chrysoglossum reticulatum Carr, Gardens' Bull. 8: 197 (1935). Plate 20C. Type: TENOMPOK, 1500 m, *Carr 3314, SFN 27060* (syntype SING!; isosyntype K!); TENOMPOK/TOMIS, 1600 m, *Clemens s.n.* (syntype BM!).

Epiphyte. Lower montane forest. Elevation: 1500–1600 m.

General distribution: Sabah, Sarawak.

55.25. **CLADERIA** Hook. f.

Fl. Brit. Ind. 5: 810 (1890).

Terrestrial herbs. *Rhizome* creeping, long, slender, bearing leafy shoots 20 cm or more apart, climbing a little way up the bases of tree trunks where the inflorescence is produced. *Leaves* 4–6 per shoot, close together, elliptic, plicate. *Inflorescence* racemose, bearing a succession of flowers one or two at a time; peduncle up to 15 cm long; rachis slowly elongating, pubescent; floral bracts stiff, pubescent. *Flowers* resupinate, green, lip paler green with darker green veins. *Sepals* and *petals* free, the sepals pubescent. *Dorsal sepal* erect or curved over column. *Lateral sepals* spreading horizontally, curved toward each other. *Lip* 3-lobed, saccate; side-lobes large, obtuse, erect; mid-lobe wider than long, apex reflexed, retuse; disc with 2 low rounded hairy keels between side-lobes. *Column* long, arcuate. *Pollinia* 2, deeply cleft.

Two species distributed in Thailand, Peninsular Malaysia, Sumatra, Bangka, Mentawai, Sulawesi and New Guinea.

55.25.1. Claderia viridiflora Hook. f., Fl. Brit. Ind. 5: 810 (1890). Plate 21A.

Climber. Hill forest. Elevation: 800 m.

General distribution: Sabah, Brunei, Kalimantan, Sarawak, Peninsular Malaysia, Thailand, Sumatra, Sulawesi, Bangka, Mentawai.

Collection. KEKEHITAN HILL: 800 m, *Carr 3423, SFN 27339* (SING).

55.26. **CLEISOCENTRON** Brühl

Guide Orchids Sikkim, 136 (1926).

Monopodial epiphytes. *Stems* long, usually pendent, up to 1 m. *Leaves* strap-shaped to terete, unequally bilobed or acute. *Inflorescences* axillary, simple or branched, short, few- to many-flowered. *Flowers* pinkish white to lavender-blue; *sepals* and *petals* spreading; *lip* immobile, 3-lobed, with a gently curving cylindrical spur, inside which is either an upward-pointing central protruberance on the back-wall (*C. merrillianum*), or a decurved shelf-like back-wall callus, front-wall callus flap-like, median septum absent; *column* erect, cylindrical, with a foot decurrent on the dorsal wall of the lip, or free; *stipes* long and slender; *viscidium* relatively large; *pollinia* 4, appearing as 2 pollen masses each completely divided into free halves.

Five species distributed in the Himalayan region, Burma, Vietnam and Borneo.

55.26.1. Cleisocentron merrillianum (Ames) Christenson, Amer. Orchid Soc. Bull. 61: 246 (1992). Plate 21B.

Sarcanthus merrillianus Ames, Orch. 6: 230 (1920). Type: MARAI PARAI SPUR, *Clemens s.n.* (holotype AMES mf; isotypes BM!, K!).

Robiquetia merrilliana (Ames) Lueckel, M. Wolff & J. J. Wood, Orchidee 40: 109 (1989).

Epiphyte. Lower montane forest, upper montane forest, sometimes on ultramafic substrate; on moss-covered trees. Elevation: 1100–3000 m.

Endemic to Mount Kinabalu.

Additional collections. BUNDU TUHAN: 1100 m, *Carr SFN 27893* (K), 1200 m, *Clemens 40112* (BM); EASTERN SHOULDER, CAMP 3/CAMP 4: 2400–3000 m, *Collenette 756* (K); LITTLE MAMUT RIVER: 1400 m, *Collenette 1020* (K); MARAI PARAI: 1500 m, *Collenette A 135* (BM), 1500 m, *Lamb SAN 89678* (K); MARAI PARAI SPUR: 1600 m, *Bailes & Cribb 819* (K), 1700 m, *855* (K), *Clemens 267* (AMES), *368* (AMES); MESILAU CAVE: 2000 m, *Collenette 597* (K); MESILAU TRAIL: 1500 m, *Sadau SAN 49690* (K); PENIBUKAN: 1200 m, *Clemens 40634* (BM); PINOSUK PLATEAU: 1500 m, *Lamb SAN 89677* (K); TAHUBANG RIVER: 1400 m, *Clemens 40344* (BM).

55.27. **CLEISOSTOMA** Blume

Bijdr., 362 (1825).

Seidenfaden, G. (1975). Orchid genera in Thailand II. Dansk Bot. Ark. 29, 3: 1–80.

Small to medium-sized monopodial epiphytes. *Stems* up to 60 cm long. *Leaves* strap-shaped or terete, to 30 × 4.5 cm, apex usually unequally bilobed. *Inflorescences* usually branched, many-flowered, erect, horizontal or pendulous. *Flowers* small, subtended by rather small bracts; *sepals* and *petals* free, spreading, usually the same size; *lip* 3-lobed, saccate or spurred, always with a callus on the back-wall and often with outgrowths on the front-wall closing the entrance, saccate base or spur usually having a longitudinal internal septum; *column* short and stout, usually without a foot; *viscidium* ranging from small and subglobose to broad and horseshoe-shaped; *pollinia* 4, appearing as 2 unequal masses.

Between 80 and 100 species widespread from India throughout Asia to NE Australia, New Guinea and the Pacific islands.

55.27.1. Cleisostoma discolor Lindl., Bot. Reg. 31: misc. 59 (1845).

Epiphyte. Hill forest on ultramafic substrate. Elevation: 500 m.

General distribution: Sabah, Sarawak, India, Peninsular Malaysia, Thailand, Cambodia, Sumatra, Java.

Collections. LOHAN RIVER: *Clements 3393* (K), 500 m, *Lamb AL 75/83* (K); PORING: *Lohok 22* (K).

55.27.2. Cleisostoma ridleyi Garay, Bot. Mus. Leafl. Harvard Univ. 23, 4: 174 (1972).

Epiphyte. Hill forest? Elevation: 600 m.

General distribution: Sabah, Sarawak.

Collection. MOUNT KINABALU: 600 m, *Lamb SAN 93453* (K).

55.27.3. Cleisostoma aff. **sagittatum** Blume, Bijdr., 363, t. 27 (1825).

Epiphyte. Hill forest, lower montane forest, especially on ultramafic substrate. Elevation: 600–1700 m.

Collections. GOLF COURSE SITE: 1600 m, *Bailes & Cribb 728* (K); LOHAN RIVER: 600–800 m, *Lamb SAN 91516* (K); MARAI PARAI SPUR: 1700 m, *Bailes & Cribb 853* (K).

55.27.4. Cleisostoma striatum (Rchb. f.) Garay, Bot. Mus. Leafl. Harvard Univ. 23: 175 (1972).

Echioglossum striatum Rchb. f., Gard. Chron. ser. 2, 12: 390 (1879).

Epiphyte. Lowlands. Elevation: 400–500 m.

General distribution: Sabah, China, India, Peninsular Malaysia, Vietnam.

Collections. KAUNG: 500 m, *Carr SFN 27805* (K, SING), 400 m, *Darnton 386* (BM).

55.27.5. Cleisostoma suaveolens Blume, Bijdr., 363 (1825).

Epiphyte. Cultivated at Poring, of local origin.

General distribution: Sabah, Sarawak, Sumatra, Java, Bali.

Collection. PORING ORCHID GARDEN: *Cribb 89/50* (K).

55.27.6. Cleisostoma sp. 1

Epiphyte. Upper montane forest?

Collection. UPPER KINABALU: *Clemens 29850* (BM, K).

55.27.7. Cleisostoma sp. 2

Epiphyte. Low-stature hill forest on ultramafic substrate. Elevation: 800–1000 m.

Collection. LOHAN RIVER: 800–1000 m, *Beaman 10009* (K).

55.27.8. Cleisostoma sp. 3

Epiphyte. Lower montane forest. Elevation: 1600 m.

Collection. MENTEKI RIVER: 1600 m, *Beaman 10773* (K).

55.27.9. **Cleisostoma** indet.

Collections. BUNDU TUHAN: 1200 m, *Carr SFN 27422* (SING); LANGANAN RIVER: *Lohok 9* (K); MESILAU BASIN: 2100–2400 m, *Clemens 29707* (BM).

55.28. **COELOGYNE** Lindl.

Coll. Bot., sub t. 33 (1821?), corr. Lindl., op. cit. sub t. 37 (1826).

Seidenfaden, G. (1975). Orchid genera in Thailand III. Dansk Bot. Ark. 29, 4: 1–94. De Vogel, E. F. (1992). Revisions in Coelogyninae (Orchidaceae) IV. *Coelogyne* section *Tomentosae*. Orchid Monogr. 6: 1–42.

Epiphytic, rarely lithophytic or terrestrial herbs. *Pseudobulbs* ovoid, conical or cylindrical, crowded or remote on rhizome, 1- or 2-leaved at apex. *Leaves* plicate, broad or narrow. *Inflorescences* erect or pendulous, hysteranthous, synanthous, protantherous or heteranthous, 1- to many-flowered. *Flowers* resupinate, small to large and showy, opening simultaneously or one at a time. *Sepals* free, often strongly concave and carinate. *Petals* free, usually narrower than sepals and often rolled backward at apex. *Lip* usually 3-lobed, rather broad and concave at base, rarely saccate; side-lobes erect, broad and widening gradually from base; mid-lobe spreading; disc keeled, keels often extending onto mid-lobe. *Column* long, with an apical wing. *Pollinia* 4.

About 300 species distributed from tropical Asia to the Pacific islands, particularly well represented in mainland Asia, Malaysia and Indonesia.

55.28.1. Coelogyne clemensii Ames & C. Schweinf., Orch. 6: 23 (1920).

a. var. **clemensii** Plate 22A, 22B. Type: MARAI PARAI SPUR, *Clemens 227* (holotype AMES mf; isotypes BM!, K!).

Epiphyte. Hill forest, lower montane forest, upper montane forest, sometimes on ultramafic substrate. Elevation: 900–2200 m.

General distribution: Sabah.

Additional collections. DALLAS: 900 m, *Clemens 27236* (BM); GURULAU SPUR: 1800 m, *Clemens 50408* (K), 1800–2100 m, *50737* (BM, K); KEMBURONGOH: 2200 m, *Carr s.n.* (SING); KIAU VIEW TRAIL: 1800 m, *Gunsalam 7* (K); LUMU-LUMU: 2100 m, *Clemens 26952* (BM), 2000 m, *27167* (BM); MT. NUNGKEK: 1400 m, *Darnton 602* (BM); PARK HEADQUARTERS: 1500 m, *Chan & Lamb AL 53/83* (K); PENIBUKAN: 1500 m, *Clemens 31000* (BM); TENOMPOK: 1500 m, *Clemens 27004* (BM, K), 1500 m, *28482* (BM).

b. var. **angustifolia** Carr, Gardens' Bull. 8: 212 (1935). Type: PENIBUKAN, 1400 m, *Carr 3091, SFN 26453* (holotype SING n.v.; isotype K!).

Epiphyte. Lower montane forest, upper montane forest, mossy forest, on ultramafic substrate. Elevation: 1400–2400 m.

General distribution: Sabah.

Additional collections. MARAI PARAI SPUR: 1400–1600 m, *Bailes & Cribb 833* (K), 1700 m, *Collenette A 42* (BM); PAKA-PAKA CAVE: 2400 m, *Meijer SAN 22073* (K); PENATARAN BASIN: 1500–2100 m, *Clemens 34333* (BM).

c. var. **longiscapa** Ames & C. Schweinf., Orch. 6: 25 (1920). Type: MOUNT KINABALU, *Haslam s.n.* (holotype AMES mf).

Epiphyte. Known only from the type collection.

Endemic to Mount Kinabalu.

55.28.2. Coelogyne compressicaulis Ames & C. Schweinf., Orch. 6: 25 (1920). Plate 23A. Type: MOUNT KINABALU, *Clemens s.n.* (holotype AMES mf).

Epiphyte. Low mossy and xerophyllous scrub forest, upper montane forest, on extreme ultramafic substrate. Elevation: 1200–2300 m.

Endemic to Mount Kinabalu.

Additional collections. KEMBURONGOH: 2200 m, *Carr 3270, SFN 27501* (BM), 2100 m, *3499, SFN 27466* (SING); KINATEKI RIVER: 1200 m, *Collenette A 105* (BM); LUBANG: 1500 m, *Clemens 122* (AMES), *217* (AMES), *293* (AMES); MARAI PARAI: 2100 m, *Carr s.n.* (SING); *Clemens 35192* (BM), 1800 m, *35194* (BM); MARAI PARAI SPUR: 2100 m, *Carr 3566, SFN 27860* (K, SING p.p.); MOUNT KINABALU: *Haslam s.n.* (AMES, BM, K); PENATARAN BASIN: 1800 m, *Clemens s.n.* (BM); PIG HILL: 2000–2300 m, *Beaman 9891* (K).

55.28.3. Coelogyne craticulaelabris Carr, Gardens' Bull. 8: 214 (1935). Fig. 13, Plate 23B. Type: LUMU-LUMU, 1700 m, *Carr 2665, SFN 27965* (holotype SING n.v.).

Pendulous epiphyte. Hill forest?, lower montane forest, upper montane forest. Elevation: 900–2400 m.

General distribution: Sabah, Brunei, Kalimantan, Sarawak.

Additional collections. BAMBANGAN RIVER: 1500 m, *RSNB 4453* (K), 1500 m, *RSNB 4556* (K); DALLAS: 900 m, *Clemens s.n.* (BM); GURULAU SPUR: *Clemens 50416* (BM), 1700 m, *50699* (BM, K); KEMBURONGOH/LUMU-LUMU: 1800–2400 m, *Clemens 27759* (BM); LUMU-LUMU: 2100 m, *Clemens s.n.* (BM); MT. NUNGKEK: 900–1200 m, *Clemens 32780* (BM).

55.28.4. Coelogyne cumingii Lindl., Bot. Reg. 26: misc. 76, no. 178 (1840). Plate 24A.

Epiphyte. Hill forest on ultramafic substrate. Elevation: 900–1000 m.

General distribution: Sabah, Peninsular Malaysia, Singapore, Thailand, Laos, Sumatra, Riau Archipelago.

Collection. MELANGKAP TOMIS: 900–1000 m, *Beaman 8993* (K).

55.28.5. Coelogyne cuprea H. Wendl. & Kraenzl., Gard. Chron., ser. 3, 11: 619 (1892).

a. var. **cuprea**

Epiphyte. Lower montane forest. Elevation: 1500 m.

Collection. TENOMPOK: 1500 m, *Clemens s.n.* (BM).

b. var. **planiscapa** J. J. Wood & C. L. Chan, Lindleyana 5: 84, f. 3 (1990).

Epiphyte. Lower montane forest, sometimes on ultramafic substrate. Elevation: 1200–1800 m.

General distribution: Sabah.

Collections. GOLF COURSE SITE: 1600 m, *Bailes & Cribb 718* (K), 1700–1800 m, *Beaman 7218* (K); GURULAU SPUR: 1800 m, *Clemens 50411* (BM, K); MAHANDEI RIVER HEAD: 1200 m, *Carr 3075, SFN 26454* (K, SING); MARAI PARAI SPUR: 1500–1600 m, *Bailes & Cribb 820* (K); PINOSUK PLATEAU: 1600 m, *Bailes & Cribb 534* (K); TENOMPOK: 1500 m, *Carr SFN 27234* (SING), 1200 m, *Clemens s.n.* (BM, BM, BM).

Fig. 13. Coelogyne craticulaelabris. A, habit; **B**, floral bract and flower (side view); **C**, dorsal sepal; **D**, lateral sepal; **E**, petal; **F**, lip (front view); **G**, pedicel-with-ovary, column and lip (side view); **H**, column with anther-cap (front view); **J**, anther-cap (back view); **K**, pollinia. **A** from *Chew & Corner 4556*; **B–K** from *Wood 770*, Mt. Alab, Sabah. Scale: single bar = 1 mm; double bar = 1 cm. Drawn by Eleanor Catherine.

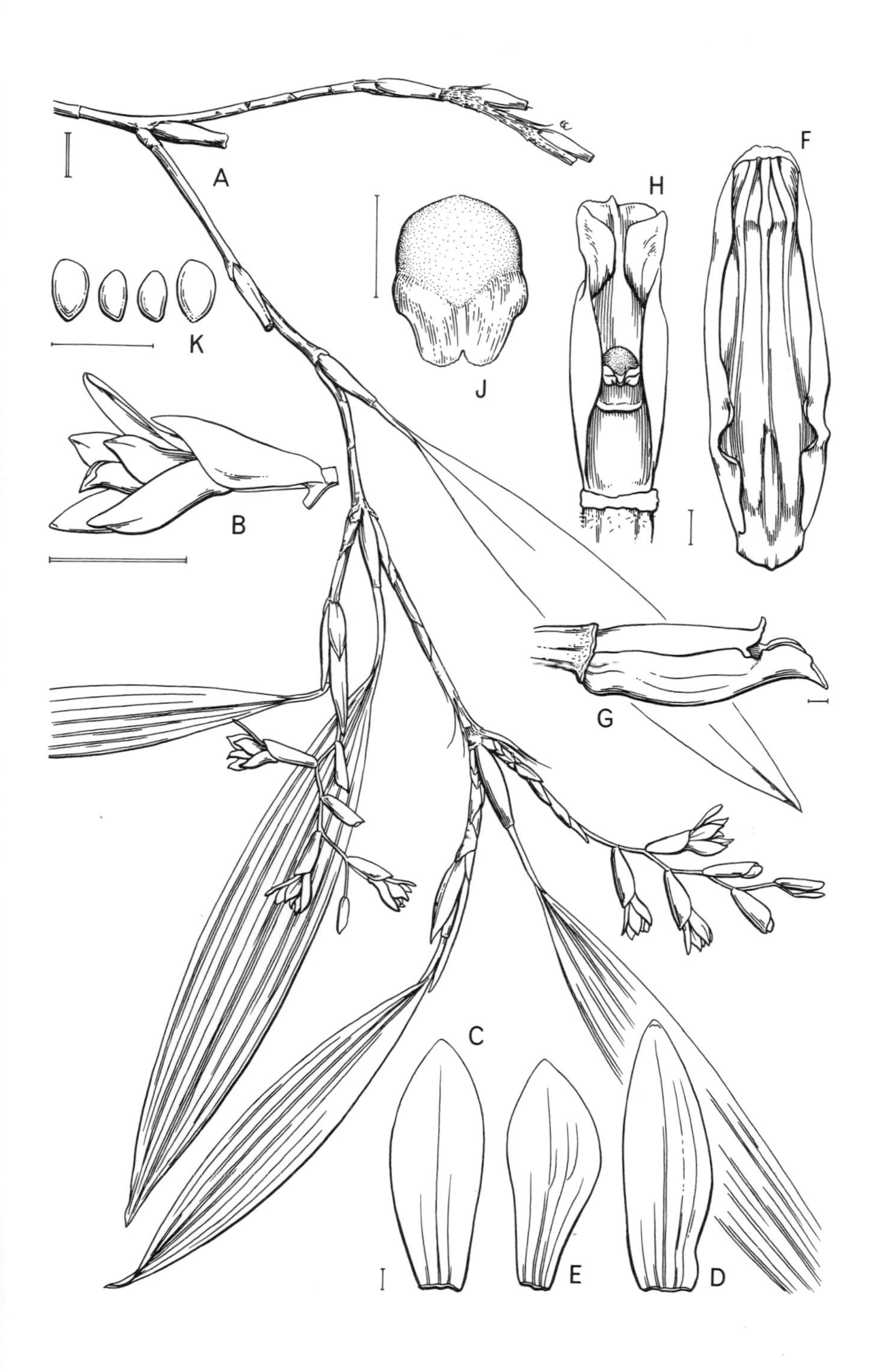
A
B
C
D
E
F
G
H
J
K

55.28.6. Coelogyne dayana Rchb. f., Gard. Chron. ser. 2, 21: 826 (1884).

Coelogyne pulverula Teijsm. & Binn., Natuurk. Tijdschr. Ned.-Ind. 24: 306 (1862).

Epiphyte or lithophyte. Lowlands, hill forest, sometimes on ultramafic substrate. Elevation: 300–1100 m.

General distribution: Sabah, Sarawak, Peninsular Malaysia, Thailand?, Sumatra, Java.

Collections. BUNDU TUHAN: 1100 m, *Carr 3714* (SING); DALLAS: 900 m, *Clemens 26044* (BM, K, SING), 900 m, *26521* (BM); KAUNG: 400 m, *Carr 3005, SFN 26261* (K, SING), 300 m, *Gibbs 4339* (BM); KIAU: 1100 m, *Clemens 40451* (BM, K); KINATEKI RIVER: 500 m, *Collenette A 128* (BM); MOUNT KINABALU: *Haslam s.n.* (AMES, K); PENATARAN RIVER: 500 m, *Beaman 8854* (K); PINAWANTAI: *Shea & Aban SAN 76860* (K).

55.28.7. Coelogyne exalata Ridl., J. Straits Branch Roy. Asiat. Soc. 49: 29 (1908).

Epiphyte. Hill forest, lower montane forest, upper montane forest. Elevation: 900–2700 m.

General distribution: Sabah, Brunei, Sarawak.

Collections. DALLAS: 900 m, *Clemens s.n.* (BM); EAST MESILAU/MENTEKI RIVERS: 1700–2000 m, *Beaman 9601* (K); 1700 m, *Wood 824* (K); PAKA-PAKA CAVE: 2700 m, *Gibbs 4215* (BM).

55.28.8. Coelogyne foerstermannii Rchb. f., Gard. Chron. ser. 2, 26: 262 (1886).

Epiphyte on tall dipterocarps and other trees. Lowland dipterocarp forest. Sight record of A. Lamb; cf. Lamb & Chan (1978: 238); no specimens seen.

General distribution: Sabah, Kalimantan, Sarawak, Peninsular Malaysia, Sumatra.

55.28.9. Coelogyne genuflexa Ames & C. Schweinf., Orch. 6: 28 (1920). Type: MARAI PARAI SPUR, *Clemens 251* (holotype AMES mf).

Epiphyte. Upper montane mossy forest on ultramafic substrate.

Endemic to Mount Kinabalu.

Additional collections. PAKA-PAKA CAVE: *Clemens 200A* (AMES), *207* (AMES).

55.28.10. Coelogyne hirtella J. J. Sm., Bull. Jard. Bot. Buit., ser. 3, 11: 105 (1931). Plate 25A.

Epiphyte or lithophyte. Lower montane forest, sometimes on ultramafic substrate. Elevation: 1200–2000 m.

General distribution: Sabah, Brunei, Kalimantan, Sarawak.

Collections. BAMBANGAN RIVER: 1600 m, *RSNB 4951* (K); EAST MESILAU/MENTEKI RIVERS: 1700 m, *Beaman 9387* (K); KINATEKI RIVER: 1200 m, *Collenette A 106* (BM); KUNDASANG: 1400 m, *Collenette 592* (K); MAMUT COPPER MINE: 1600–1700 m, *Beaman 9941* (K); MARAI PARAI: 1400 m, *Clemens 32359* (BM); MARAI PARAI SPUR: 1400–1600 m, *Bailes & Cribb 822* (K); MESILAU CAMP: *Poore H 424* (K); MESILAU CAVE: 1600 m, *Bailes & Cribb 697* (K), 1900 m, *Collenette 596* (K); MESILAU CAVE TRAIL: *Wood 852* (K); MT. NUNGKEK: 1200 m, *Clemens 32547* (BM, K); NUMERUK RIDGE: 1400 m, *Clemens 40094* (BM); PENIBUKAN: 1400 m, *Carr 3122, SFN 26876* (SING); PIG HILL: 1700–2000 m, *Sutton 6* (K); TENOMPOK: 1500 m, *Clemens 29478* (BM, K), 1500 m, *29638* (BM, K); WEST MESILAU RIVER: 1600 m, *Beaman 9465* (K).

55.28.11. Coelogyne kinabaluensis Ames & C. Schweinf., Orch. 6: 30 (1920). Type: MARAI PARAI SPUR, *Clemens 229* (holotype AMES mf).

Epiphyte. Lower montane forest, sometimes on ultramafic substrate. Elevation: 900–1800 m.

General distribution: Sabah.

Additional collections. BAMBANGAN RIVER: 1600 m, *RSNB 4972* (K); DALLAS: 900 m, *Clemens 26518* (BM, K); EAST MESILAU/MENTEKI RIVERS: 1700 m, *Beaman 8761* (K); GOLF COURSE SITE: 1700–1800 m, *Beaman 7221* (K), 1700–1800 m, *7222* (K); KIAU: *Clemens 29* (AMES), 900 m, *161* (AMES); LUBANG: 1500 m, *Clemens 105* (AMES); LUMU-LUMU: 1500 m, *Carr 3328* (BM, K, SING); MAMUT COPPER MINE: 1600–1700 m, *Beaman 9948* (K), 1400–1500 m, *10349* (K); MARAI PARAI: 1200 m, *Clemens s.n.* (BM), 1400 m, *32355* (BM); MARAI PARAI SPUR: *Clemens 236* (AMES), *369* (AMES); MARAI PARAI/KIAU TRAIL: 1500 m, *Bailes & Cribb 877* (K); MOUNT KINABALU: *Clemens s.n.* (AMES), *Haslam s.n.* (AMES); PENIBUKAN: 1200–1500 m, *Clemens s.n.* (BM), 1200 m, *30591* (BM), 1200 m, *30780* (BM), 1200 m, *30786* (BM), 1500 m, *30994* (BM), 1200 m, *31047* (BM), 1200 m, *50113* (BM, K); TAHUBANG FALLS: 1200 m, *Clemens 30706* (BM), 1200 m, *40335* (BM).

55.28.12. Coelogyne latiloba de Vogel, Orchid Monogr. 6: 20 (1992). Type: MARAI PARAI, 1600 m, *Collenette A 38* (holotype BM!).

Epiphyte. Lower montane forest on ultramafic substrate. Elevation: 1600 m. Known only from the type.

Endemic to Mount Kinabalu.

55.28.13. Coelogyne longibulbosa Ames & C. Schweinf., Orch. 6: 33 (1920). Type: KIAU, 900 m, *Clemens 79* (holotype AMES mf; isotypes BM!, K!).

Epiphyte. Hill forest, lower montane forest, sometimes on ultramafic substrate. Elevation: 900–1500 m.

General distribution: Sabah, Brunei, Sarawak.

Additional collections. KIAU: 900 m, *Clemens 78* (AMES), 900 m, *173* (BM, K, SING), 900 m, *175* (AMES); MAHANDEI RIVER: 1100 m, *Carr 3479* (SING); MAMUT RIVER: 1400 m, *Collenette 1032* (K); MOUNT KINABALU: *Clemens s.n.* (BM, K); PENIBUKAN: 1200 m, *Clemens 30798* (BM), 1200–1500 m, *31627* (BM), 1200–1500 m, *35191* (BM, K), 1400 m, *40254* (BM), 1100 m, *Kanis & Meijer SAN 51454* (K); TAHUBANG RIVER: *Clemens 31341* (BM), 1200–1500 m, *31655* (BM); 1100 m, *40224* (BM), 1200 m, *40332* (BM).

55.28.14. Coelogyne monilirachis Carr, Gardens' Bull. 8: 206 (1935). Plate 25B. Type: TENOMPOK, 1500 m, *Carr 3366, SFN 27230* (holotype SING!; isotypes AMES mf, K!).

Epiphyte. Lower montane forest. Elevation: 1200–1600 m.

General distribution: Sabah, Sarawak.

Additional collections. LIWAGU RIVER TRAIL: 1600 m, *Sands 3882* (K); MARAI PARAI/ KILEMBUN RIVER: *Bailes & Cribb 867* (K); MENTEKI RIVER: 1600 m, *Beaman 10772* (K); MESILAU CAMP: 1500 m, *RSNB 6007* (K, SING); TENOMPOK: 1200 m, *Clemens 26127* (BM, K), 1500 m, *27166* (BM, K), 1500 m, *28294* (BM, K), 1500 m, *28316* (BM, K), 1500 m, *28454* (BM, K), 1500 m, *28951* (K); WEST MESILAU RIVER: 1600 m, *Beaman 8997* (K).

55.28.15. Coelogyne moultonii J. J. Sm., Bull. Jard. Bot. Buit., ser. 2, 111: 54 (1912). Plate 25C, 25D.

Epiphytic or terrestrial. Lower montane forest. Elevation: 1400–2200 m.

General distribution: Sabah, Sarawak.

Collections. GOLF COURSE SITE: 1600 m, *Bailes & Cribb 703* (K); KILEMBUN BASIN: 1400 m, *Clemens 33176* (BM); MARAI PARAI: 2100 m, *Clemens 32286* (BM); MESILAU: 1500 m, *RSNB 4237* (K); MESILAU CAVE: 1900–2200 m, *Beaman 9573* (K); MESILAU CAVE TRAIL: *Wood 850* (K); MESILAU RIVER: 1500 m, *RSNB 4151* (K), 2100 m, *Clemens 51736* (BM); MOUNT KINABALU: 1800 m, *Smith & Everard 142* (K); TENOMPOK: 1500 m, *Carr 3262, SFN 27112* (K, SING), 1500 m, *Clemens 29296* (BM, K), 1500 m, *30157* (K), 1500 m, *30158A* (K); WEST MESILAU RIVER: 1600 m, *Beaman 9467* (K).

55.28.16. Coelogyne obtusifolia Carr, Gardens' Bull. 8: 205 (1935). Type: BUNDU TUHAN, 600 m, *Carr 3149, SFN 27897* (syntype SING!; isosyntypes AMES mf, K!); TENOMPOK, 1500 m, *Clemens 26125* (syntype BM? n.v.; isosyntype K!).

Epiphyte. Hill forest, lower montane forest. Elevation: 600–1500 m.

General distribution: Sabah.

Additional collection. MESILAU TRAIL: *Chow & Leopold SAN 74529* (K).

55.28.17. Coelogyne pandurata Lindl., Gard. Chron. 1853: 791 (1853). Plate 26A.

Epiphytic or terrestrial. Hill forest, steep rocky ridge, on ultramafic substrate. Elevation: 800 m.

General distribution: Sabah, Sarawak.

Collections. HEMPUEN HILL: *Lamb AL 312/85* (K); LOHAN RIVER: 800 m, *Lamb K 57* (K).

55.28.18. Coelogyne papillosa Ridl. in Stapf, Trans. Linn. Soc. Bot. 4: 238 (1894). Fig. 14, Plate 26B, 26C. Type: MOUNT KINABALU, 3200 m, *Haviland 1098* (holotype SING!; isotype K!).

Terrestrial, lithophyte. Upper montane forest, among rocks, in the open or in scrub, on either granitic or ultramafic substrate. Elevation: 1800–3700 m.

Endemic to Mount Kinabalu.

Additional collections. GURULAU SPUR: 2400–2700 m, *Clemens 50662* (K); KEMBURONGOH/ PAKA-PAKA CAVE: 2900 m, *Sinclair et al. 9195* (K, SING); LUMU-LUMU: 1800 m, *Clemens 27872* (BM, K); MARAI PARAI: 2400–3000 m, *Clemens 32941* (K); MESILAU: 2100 m, *Clemens 29293* (BM, K); MOUNT KINABALU: 2400–2700 m, *Gibbs 4261* (K), *Haslam s.n.* (AMES), 3200 m, *Richards 111* (K); PAKA-PAKA CAVE: 3400 m, *Carr 3522, SFN 27532* (K, SING), 2900 m, *Clemens 199* (AMES), 3000 m, *27172* (BM), 3700 m, *27862* (BM, K); PANAR LABAN: 3300 m, *Beaman 8107* (K); SUMMIT TRAIL: 2700–3000 m, *Wood 607* (K); UPPER KINABALU: *Clemens 27154* (K).

55.28.19. Coelogyne planiscapa Carr, Gardens' Bull. 8: 74 (1935).

a. var. **planiscapa**

Epiphyte. Lower montane forest? Noted by Carr (1935) as being less common than var. *grandis*, but no specimens seen by us.

General distribution: Sabah, Sarawak.

b. var. **grandis** Carr, Gardens' Bull. 8: 202 (1935). Plate 27A. Type: PENIBUKAN, 1400 m, *Carr 3120, SFN 27464* (holotype SING!; isotypes AMES mf, K!).

Epiphyte. Lower montane forest, upper montane forest, sometimes on ultramafic substrate. Elevation: 1400–2000 m.

General distribution: Sabah.

Additional collections. EAST MESILAU/MENTEKI RIVERS: 1700–2000 m, *Beaman 9612* (K); KEMBURONGOH: 1900 m, *Carr 3120A* (SING); LUMU-LUMU: 1800 m, *Clemens s.n.* (BM); MARAI PARAI SPUR: 1400–1600 m, *Bailes & Cribb 824* (K); UPPER KINABALU: *Clemens 30015* (K).

55.28.20. Coelogyne plicatissima Ames & C. Schweinf., Orch. 6: 35 (1920). Fig. 15, Plate 27B. Type: PAKA-PAKA CAVE, 1500–3000 m, *Clemens 204* (holotype AMES mf; isotypes BM!, K!).

Epiphyte. Lower montane forest, upper montane forest and scrub, mossy ericaceous forest, frequently on ultramafic substrate. Elevation: 1200–3400 m.

General distribution: Sabah.

Additional collections. EAST MESILAU/MENTEKI RIVERS: 1700 m, *Wood 833* (K); GURULAU SPUR: 2400–2700 m, *Clemens 50655* (BM, K), 2400–2700 m, *50680* (BM); KEMBURONGOH: 2200 m, *Carr 3516, SFN 27504* (K, SING), 2400 m, *3712* (SING), 2400 m, *Clemens s.n.* (BM); KEMBURONGOH/PAKA-PAKA CAVE: 2900 m, *Sinclair et al. 9197* (K); KINATEKI RIVER: 1200–1500 m, *Clemens 31045* (BM); KINATEKI RIVER HEAD: 2700 m, *Clemens 31907* (BM); LUMU-LUMU: 2100 m, *Clemens 27185* (BM, K); MARAI PARAI: 2400 m, *Clemens 31682* (BM, K), 3400 m, *32354* (BM), 1500 m, *33153* (BM, K); MARAI PARAI SPUR: 2000 m, *Collenette A 56* (BM); MOUNT KINABALU: *Haslam s.n.* (AMES); MURU-TURA RIDGE: 1500 m, *Clemens 34359* (BM); PAKA-PAKA CAVE: 2900 m, *Carr s.n.* (SING), 3000 m, *Clemens 29955* (BM, K), 2400 m, *Mikil SAN 46555* (K); PIG HILL: 1700–2000 m, *Sutton 10* (K), 1700–2000 m, *13* (K); SUMMIT TRAIL: 2100 m, *Bailes & Cribb 773* (K), *Wood 609* (K); UPPER KINABALU: *Clemens 30023* (K).

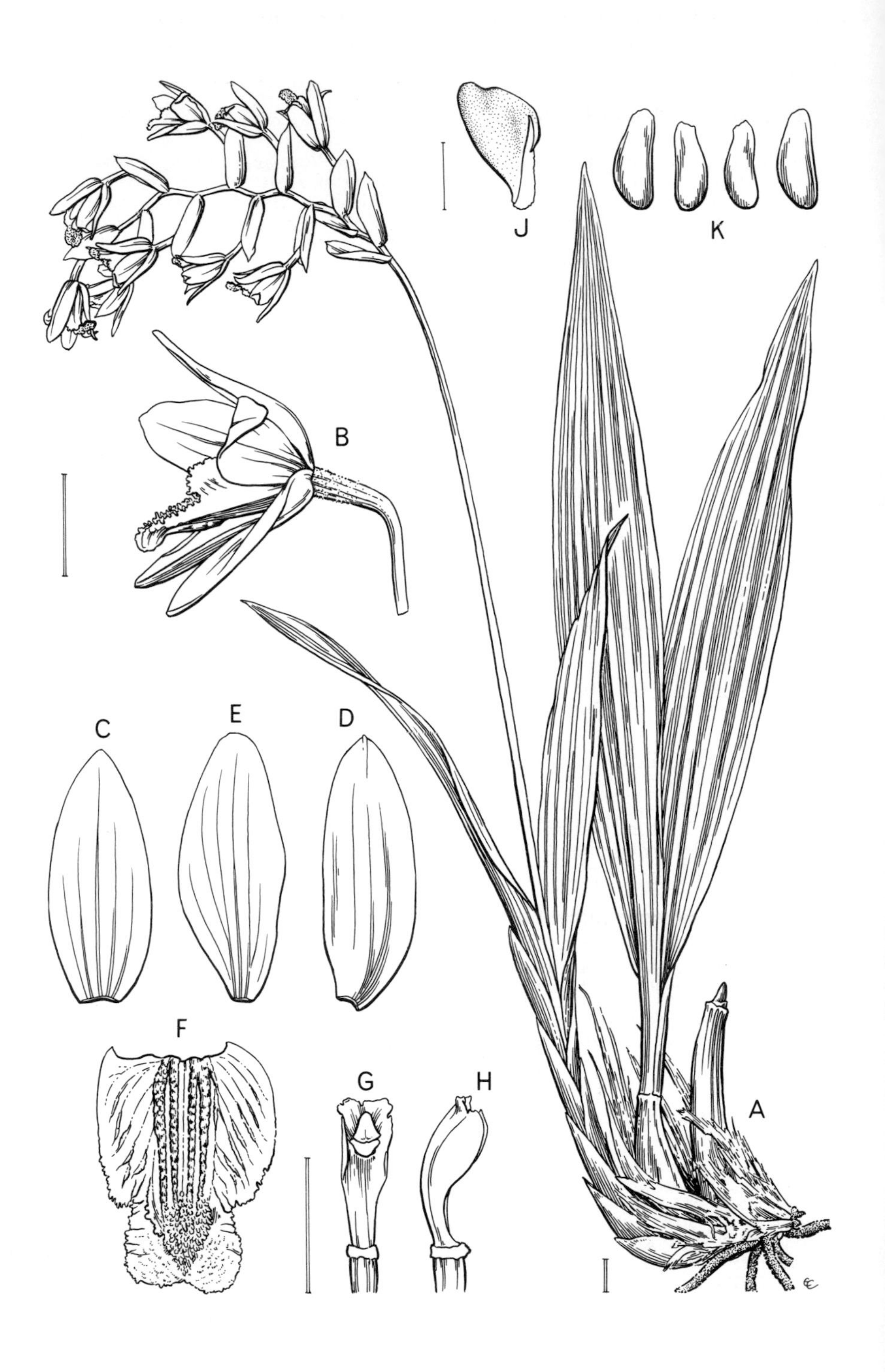
J
K
B
C
E
D
F
G
H
A

55.28.21. Coelogyne prasina Ridl., J. Linn. Soc. Bot. 32: 326 (1896).

Epiphyte. Hill forest.

General distribution: Sabah, Peninsular Malaysia, Sulawesi, Maluku.

Collection. LUGAS HILL (cult. Mountain Garden, Park Headquarters): *Wood 844* (K).

55.28.22. Coelogyne radioferens Ames & C. Schweinf., Orch. 6: 38 (1920). Plate 27C, 27D. Type: PAKA-PAKA CAVE, *Clemens 200* (holotype AMES mf).

Epiphyte, sometimes terrestrial. Lower montane forest, upper montane forest, frequently on ultramafic substrate. Elevation: 1400–2700 m.

General distribution: Sabah, Brunei, Kalimantan, Sarawak.

Additional collections. BAMBANGAN RIVER: 1500 m, *RSNB 4485* (K); GURULAU SPUR: 2400–2700 m, *Clemens 50664* (BM, K), 2100–2700 m, *50767* (BM); KEMBURONGOH: 2100 m, *Carr 3496, SFN 27458* (K, SING); MARAI PARAI: 1500 m, *Carr 3121, SFN 27458* (SING), 1500 m, *Clemens 32230* (BM), 1500 m, *Collenette A 35* (BM), 1700 m, *A 43* (BM), 1800 m, *A 73* (BM); MARAI PARAI SPUR: 1400–1600 m, *Bailes & Cribb 821* (K), 1700 m, *Collenette A 45* (BM), 1800 m, *A 72* (BM); MESILAU CAVE TRAIL: 1700–1900 m, *Beaman 7990* (K); MOUNT KINABALU: 2400–2700 m, *Burbidge s.n.* (BM), *Haslam s.n.* (AMES); SUMMIT TRAIL: 2400 m, *Wood 612* (K); TAHUBANG RIVER HEAD: 2100 m, *Clemens s.n.* (BM).

55.28.23. Coelogyne reflexa J. J. Wood & C. L. Chan, Lindleyana 5: 87, f. 5 (1990). Fig. 16.

Epiphyte. Lower montane forest, sometimes on ultramafic substrate. Elevation: 1200–1900 m.

General distribution: Sabah.

Collections. EAST MESILAU RIVER: 1700 m, *Beaman 9047* (K); KEMBURONGOH: *Cribb s.n.* (K, photo); KINATEKI RIVER: 1200–1500 m, *Clemens 31455* (BM); MARAI PARAI SPUR, 2100 m, *Carr 3566, SFN 27860* (SING, p.p.); MESILAU CAVE: 1900 m, *Collenette 629* (K).

55.28.24. Coelogyne rhabdobulbon Schltr., Notizbl. Bot. Gart. Berlin-Dahlem 8: 15 (1921). Plate 28.

Coelogyne pulverula auct. non Teijsm. & Binn.

Epiphyte. Hill forest, lower montane forest. Elevation: 800–1500 m.

Fig. 14. Coelogyne papillosa. A, habit; **B**, flower (side view); **C**, dorsal sepal; **D**, lateral sepal; **E**, petal; **F**, lip (front view); **G**, column (front view); **H**, column (side view); **J**, anther-cap (side view), **K**, pollinia. **A**, from *Carr 3522*, **B–K** from *Wood 607*. Scale: single bar = 1 mm; double bar = 1 cm. Drawn by Eleanor Catherine.

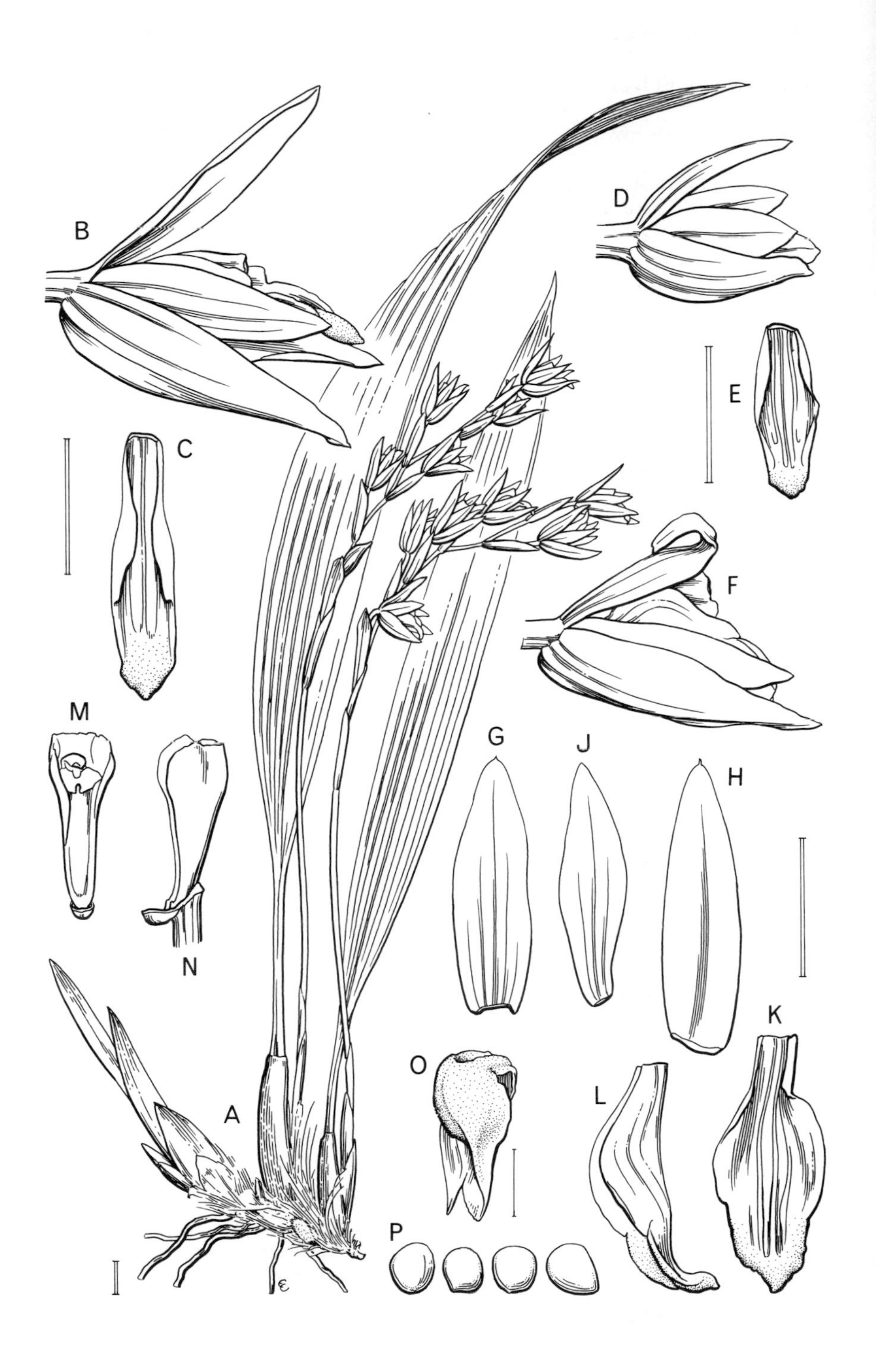
B
D
E
C
F
M
G
J
H
N
K
O
A
L
P

General distribution: Sabah, Sarawak.

Collections. PARK HEADQUARTERS: 1500 m, *Wood 623* (K); PINOSUK: 800 m, *Carr 3678* (SING).

55.28.25. Coelogyne rigidiformis Ames & C. Schweinf., Orch. 6: 40 (1920). Type: KIAU, 900 m, *Clemens 71* (holotype AMES mf).

Epiphyte. Hill forest, lower montane forest. Elevation: 900–1600 m.

Endemic to Mount Kinabalu.

Additional collection. TENOMPOK: 1600 m, *Carr 3244, SFN 26988* (BM, K, SING).

55.28.26. Coelogyne rochussenii De Vriese, Ill. Orch. Ind. Orient., t. 2, t. 11, f. 6 (1854). Plate 29.

Epiphyte. Lowlands, hill forest. Elevation: 400–900 m.

General distribution: Sabah, Kalimantan, Sarawak, Peninsular Malaysia, Thailand, Sumatra, Java, Sulawesi, Maluku, Philippines.

Collections. DALLAS: 900 m, *Clemens 27616* (BM, SING); KAUNG: 400 m, *Carr 3008, SFN 26264* (SING), 400 m, *Darnton 320* (BM); KIAU: 900 m, *Clemens 174* (AMES).

55.28.27. Coelogyne rupicola Carr, Gardens' Bull. 8: 210 (1935). Plate 30A. Type: KEMBURONGOH, 2700 m, *Carr 3552, SFN 27793* (holotype SING!; isotype K!).

Lithophyte. Lower montane forest, upper montane scrub, among rocks on ultramafic substrate. Elevation: 1500–2700 m.

Endemic to Mount Kinabalu.

Additional collections. GURULAU SPUR: 2400 m, *Clemens 50362* (BM, K), 2400 m, *50953* (BM, K); MARAI PARAI: 1700 m, *Clemens 40229* (BM), 1500 m, *Collenette A 22A* (BM), *A 22B* (BM), 1500 m, *A 23* (BM); MARAI PARAI SPUR: 1700 m, *Clemens 34374* (BM).

55.28.28. Coelogyne sanderiana Rchb. f., Gard. Chron. ser. 3, 1: 764 (1887).

Epiphytic or terrestrial. Secondary hill forest. Elevation: 800 m.

Fig. 15. Coelogyne plicatissima. A, habit; **B**, flower (side view); **C**, lip (front view); **D**, flower (side view); **E**, lip (front view); **F**, flower (side view); **G**, dorsal sepal; **H**, lateral sepal; **J**, petal; **K**, lip (front view); **L**, lip (side view); **M**, column (front view); **N**, column (side view); **O**, anther-cap (side view); **P**, pollinia. **A–C** from *Wood 900*, Mt. Trus Madi, Sabah; **D–E** from *Bailes & Cribb 773*; **F–M** from *Wood 824*; **O–P** from *Wood 900*. Scale: single bar = 1 mm; double bar = 1 cm. Drawn by Eleanor Catherine.

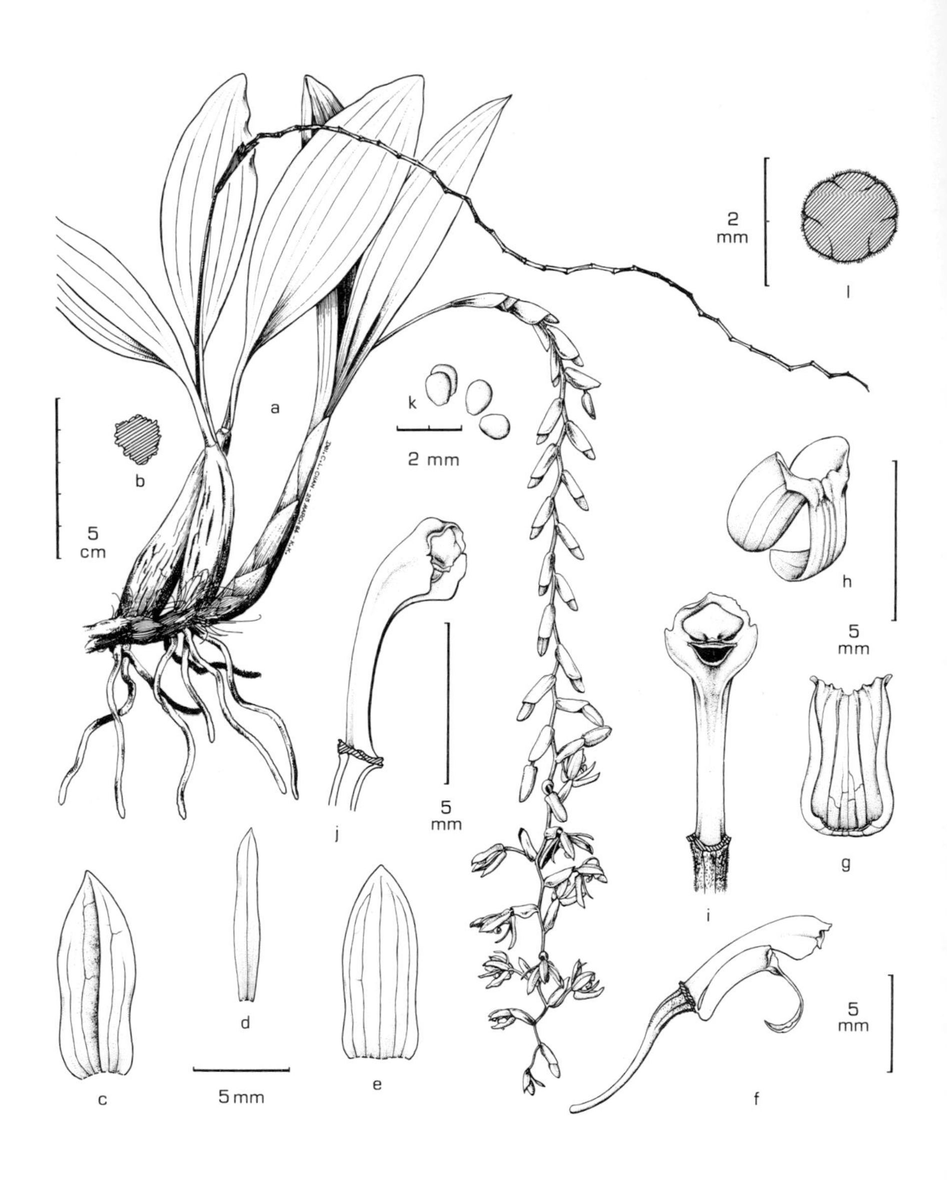

Fig. 16. Coelogyne reflexa **a**, habit; **b**, pseudobulb (transverse section); **c**, lateral sepal; **d**, petal; **e**, dorsal sepal; **f**, pedicel-with-ovary, column and lip (side view); **g**, lip hypochile; **h**, whole lip (oblique view); **i**, column (front view); **j**, column (side view); **k**, pollinia; **l**, ovary (transverse section). All from *Beaman 9047*. Drawn by Chan Chew Lun.

General distribution: Sabah, Sarawak, Sumatra.

Collection. KIAU/KAUNG: 800 m, *Gibbs 3971* (K).

55.28.29. Coelogyne septemcostata J. J. Sm., Icon. Bogor. 2: 23, t. 106A (1903). Plate 30B.

Epiphyte. Mossy forest on ultramafic substrate. Elevation: 800–900 m.

General distribution: Sabah, Brunei, Sarawak, Sumatra.

Collections. PENATARAN RIVER: *Lamb s.n.* (K photo), 800–900 m, *SAN 93361* (K).

55.28.30. Coelogyne swaniana Rolfe, Kew Bull. 1894: 183 (1894).

Epiphyte. Hill forest, sometimes on ultramafic substrate. Elevation: 700–1000 m.

General distribution: Sabah, Sarawak, Peninsular Malaysia, Sumatra.

Collections. LOHAN RIVER: 800 m, *Bailes & Cribb 664* (K), 700–900 m, *Beaman 9253* (K); SAYAP: 800–1000 m, *Beaman 9781* (K).

55.28.31. Coelogyne tenompokensis Carr, Gardens' Bull. 8: 203 (1935). Plate 30C. Type: TENOMPOK, 1600 m, *Carr 3270, SFN 27501* (syntype SING!; isosyntypes AMES mf, K!), 1500 m, *Clemens 27191* (syntype BM!), 1500 m, *29126* (syntype BM!).

Epiphyte. Lower montane forest, sometimes on ultramafic substrate. Elevation: 1200–2100 m.

General distribution: Sabah, Sarawak.

Additional collections. EAST MESILAU RIVER: 1500–1600 m, *Bailes & Cribb 510* (K), 1500–1600 m, *512* (K); GOLF COURSE SITE: 1600 m, *Bailes & Cribb 729* (K), 1700–1800 m, *Beaman 7231* (K); KILEMBUN BASIN: 1400 m, *Clemens 34399* (BM), 1200 m, *34487* (BM); KINATEKI RIVER HEAD: 1200 m, *Clemens 31040* (BM); MARAI PARAI: 1700 m, *Clemens s.n.* (BM), 1500 m, *32248* (BM); MESILAU CAVE TRAIL: 1700–1900 m, *Beaman 8002* (K), *Wood 847* (K); PORING ORCHID GARDEN: *Lohok 19* (K); SUMMIT TRAIL: 2100 m, *Bailes & Cribb 772* (K); TINEKUK FALLS: 2000 m, *Clemens 40910* (BM, K).

55.28.32. Coelogyne aff. **tenompokensis** Carr, Gardens' Bull. 8: 203 (1935).

Epiphyte. Lower montane forest. Elevation: 1200–1800 m.

Collections. BAMBANGAN RIVER: 1200 m, *Clemens 35193* (BM); MESILAU CAVE TRAIL: 1800 m, *Beaman 7482* (K); TINEKUK FALLS: 1500 m, *Clemens 40820* (BM).

55.28.33. Coelogyne venusta Rolfe, Gard. Chron. ser. 3, 35: 259 (1904). Plate 31A.

Epiphyte. Hill forest, lower montane forest, frequently on ultramafic substrate. Elevation: 900–2100 m.

General distribution: Sabah, Brunei, Sarawak.

Collections. BUNDU TUHAN: 1400 m, *Darnton 228* (BM); DALLAS: 900 m, *Clemens 27418* (BM); HEMPUEN HILL: 1200 m, *Madani SAN 89543* (K); KEMBURONGOH/LUMU-LUMU: 2100 m, *Clemens 27154* (BM); KIAU: 900 m, *Clemens 28* (AMES), 900 m, *30* (AMES, BM), 900 m, *131* (AMES), 900 m, *142* (AMES, BM, K), *315* (AMES), *357* (AMES); LUBANG: *Clemens 300* (AMES); MAHANDEI RIVER: 1100 m, *Carr 3111, SFN 26530* (K, SING); MARAI PARAI SPUR: *Clemens 294* (AMES), *299* (AMES); PENIBUKAN: 1200 m, *Clemens 32061* (BM, K), 1200 m, *40630* (BM, K), 1200 m, *40854* (BM, K), 1200 m, *40854a* (BM); TENOMPOK: 1500 m, *Clemens s.n.* (BM, SING), 1500 m, *27002* (BM, K), 1500 m, *28417* (BM, K), 1500 m, *29407* (BM, K), 1500 m, *Collenette 591* (K); TENOMPOK ORCHID GARDEN: *Clemens 50171* (BM); UPPER KINABALU: *Clemens 30088* (K).

55.28.34. Coelogyne vermicularis J. J. Sm., Icon. Bogor. 3: 9, t. 204 (1906).

Epiphyte. Lower montane forest. Elevation: 1400 m.

General distribution: Sabah, Kalimantan.

Collection. PARK HEADQUARTERS: 1400 m, *Price 122* (K).

55.28.35. Coelogyne sp. 1

Epiphyte. Lower montane forest, on ultramafic substrate. Elevation: 1200–1400 m.

Collections. KINATEKI RIVER/MARAI PARAI: 1200 m, *Collenette A 19* (BM); LITTLE MAMUT RIVER: 1400 m, *Collenette 1011* (K).

55.28.36. Coelogyne indet.

Collections. BAMBANGAN RIVER: 1500 m, *RSNB 4453* (K); EAST MESILAU/MENTEKI RIVERS: 1700 m, *Beaman 9389* (K); TENOMPOK: 1500 m, *Carr SFN 27230* (K); TINEKUK FALLS: 2000 m, *Clemens 40901* (K); UPPER KINABALU: *Clemens 30158* (K).

55.29. CORDIGLOTTIS J. J. Sm.

Bull. Jard. Bot. Buit., ser. 3, 5: 95 (1922).

Small monopodial epiphytes. *Stems* very short, usually pendulous. *Leaves* terete or laterally compressed. *Inflorescences* usually with a peduncle and a much shorter rachis, few- to many-flowered. *Flowers* small, lasting only one day; *dorsal sepal* to 1.4 × 0.5 mm; *lip* mobile, slightly saccate, mid-lobe somewhat fleshy, powdery or hairy; *column* short, with a distinct foot; *stipes* broad and spathulate; *viscidium* large; *pollinia* 4, appearing as 2 unequal masses; fruits long and slender as in *Thrixspermum.*

Seven species distributed from S Thailand to Sumatra and Borneo, absent from Java. The centre of distribution is Peninsular Malaysia.

55.29.1. Cordiglottis sp. 1

Epiphyte. Lower montane forest. Elevation: 1200 m. The specimen lacks flowers and cannot readily be determined to species.

Collection. PENIBUKAN: 1200 m, *Clemens 31115* (BM).

55.30. CORYBAS Salisb.

Parad. Lond., t. 83 (1807).

Dransfield, J., Comber, J. B. & Smith, G. (1986). *Corybas* west of Wallace's Line. Kew Bull. 41(3): 575–613.

Dwarf terrestrial, lithophytic, rarely epiphytic herbs arising from small tuberoids. *Stem* erect, very short and slender. *Leaf* solitary, ovate, orbicular or cordate, borne horizontally above the ground, often with red or white veins. *Inflorescence* terminal, erect, 1-flowered. *Flower* sessile or shortly stalked, 1–2 cm high, helmet-shaped, large for size of plant, resupinate; *dorsal sepal* large, hooded; *lateral sepals* and *petals* linear, often filiform; *lip* large, recurved, entire, tubular below, with 2 short basal spurs, embracing column at base, apex rounded, often fimbriate; *column* small, erect; *pollinia* 2, 2-lobed, granular.

About 100 species distributed from India, S China, Taiwan, and SE Asia eastward to New Guinea, Australia, New Zealand and the SW Pacific.

55.30.1. Corybas carinatus (J. J. Sm.) Schltr., Feddes Repert. 19: 19 (1924).

Corysanthes carinata J. J. Sm., Bull. Dép. Agric. Indes Néerl. 13: 8 (1907).

Terrestrial. Lower montane forest; in moss carpets on the lips of podzolized ridges. Elevation: 1500 m.

General distribution: Sabah, Kalimantan, Sarawak, Peninsular Malaysia, Sumatra, Java.

Collection. MAMUT COPPER MINE: 1500 m, *Collenette 1047* (K).

55.30.2. Corybas kinabaluensis Carr, Gardens' Bull. 8: 173 (1935). Type: MAHANDEI RIVER HEAD, 1200 m, *Carr 3067, SFN 26395* (holotype SING!; isotype K!).

Terrestrial. Lower montane forest; in mossy areas on ultramafic substrate. Elevation: 1200 m. Known only from the type.

Endemic to Mount Kinabalu.

55.30.3. Corybas pictus (Blume) Rchb. f., Beitr. Syst. Pflanzenk., 67 (1871). Plate 31B.

Calcearia picta Blume, Bijdr., 418, t. 33 (1825).

Terrestrial. Lower montane forest or hill forest; mossy, well drained banks, bases of mossy tree trunks, infrequently on ultramafics. Elevation: 1000–1700 m.

General distribution: Sabah, Sumatra, Java.

Collections. EASTERN SHOULDER: 1100 m, *Collenette 821* (K); HEMPUEN HILL: 1000 m, *Cribb 89/32* (K); LIWAGU RIVER TRAIL: 1400 m, *Dransfield JD 5709* (K); MAMUT COPPER MINE: 1300 m, *Collenette 1046* (K); MINITINDUK/KINATEKI DIVIDE: 1100 m, *Carr 3205, SFN 26841* (SING); MT. NUNGKEK: 1200 m, *Clemens 32540* (K); PARK HEADQUARTERS: 1500 m, *Wood 616* (K); PENIBUKAN: 1500 m, *Clemens 51723* (BM); POWER STATION: 1700 m, *Dransfield JD 5558* (K).

55.31. CORYMBORKIS Thouars

Hist. Orchid. t. 37, 38 (1822).

Rasmussen, F. (1977). The genus *Corymborkis* Thou., Orchidaceae: A taxonomic revision. Bot. Tidsskr. 71: 161–192.

Terrestrial herbs with subterranean rhizomes and persistent roots. *Stems* erect, slender, woody, unbranched. *Leaves* narrowly elliptic to ovate, plicate, scattered along upper part of stem, sheathing at base, thin, but tough-textured. *Inflorescences* lateral, paniculate, corymbose, many-flowered. *Flowers* white to greenish white; *sepals* and *petals* long and spathulate, of equal length; *lip* equal to sepals and petals, narrow, apex expanded and flabellate, disc with 2 obscure longitudinal keels; *column* straight, long and slender, apex dilated; *rostellum* erect; *pollinia* 2.

A pantropical genus containing 5 species.

55.31.1. Corymborkis veratrifolia (Reinw.) Blume, Fl. Javae, ser. 2, 1, Orch.: 125, pl. 42–43 (1859).

a. var. **veratrifolia**

Hysteria veratrifolia Reinw., Syll. Pl. Nov. 2: 5 (1825–26).

Terrestrial. Hill forest, lower montane forest. Elevation: 900–1500 m.

General distribution: Sabah, Sarawak, Brunei, Kalimantan, widespread from India & Sri Lanka to Samoa.

Collections. DALLAS: 900 m, *Clemens 26800* (BM); LUBANG: 1200 m, *Carr 3218, SFN 27310* (K, SING), *Clemens 123* (AMES, K, SING), *200* (AMES); PENIBUKAN: 1200–1500 m, *Clemens s.n.* (K), 1200 m, *40580* (BM), 1200 m, *50142* (BM).

55.32. CRYPTOSTYLIS R. Br.

Prodr., 317 (1810).

Terrestrial herbs. *Rhizome* short, thick. *Stem* erect, slender. *Leaves* 1-4, ovate, petiolate, with dark reticulate veins, arising from ground level. *Inflorescence* terminal, arising separately on rhizome, lax or subdense, many-flowered. *Flowers* non-resupinate, lip uppermost; *sepals* and *petals* free, sepals usually longer, both very narrow, spreading; *dorsal sepal* revolute; *lip* entire, not spurred, strongly concave at base and enclosing column, tapered at apex; *column* very short, with lateral auricles; *pollinia* 4.

About 20 species distributed from SE Asia to New Guinea, Australia, and the Pacific islands.

55.32.1. Cryptostylis acutata J. J. Sm., Bull. Jard. Bot. Buit., ser. 3, 3: 243 (1921). Plate 31C.

Terrestrial. Hill forest, lower montane forest, sometimes on ultramafic substrate. Elevation: 900–2700 m.

General distribution: Sabah, Sumatra, Java.

Collections. EASTERN SHOULDER, CAMP 1/CAMP 2: 1400 m, *Collenette 752* (K); GURULAU SPUR: 1500 m, *Clemens 50542* (BM); HEMPUEN HILL: 900 m, *Wood 839* (K); KEMBURONGOH: 2100 m, *Clemens s.n.* (BM); LOHAN/MAMUT COPPER MINE: 1500 m, *Collenette 1055* (K); MAMUT COPPER MINE: 1400–1500 m, *Beaman 10346* (K); MESILAU RIVER: 2100–2700 m[?], *Clemens 51427* (BM); MINITINDUK/KINATEKI DIVIDE: 1100 m, *Carr 3208, SFN 27132* (K, SING); MT. NUNGKEK: 1100 m, *Darnton 486* (BM); TENOMPOK: 1200 m, *Clemens s.n.* (BM).

55.32.2. Cryptostylis arachnites (Blume) Hassk., Cat. Hort. Bot. Bog., 48 (1844). Plate 31D.

Zosterostylis arachnites Blume, Bijdr., 419, t. 32 (1825).

Terrestrial. Hill forest, lower montane forest, sometimes on ultramafic substrate. Elevation: 900–2200 m.

General distribution: Sabah, Sarawak, widespread from Sri Lanka & India through Thailand to Indonesia, Philippines & the Pacific islands.

Collections. KEMBURONGOH: 2200 m, *Carr 3508, SFN 27978* (K, SING), 2200 m, *3508A* (K); LUMU-LUMU: 1800 m, *Clemens s.n.* (BM); MELANGKAP TOMIS: 900–1000 m, *Beaman 8992* (K); MESILAU CAVE: 1900–2200 m, *Beaman 9570* (K); MESILAU CAVE TRAIL: 1800 m, *Wood 845* (K); PENIBUKAN: *Clemens s.n.* (BM), *35175* (BM); SUMMIT TRAIL: 1800 m, *Darnton 512* (BM).

55.32.3. Cryptostylis clemensii (Ames & C. Schweinf.) J. J. Sm., Mitt. Inst. Allg. Bot. Hamburg 7: 17 (1927).

Chlorosa clemensii Ames & C. Schweinf., Orch. 6: 9 (1920). Type: MARAI PARAI SPUR, *Clemens 399* (holotype AMES mf).

Cryptostylis tridentata Carr, Gardens' Bull. 8: 174 (1935). Type: PENIBUKAN, 1400 m, *Carr 3089, SFN 26556* (holotype SING!; isotype K!).

Terrestrial. Lower montane forest on ultramafic substrate. Elevation: 1100–1700 m.

General distribution: Sabah.

Additional collections. KINATEKI RIVER/MARAI PARAI: 1100 m, *Collenette A 5* (BM); MARAI PARAI SPUR: 1500–1700 m, *Bailes & Cribb 836* (K); MOUNT KINABALU: 1500 m, *Clemens s.n.* (BM).

55.33. CYMBIDIUM Sw.

Nova Acta Regiae Soc. Sci. Upsal., ser. 2, 6: 70 (1799).

Du Puy, D. & Cribb, P. (1988). The Genus *Cymbidium.* Christopher Helm/Timber Press.

Epiphytic, lithophytic or terrestrial herbs. *Pseudobulbs* short, rarely elongate, occasionally absent and replaced by a slender stem, ovoid to spindle-shaped, often enclosed in sheathing leaf bases. *Leaves* up to 13, distichous, linear-elliptic or narrowly ligulate to elliptic, acuminate to strongly bilobed at apex, articulate to sheaths. *Inflorescences* racemose, dense or lax, erect, porrect or pendulous, 2- to many-flowered; peduncle loosely covered by inflated cymbiform sheaths. *Flowers* resupinate, often large and showy, up to 12 cm in diameter (Bornean species smaller), sometimes fragrant. *Sepals* and *petals* free, spreading or erect. *Lip* 3-lobed, free or fused at base to base of column; side-lobes erect around column; mid-lobe often recurved; disc with usually 2 pubescent parallel ridges. *Column* long, somewhat arcuate. *Pollinia* usually 2, deeply cleft, sometimes 4 in 2 unequal pairs.

Du Puy & Cribb recognise 44 species widely distributed from NW Himalaya to Japan, south through Indochina, Malaysia & Indonesia to the Philippines, New Guinea and Australia.

55.33.1. Cymbidium atropurpureum (Lindl.) Rolfe, Orchid Rev. 11: 190 (1903).

Cymbidium pendulum (Roxb.) Sw. var. *atropurpureum* Lindl., Gard. Chron. 1854: 287 (1854).

Epiphyte. Lowlands. No specimens seen. Noted by Lamb and Chan (1978: 236) as occurring at Poring.

General distribution: Sabah, Peninsular Malaysia, Thailand, Sumatra, Java, Philippines.

55.33.2. Cymbidium bicolor Lindl.

a. subsp. **pubescens** (Lindl.) D. J. Du Puy & P. J. Cribb, The Genus *Cymbidium,* 73 (1988). Plate 32C.

Cymbidium pubescens Lindl., Bot. Reg. 26: misc. 75, no. 177 (1840).

Epiphyte. Lowlands, hill forest. Elevation: 400–900 m.

General distribution: Sabah, Sarawak, Peninsular Malaysia, Sumatra, Java, Sulawesi, Philippines.

Collections. DALLAS: 900 m, *Clemens s.n.* (BM); KAUNG: *Darnton 309* (BM), 400 m, *314* (BM); MOUNT KINABALU: *Haslam s.n.* (AMES).

55.33.3. Cymbidium borneense J. J. Wood, Kew Bull. 38: 69, t. 1 (1983). Plate 32A, 32B.

Terrestrial. Hill forest, with *Gymnostoma sumatrana* on ultramafic substrate (serpentine boulders), in shade. Elevation: 1200 m.

General distribution: Sabah, Sarawak.

55.33.4. Cymbidium chloranthum Lindl., Bot. Reg. 29: misc. 68, no. 102 (1843).

Epiphyte. Lowlands, hill forest. Elevation: 400–900 m.

General distribution: Sabah, Sarawak, Peninsular Malaysia, Sumatra, Java.

Collections. KAUNG: 400 m, *Darnton 310* (BM); KIAU: 900 m, *Carr 3136, SFN 26596* (SING), 900 m, *Clemens 50* (AMES), 900 m, *51* (AMES), 900 m, *40456* (BM, K).

55.33.5. Cymbidium dayanum Rchb. f., Gard. Chron. 1869: 710 (1869). Plate 33A.

Cymbidium angustifolium Ames & C. Schweinf., Orch. 6: 212 (1920). Type: KIAU, 900 m, *Clemens 74* (holotype AMES mf).

Epiphytic or terrestrial. Hill forest, lower montane forest, *Agathis-Podocarpus*-oak forest, often on mossy banks and stumps. Elevation: 900–1700 m.

General distribution: Sabah, China, Taiwan, Japan, Ryukyu Islands, India, Peninsular Malaysia, Thailand, Cambodia, Sumatra, Philippines.

Additional collections. DALLAS: 900 m, *Clemens 26691* (BM), 1200 m, *26771* (BM); KIAU: 900 m, *Clemens 39* (AMES), 900 m, *82* (AMES); KINATEKI RIVER: 1200 m, *Clemens 40126* (BM); MINITINDUK/KINATEKI DIVIDE: 1100 m, *Carr SFN 26799* (AMES, SING); PARK HEADQUARTERS: 1700 m, *Bailes & Cribb 751* (K); TAHUBANG RIVER: *Clemens 40347* (BM); TENOMPOK: 1200 m, *Clemens 26124* (BM), 1500 m, *29453* (BM), 1200 m, *40113* (BM).

55.33.6. Cymbidium elongatum J. J. Wood, D. J. Du Puy & Shim in Du Puy & Cribb, The Genus *Cymbidium*, 103, f. 22.1 (1988). Plate 33B. Type: MOUNT KINABALU, 1600 m, *Collenette A 47* (holotype BM!; isotype K!).

Terrestrial or epiphyte. Lower montane forest on ultramafic substrate; marshy areas among scrub, stunted trees, rattans and sedges. Elevation: 1200–1800 m.

General distribution: Sabah, Sarawak.

55.33.7. Cymbidium ensifolium (L.) Sw.

a. subsp. **haematodes** (Lindl.) D. J. Du Puy & P. J. Cribb, The Genus *Cymbidium*, 161 (1988).

Cymbidium haematodes Lindl., Gen. Sp. Orch., 162 (1833).

Terrestrial. Hill forest, lower montane forest, sometimes on ultramafic substrate. Elevation: 500–1700 m.

General distribution: Sabah, India, Sri Lanka, Peninsular Malaysia, Thailand, Sumatra, Java, New Guinea.

Collections. EAST MESILAU RIVER: 1600–1700 m, *Bailes & Cribb 746* (K); HEMPUEN HILL: *Lamb AL 877/87* (K), 500–600 m, *Wood 602* (K); MESILAU RIVER: 900 m, *Lamb SAN 91507* (K); WEST MESILAU RIVER: 1500 m, *Bailes & Cribb 528* (K).

55.33.8. Cymbidium finlaysonianum Lindl., Gen. Sp. Orch., 164 (1833).

Epiphyte. Hill forest. Elevation: 1200 m.

General distribution: Sabah, Sarawak, Peninsular Malaysia, Thailand, Cambodia, Vietnam, Sumatra, Java, Sulawesi, Philippines.

Collection. BUNDU TUHAN: 1200 m, *Carr SFN 27414* (SING).

55.33.9. Cymbidium lancifolium Hook., Exot. Fl. 1: t. 51 (1823).

Epiphytic or rarely terrestrial. Hill forest, lower montane forest. Elevation: 900–1500 m.

General distribution: Sabah, widespread from Nepal and India to Japan and Taiwan through Indochina, Peninsular Malaysia, Sumatra, Java & the Philippines to New Guinea.

Collections. DALLAS: 900 m, *Clemens 26813* (BM); TENOMPOK: 1500 m, *Clemens s.n.* (BM).

55.34. CYRTOSIA Blume

Bijdr., 396 (1825).

Leafless saprophytic herbs. *Rhizome* stout, erect, bearing several fleshy roots. *Stems* several from each rhizome, simple or branched. *Inflorescence* terminal. *Flowers* non-resupinate, not fully opening; *sepals* and *petals* connivent; *lip* spurless, entire, erect, fused with base of column; *column* slightly curved, shortly clawed below, flabellate above, foot absent; *anther-cap* terminal; *pollinia* 2; *fruits* succulent, indehiscent, *seeds* exalate.

Five species distributed from Sri Lanka, Thailand and Indochina to Malaysia and Indonesia, with one species native to Japan.

55.34.1. Cyrtosia javanica Blume, Bijdr., 396, t. 6 (1825). Plate 33C.

Saprophyte. Lower montane forest. Elevation: 1100–2700 m.

General distribution: Sabah, Java, Sulawesi.

Collections. BUNDU TUHAN: 1200 m, *Carr 3199, SFN 26778A* (K, SING); DALLAS/TENOMPOK: 1400 m, *Clemens 27024* (BM); KEGIITAN AGAYE HEAD: 1100 m, *Carr 3199, SFN 26778* (SING); MESILAU RIVER: 2100–2700 m, *Clemens 51425* (BM, K); PENATARAN BASIN: 1200 m, *Clemens 34040* (BM); TENOMPOK: 1500 m, *Clemens 27024* (K), 1200 m, *29639* (BM); WEST MESILAU RIVER: 1600–1700 m, *Beaman 7426* (K).

55.35. CYSTORCHIS Blume

Fl. Javae ser. 2, 1, Orch.: 87 (1859).

Terrestrial herbs. *Stem* with a few leaves or, in saprophytic species, replaced by brown scale-leaves. *Leaves* green, purplish, occasionally variegated. *Inflorescence* lax or rather dense, few- to many-flowered. *Flowers* small, resupinate; *dorsal sepal* and *petals* connivent, forming a hood; *lateral sepals* partially or entirely enclosing base of lip; *lip* divided into a hypochile and epichile; *hypochile* swollen at base into twin lateral sacs, each usually containing a globular sessile gland; *epichile* with fleshy involute margins, forming a tube; *column* very short; *rostellum* rather large (absent in one species); *stigma* on front of column, large, rounded; *pollinia* 2.

About 20 species distributed in China, India, Malaysia, Indonesia, the Philippines, New Guinea and Vanuatu.

55.35.1. Cystorchis aphylla Ridl., J. Linn. Soc. Bot. 32: 400 (1896).

Saprophyte. Hill forest, sometimes on ultramafic substrate. Elevation: 800–1300 m.

General distribution: Sabah, Sarawak, Peninsular Malaysia, Thailand, Sumatra, Java, Buru, Philippines.

Collections. HEMPUEN HILL: 800–1000 m, *Beaman 7409a* (K), 800–1000 m, *Cribb 89/27* (K); LUGAS HILL: 1300 m, *Beaman 8479* (K), 1300 m, *10537* (K).

55.35.2. Cystorchis variegata Blume, Fl. Javae, ser. 2, 1, Orch.: 89, t. 24, f. 3 & 36c (1859).

Terrestrial. Hill forest, lower montane forest. Elevation: 600–1500 m.

General distribution: Sabah, Peninsular Malaysia, Sumatra, Java, Vanuatu.

Collections. MELANGKAP KAPA: 600–700 m, *Beaman 8603* (K); PENIBUKAN: 1200–1500 m, *Clemens 31136* (BM), 1200–1500 m, *35196* (BM); TAHUBANG RIVER: *Collenette A 9* (BM).

55.36. DENDROBIUM Sw.

Nova Acta Regiae Soc. Sci. Upsal., ser. 2, 6: 82 (1799).

Seidenfaden, G. (1985). Orchid genera in Thailand XII. *Dendrobium* Sw. Opera Bot. 83: 1–295.

Epiphytic, lithophytic, rarely terrestrial, polymorphic herbs. *Stems* either 1) rhizomatous, 2) erect and many-noded, 3) erect and 1-noded or several-noded from a many-noded rhizome, or 4) without a rhizome, the new stems of many nodes arising from base of old ones; 1 or 2 cm to 5 m long, tough or fleshy, swollen at base or along the whole length, often pseudobulbous, more or less covered with sheathing leaf-bases and cataphylls. *Leaves* 1 to many, borne at apex or distichously along stem, linear, lanceolate, oblong or ovate, papery or coriaceous, usually bilobed or emarginate at apex. *Inflorescences* racemose, 1- to many-flowered, erect, horizontal or pendulous, either lateral or terminal. *Flowers* often showy, small to large, resupinate or non-resupinate, ephemeral or long-lived. *Sepals* short to filiform. *Lateral sepals* adnate to the elongated column-foot forming a mentum, often spur-like at the base, 0.1–3 cm long. *Petals* narrower or broader than sepals. *Lip* entire to 3-lobed, base joined to column-foot, often forming a closed spur with the lateral sepals to which it may be joined laterally for a short distance; disc with 1–7 keels; calli rarely present. *Column* short, foot long, apical stelidia present. *Pollinia* 4 in appressed pairs, naked, i.e., without caudicles or stipes.

Approximately 1400 species distributed from India across to Japan, south to Malaysia, Indonesia, east to New Guinea, Australia, New Zealand and the Pacific islands. The second largest orchid genus in Borneo and on Mount Kinabalu.

55.36.1. Dendrobium aff. **acerosum** Lindl., Bot. Reg. 27: misc. 43, no. 86 (1841).

Epiphyte. Hill forest. Elevation: 900 m.

Collection. BUNDU TUHAN: 900 m, *Carr SFN 27975* (SING).

55.36.2. Dendrobium acuminatissimum (Blume) Lindl., Gen. Sp. Orch., 86 (1830).

Grastidium acuminatissimum Blume, Bijdr., 333 (1825).

Epiphyte. Lower montane forest. Elevation: 900–1500 m.

General distribution: Sabah, Sumatra, Java, Philippines.

Collections. BUNDU TUHAN: 1200 m, *Carr SFN 27405* (SING), 900 m, *SFN 27943* (SING), 1400 m, *Darnton 223* (BM); TENOMPOK: 1500 m, *Clemens 29840* (BM, G, K), 1500 m, *30161* (G, K).

55.36.3. Dendrobium alabense J. J. Wood, Lindleyana 5: 90, f. 6 (1990). Fig. 17, Plate 34A.

Epiphyte. Lower montane forest, sometimes on ultramafic substrate. Elevation: 1500–2400 m.

General distribution: Sabah.

Collections. EAST MESILAU RIVER: 1900 m, *Wood 851* (K); GURULAU SPUR: 1800–2100 m, *Clemens 50819* (BM); KEMBURONGOH: 2000 m, *Carr 3123, SFN 26557A* (SING); KILEMBUN RIVER: 1500–1800 m, *Clemens 32608* (BM); LUMU-LUMU: 2000 m, *Clemens s.n.* (BM); MARAI PARAI: 1600 m, *Carr 3123, SFN 26557* (SING), 2100–2400 m, *Clemens s.n.* (BM); PIG HILL: 2000–2300 m, *Beaman 9890* (K), 1700–2000 m, *Sutton 20* (K); TENOMPOK: 1500 m, *Clemens s.n.* (BM); TENOMPOK ORCHID GARDEN: 1500 m, *Clemens 51734* (BM).

55.36.4. Dendrobium aloifolium (Blume) Rchb. f., Walp. Ann. Bot. Syst. 6: 279 (1861).

Macrostomium aloefolium Blume, Bijdr., 335 (1825).

Epiphyte. Hill forest on ultramafic substrate. Elevation: 500–600 m.

General distribution: Sabah, Burma, Peninsular Malaysia, Singapore, Laos, Vietnam, Sumatra, Java, Anambas Islands, Bali, Bangka, Lingga Archipelago, Riau Archipelago.

Collection. LOHAN RIVER: 500–600 m, *Clements 3317* (K).

55.36.5. Dendrobium anosmum Lindl., Bot. Reg. 31: misc. 32, no. 41 (1845).

Epiphyte. Lower montane forest. Elevation: 1500 m.

General distribution: Sabah, Kalimantan, Sarawak, widespread from Sri Lanka to Indochina & Malaysia through Indonesia to the Philippines, New Guinea & the Solomon Islands.

Collection. TENOMPOK: 1500 m, *Clemens 28865* (BM).

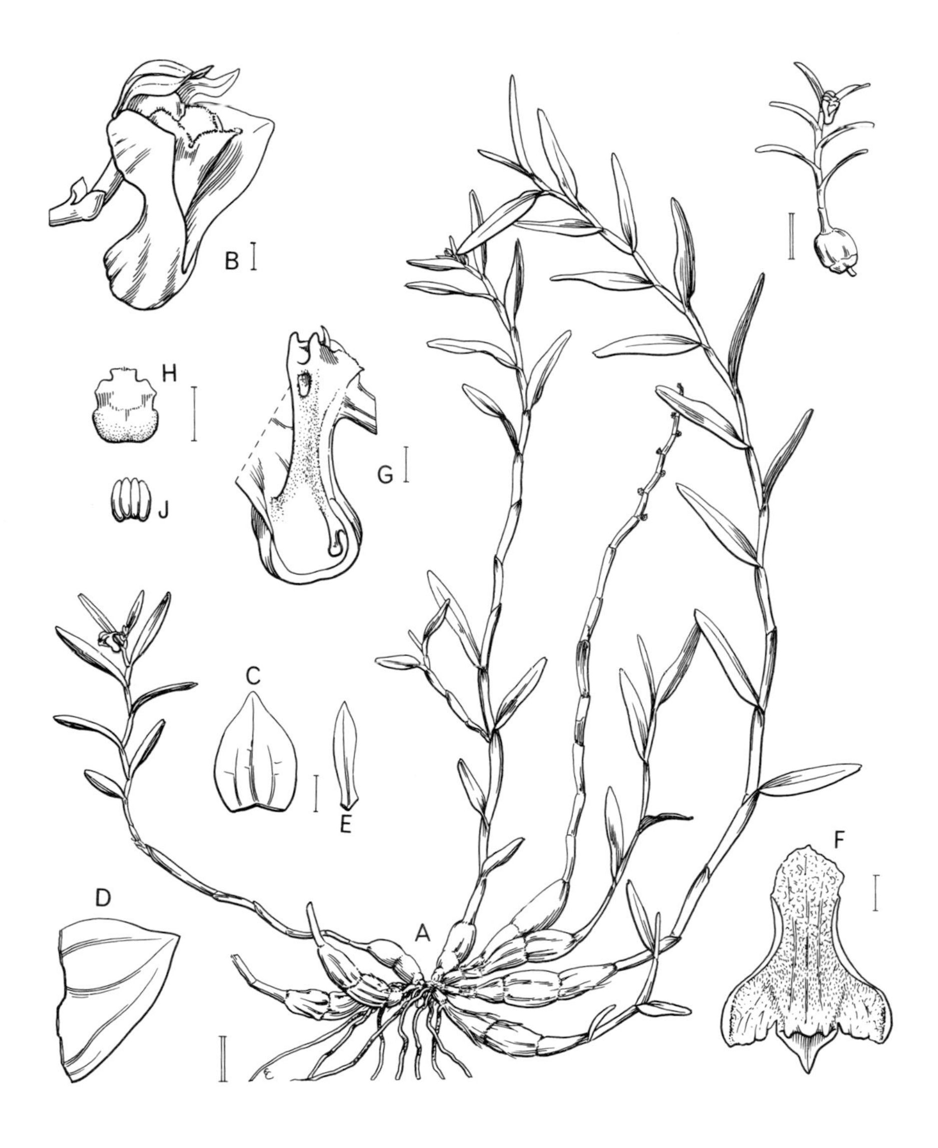

Fig. 17. Dendrobium alabense. **A**, habit; **B**, flower (oblique view); **C**, dorsal sepal; **D**, lateral sepal; **E**, petal; **F**, lip (front view); **G**, column and longitudinal section through mentum; **H**, anther-cap (back view); **J**, pollinia. All from *Wood 777* (holotype) from Mt. Alab, Sabah. Scale: single bar = 1 mm; double bar = 1 cm. Drawn by Eleanor Catherine.

55.36.6. Dendrobium beamanianum J. J. Wood & A. Lamb, **sp. nov.** Fig. 18.

Dendrobio corrugatilobo J. J. Sm., species javanica, accedens, sed foliis longioribus, floribus majoribus, sepalis longioribus, parte labii distali carina distincta carnosa apicali instructa differt.

Type: EAST MALAYSIA, SABAH, MOUNT KINABALU, PINOSUK PLATEAU: 1380 m, 28 July 1984, *Beaman 10722* (holotype K!, herbarium and spirit material).

Epiphytic herb. *Stems* 30–60 cm long, 0.3–0.5 cm wide, erect to spreading. *Leaves* linear-ligulate, acutely unequally bilobed, blade 3.5–8(–9) × 0.3–0.6(–0.8) cm, sheaths 1.5–2 cm long, densely black-hirsute, the lowermost becoming glabrous. *Inflorescences* 1–2(–3)-flowered, usually only 1 open at a time, emerging from below top of leaf sheath opposite blade; *peduncle* and *rachis* (0.5–)1–1.5 cm long, glabrous; *floral bracts* 1 mm long, triangular-ovate, acute, yellowish. *Flowers* 2.5–2.8 cm across, non-resupinate. *Pedicel-with-ovary* 1–1.2 cm long, straight, orange and greenish. *Sepals* spreading, acute to acuminate, stiff, pale brown or pale creamy-yellow, with pale orange or orange-brown veins. *Dorsal sepal* 1.1–1.4 × 0.45–0.5 cm, oblong-elliptic. *Lateral sepals* 1.7–1.8 × 0.4 cm, obliquely narrowly elliptic. *Mentum* 4–5 mm long, conical. *Petals* 1–1.5 × 0.2–0.25 cm, linear-spathulate, acute, very minutely erose, pale brown or pale creamy-yellow. *Lip* 1.5–1.6 cm long, 3-lobed, fleshy, slightly to densely white farinose, divided into a hypochile and epichile, pale yellow or pale brown with white keel and brownish yellow wrinkled epichile; *hypochile* 8–9 × 3 mm, lobes erect, rounded, 1.5–2 mm high, disc with 2 fleshy keels extending from base and terminating on base of epichile, between which is a shorter apical median keel, outer keels farinose, particularly near the base; *epichile* 6–8 × 3–5 mm, oblong, obtuse, margin at first conduplicate, then parting, very fleshy and rugose, with a variably sized but usually prominent erect fleshy apical keel in between. *Column* 2 mm long, oblong, with obtuse oblong arms c. 1 mm long, pale green, foot 4–5 mm long, sulcate. *Anther-cap* 2 × 1 mm, cucullate, white.

Dendrobium beamanianum has so far only been recorded from lower montane forest on ultramafic substrate. It is closely related to the Javan *D. corrugatilobum* J. J. Sm. (section *Conostalix*), but is distinguished by its longer leaves, larger flowers with longer sepals and a lip epichile with a distinct fleshy apical keel.

The specific epithet honours the Beaman family, Professor John H. Beaman, Dr. Teofila E. Beaman and Reed S. Beaman, who all have contributed significantly to our knowledge of the orchids of Mount Kinabalu.

Elevation: 1200–1700 m.

Endemic to Mount Kinabalu.

Additional collections: MARAI PARAI: 1700 m, *Clemens 32780* (BM); MARAI PARAI SPUR: 1600 m, *Lamb AL 52/83* (K, herbarium and spirit material); PENIBUKAN: 1400 m, *Carr 3088, SFN 26548* (SING), 1200 m, *Clemens 30776* (BM, K), 1500 m, *30998* (BM,); LOCALITY UNKNOWN, cult. Schelpe, Cape Town, S. Afr: *Lamb s.n.* (K).

A
B
C
D
E
F
G
H
J
K
L

55.36.7. Dendrobium bifarium Lindl., Gen. Sp. Orch., 81 (1830). Plate 34B.

Epiphyte. Lower montane forest, sometimes on ultramafic substrate. Elevation: 1000–1800 m.

General distribution: Sabah, Kalimantan, Sarawak, Peninsular Malaysia, Singapore, Thailand, Maluku.

Collections. GOLF COURSE SITE: 1700–1800 m, *Beaman 7237* (K), 1700 m, *8484* (K), 1700 m, *8558* (K), 1700–1800 m, *10670* (K); GURULAU SPUR: 1800 m, *Clemens 50415* (BM); HEMPUEN HILL: 1000 m, *Cribb 89/39* (K); LUMU-LUMU: 1600 m, *Carr SFN 27029* (SING); MARAI PARAI: 1500 m, *Clemens 32310* (BM, K), *32453* (BM), 1500 m, *32772* (G, K), 1500 m, *33003* (BM, K); MESILAU RIVER: 1700 m, *Held s.n.* (K); MINITINDUK/KINATEKI DIVIDE: 1100 m, *Carr SFN 27052* (SING); PORING ORCHID GARDEN: *Lohok 25* (K); TENOMPOK ORCHID GARDEN: 1500 m, *Clemens 50494* (BM); WEST MESILAU RIVER: 1600 m, *Beaman 9036a* (K), 1700 m, *Brentnall 138* (K), 1700 m, *152* (K).

55.36.8. Dendrobium cinereum J. J. Sm., Bull. Jard. Bot. Buit., ser. 3, 2: 78 (1920). Plate 34C.

Epiphyte. Cultivated at Poring, of local origin.

General distribution: Sabah, Kalimantan.

Collection. PORING ORCHID GARDEN: *Lohok 15* (K).

55.36.9. Dendrobium concinnum Miq., Fl. Ned. Ind. 3: 644 (1859).

Epiphyte. Lower montane forest. Elevation: 1200 m.

General distribution: Sabah, Sarawak, Burma, Peninsular Malaysia, Thailand, Cambodia, Vietnam, Sumatra, Java, Riau Archipelago.

Collection. PENIBUKAN: 1200 m, *Carr SFN 27894* (K).

55.36.10. Dendrobium connatum (Blume) Lindl., Gen. Sp. Orch., 89 (1830).

Onychium connatum Blume, Bijdr., 328 (1825).

Epiphyte.

General distribution: Sabah, Sumatra, Java.

Collection. MOUNT KINABALU: *Haslam s.n.* (AMES).

Fig. 18. Dendrobium beamanianum. A, habit; **B**, flower (side view); **C**, dorsal sepal; **D**, lateral sepal; **E**, petal; **F**, lip (front view); **G**, column (front view); **H**, pedicel-with-ovary and column (side view); **J**, anther-cap (front view); **K**, anther-cap (side view); **L**, pollinia. **A** from *Beaman 10722*; **B–E** from *Lamb AL 52/83*; **F-L** from *Beaman 10722*. Scale: single bar = 1 mm; double bar = 1 cm. Drawn by Eleanor Catherine.

55.36.11. Dendrobium crumenatum Sw., J. Bot. (Schrader) 2: 237 (1799).

Dendrobium crumenatum Sw. var. *parviflorum* Ames & C. Schweinf., Orch. 6: 100 (1920). Type: KIAU, 900 m, *Clemens 188* (holotype AMES mf; isotypes BM!, K!, SING!).

Epiphyte. Hill forest. Elevation: 700–900 m. Pigeon orchid.

General distribution: Sabah, Kalimantan, Sarawak, China, India, Sri Lanka, SE Asia.

Additional collections. BUNDU TUHAN: 900 m, *Carr SFN 27795* (SING); DALLAS: 900 m, *Clemens 26726* (K), 900 m, *27630* (K); KADAMAIAN RIVER: 700 m, *Carr 3173, SFN 26693* (SING).

55.36.12. Dendrobium cumulatum Lindl., Gard. Chron. 1855: 756 (1855). Fig. 19, Plate 35A.

Epiphyte. Lower montane forest. Elevation: 1500 m.

General distribution: Sabah, Bhutan, Nepal, India, Burma, Thailand, Cambodia, Laos, Vietnam.

Collections. GURULAU SPUR: 1500 m, *Clemens 50376* (BM, K), 1500 m, *50481* (BM); MT. NUNGKEK: 1500 m, *Sands 3976* (K); TENOMPOK: 1500 m, *Clemens 29590* (BM, K), 1500 m, *29631* (E, G), *29901* (BM, E, G).

55.36.13. Dendrobium cymbulipes J. J. Sm., Mitt. Inst. Allg. Bot. Hamburg 7: 51, t. 8, f. 41 (1927). Plate 35B.

Epiphyte. Lower montane forest, sometimes on ultramafic substrate. Elevation: 900–2400 m.

General distribution: Sabah, Kalimantan, Sarawak.

Collections. BAMBANGAN RIVER: 1500 m, *RSNB 4527* (K); DALLAS: 900 m, *Clemens s.n.* (BM); EAST MESILAU RIVER: 1500–1600 m, *Bailes & Cribb 524* (K); EAST MESILAU/MENTEKI RIVERS: 1700 m, *Beaman 8764* (K); KEMBURONGOH: 2100 m, *Clemens s.n.* (BM, BM); KILEMBUN BASIN: 1400 m, *Clemens 35166* (BM, G, K); KILEMBUN RIVER: 1400 m, *Clemens 34109* (BM, G, K); MARAI PARAI: 1400 m, *Clemens 32354* (BM), 1800 m, *32609* (BM); MESILAU BASIN: 2100 m, *Clemens 29101* (BM); MURU-TURA RIDGE: 1500 m, *Clemens 34366* (BM, G, K); PARK HEADQUARTERS: 1700 m, *Gunsalam 6* (K), 1600 m, *Lamb AL 19/82* (K); PENIBUKAN: 1200–1500 m, *Clemens 31512* (BM); TENOMPOK: 1500 m, *Carr SFN 27024* (SING), 1500 m, *Clemens s.n.* (BM, BM, BM, BM, BM), 1500 m, *27240* (BM), 1500 m, *27768* (BM, K), 1500 m, *28204* (BM, K), 1500 m, *28636* (BM), 1500 m, *30165* (G, K); TENOMPOK ORCHID GARDEN: 1500 m, *Clemens 50237* (BM); WEST MESILAU RIVER: 1600–1700 m, *Beaman 7458* (K), 1600–1700 m, *8667* (K), 1600 m, *9189* (K).

55.36.14. Dendrobium gracile (Blume) Lindl., Gen. Sp. Orch., 91 (1830).

Onychium gracile Blume, Bijdr., 327 (1825).

Epiphyte. Lower montane forest. Elevation: 1500–1700 m.

General distribution: Sabah, Kalimantan, Sarawak, Sumatra, Java.

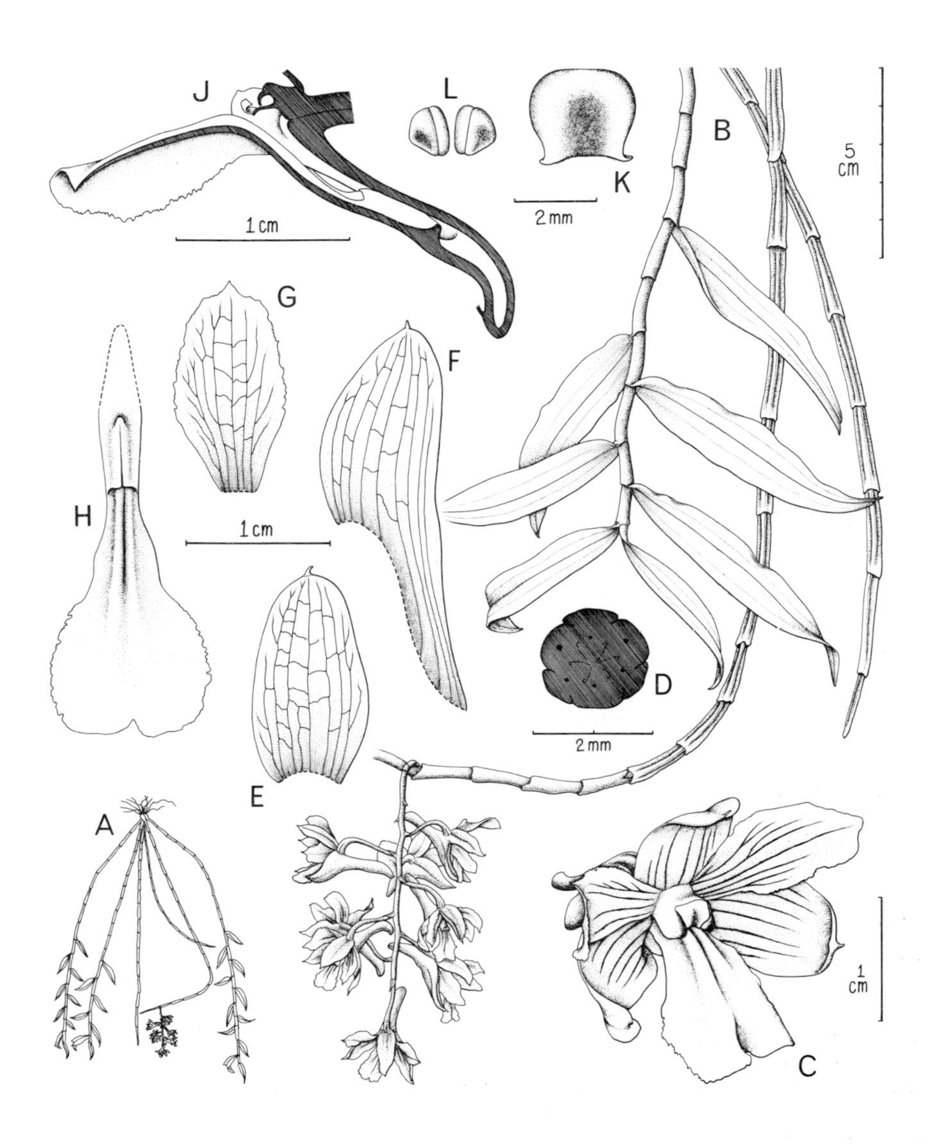

Fig. 19. Dendrobium cumulatum. **A**, habit; **B**, leafy shoot and inflorescence; **C**, flower (oblique view); **D**, ovary (transverse section); **E**, dorsal sepal; **F**, lateral sepal; **G**, petal; **H**, lip (front view); **J**, pedicel-with-ovary, column, lip and mentum (longitudinal section); **K**, anther-cap (back view); **L**, pollinia. Crocker Range. Drawn by Chan Chew Lun.

Collections. EAST MESILAU RIVER: 1500–1600 m, *Bailes & Cribb 505* (K); MARAI PARAI: 1500 m, *Clemens 32245* (BM); PINOSUK PLATEAU: 1600 m, *Bailes & Cribb 716* (K); WEST MESILAU RIVER: 1700 m, *Brentnall 125* (K).

55.36.15. Dendrobium grande Hook. f., Fl. Brit. Ind. 5: 724 (1890).

Epiphyte. Low-stature hill forest on ultramafic substrate. Elevation: 800–1000 m.

General distribution: Sabah, Sarawak, Andaman Islands, Burma, Peninsular Malaysia, Thailand.

Collection. LOHAN RIVER: 800–1000 m, *Beaman 10008* (K).

55.36.16. Dendrobium hamaticalcar J. J. Wood & Dauncey, **sp. nov.** Fig. 20, Plate 35C.

Dendrobio sarawakensi Ames, species sarawakensis, affinis sed mento fortuis hamato, lamina labello angustiore callo basali epsiliformi neque carinis duobus vel tribus longitudinalibus instructa distingui possit.

Type: EAST MALAYSIA, SABAH, FOOTHILLS OF MT. TRUS MADI, NEAR KAINGARAN, 750–900 m, October/November 1968, cult. Edinburgh Botanic Garden no. C6614, flowered in cult. October 1973, *Bacon 142* (holotype E!, herbarium material only; isotype K!, herbarium material only).

Epiphytic herb. *Stems* 60–100 cm long, internodes c. 2 cm long, c. 1 cm thick, arching, somewhat thicker toward apex, sheaths flushed violet. *Leaves* 6–10 × 2.5–3.5 cm, thin-textured. *Inflorescence* pendulous, more or less perpendicular to stem, from upper nodes of leafy and leafless stems, 4–9-flowered; peduncle c. 1.2–2.1 cm long, c. 1 mm thick; rachis c. 0.8–1.6 cm long, pink; *floral bracts* 2–3 mm long. *Flowers* medium-sized, 2.4–3.6 cm long, pale dull yellow, reverse of sepals and spur darker coloured, veins magenta especially on lip, lip callus yellow, column with dark magenta line on each side. *Pedicel-with-ovary* 1.1–1.5 cm long, 1.8–2.6 mm thick, straight or gently curving, perpendicular to mentum. *Dorsal sepal* 10–14 × 5.5–8 mm, ovate-elliptic, obtusely rounded. *Lateral sepals* obliquely ovate, apex obtuse, apiculate, free distal portion 10.5–14 × 7–9 mm, lower portion extended into a narrow mentum 24–30 mm long, 2–3.5 mm thick, curving through 180–300°, fused at base for 2.5–9.5 mm. *Petals* 10–14.5 × 6–8.5 mm, oblong-spathulate, apex obtusely rounded, somewhat retuse. *Lip* 18–23.5 × 2–2.5 mm, spathulate, claw narrow, linear, with callus 5.5–10 mm from base, abruptly expanding into an oblong lamina, apex obtusely rounded, somewhat prominently retuse, apical margin coarsely dentate, callus thick, U-shaped, at base of lamina, 10.5–14.5 mm long, 6–10.5 mm wide. *Column* short, with prominent rhombic stelidia, anther-filament broadly triangular, with a somewhat triangular dorsal protrusion, foot 26–30 × 2–2.5 mm, curved, with a nectary c. 2 mm from apex, rostellum absent, stigma narrow. *Anther-cap* 3.5–4.2 × 2.5–3.2 mm, somewhat dorso-ventrally flattened, lower edge concave. *Pollinia* 2 mm long.

This is a striking species, belonging to section *Calcarifera*. It has a long mentum which forms a semicircular hook or may even be curved into an almost complete circle. It is closely related to *D. sarawakense* Ames (syn. *D.*

multiflorum Ridl.), which has a broader lip with 2 to 3 longitudinal keels and a shorter, less hooked mentum. *Dendrobium hamaticalcar* has a distinct U-shaped basal callus on the lip. It has been in cultivation at Edinburgh Botanic Gardens for a number of years.

The specific epithet is derived from the Latin *hamatus*, hooked at the tip, and *calcar*, spur, in reference to the hooked, spur-like mentum.

Lowlands and hill forest. Elevation: 400–900 m.

General distribution: Sabah.

Additional collections: KAUNG: 400 m, *Carr 3019* (K, SING); MT. TRUS MADI: *Bacon 110* (E).

55.36.17. Dendrobium indivisum (Blume) Miq.

a. var. **pallidum** Seidenf., Opera Bot. 83: 216, f. 146 (1985).

Epiphyte. Hill forest. Elevation: 600 m.

General distribution: Sabah, Burma, Peninsular Malaysia, Thailand.

Collections. BUNDU TUHAN: 600 m, *Carr SFN 27955* (SING), 600 m, *SFN 28024* (SING); RANAU: 600 m, *Collenette 70* (BM).

55.36.18. Dendrobium kentrophyllum Hook. f., Fl. Brit. Ind. 5: 725 (1890).

Epiphyte. Hill forest. Elevation: 600–1200 m.

General distribution: Sabah, Kalimantan, Sarawak, Peninsular Malaysia, Thailand, Sumatra.

Collections. BUNDU TUHAN: 1200 m, *Carr SFN 27818* (SING); DALLAS: 900 m, *Clemens 26515* (BM, G, K); RANAU: 600 m, *Collenette 69* (BM).

55.36.19. Dendrobium kiauense Ames & C. Schweinf., Orch. 6: 103 (1920). Fig. 21, Plate 35D. Type: KIAU, 900 m, *Clemens 176* (holotype AMES mf; isotype BM!).

Epiphyte. Hill forest, lower montane forest, sometimes on ultramafic substrate. Elevation: 800–1500 m.

General distribution: Sabah, Kalimantan.

Additional collections. BUNDU TUHAN: 1400 m, *Carr SFN 27235* (SING), 1200 m, *SFN 27967* (SING); GURULAU SPUR: *Clemens 304* (AMES, BM, K); HEMPUEN HILL: 800–1000 m, *Cribb 89/31* (K); LUBANG: *Clemens 128* (AMES, BM, K); PENATARAN RIVER: 1200 m, *Clemens 34051* (BM); PENIBUKAN: 1200–1500 m, *Clemens 35164* (BM); TENOMPOK: 1500 m, *Clemens 28396* (BM, G, K), 1500 m, *29317* (BM, G, K).

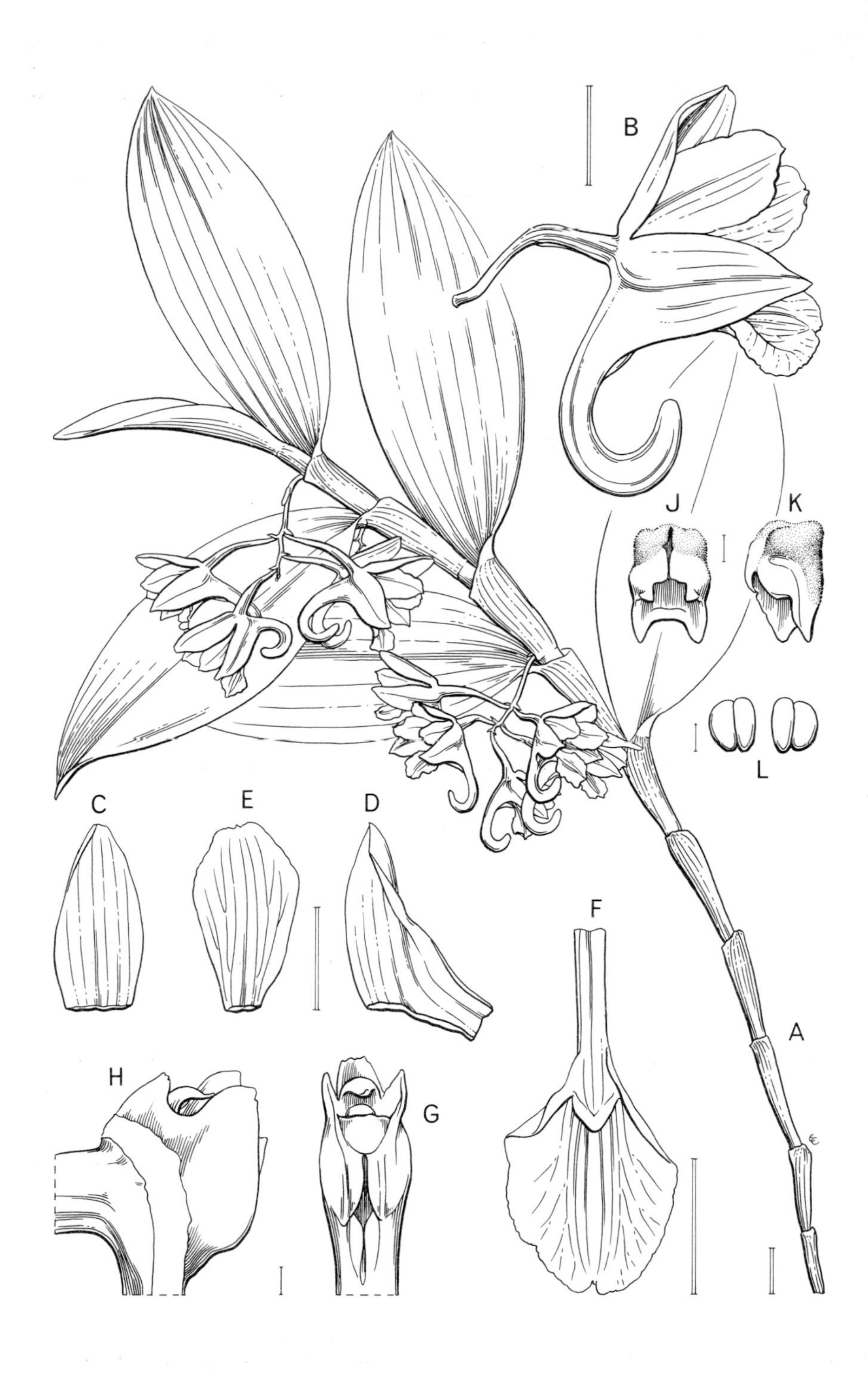
B
J
K
L
C
E
D
F
A
H
G

55.36.20. Dendrobium lamellatum (Blume) Lindl., Gen. Sp. Orch., 89 (1830).

Onychium lamellatum Blume, Bijdr., 326, t. 10 (1825).

Epiphyte. Hill forest on ultramafic substrate.

General distribution: Sabah, Kalimantan, Sarawak, Burma, Peninsular Malaysia, Thailand, Sumatra, Java, Natuna Islands, Riau Archipelago.

Collection. LOHAN RIVER: *Clements 3402* (K).

55.36.21. Dendrobium lamelluliferum J. J. Sm., Mitt. Inst. Allg. Bot. Hamburg 7: 52, t. 8, f. 42 (1927).

Epiphyte. Lower montane forest. Elevation: 1200–2200 m.

General distribution: Sabah, Kalimantan, Sarawak.

Collections. GOLF COURSE SITE: 1700 m, *Beaman 8501* (K); KEMBURONGOH: 2200 m, *Carr SFN 27831* (SING); KUNDASANG: 1400 m, *Kidman Cox 2502* (K); MARAI PARAI: 1500 m, *Collenette A 94* (BM); MESILAU RIVER: 1800 m, *Clemens s.n.* (BM); TENOMPOK: 1500 m, *Carr SFN 27053* (SING), 1500 m, *Clemens s.n.* (BM), 1200 m, *28039* (BM); TENOMPOK ORCHID GARDEN: 1500 m, *Clemens 50176* (BM); WEST MESILAU RIVER: 1700 m, *Brentnall 123* (K).

55.36.22. Dendrobium aff. **lamelluliferum** J. J. Sm., Mitt. Inst. Allg. Bot. Hamburg 7: 52, t. 8, f. 42 (1927).

Epiphyte. Lower montane forest, upper montane forest, sometimes on ultramafic substrate. Elevation: 1200–2300 m.

Collections. BUNDU TUHAN: 1200 m, *Carr SFN 27427* (SING); PIG HILL: 2000–2300 m, *Beaman 9886* (K).

55.36.23. Dendrobium aff. **lancilobum** J. J. Wood, Lindleyana 5: 90, f. 7 (1990).

Epiphyte. Hill forest. Elevation: 900 m.

Collection. BUNDU TUHAN: 900 m, *Carr SFN 27764* (SING).

55.36.24. Dendrobium linguella Rchb. f., Gard. Chron. ser. 2, 18: 552 (1882). Plate 35E.

Epiphyte. Lowlands, hill forest. Elevation: 300–900 m.

Fig. 20. Dendrobium hamaticalcar. A, habit; **B**, flower (side view); **C**, dorsal sepal; **D**, lateral sepal; **E**, petal; **F**, lip (front view); **G**, column (front view); **H**, column (side view); **J**, anther-cap (front view); **K**, anther-cap (side view); **L**, pollinia. All from *Bacon 142.* Scale: single bar = 1 mm; double bar = 1 cm. Drawn by Eleanor Catherine.

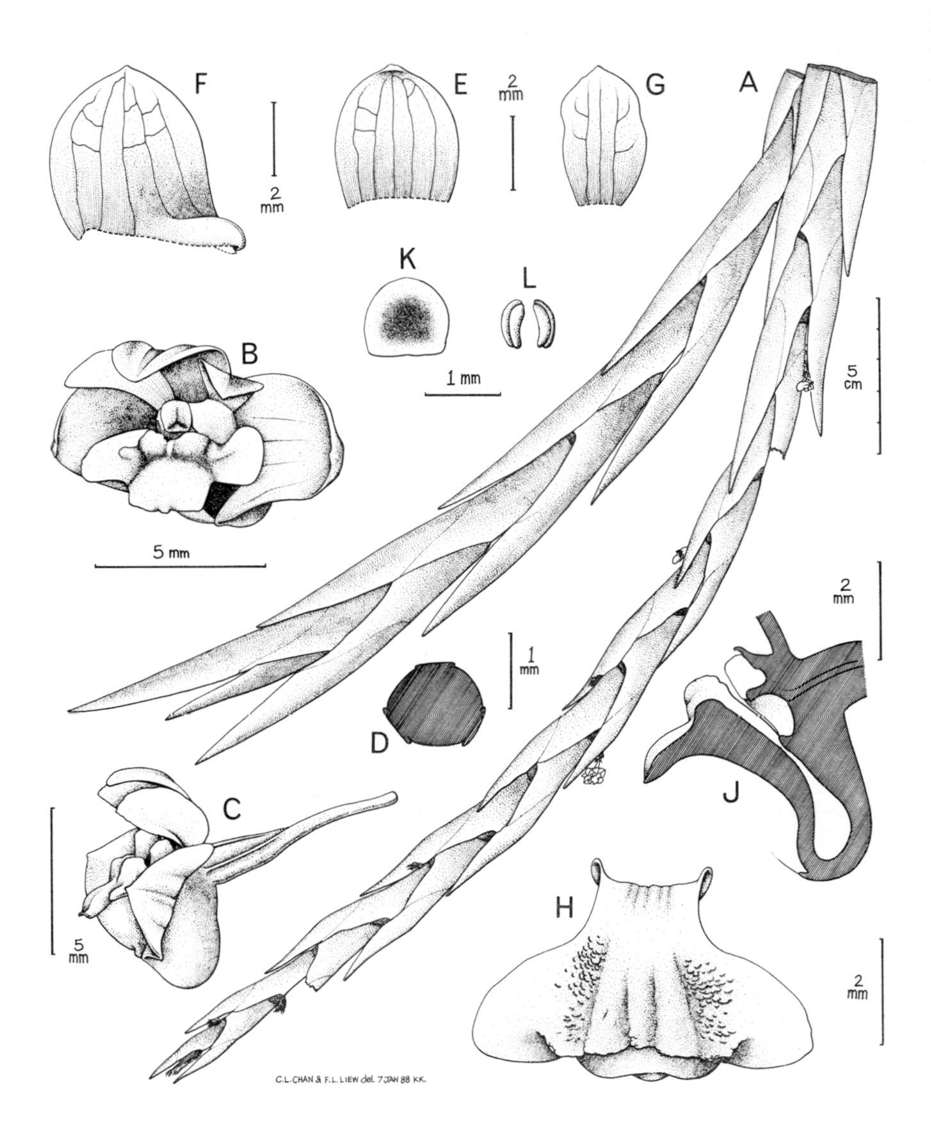

Fig. 21. Dendrobium kiauense. **A**, habit; **B**, flower (front view); **C**, flower (side view); **D**, ovary (transverse section); **E**, dorsal sepal; **F**, lateral sepal; **G**, petal; **H**, lip (front view); **J**, pedicel-with-ovary, column and lip (longitudinal section); **K**, anther-cap (back view); **L**, pollinia. From plant cultivated in Tenom Orchid Centre, ex Lohan River. Drawn by Chan Chew Lun and F. L. Liew.

General distribution: Sabah, Peninsular Malaysia, Thailand, Vietnam, Sumatra, Anambas Islands.

Collections. BUNDU TUHAN: 900 m, *Carr SFN 27953* (AMES, SING); KAUNG: 300 m, *Collenette 2* (BM).

55.36.25. Dendrobium maraiparense J. J. Wood & C. L. Chan, ined. Type: BAMBANGAN RIVER, 1600 m, *RSNB 4962* (holotype K!; isotype SING!).

Epiphyte, stiffly erect. Lower montane forest on ultramafic substrate. Elevation: 1200–2000 m.

Endemic to Mount Kinabalu.

Additional collections. MARAI PARAI: 1500 m, *Clemens 32242* (BM), 1400 m, *32358* (BM), 1200 m, *Collenette A 3* (BM), 1500 m, *A 81* (BM); MARAI PARAI SPUR: 1400–1600 m, *Bailes & Cribb 825* (K), 1600 m, *Lamb SAN 93368* (K); PENIBUKAN: 1400 m, *Carr 3069, SFN 26455* (SING), 1200 m, *Clemens 30776* (BM), 1200 m, *30971* (BM), 1200–1500 m, *31255* (BM); PIG HILL: 1700–2000 m, *Sutton 4* (K).

55.36.26. Dendrobium microglaphys Rchb. f., Gard. Chron. 1868: 1014 (1868). Plate 36A.

Epiphyte. Hill forest on ultramafic substrate. Elevation: 800 m.

General distribution: Sabah, Kalimantan, Sarawak, Singapore, Thailand.

Collection. MELANGKAP KAPA: 800 m, *Lamb AL 972/88* (K).

55.36.27. Dendrobium minimum Ames & C. Schweinf., Orch. 6: 107, pl. 91 (1920). Type: MARAI PARAI SPUR, *Clemens 287* (holotype AMES mf; isotypes BM!, K!, SING!).

Epiphyte. Lower montane forest, frequently on ultramafic substrate. Elevation: 1200–2000 m.

General distribution: Sabah.

Additional collections. EAST MESILAU/MENTEKI RIVERS: 1700–2000 m, *Beaman 9604* (K), 1700 m, *Wood 831* (K); LIWAGU RIVER: 1600 m, *Darnton 572* (BM); LIWAGU RIVER TRAIL: 1700 m, *Collenette 2372* (K); LUMU-LUMU: *Clemens 27857* (BM, G, K); MAMUT COPPER MINE: 1600–1700 m, *Beaman 9949* (K); MARAI PARAI: 1500 m, *Clemens 32353* (BM); MARAI PARAI SPUR: *Clemens 403* (AMES); PARK HEADQUARTERS: 1700 m, *Gunsalam 8* (K); PENIBUKAN: 1400 m, *Carr 3087, SFN 26555* (SING), 1200–1500 m, *Clemens 3827?* (K), 1400 m, *30810* (BM), 1200 m, *30827* (BM); TENOMPOK: 1500 m, *Carr SFN 27092* (SING), 1500 m, *Clemens s.n.* (BM), 1500 m, *28582* (BM, K), 1500 m, *28821* (G, K).

55.36.28. Dendrobium oblongum Ames & C. Schweinf., Orch. 6: 108 (1920). Fig. 22. Type: MOUNT KINABALU, *Haslam s.n.* (holotype AMES mf).

Epiphyte. Hill forest, lower montane forest, sometimes on ultramafic substrate. Elevation: 1100–1700 m.

General distribution: Sabah, Kalimantan, Sarawak.

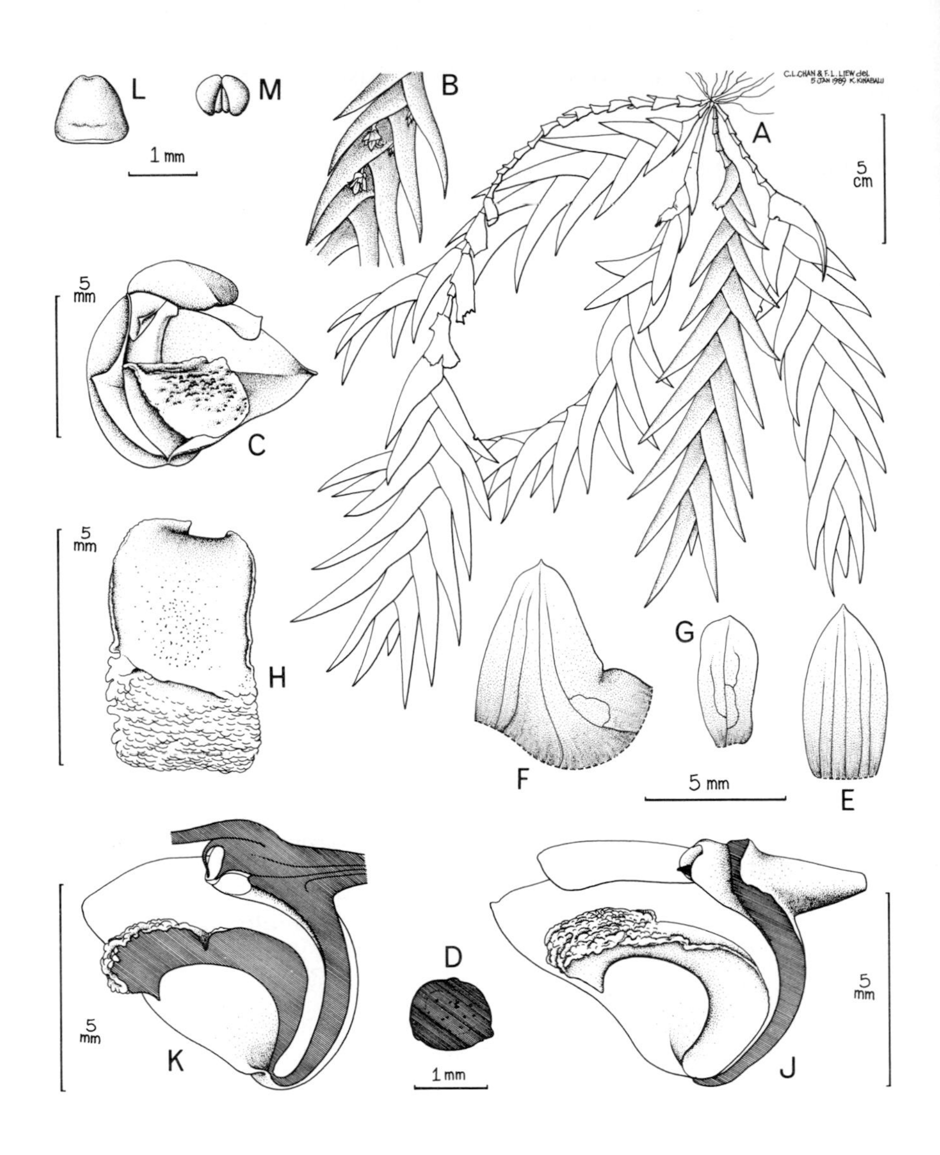

Fig. 22. Dendrobium oblongum. **A**, habit; **B**, part of flowering shoot; **C**, flower (oblique view); **D**, ovary (transverse section); **E**, dorsal sepal; **F**, lateral sepal; **G**, petal; **H**, lip (front view); **J**, pedicel-with-ovary, column and lip (with dorsal sepal, lateral sepal and petal removed); **K**, pedicel-with-ovary, column and lip (longitudinal section, with one lateral sepal attached); **L**, anther-cap (back view); M, pollinia. All from *Chan & Gunsalam 128.* Drawn by Chan Chew Lun and F. L. Liew.

Additional collections. GURULAU SPUR: 1700 m, *Clemens 50700* (BM); KIAU: *Clemens 334* (AMES); MAHANDEI RIVER: 1100 m, *Carr 3037, SFN 26372* (BM, SING); MARAI PARAI: 1500 m, *Clemens 32254* (BM); MOUNT KINABALU: 1600 m, *Chan & Gunsalam s.n.* (K); PENIBUKAN: 1200–1500 m, *Clemens s.n.* (BM), 1200–1500 m, *30586* (K), 1200–1500 m, *30686* (BM); TENOMPOK: 1500 m, *Clemens 28171* (BM, G, K); TENOMPOK ORCHID GARDEN: 1500 m, *Clemens 50238* (BM, K).

55.36.29. Dendrobium olivaceum J. J. Sm., Bull. Jard. Bot. Buit., ser. 2, 8: 41 (1912). Plate 36B.

Epiphytic or terrestrial. Lower montane forest. Elevation: 1500–1600 m.

General distribution: Sabah, Kalimantan.

Collections. PARK HEADQUARTERS: 1600 m, *Lamb LKC 3023* (K); TENOMPOK: 1500 m, *Carr SFN 28021* (SING).

55.36.30. Dendrobium aff. **olivaceum** J. J. Sm., Bull. Jard. Bot. Buit., ser. 2, 8: 41 (1912).

Epiphyte. Lower montane forest. Elevation: 1700 m.

Collection. KIAU VIEW TRAIL: 1700 m, *Lamb & Phillipps AL 161/83* (K).

55.36.31. Dendrobium orbiculare J. J. Sm., Bull. Jard. Bot. Buit., ser. 2, 13: 23 (1914). Plate 36C.

Dendrobium fuscopilosum Ames & C. Schweinf., Orch. 6: 101 (1920), **syn. nov.** Type: GURULAU SPUR, *Clemens 300* (holotype AMES mf; isotypes BM!, K!, SING!).

Epiphytic and terrestrial. Lower montane forest on ultramafic substrate. Elevation: 900–1700 m.

General distribution: Sabah, Kalimantan, Sarawak.

Additional collections. GURULAU SPUR: 1500 m, *Clemens 50442* (BM), 1700 m, *Gibbs 4012* (BM, K); MAHANDEI RIVER HEAD: 1200 m, *Carr 3068, SFN 26430* (SING); MARAI PARAI SPUR: *Clemens 284* (AMES, BM, K, SING); MOUNT KINABALU: *Haviland 1299* (K); PENIBUKAN: 1200–1500 m, *Clemens 30783* (BM, G, K), 1200–1500 m, *31212* (G, K), 1200–1500 m, *31510* (BM), 1400 m, *40693* (BM, K), 1200 m, *40858* (BM), 900 m, *50145* (BM, K); TAHUBANG RIVER: 1100 m, *Clemens 40341* (BM).

55.36.32. Dendrobium pachyanthum Schltr., Feddes Repert. 9: 290 (1911).

Epiphyte. Hill forest. Elevation: 600–1200 m.

General distribution: Sabah, Sarawak.

Collection. PORING HOT SPRINGS: 600–1200 m, *Lamb SAN 93360* (K).

55.36.33. Dendrobium pachyphyllum (Kuntze) Bakh. f. in Bakh. f. & J. Kost., Blumea 12: 69 (1963).

Callista pachyphylla Kuntze, Revis. Gen. Pl. 2: 654 (1891).

Epiphyte. Hill forest, lower montane forest, sometimes on ultramafic substrate. Elevation: 600–2100 m.

General distribution: Sabah, Kalimantan, Sarawak, India, Burma, Peninsular Malaysia, Singapore, Thailand, Vietnam, Sumatra, Java, Mentawai, Riau Archipelago.

Collections. BUNDU TUHAN: 600 m, *Carr SFN 27918* (SING); GURULAU SPUR: 1500–2100 m, *Clemens 50738* (BM); HEMPUEN HILL: *Clements 3397* (K).

55.36.34. Dendrobium pandaneti Ridl., J. Linn. Soc. Bot. 32: 257 (1896).

Epiphyte. Lowlands. Elevation: 400 m.

General distribution: Sabah, Peninsular Malaysia, Singapore, Thailand, Sumatra, Java, Lingga Archipelago.

Collections. KAUNG: 400 m, *Carr SFN 27336* (SING), 400 m, *Darnton 308* (BM).

55.36.35. Dendrobium panduriferum Hook. f., Fl. Brit. Ind. 6: 186 (1890). Plate 37A.

Epiphyte. Hill forest on ultramafic substrate. Elevation: 600 m.

General distribution: Sabah, Burma, Peninsular Malaysia, Thailand.

Collection. LOHAN RIVER: 600 m, *Lamb s.n.* (K).

55.36.36. Dendrobium parthenium Rchb. f., Gard. Chron. ser. 2, 24: 489 (1885). Plate 37B.

Epiphyte. Hill forest on ultramafic substrate. Elevation: 600–900 m.

Endemic to Mount Kinabalu.

Collections. HEMPUEN HILL: 900 m, *Cribb 89/59* (K), 600–900 m, *Lamb AL 311/85* (K), *Wood 837* (K); LOHAN RIVER: 800 m, *Bailes & Cribb 671* (K), 800–900 m, *Lamb LKC 3164* (K).

55.36.37. Dendrobium patentilobum Ames & C. Schweinf., Orch. 6: 110 (1920). Fig. 23, Plate 37C. Type: MARAI PARAI SPUR, *Clemens 366* (holotype AMES mf; isotypes BM!, K!, SING!).

Epiphyte. Hill forest, lower montane forest on ultramafic substrate. Elevation: 800–1500 m.

General distribution: Sabah.

Additional collections. HEMPUEN HILL: 800–1200 m, *Beaman 7715* (K); MAMUT COPPER MINE: 1400–1500 m, *Beaman 10321* (K); MARAI PARAI: 1200 m, *Collenette A 4* (BM); MARAI PARAI SPUR: *Clemens 226* (AMES, BM, K, SING); MARAI PARAI/KIAU TRAIL: 1400 m, *Bailes & Cribb 800* (K); MELANGKAP TOMIS: 900–1000 m, *Beaman 8401* (K), 900–1000 m, *8620* (K), 900–1000 m, *8982* (K); MOUNT KINABALU: *Chan s.n.* (K), *Clemens s.n.* (BM, K); NUMERUK RIDGE: 1400 m,

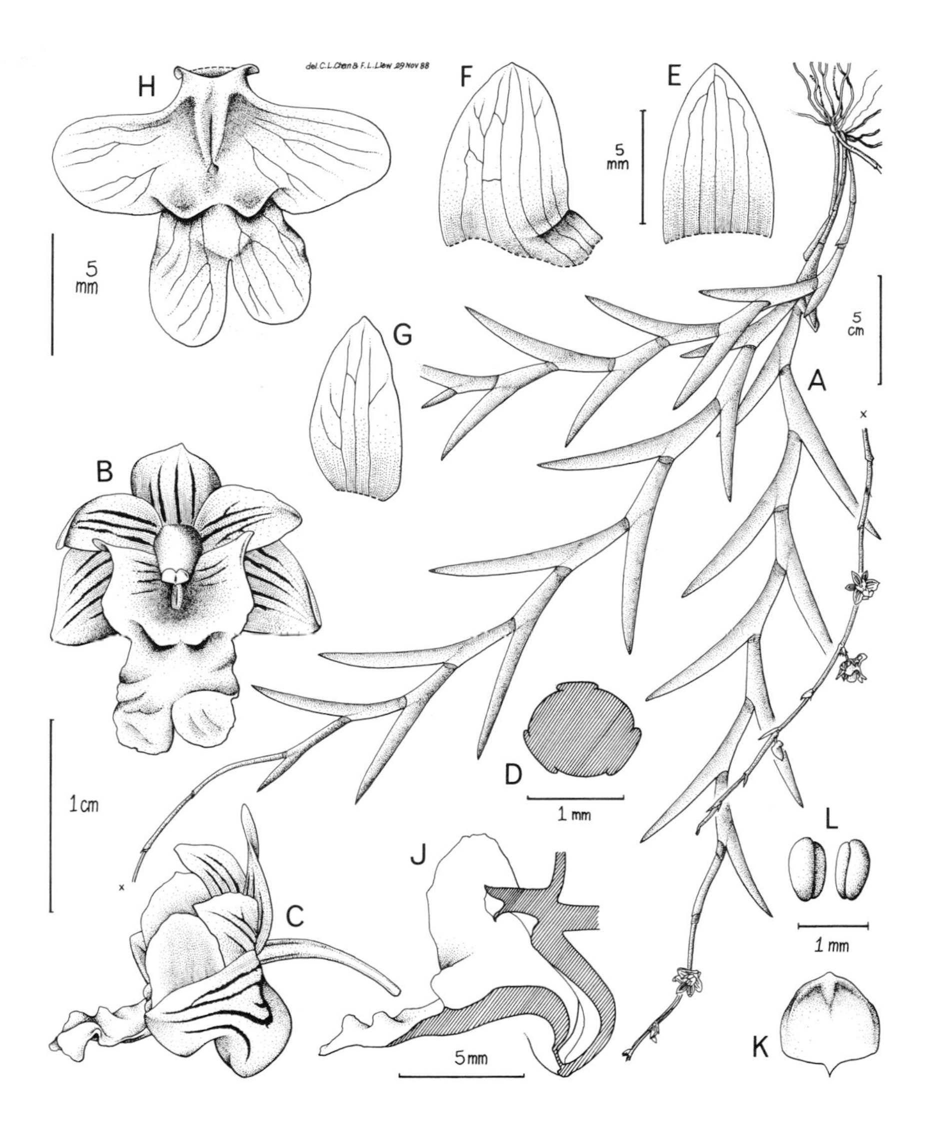

Fig. 23. Dendrobium patentilobum. **A**, habit; **B**, flower (front view); **C**, flower (side view); **D**, ovary (transverse section); **E**, dorsal sepal; **F**, lateral sepal; **G**, petal; **H**, lip (flattened, front view); **J**, pedicel-with-ovary, column and lip (longitudinal section); **K**, anther-cap (back view); **L**, pollinia. All from *Lamb AL 1117/89.* Drawn by Chan Chew Lun and F. L. Liew.

Clemens 40060 (BM); PENIBUKAN: *Carr 3044, SFN 26367* (SING), 1200–1500 m, *Clemens 30472* (BM, K), 1200 m, *30587* (BM, G, K), 1500 m, *34332* (BM), 1500 m, *50343* (BM, K).

55.36.38. Dendrobium pensile Ridl., J. Linn. Soc. Bot. 32: 253 (1896).

Pendulous epiphyte. Hill forest. Elevation: 900 m.

General distribution: Sabah, Peninsular Malaysia, Sumatra.

Collection. DALLAS: 900 m, *Clemens 26998* (BM, G, K).

55.36.39. Dendrobium piranha C. L. Chan & P. J. Cribb, ined. Plate 37D. Type: MARAI PARAI SPUR, 1600 m, *Bailes & Cribb 815* (holotype K!).

Epiphytic or terrestrial. Lower montane forest, low scrubby woodland on ultramafic substrate. Elevation: 1400–2400 m.

Endemic to Mount Kinabalu.

Additional collections. MARAI PARAI: 1500 m, *Clemens 32261* (BM), 1800–2400 m, *33135* (BM); MARAI PARAI SPUR: 1400–1700 m, *Lamb SAN 89671* (K).

55.36.40. Dendrobium prostratum Ridl., J. Linn. Soc. Bot. 32: 248 (1896).

Epiphyte. Hill forest on ultramafic substrate. Elevation: 600 m.

General distribution: Sabah, Sarawak, Peninsular Malaysia, Singapore.

Collection. HEMPUEN HILL: 600 m, *Lamb AL 1402/92* (K).

55.36.41. Dendrobium salaccense (Blume) Lindl., Gen. Sp. Orch., 86 (1830).

Grastidium salaccense Blume, Bijdr., 333 (1825).

Epiphyte. Lowlands. Elevation: 500 m.

General distribution: Sabah, Sarawak, China, Burma, Peninsular Malaysia, Thailand, Laos, Vietnam, Sumatra, Java.

Collection. KAUNG: 500 m, *Carr SFN 27389* (SING).

55.36.42. Dendrobium sanguinolentum Lindl., Bot. Reg. 28: misc. 62, no. 73 (1842).

Epiphyte. Hill forest. Apparently not collected on Mount Kinabalu. Sight record of A. Lamb (pers. comm.).

General distribution: Sabah, Peninsular Malaysia, Thailand, Sumatra, Java, Philippines.

55.36.43. Dendrobium secundum (Blume) Lindl., Bot. Reg. 15: t. 1291 (1829). Plate 39.

Pedilonum secundum Blume, Bijdr., 322 (1825).

Epiphyte. Hill forest. Elevation: 600 m.

General distribution: Sabah, Sarawak, Kalimantan, widespread from Burma to New Guinea.

Collection. BUNDU TUHAN: 600 m, *Carr 3692* (SING).

55.36.44. Dendrobium setifolium Ridl., J. Linn. Soc. Bot. 31: 270 (1896).

Epiphyte. Lower montane forest. Elevation: 1400–1600 m.

General distribution: Sabah, Sarawak, Peninsular Malaysia, Singapore, Thailand, Riau Archipelago.

Collection. PINOSUK PLATEAU: 1400–1600 m, *Lamb AL 17/82* (K).

55.36.45. Dendrobium aff. **setifolium** Ridl., J. Linn. Soc. Bot. 31: 270 (1896).

Epiphyte. Lowlands?, lower montane forest, sometimes on ultramafic substrate. Elevation: 300–1800 m.

Collections. GOLF COURSE SITE: 1700–1800 m, *Beaman 7219* (K); LUBANG: 1400 m, *Carr SFN 26855* (K, SING); MELANGKAP KAPA: 300 m, *Carr SFN 28024* (SING); PIG HILL: *Sutton 8* (K); TENOMPOK: 1500 m, *Clemens 28134* (BM, K).

55.36.46. Dendrobium singkawangense J. J. Sm., Gardens' Bull. 9: 91 (1935).

Epiphyte. Lower montane forest. Elevation: 1400 m.

General distribution: Sabah, Kalimantan.

Collection. GURULAU SPUR: 1400 m, *Carr 3144, SFN 26592* (L, SING).

55.36.47. Dendrobium singulare Ames & C. Schweinf., Orch. 6: 112 (1920). Type: LUBANG, *Clemens 118* (holotype AMES mf).

Epiphyte. Lower montane forest.

General distribution: Sabah.

Additional collection. LUBANG: *Clemens 222A* (AMES).

55.36.48. Dendrobium spectatissimum Rchb. f., Linnaea 41: 41 (1877). Plate 38. Type: BORNEO, *Lobb s.n.* (holotype W n.v.).

Dendrobium speciosissimum Rolfe, Orch. Rev. 3: 119, 240 (1895). Type: MOUNT KINABALU, *Low s.n.* (holotype K!).

Dendrobium reticulatum J. J. Sm., Bull. Jard. Bot. Buit., ser. 2, 13: 18 (1914). Type: MOUNT KINABALU, *Dumas s.n.* (holotype BO n.v.).

Epiphytic, 1-3 m above ground on small bushes. Lower montane forest; *Leptospermum* scrub on ultramafic substrate. Elevation: 1600–1700 m.

Endemic to Mount Kinabalu.

55.36.49. Dendrobium spegidoglossum Rchb. f., Bonplandia 2: 88 (1854).

Epiphyte. Lower montane forest. Elevation: 1200–1500 m.

General distribution: Sabah, Kalimantan, Sarawak, Burma, Peninsular Malaysia, Singapore, Thailand, Sumatra, Java, Riau Archipelago.

Collections. PENIBUKAN: 1200 m, *Clemens 30585* (BM), 1200 m, *30606* (BO); TENOMPOK ORCHID GARDEN: 1500 m, *Clemens 50375* (BM).

55.36.50. Dendrobium spurium (Blume) J. J. Sm., Orch. Java, 343 (1905).

Dendrocolla spuria Blume, Bijdr., 290 (1825).

Epiphyte. Hill forest on ultramafic substrate. Elevation: 800–900 m.

General distribution: Sabah, Kalimantan, Sarawak, Peninsular Malaysia, Sumatra, Java, Bali, Philippines.

Collections. HEMPUEN HILL: 800–900 m, *Lamb SAN 89687* (K), *Lohok 17* (K).

55.36.51. Dendrobium cf. **subulatoides** Schltr., Feddes Repert. 9: 290 (1911).

Epiphyte. Hill forest. Elevation: 900 m.

Collection. BUNDU TUHAN: 900 m, *Carr SFN 27767* (SING).

55.36.52. Dendrobium tetrachromum Rchb. f., Gard. Chron. ser. 2, 13: 712 (1880). Fig. 24, Plate 39B.

Fig. 24. Dendrobium tetrachromum. **A**, habit; **B**, flower (side view); **C**, dorsal sepal; **D**, lateral sepal; **E**, petal; **F**, lip (front view, flattened); **G**, pedicel-with-ovary, column and mentum (oblique view); **H**, anther-cap with pollinia (front view); **J**, pollinia. **A** from *Lamb AL 30/82* and *Collenette s.n.* (locality unknown); **B–J** from *Collenette s.n.* Scale: single bar = 1 mm; double bar = 1 cm. Drawn by Eleanor Catherine.

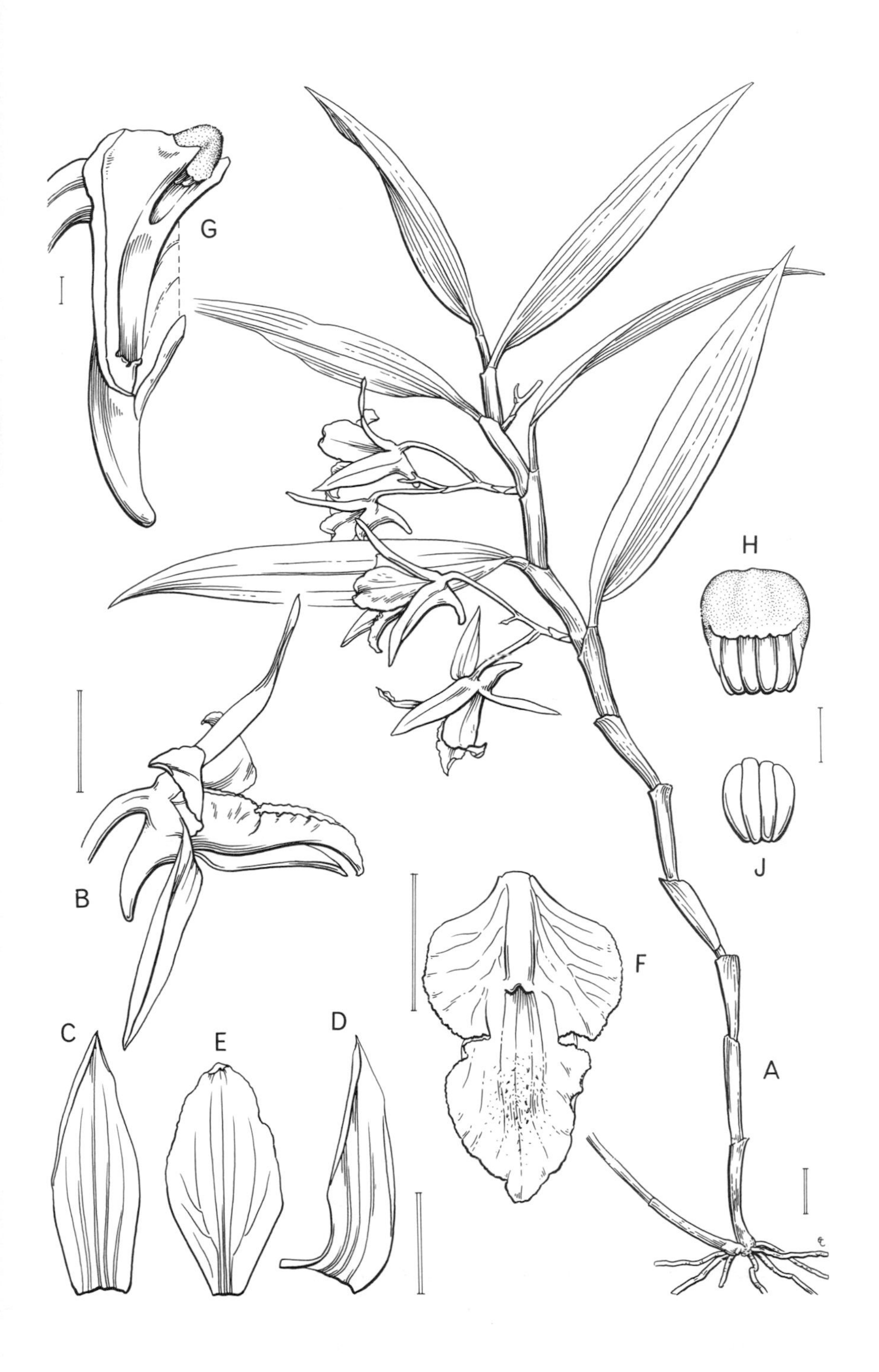
G
H
J
B
F
C
E
D
A

Epiphyte. Hill forest. Elevation: 500 m.

General distribution: Sabah, Kalimantan.

Collection. KINATEKI RIVER: 500 m, *Collenette A 127* (BM).

55.36.53. Dendrobium tridentatum Ames & C. Schweinf., Orch. 6: 115 (1920). Fig. 25, Plate 39C. Type: MARAI PARAI SPUR, *Clemens 257* (holotype AMES mf).

Epiphyte. Lower montane forest, upper montane forest, on ultramafic substrate. Elevation: 1500–2200 m.

General distribution: Sabah.

Additional collections. LAYANG-LAYANG/MESILAU CAVE ROUTE: 2100 m, *Collenette 892* (K); MARAI PARAI: 1500 m, *Carr SFN 27976* (SING), 1700 m, *Clemens 40111* (BM), 1700 m, *Collenette A 40* (BM), 1500 m, *A 82* (BM); MARAI PARAI SPUR: 1600 m, *Collenette A 74* (BM); MESILAU CAVE: 1900–2200 m, *Beaman 9574* (K); PIG HILL: *Lamb AL 733/87* (K), *Sutton 17* (K); TAHUBANG RIVER HEAD: 1500 m, *Clemens 40702* (BM).

55.36.54. Dendrobium uncatum Lindl., J. Linn. Soc. Bot. 3: 5 (1859).

Epiphyte. Lower montane forest. Elevation: 1400 m.

General distribution: Sabah, Java, Krakatau.

Collection. BUNDU TUHAN: 1400 m, *Carr SFN 27404* (SING).

55.36.55. Dendrobium uniflorum Griff., Not. 3: 305 (1851). Plate 40A.

Epiphyte. Hill forest. Elevation: 800–900 m.

General distribution: Sabah, Sarawak, Peninsular Malaysia, Thailand, Vietnam, Philippines.

Collections. BUNDU TUHAN: 900 m, *Carr SFN 27743* (SING); EASTERN SHOULDER: 800 m, *Lamb SAN 93472* (K); MOUNT KINABALU: *Lamb SAN 91573* (K).

55.36.56. Dendrobium ustulatum Carr, Gardens' Bull. 7: 13 (1932).

Epiphyte. Hill forest. Elevation: 900 m.

General distribution: Sabah, Sarawak, Peninsular Malaysia.

Collection. BUNDU TUHAN: 900 m, *Carr SFN 27958* (SING).

55.36.57. Dendrobium ventripes Carr, Gardens' Bull. 8: 103 (1935). Fig. 26, Plate 40B.

Epiphyte. Lower montane forest, upper montane forest on ultramafic substrate. Elevation: 1200–2600 m.

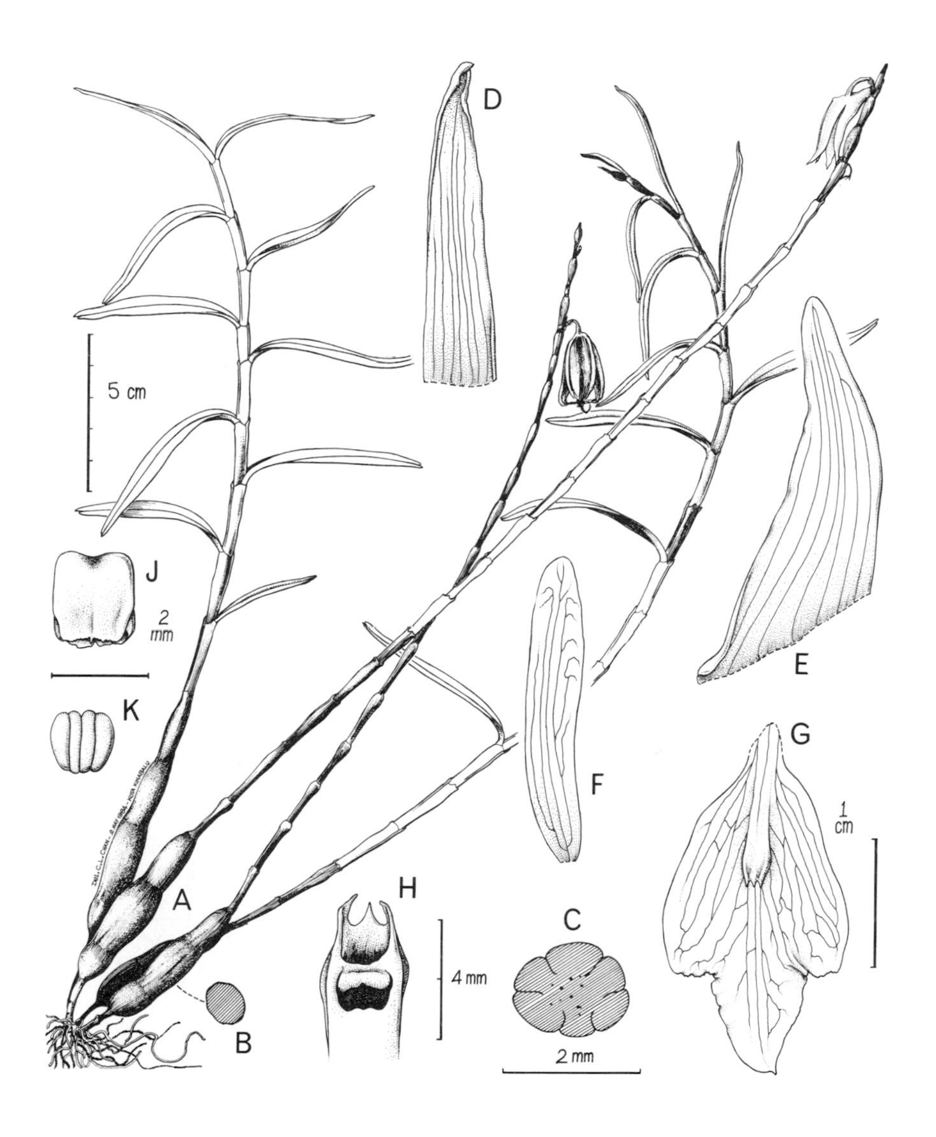

Fig. 25. Dendrobium tridentatum. **A**, habit; **B**, pseudobulb (transverse section); **C**, ovary (transverse section); **D**, dorsal sepal; **E**, lateral sepal; **F**, petal; **G**, lip (flattened, front view); **H**, column (front view); **J**, anther-cap (back view); **K**, pollinia. All from *Beaman 9574*. Drawn by Chan Chew Lun.

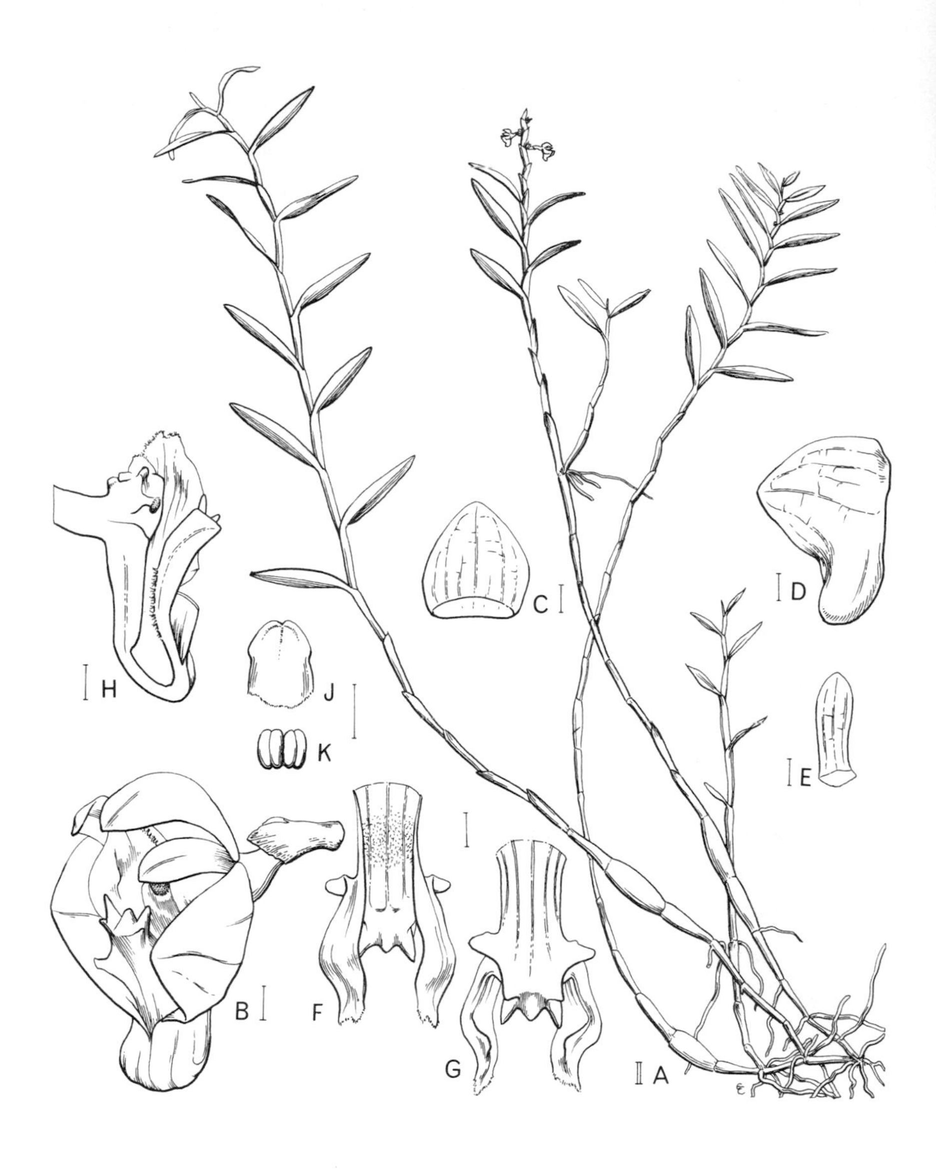

Fig 26. Dendrobium ventripes. **A**, habit; **B**, flower (oblique view); **C**, dorsal sepal; **D**, lateral sepal; **E**, petal; **F**, lip (front view); **G**, lip (back view); **H**, pedicel-with-ovary, column and lip (longitudinal section); **J**, anther-cap (back view); **K**, pollinia. **A** from *Synge 461*, Mt. Dulit, Sarawak; **B–K** from *Wood 879*, Mt. Trus Madi, Sabah. Scale: single bar = 1 mm; double bar = 1 cm. Drawn by Eleanor Catherine.

General distribution: Sabah, Sarawak.

Collections. KEMBURONGOH: 2100 m, *Carr SFN 27460* (SING), 2400 m, *Clemens 29122* (BM); MINETUHAN SPUR: 2600 m, *Clemens 33792* (BM); PENIBUKAN: 1200 m, *Clemens 35164* (BM).

55.36.58. **Dendrobium** aff. **xiphophyllum** Schltr., Feddes Repert. 9: 291 (1911).

Epiphyte. Low-stature hill forest on ultramafic substrate. Elevation: 700–1000 m.

Collections. LOHAN RIVER: 800–1000 m, *Beaman 9068* (K), 700–900 m, *9259* (K), *Lamb AL 500/85* (K).

55.36.59. Dendrobium sp. 1, sect. Aporum

Epiphyte. Lowlands. Elevation: 400–500 m.

Collection. MEKEDEU RIVER: 400–500 m, *Beaman 9649* (K).

55.36.60. Dendrobium sp. 2, sect. Calcarifera

Epiphyte. Hill forest, lower montane forest, sometimes on ultramafic substrate. Elevation: 600–1500 m.

Collections. BUNDU TUHAN: 600 m, *Carr SFN 27881* (SING); MAHANDEI RIVER: 1100 m, *Carr 3117, SFN 26535* (SING).

55.36.61. Dendrobium sp. 3, sect. Rhopalanthe

Epiphyte. Lower montane ridge forest on ultramafic substrate. Elevation: 1700 m.

Collection. MARAI PARAI SPUR: 1700 m, *Bailes & Cribb 839* (K).

55.36.62. **Dendrobium** indet.

Collections. KAUNG: 500 m, *Carr SFN 28003* (SING); MAMUT COPPER MINE: 1400–1500 m, *Beaman 10345* (K); MENSAHABAN RIVER: *Tiong SAN 88658* (K).

55.37. **DENDROCHILUM** Blume

Bijdr., 398 (1825).

Epiphytic, lithophytic, rarely terrestrial herbs. *Pseudobulbs* tufted, narrow, fusiform or ovoid, 1-leaved. *Leaves* flat, linear to narrowly elliptic, coriaceous. *Inflorescence* lateral, slender, suberect to pendulous, spicate or racemose, usually densely many-flowered, synanthous (in sect. *Platyclinis*) and heteranthous (in sect. *Dendrochilum*). *Flowers* small, usually resupinate, thin-textured. *Sepals* subequal, spreading. *Lateral sepals* adnate to base of column. *Petals* smaller than sepals, often erose. *Lip* 3-lobed or entire, usually with small side-lobes and a large mid-lobe, disc 2- or 3-keeled. *Column* usually short, curved, without a foot, with narrow lateral arms and an often toothed apical wing around anther. *Pollinia* 4.

Between 100 and 150 species distributed from mainland Asia east to the Philippines and New Guinea, particularly well represented in the montane areas of Sumatra and Borneo.

55.37.1. Dendrochilum acuiferum Carr, Gardens' Bull. 8: 227 (1935). Type: PAKA-PAKA CAVE, 3000 m, *Carr 3550, SFN 27645* (holotype SING!; isotypes AMES mf, K!).

Terrestrial. Upper montane forest, under dwarf *Leptospermum*, or in the open, on ultramafic substrate. Elevation: 3000 m. Known only from the type.

Endemic to Mount Kinabalu.

55.37.2. Dendrochilum alatum Ames, Orch. 6: 45, pl. 82, 3 (1920). Plate 40C, 40D. Type: MARAI PARAI SPUR, *Clemens 383* (holotype AMES mf; isotypes BM!, K!, SING!).

Epiphyte or lithophyte. Lower montane forest, upper montane forest, in scrub on ridges, on ultramafic and granitic substrates. Elevation: 1700–3200 m.

General distribution: Sabah, Sarawak.

Additional collections. GURULAU SPUR: 2400–2700 m, *Clemens 50652* (BM, K), 2100–2700 m, *50777* (BM, K); KEMBURONGOH: 2000 m, *Carr 3715* (SING); MESILAU CAVE TRAIL: 1700–1900 m, *Beaman 8005* (K); PANAR LABAN: 3200 m, *Smith & Everard 155* (K); TENOMPOK: 1700 m, *Clemens s.n.* (BM).

55.37.3. Dendrochilum alpinum Carr, Gardens' Bull. 8: 235 (1935). Fig. 27, Plate 41A. Type: SAYAT-SAYAT, 3500 m, *Carr 3545, SFN 27624* (holotype SING!; isotypes AMES mf, K!).

Lithophyte. Upper montane forest, summit area, on granitic rocks sheltered by stunted trees. Elevation: 2400–3700 m.

Endemic to Mount Kinabalu.

Additional collections. GURULAU SPUR: 2400–2700 m, *Clemens 50669* (BM); PANAR LABAN/SAYAT-SAYAT: 3400–3700 m, *Sato 759* (UKMS), 3400–3700 m, *762* (UKMS); SUMMIT AREA: 3500 m, *Sato et al. 1375* (UKMS).

55.37.4. Dendrochilum angustilobum Carr, Gardens' Bull. 8: 222 (1935). Type: TENOMPOK, 1500 m, *Carr 3233, SFN 26874* (holotype SING!; isotypes AMES mf, K!).

Epiphytic or terrestrial. Lower montane forest. Elevation: 1400–1600 m.

Fig. 27. Dendrochilum alpinum. **A**, habit; **B**, inflorescence; **C**, flower (front and side views), **D**, dorsal sepal; **E**, lateral sepal; **F**, petal; **G**, lip (front view); **H**, column and top of pedicel-with-ovary (front view); **J**, anther-cap (back view). All from *Carr 3545*. Scale: single bar = 1 mm; double bar = 1 cm. Drawn by Eleanor Catherine.

General distribution: Sabah.

Additional collections. KINATEKI RIVER HEAD: 1500 m, *Clemens s.n.* (BM); PARK HEADQUARTERS: 1400 m, *Price 217* (K); TENOMPOK: 1500 m, *Clemens 28949* (BM), 1500 m, *29295* (BM), 1500 m, *29361* (BM, K); TENOMPOK/TOMIS: 1600 m, *Clemens 29412* (BM).

55.37.5. Dendrochilum angustipetalum Ames, Orch. 6: 47, pl. 83b (1920). Type: MARAI PARAI SPUR, *Clemens 270* (holotype AMES mf).

Epiphyte. Lower montane forest, on ultramafic substrate. Elevation: 1200–2000 m.

Endemic to Mount Kinabalu.

Additional collections. KINATEKI RIVER: 1200 m, *Collenette A 104* (BM); MARAI PARAI SPUR: 2000 m, *Carr 3684, SFN 28019* (K, SING); PENIBUKAN: 1400 m, *Clemens 40255* (BM), 1400 m, *40467* (BM, K), 1400 m, *51714* (BM, K).

55.37.6. Dendrochilum anomalum Carr, Gardens' Bull. 8: 87 (1935).

Epiphyte. Hill forest on ultramafic substrate. Elevation: 600 m.

General distribution: Sabah, Sarawak.

Collection. HEMPUEN HILL: 600 m, *Lamb AL 1365/91* (K).

55.37.7. Dendrochilum conopseum Ridl. in Stapf, Trans. Linn. Soc. Bot. 4: 236 (1894). Type: MARAI PARAI SPUR, 1700 m, *Haviland s.n.* (holotype? SAR n.v.).

Epiphyte. Lowlands, hill forest, lower montane forest, sometimes on ultramafic substrate. Elevation: 400–1700 m.

General distribution: Sabah, Sarawak.

Additional collections. DALLAS: 900 m, *Clemens 26899* (BM, K); KAUNG: 400 m, *Carr 3412, SFN 27315* (K, SING), *Haviland 1381* (K); MAHANDEI RIVER: 1100 m, *Carr 3186, SFN 26759* (K, SING); MARAI PARAI: 1200 m, *Clemens 32556* (BM); PENIBUKAN: 1200–1500 m, *Clemens 31337* (BM), 1200 m, *40633* (BM, K).

55.37.8. Dendrochilum corrugatum (Ridl.) J. J. Sm., Recueil Trav. Bot. Néerl. 1: 65 (1904).

Platyclinis corrugata Ridl. in Stapf, Trans. Linn. Soc. Bot. 4: 233 (1894). Type: MARAI PARAI SPUR, 1700 m, *Haviland s.n.* (holotype? SING!).

Epiphyte. Lower montane forest, on ultramafic substrate. Elevation: 1700 m. Known only from the type.

Endemic to Mount Kinabalu.

55.37.9. Dendrochilum crassifolium Ames, Orch. 6: 49, pl. 84a (1920). Type: MOUNT KINABALU, *Haslam s.n.* (holotype AMES mf).

Epiphyte. Lower montane forest. Elevation: 1500–1800 m.

General distribution: Sabah.

Additional collections. TENOMPOK: 1600 m, *Carr 3671, SFN 28047* (K, SING), *Clemens 40434* (BM), 1500 m, *50234* (SING), *50258* (BM); TENOMPOK ORCHID GARDEN: 1500 m, *Clemens 40434* (K), 1500 m, *50235* (BM, K), 1500 m, *50258* (K); TINEKUK FALLS: 1800 m, *Clemens 40931* (BM).

55.37.10. Dendrochilum crassum Ridl., J. Linn. Soc. Bot. 32: 288 (1896). Fig. 28, Plate 41B, 41C.

Lithophyte. Hill forest; rocky banks, cliffs. Elevation: 900 m.

General distribution: Sabah, Peninsular Malaysia.

Collection. MINITINDUK GORGE: 900 m, *Carr 3172, SFN 26668* (K, SING).

55.37.11. Dendrochilum dewindtianum W. W. Sm., Notes Roy. Bot. Gard. Edinburgh 8: 321 (1915). Fig. 29, Plate 41D. Type: MOUNT KINABALU, *Native collector 68* (syntype E n.v.), 2200 m, *99* (syntype E n.v.; isosyntype K!).

Dendrochilum perspicabile Ames, Orch. 6: 62, Pl. 82, IV, 5 (1920). Type: MOUNT KINABALU, 1500–3000 m, *Clemens 202* (holotype AMES mf; isotypes BM!, K!, SING!).

Epiphytic or terrestrial. Lower montane forest, upper montane forest, mostly on ultramafic substrate. Elevation: 1500–3000 m.

General distribution: Sabah.

Additional collections. GURULAU SPUR: 2100–2700 m, *Clemens 50774* (BM, K); KEMBURONGOH: 2200 m, *Carr SFN 27562* (K), 2400 m, *Clemens 27140* (BM, K, SING), 2700 m, *27145* (BM, K), 2300 m, *27148* (BM, K), 2300 m, *Meijer SAN 20367* (K); LUMU-LUMU: 2000 m, *Carr 3533, SFN 27562* (K, SING); MARAI PARAI SPUR: 2000 m, *Carr 3534* (SING); MOUNT KINABALU: *Haslam s.n.* (BM, K), 2200 m, *Native collector s.n.* (K).

55.37.12. Dendrochilum exasperatum Ames, Orch. 6: 50 (1920). Fig. 30. Type: MARAI PARAI SPUR, *Clemens 396* (holotype AMES mf).

Epiphyte. Hill forest, lower montane forest. Elevation: 900–1600 m.

General distribution: Sabah, Sarawak.

Additional collections. DALLAS: 900 m, *Clemens 26784A* (BM), 900 m, *27256* (BM, K); PENIBUKAN: 1200–1500 m, *Clemens s.n.* (BM), 1200 m, *30640* (BM), 1200 m, *30799* (BM); TENOMPOK: 1600 m, *Carr 3668, SFN 28029* (K, SING), 1500 m, *3718, SFN 28049* (SING), *Clemens 50246* (BM), 1500 m, *50256* (SING); TENOMPOK ORCHID GARDEN: 1500 m, *Clemens 50235* (K), 1500 m, *50246* (K).

Fig. 28. Dendrochilum crassum. **A**, habit; **B**, flower; **C**, dorsal sepal; **D**, lateral sepal; **E**, petal; **F**, lip (front view); **G**, column (side view); **H**, anther-cap (back view); **J**, pollinia. **A** from *Vermeulen & Duistermaat 693*, Crocker Range, Sabah; **B–J** from *Wood 733*, Crocker Range, Sabah. Scale: single bar = 1 mm; double bar = 1 cm. Drawn by Eleanor Catherine.

55.37.13. Dendrochilum fimbriatum Ames, Orch. 6: 51 (1920). Type: MARAI PARAI SPUR, *Clemens 248* (holotype AMES mf).

Epiphyte. Lower montane forest, ridge scrub, on ultramafic substrate. Elevation: 1500–2100 m.

Endemic to Mount Kinabalu.

Additional collections. MARAI PARAI: 1700 m, *Carr 3128, SFN 27428* (BM, K, SING), 1500 m, *Clemens 32244* (BM), 1500 m, *Collenette A 31* (BM); MARAI PARAI SPUR: 1700 m, *Bailes & Cribb 841* (K); PENATARAN BASIN: 1700 m, *Clemens 40135* (BM); TAHUBANG RIVER HEAD: 2100 m, *Clemens 32548* (BM).

55.37.14. Dendrochilum gibbsiae Rolfe in Gibbs, J. Linn. Soc. Bot. 42: 147 (1914). Fig. 31, Plate 42A. Type: MARAI PARAI SPUR, 2100 m, *Gibbs 4087* (holotype BM!; isotype K!).

Dendrochilum kinabaluense Rolfe in Gibbs, J. Linn. Soc. Bot. 42: 148 (1914). Type: MARAI PARAI SPUR, 2100 m, *Gibbs 4085* (holotype BM!; isotype K!).

Dendrochilum quinquelobum Ames, Orch. 6: 63, Pl. 82, II, 2 (1920). Type: KIAU, *Clemens 361* (holotype AMES mf; isotypes BM!, K!, SING!).

Epiphytic, terrestrial. Hill forest, lower montane forest, upper montane forest, frequently on ultramafic substrate. Elevation: 900–2400 m.

General distribution: Sabah, Sarawak.

Additional collections. BUNDU TUHAN: 900 m, *Carr 3620, SFN 27884* (K, SING); DALLAS: 900 m, *Clemens s.n.* (BM), 900 m, *26784* (BM), 900 m, *26841* (BM, K), 900 m, *26930* (BM); GOLF COURSE SITE: 1600 m, *Bailes & Cribb 722* (K), 1700–1800 m, *Beaman 7239* (K); KADAMAIAN RIVER: 1800 m, *Carr 3709, SFN 28034* (SING); KIAU: 900 m, *Clemens 146* (AMES, BM, K, SING), *178* (AMES, BM, K); KILEMBUN RIVER: 1400 m, *Clemens 32432* (BM, K); LUBANG: *Clemens 289* (AMES, BM, K); MAHANDEI RIVER: 1100 m, *Carr 3134, SFN 26745* (SING); MARAI PARAI: 1500 m, *Carr 3669, SFN 27970* (SING); MARAI PARAI/NUNGKEK: 1400 m, *Clemens 32552* (BM); PARK HEADQUARTERS: 1500 m, *Wood 621* (K); PENIBUKAN: 1200 m, *Clemens s.n.* (BM), 1200–1500 m, *30573* (BM, K), 1200–1500 m, *31457* (BM, K); PIG HILL: 2000–2300 m, *Beaman 9889* (K); TAHUBANG FALLS: 1200 m, *Clemens 30702* (BM); TENOMPOK: 1500 m, *Clemens 29235* (BM, K), 1500 m, *50184* (BM), 1500 m, *Meijer SAN 20326* (K); TENOMPOK ORCHID GARDEN: 1500 m, *Clemens 50177* (BM), 1500 m, *50184* (K); TINEKUK FALLS: 1500–2400 m, *Clemens 40925* (BM, K).

55.37.15. Dendrochilum gracile (Hook. f.) J. J. Sm., Recueil Trav. Bot. Néerl. 1: 69 (1904).

Platyclinis gracilis Hook. f., Fl. Brit. Ind. 5: 708 (1890).

Epiphyte. Low-stature lower montane ridge forest on ultramafic substrate. Elevation: 1400–1500 m.

General distribution: Sabah, Sarawak, Peninsular Malaysia, Sumatra, Java.

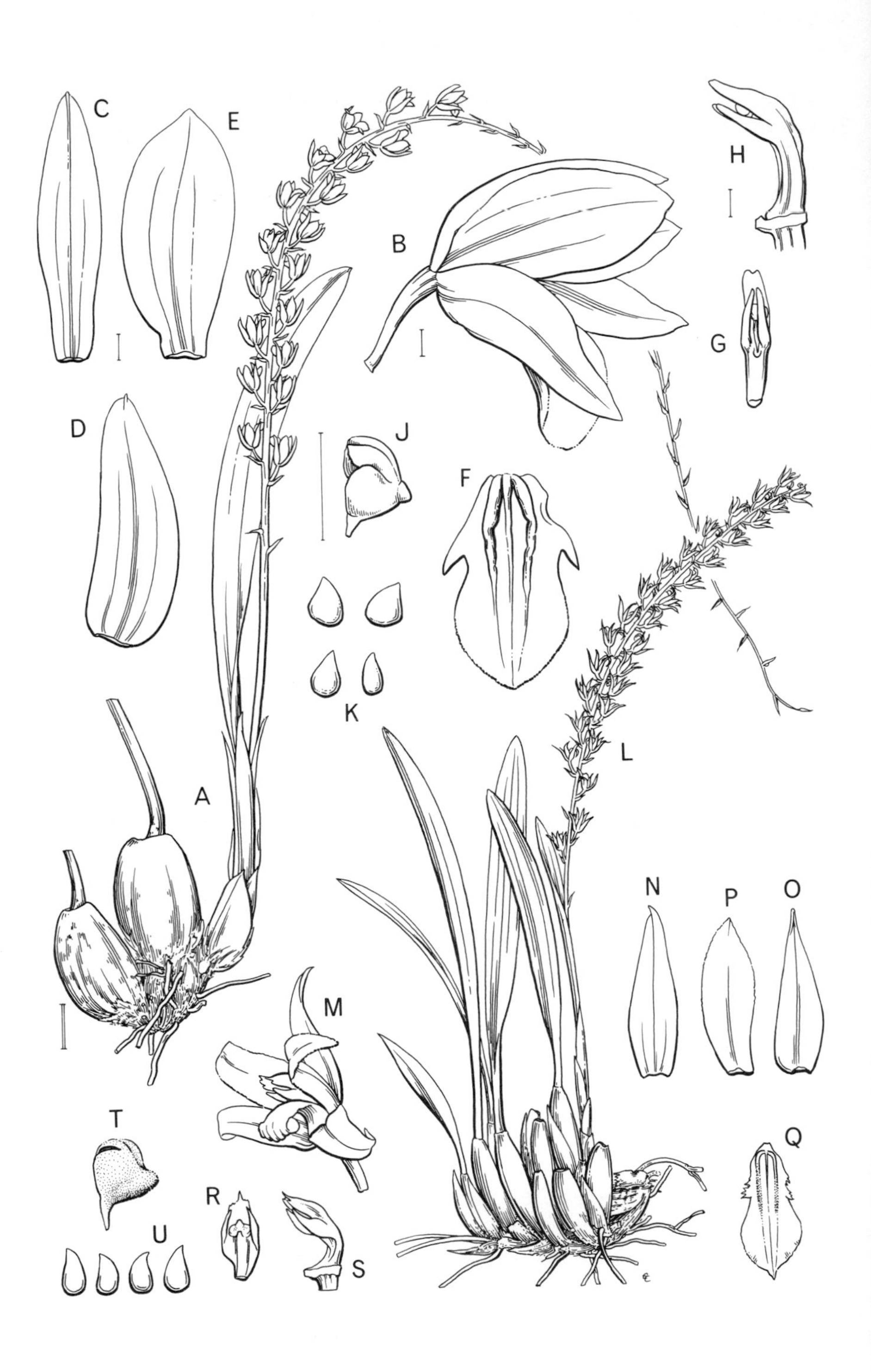
C
E
H
B
G
D
J
F
K
L
A
N
P
O
M
T
Q
R
U
S

Collection. MAMUT COPPER MINE: 1400–1500 m, *Beaman 10347* (K).

55.37.16. Dendrochilum graminoides Carr, Gardens' Bull. 8: 229 (1935). Type: PINOSUK, 900 m, *Carr 3680, SFN 28006* (holotype SING!; isotypes AMES mf, K!).

Epiphyte. Hill forest, lower montane forest, sometimes on ultramafic substrate. Elevation: 900–2700 m.

General distribution: Sabah, Sarawak.

Additional collections. GURULAU SPUR: 2100–2700 m, *Clemens 50774* (BM); KADAMAIAN RIVER: 2000 m, *Carr 3675, SFN 28004* (K, SING); KINATEKI RIVER HEAD: 1500 m, *Clemens s.n.* (BM); MARAI PARAI SPUR: 1700 m, *Collenette A 41* (BM); PENATARAN BASIN: 900–1200 m, *Clemens 34329* (BM), 900 m, *40134* (BM); PENIBUKAN: 1200–1500 m, *Clemens s.n.* (BM, BM, BM), 1200–1500 m, *30826* (BM, K), 1200 m, *32220* (BM), 1500 m, *50322* (BM, K); TAHUBANG RIDGE: 1200 m, *Clemens 30471* (BM); TINEKUK FALLS: 1800 m, *Clemens 50278* (BM, K).

55.37.17. Dendrochilum grandiflorum (Ridl.) J. J. Sm., Recueil Trav. Bot. Néerl. 1: 66 (1904). Fig. 32, Plate 42B.

Platyclinis grandiflora Ridl. in Stapf, Trans. Linn. Soc. Bot. 4: 233 (1894). Type: MOUNT KINABALU, 3200 m, *Haviland s.n.* (holotype unlocated).

Epiphytic or terrestrial. Upper montane forest on ultramafic substrate. Elevation: 1500–3800 m.

Endemic to Mount Kinabalu.

Additional collections. GURULAU SPUR: 2400–2700 m, *Clemens 50661* (BM, K), 2100–2700 m, *50768* (BM), 2700 m, *50801* (BM, K), 3800 m, *51534* (BM, K); KADAMAIAN RIVER, JUST BELOW PAKA-PAKA CAVE: 2900 m, *Sinclair et al. 9180* (K, SING); KEMBURONGOH: 2100 m, *Carr 3476, SFN 27430* (K, SING), 2400–3000 m, *Clemens 209* (AMES, BM, K, SING), 2700 m, *27142* (K, SING), 2700 m, *27149* (K, SING), 2600 m, *27867* (BM, K, SING), 2600 m, *Meijer SAN 29165* (K); KEMBURONGOH/PAKA-PAKA CAVE: 2700 m, *Clemens 27866* (BM, SING), 2700 m, *29128* (BM, K); KINATEKI RIVER HEAD: 2400 m, *Clemens 31835* (BM); LAYANG-LAYANG: 2400–2700 m, *Sato 699* (UKMS); LIWAGU RIVER HEAD: 2300 m, *Meijer SAN 24131* (K); MARAI PARAI: 2400 m, *Clemens 31683* (BM), 2400–3000 m, *33173* (BM), 2400–3000 m, *33178* (BM, K), 2400–3000 m, *33180* (BM); MOUNT KINABALU: *Haslam s.n.* (BM, K); PAKA-PAKA CAVE: 3100 m, *Carr 3476, SFN 27430A* (K, SING), 2700–3000 m, *Gibbs 4250* (BM); SUMMIT TRAIL: 2700–3000 m, *Sato 1223* (UKMS), 2800 m, *Sato et al. 970* (UKMS), 2800 m, *972* (UKMS), 3000 m, *974* (UKMS), 2600 m, *1046* (UKMS), 2600 m, *1047* (UKMS), 2600 m, *1080* (UKMS), 2700 m, *1081* (UKMS), 2600 m, *1082* (UKMS), 2700–3000 m, *Wood 608* (K); TENOMPOK: 1500 m, *Clemens 30101* (K), 1500 m, *30147* (K, SING); UPPER KINABALU: *Clemens 30103* (K).

Fig. 29. Dendrochilum dewindtianum. – Showing two extremes in variation. **A**, habit; **B**, flower (side view); **C**, dorsal sepal; **D**, lateral sepal; **E**, petal; **F**, lip (front view); **G**, column (front view); **H**, column (side view); **J**, anther-cap (side view); **K**, pollinia; **L**, habit; **M**, flower (side view); **N**, dorsal sepal; **O**, lateral sepal; **P**, petal; **Q**, lip (front view); **R**, column (front view); **S**, column (side view); **T**, anther-cap (side view); **U**, pollinia. **A–K** from *Native Collector 99;* **L** from *Carr SFN 27562*; **M–U** from *Wood 583*, Crocker Range, Sabah. Scale: single bar = 1 mm; double bar = 1 cm. Drawn by Eleanor Catherine.

Fig. 30. Dendrochilum exasperatum. **A**, habit; **B**, flower (side view); **C**, dorsal sepal; **D**, lateral sepal; **E**, petal; **F**, lip (front and side views); **G**, column (front view); **H**, anther-cap (back view); **J**, pollinia. All from *Comber 122*, Mt. Lumaku, Sabah. Scale: single bar = 1 mm; double bar = 1 cm. Drawn by Eleanor Catherine.

Fig. 31. Dendrochilum gibbsiae. **A** and **B**, habit; **C**, flower (front view); **D**, flower (side view); **E**, pedicel-with-ovary, column and longitudinal section through lip; **F**, dorsal sepal; **G**, lateral sepal; **H**, petal; **J**, lip (front view); **K**, anther-cap (back view); **L**, pollinia. **A** from *Wood 578*, Crocker Range, Sabah; **B** from *Carr 3620*; **C–L** from *Bailes & Cribb 722.* Scale: single bar = 1 mm; double bar = 1 cm. Drawn by Eleanor Catherine.

Fig. 32. Dendrochilum grandiflorum. **A**, habit; **B**, flower; **C**, dorsal sepal; **D**, lateral sepal; **E**, petal; **F**, lip (front view); **G**, pedicel-with-ovary and column (front view); **H**, anther-cap (back view); **J**, pollinia. **A** from *Carr 3476* and *Clemens 27867*; **B–J** from *Carr 3476*. Scale: single bar = 1 mm; double bar = 1 cm. Drawn by Eleanor Catherine.

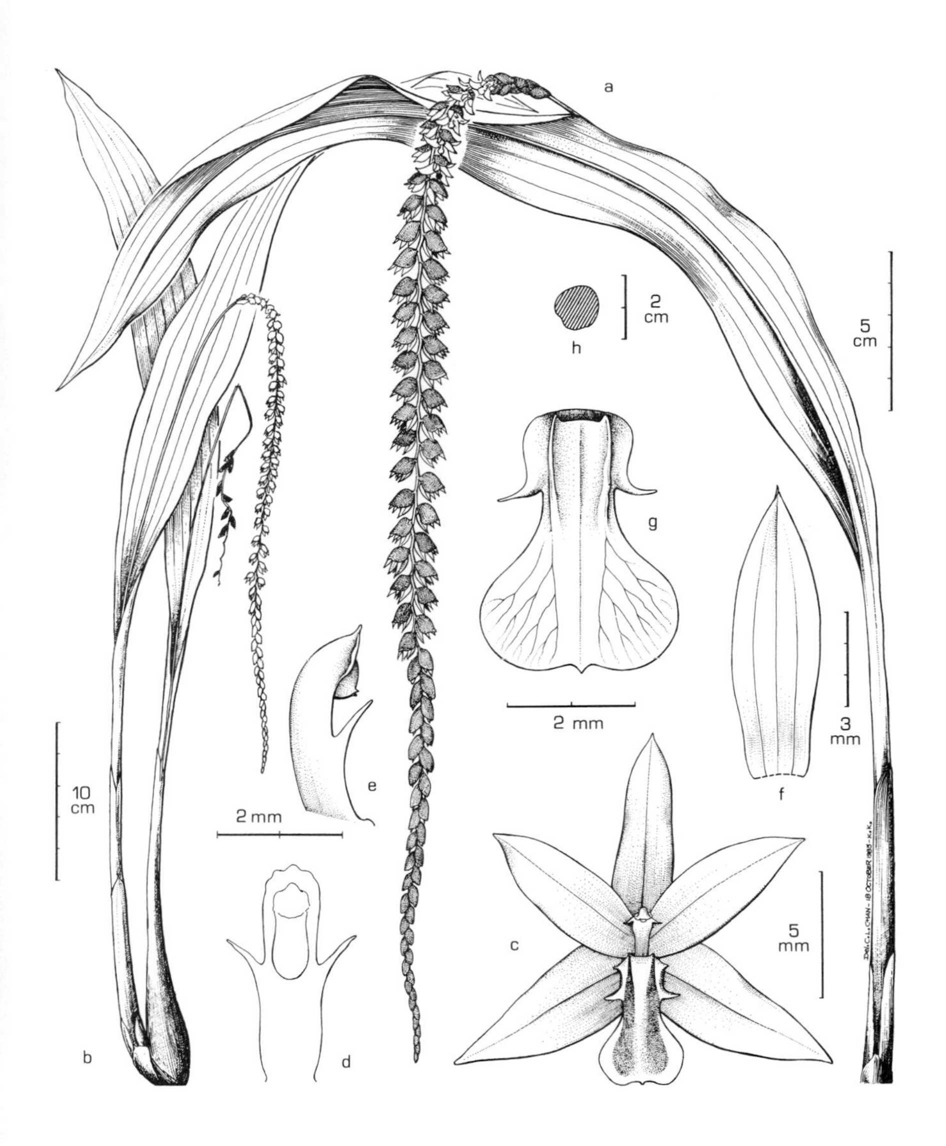

Fig. 33. Dendrochilum imbricatum. **a**, inflorescence and leaf; **b**, habit; **c**, flower (front view); **d**, column (front view); **e**, column (side view); **f**, dorsal sepal; **g**, lip (front view); **h**, pseudobulb (transverse section). All from *Phillipps & Lamb s.n.*, Liwagu River. Drawn by Chan Chew Lun.

55.37.18. Dendrochilum haslamii Ames, Orch. 6: 53, pl. 85 (1920). Plate 42C. Type: MOUNT KINABALU, *Haslam s.n.* (holotype AMES mf).

Epiphyte. Upper montane ridge forest, frequently on ultramafic substrate. Elevation: 2400–3100 m.

General distribution: Sabah, Sarawak.

Additional collections. GURULAU SPUR: 2400–2700 m, *Clemens 50650* (BM, K); JANET'S HALT/SHEILA'S PLATEAU: 2900 m, *Collenette 21535* (K); KEMBURONGOH: 2400 m, *Clemens s.n.* (BM); KEMBURONGOH/PAKA-PAKA CAVE: 2900 m, *Carr 3528, SFN 27864* (K, SING); KINATEKI RIVER HEAD: 2700 m, *Clemens 31830* (BM); PAKA-PAKA CAVE: 3100 m, *Carr SFN 36565* (SING); SUMMIT TRAIL: 2700 m, *Gunsalam 3* (K).

55.37.19. Dendrochilum imbricatum Ames, Orch. 6: 54, pl. 82 I (1920). Fig. 33, Plate 42D, 42E. Type: KIAU, 900 m, *Clemens 179* (holotype AMES mf; isotypes BM!, K!, SING!).

Epiphyte. Hill forest, lower montane forest. Elevation: 900–1500 m.

General distribution: Sabah, Sarawak.

Additional collections. KIAU: *Clemens 318* (AMES); MAMUT RIVER: 1400 m, *Collenette 1042* (K); PARK HEADQUARTERS: 1500 m, *Wood 618* (K); PENIBUKAN: 1200 m, *Clemens 30578* (BM); TENOMPOK: 1500 m, *Carr 3653, SFN 27998* (SING); TENOMPOK ORCHID GARDEN: 1500 m, *Clemens 50162* (BM, K).

55.37.20. Dendrochilum joclemensii Ames, Orch. 6: 55, pl. 83 (1920). Type: MARAI PARAI SPUR, *Clemens 247* (holotype AMES mf).

Epiphyte. Lower montane forest, sometimes on ultramafic substrate. Elevation: 1800–2000 m.

Endemic to Mount Kinabalu.

Additional collections. KADAMAIAN RIVER: 2000 m, *Carr 3710* (SING); KIAU VIEW TRAIL: 1800 m, *Gunsalam 10* (K).

55.37.21. Dendrochilum kamborangense Ames, Orch. 6: 57, pl. 84 (1920). Type: KEMBURONGOH, *Clemens 205* (holotype AMES mf; isotypes BM!, K!, SING!).

Epiphyte. Upper montane mossy forest, often in exposed places on ultramafic substrate. Elevation: 1500–2900 m.

Endemic to Mount Kinabalu.

Additional collections. EASTERN SHOULDER: 2400 m, *RSNB 182* (SING); GURULAU SPUR: 2400–2700 m, *Clemens 50654* (BM, K); JANET'S HALT/SHEILA'S PLATEAU: 2900 m, *Collenette 21534* (K); KEMBURONGOH: 2300 m, *Carr 3622, SFN 27908* (K, SING), 2100 m, *3751* (SING), 2200 m, *SFN 36567* (SING), 2700 m, *Clemens 27143* (BM); LUMU-LUMU: 2100 m, *Clemens 27860* (BM); MARAI PARAI SPUR: *Clemens 385* (AMES, BM, K, SING); MESILAU CAVE: 1600 m, *Bailes & Cribb 695* (K); MESILAU CAVE/JANET'S HALT: 2400 m, *Fuchs & Collenette 21404* (K); MOUNT KINABALU: 2400 m, *Carr 3742* (SING), *Jumaat 3397* (UKMS); PAKA-PAKA CAVE: 2600 m, *Clemens 27146* (BM, K); SUMMIT TRAIL: 2200 m, *Sato 2100* (UKMS), 2300 m, *2158* (UKMS), 2400 m, *Sato et al. 1436*

(UKMS); TENOMPOK: 1500 m, *Clemens 27860* (K); TENOMPOK ORCHID GARDEN: 1500 m, *Clemens 50235* (BM).

55.37.22. Dendrochilum lacteum Carr, Gardens' Bull. 8: 223 (1935). Fig. 34. Type: LUMU-LUMU, 1800 m, *Carr 3608, SFN 27892* (holotype SING!; isotypes AMES mf, K!).

Epiphyte. Lower montane forest. Elevation: 1400–1800 m.

General distribution: Sabah, Sarawak.

Additional collection. PARK HEADQUARTERS: 1400 m, *Price 145* (K).

55.37.23. Dendrochilum lancilabium Ames, Orch. 6: 58, pl. 83 (1920). Type: MARAI PARAI SPUR, *Clemens 280* (holotype AMES mf; isotypes BM!, K!, SING!).

Epiphyte. Upper montane forest on ultramafic substrate. Elevation: 2000–2400 m.

General distribution: Sabah, Sarawak.

Additional collections. KEMBURONGOH: 2300 m, *Carr 3752* (SING); LUMU-LUMU: 2100 m, *Clemens s.n.* (BM); MARAI PARAI SPUR: *Clemens 224A* (AMES), *242* (AMES), *386* (BM, K, SING); MESILAU RIVER: 2400 m, *Clemens 51626* (BM); MOUNT KINABALU: *Haslam s.n.* (AMES, BM, K); PAKA-PAKA CAVE: *Clemens 114* (AMES); PIG HILL: 2000–2300 m, *Beaman 9888* (K); UPPER KINABALU: *Clemens 27147* (K, SING).

55.37.24. Dendrochilum lobongense Ames, Orch. 6: 59 (1920). Type: LUBANG, *Clemens 116* (holotype AMES mf).

Epiphyte. Lower montane forest. Known only from the type.

Endemic to Mount Kinabalu.

55.37.25. Dendrochilum longirachis Ames, Orch. 6: 60 (1920). Type: KIAU, *Clemens 332* (holotype AMES mf; isotype K!).

Epiphyte. Lower montane forest, sometimes on ultramafic substrate. Elevation: 1200–1500 m.

General distribution: Sabah, Sarawak.

Additional collections. GURULAU SPUR: 1200 m, *Clemens 51080* (BM, K); MAMUT RIVER: 1200 m, *Sato 1558* (UKMS); MARAI PARAI SPUR: *Clemens 377* (AMES, BM, K, SING); MOUNT KINABALU: 1500 m, *Clemens 35205* (BM); PENIBUKAN: 1200 m, *Clemens 30601* (BM), 1500 m, *31002* (BM); TAHUBANG RIVER: *Clemens s.n.* (BM).

55.37.26. Dendrochilum planiscapum Carr, Gardens' Bull. 8: 228 (1935). Fig. 35. Type: TENOMPOK, 1500 m, *Carr 3663, SFN 28020* (holotype SING!; isotypes AMES mf, K!).

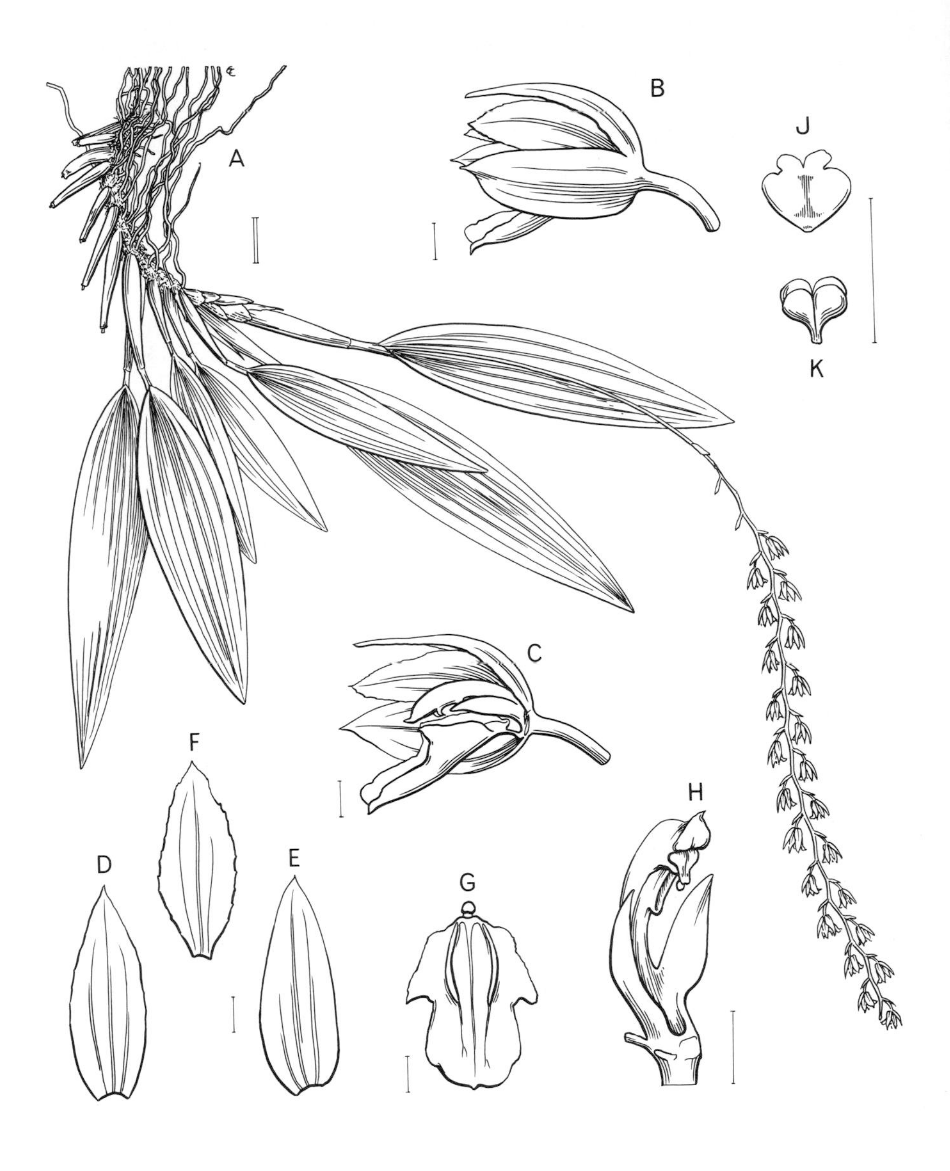

Fig. 34. Dendrochilum lacteum. **A**, habit; **B**, flower (side view); **C**, flower (side view, with lateral sepal and petal removed); **D**, dorsal sepal; **E**, lateral sepal; **F**, petal; **G**, lip (front view); **H**, column (oblique view); **J**, anther-cap (back view); **K**, pollinia. All from *Carr 3608*. Scale: single bar = 1 mm; double bar = 1 cm. Drawn by Eleanor Catherine.

Fig. 35. Dendrochilum planiscapum. A, habit; **B**, flower (side view); **C**, flower (side view, with lateral sepal and petal removed); **D**, dorsal sepal; **E**, lateral sepal; **F**, petal; **G**, lip (flattened, front view); **H**, anther-cap (back view). **A** from *Chow & Leopold SAN 74511* and *Meijer SAN 48111*; **B–H** from *Meijer SAN 48111.* Scale: single bar = 1 mm; double bar = 1 cm. Drawn by Eleanor Catherine.

Epiphyte. Lower montane forest, upper montane forest? Elevation: 1500–2400 m.

General distribution: Sabah.

Additional collections. MESILAU CAVE TRAIL: 2400 m, *Meijer SAN 48111* (K); MESILAU TRAIL: 2100 m, *Chow & Leopold SAN 74511* (K).

55.37.27. Dendrochilum pterogyne Carr, Gardens' Bull. 8: 236 (1935). Type: PAKA-PAKA CAVE, 3100 m, *Carr 3541, SFN 27597* (holotype SING!; isotypes AMES mf, K!).

Epiphyte. Upper montane forest, sometimes on ultramafic substrate. Elevation: 900–3800 m.

Endemic to Mount Kinabalu.

Additional collections. DALLAS: 900 m, *Clemens 30102* (K); GURULAU SPUR: 2400–3700 m, *Clemens 51079* (BM, K); KADAMAIAN RIVER, JUST BELOW PAKA-PAKA CAVE: 2900 m, *Sinclair et al. 9181* (K, SING); KINATEKI RIVER HEAD: 2100 m, *Clemens 31831* (BM); LAYANG-LAYANG: 2600 m, *Sands 3880* (K); MARAI PARAI: 3200 m, *Clemens 32322* (BM), 3200 m, *32324* (BM), 2400–3000 m, *33180* (BM); PAKA-PAKA CAVE: 3100 m, *Carr 3548, SFN 27635* (K, SING), 3400 m, *Clemens 27864* (BM, K); SUMMIT AREA: 3800 m, *Collenette 616* (K), 3400–3700 m, *Sato 162* (UKMS).

55.37.28. Dendrochilum scriptum Carr, Gardens' Bull. 8: 234 (1935). Type: KEMBURONGOH/PAKA-PAKA CAVE, 2600 m, *Carr 3597* (holotype SING!).

Epiphyte. Upper montane forest on ultramafic substrate. Elevation: 2600 m. Known only from the type.

Endemic to Mount Kinabalu.

55.37.29. Dendrochilum simplex J. J. Sm., Bull. Dép. Agric. Indes Néerl. 22: 13 (1909).

Epiphytic or terrestrial. Lower montane forest, mostly on ultramafic substrate. Elevation: 1200–1500 m.

General distribution: Sabah, Kalimantan.

Collections. MARAI PARAI SPUR: *Clemens 278* (AMES, BM); PENIBUKAN: 1400 m, *Clemens 40548* (BM), 1200 m, *50105* (BM, K); TENOMPOK ORCHID GARDEN: 1500 m, *Clemens 50347* (BM, K).

55.37.30. Dendrochilum stachyodes (Ridl.) J. J. Sm., Recueil Trav. Bot. Néerl. 1: 77 (1904). Fig. 36, Plate 43.

Platyclinis stachyodes Ridl. in Stapf, Trans. Linn. Soc. Bot. 4: 234 (1894). Type: MOUNT KINABALU, 3400 m, *Haviland 1097* (holotype SING!; isotype K!).

Lithophyte. Upper montane forest, summit area, in granitic rock crevices or on ultramafic substrate, in the open or among scrub. Elevation: 2400–3700 m.

Endemic to Mount Kinabalu.

Additional collections. GURULAU SPUR: 2400–2700 m, *Clemens 50653* (BM, K), 3500 m, *51012* (BM, K); KEMBURONGOH: 2700 m, *Clemens 27870* (BM, K, SING), 2400 m, *29120* (BM); KILEMBUN BASIN: 2900 m, *Clemens 33177* (BM); LAYANG-LAYANG: *Sidek bin Kiah S. 38* (L, SING); LUBANG: *Clemens 224* (AMES); MARAI PARAI: 2400–3000 m, *Clemens 33177* (K); MOUNT KINABALU: *Haslam s.n.* (BM, K); PAKA-PAKA CAVE: 3000 m, *Carr 3521, SFN 27531* (K, SING), *Clemens 115* (AMES, BM, K, SING), 3400 m, *27141* (BM, K, SING), 3200 m, *Meijer SAN 28560* (K); SUMMIT AREA: 3500 m, *Gibbs 4181* (BM, K); SUMMIT TRAIL: 3400 m, *Wood 605* (K); UPPER KINABALU: *Clemens 30142* (K).

55.37.31. Dendrochilum subintegrum Ames, Orch. 6: 65 (1920). Type: LUBANG, *Clemens 285* (holotype AMES mf).

Epiphyte. Lower montane forest.

General distribution: Sabah, Sarawak.

55.37.32. Dendrochilum tenompokense Carr, Gardens' Bull. 8: 225 (1935). Type: TENOMPOK, 1600 m, *Carr 3623, SFN 27891* (holotype SING!; isotypes AMES mf, K!).

Epiphyte. Lower montane forest. Elevation: 1500–1600 m.

General distribution: Sabah, Sarawak.

Additional collections. TENOMPOK: *Clemens 50236* (BM), *50360* (BM); TENOMPOK ORCHID GARDEN: 1500 m, *Clemens 50236* (K), 1500 m, *50360* (K).

55.37.33. Dendrochilum transversum Carr, Gardens' Bull. 8: 233 (1935). Type: MARAI PARAI, 2100 m, *Carr 3477, SFN 27431* (holotype SING!; isotypes AMES mf, K!).

Epiphyte. Upper montane forest on ultramafic substrate. Elevation: 2100–3000 m.

Endemic to Mount Kinabalu.

Additional collections. MARAI PARAI: 2100–2700 m, *Clemens 33130* (BM, K), 2400–3000 m, *33173* (BM).

55.37.34. Dendrochilum sp. 1, sect. Platyclinis

Upper montane forest. Elevation: 2700–3000 m.

Collection. SUMMIT TRAIL: 2700–3000 m, *Sato 1214* (UKMS).

55.37.35. Dendrochilum indet.

Collections. HEMPUEN HILL: 600 m, *Lamb & Surat AL 1365/91* (K); KEMBURONGOH: 2200 m, *Carr SFN 36566* (SING); MARAI PARAI SPUR: 1700 m, *Bailes & Cribb 841* (K); PAKA-PAKA CAVE: 3100 m, *Carr SFN 36570* (SING); PENIBUKAN: 1200–1500 m, *Clemens s.n.* (K), 1200 m, *50370* (K).

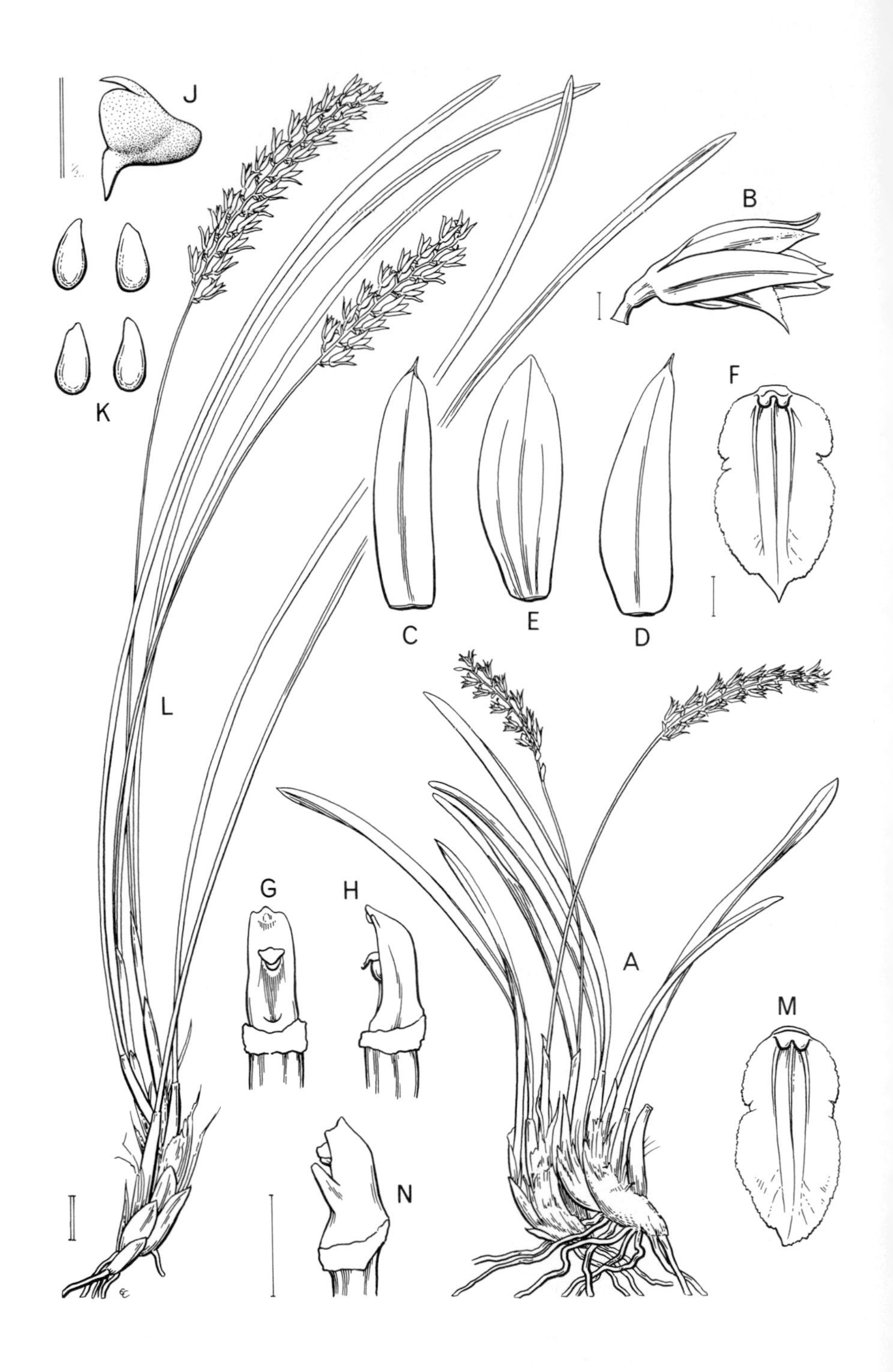
J
B
K
F
C
E
D
L
G
H
A
M
N

55.38. DIDYMOPLEXIELLA Garay

Arch. Jard. Bot. Rio de Janeiro 13: 33 (1955).

Leafless saprophytic herbs. *Rhizome* small, fleshy. *Stems* simple, erect, slender. *Inflorescence* terminal, racemose. *Flowers* small, resupinate; *sepals* and *petals* connate near the base, *lateral sepals* a little more than to the middle; *petals* connate halfway with dorsal sepal; *lip* entire and retuse at apex or 3-lobed, callose; *column* with long decurved, falcate wings (stelidia), foot absent; *stigma* near column apex; *pollinia* 4.

Six species distributed in Thailand, Peninsular Malaysia and Borneo.

55.38.1. Didymoplexiella kinabaluensis (Carr) Seidenf., Dansk Bot. Ark. 32: 175 (1978).

Didymoplexis kinabaluensis Carr, Gardens' Bull. 8: 178 (1935). Type: MINITINDUK GORGE, 900 m, *Carr 3155, SFN 26606* (holotype SING!; isotype K!).

Saprophyte. Hill forest. Elevation: 900–1300 m.

Endemic to Mount Kinabalu.

Additional collection. LUGAS HILL: 1300 m, *Beaman 9524* (K).

55.39. DILOCHIA Lindl.

Gen. Sp. Orch., 38 (1830).

Terrestrial or epiphytic herbs. *Stems* tall, stout, erect, leafy, without pseudobulbs. *Leaves* elliptic, stiff. *Inflorescences* terminal, usually branched, straight or decurved, many-flowered; floral bracts large, often concave, deciduous. *Flowers* resupinate; *sepals* and *petals* similar, free, spreading or not opening widely; *lip* 3-lobed, side-lobes erect, mid-lobe often bilobed, disc with 5 keels; *column* long, slender, curved; *pollinia* 8. Fruit globose.

About 6 or 7 species distributed from Thailand and Peninsular Malaysia, through Indonesia to the Philippines and New Guinea.

Fig. 36. Dendrochilum stachyodes. **A**, habit; **B**, flower (side view); **C**, dorsal sepal; **D**, lateral sepal; **E**, petal; **F**, lip (front view); **G**, column (front view); **H**, column (side view); **J**, anther-cap (side view); **K**, pollinia; **L**, habit; **M**, lip (front view); **N**, column (side view). **A** from *Chan & Gunsalam 53/87*; **B–K** from *Wood 605*; **L–N** from *Clemens 33177*. Scale: single bar = 1 mm; double bar = 1 cm. Drawn by Eleanor Catherine.

55.39.1. Dilochia cantleyi (Hook. f.) Ridl., J. Linn. Soc. Bot. 32: 332 (1896). Plate 44.

Arundina cantleyi Hook. f., Fl. Brit. Ind. 5: 858 (1890).

Terrestrial. Lower montane forest, upper montane forest, mostly on ultramafic substrate. Elevation: 1600–2200 m.

General distribution: Sabah, Peninsular Malaysia, Sumatra.

Collections. MARAI PARAI: 1700 m, *Carr 3129* (SING), 1600 m, *SFN 26549* (SING), *Collenette A 18* (BM); MARAI PARAI SPUR: *Clemens 266A* (AMES), *272* (AMES, BM, K, SING); MESILAU CAVE: 2000–2100 m, *Beaman 8142* (K), 2200 m, *Lamb AL 588/86* (K); MESILAU CAVE TRAIL: 1700–1900 m, *Beaman 7998* (K); TAHUBANG RIVER HEAD: 2100 m, *Clemens 34318* (BM).

55.39.2. Dilochia parviflora J. J. Sm., Bull. Jard. Bot. Buit., ser. 3, 11: 112 (1931). Plate 45A.

Terrestrial. Lower montane forest. Elevation: 1500 m.

General distribution: Sabah, Kalimantan.

Collection. TENOMPOK: 1500 m, *Carr 3246* (SING).

55.39.3. Dilochia rigida (Ridl.) J. J. Wood, **comb. nov.** Fig. 37.

Bromheadia rigida Ridl. in Stapf, Trans. Linn. Soc. Bot. 4: 239 (1894). Type: MOUNT KINABALU, 1800 m, *Haviland 1251* (holotype? K!).

Arundina gracilis Ames & C. Schweinf., Orch. 6: 96 (1920). Type: MARAI PARAI SPUR, *Clemens 370* (holotype AMES mf; isotype BM!).

Dilochia gracilis (Ames & C. Schweinf.) Carr, Gardens' Bull. 8: 91 (1935).

Epiphyte or terrestrial. Lower montane forest, upper montane forest, often along ridges, usually on ultramafic substrate. Elevation: 1400–2400 m.

General distribution: Sabah, Sarawak.

Additional collections. EASTERN SHOULDER, CAMP 2: 2300 m, *RSNB 1053* (K, SING); GURULAU SPUR: 1700 m, *Clemens 50421* (BM, K); KEMBURONGOH: 2400 m, *Clemens 29123* (BM); LUMU-LUMU: 2100 m, *Clemens s.n.* (BM); MARAI PARAI: 1500 m, *Carr 3119, SFN 26550* (SING); MARAI PARAI SPUR: *Clemens 244* (AMES), *266* (AMES, BM); MOUNT KINABALU: *Haslam s.n.* (AMES); MT. NUNGKEK: 1400 m, *Sands 3989* (K); PINOSUK PLATEAU: 1500 m, *Lamb s.n.* (K), 1500 m, *T 6* (K); TINEKUK FALLS: 2000 m, *Clemens 40915* (BM); UPPER KINABALU: *Clemens 27155* (G, K).

55.39.4. Dilochia wallichii Lindl., Gen. Sp. Orch., 38 (1830). Plate 45B.

Terrestrial, lithophytic, sometimes epiphytic. Hill forest, ridge scrub. Elevation: 400–900 m.

General distribution: Sabah, Kalimantan, Sarawak, Peninsular Malaysia, Thailand, Sumatra, Java, Bangka, New Guinea.

Fig. 37. Dilochia rigida. **A**, habit; **B**, flower (side view); **C**, dorsal sepal; **D**, lateral sepal; **E**, petal; **F**, lip (front view); **G**, column (oblique view); **H**, anther-cap (back view); **J**, pollinia. **A** from *Beaman 8243*, Mt. Alab, Sabah; **B–J** from *Wood 769*, Mt. Alab, Sabah. Scale: single bar = 1 mm; double bar = 1 cm. Drawn by Eleanor Catherine.

Collections. DALLAS: 900 m, *Clemens 27615* (BM, K), 900 m, *27728* (BM, K, SING); KAUNG: 400 m, *Darnton 358* (BM); KEKEHITAN HILL: 800 m, *Carr 3422, SFN 27338* (SING).

55.40. DIMORPHORCHIS Rolfe

Orchid Rev. 27: 149 (1919).

Tan, K. W. (1975–1976). Taxonomy of *Arachnis, Armodorum, Esmeralda* & *Dimorphorchis*, Orchidaceae, Part I & Part II. Selbyana 1(1): 1–15; 1(4): 365–373.

Large monopodial epiphytes. *Stems* long, usually pendent, leafy, up to 200 cm long. *Leaves* strap-shaped, apex unequally bilobed, arcuate, 30–70 × 1.7–6 cm long. *Inflorescences* pendent, 28–300 cm long, laxly few- to many-flowered, peduncle and rachis flexuous, tomentose, bracts 0.7–3 cm long. *Flowers* large, showy, dimorphic, resupinate, the basal 2 always strongly scented, 5–6.5 cm across, *sepals* and *petals* yellow or orange-yellow, spotted purple or with a few small red basal spots or a few red spots only on the lateral sepals, the apical flowers unscented, 5–6.5 cm across, *sepals* and *petals* yellow or white with either large irregular purple blotches or small red or purple spots; *sepals* free, spreading, usually acute, margin often undulate in apical flowers, outer surface stellate-pubescent, 2.4–3.5 × 1.3–2 cm; *lip* mobile, 3-lobed, very fleshy, sometimes L-shaped in side view, 0.8–1.3 cm long, side-lobes erect, margins often incurved, to 6 mm long, mid-lobe at an obtuse or right angle to the base of the lip, sometimes bilaterally compressed, with a keel-like callus; *column* 0.5–1.2 cm long, with a short foot; *anther* pubescent; *stipes* broad; *pollinia* 4, appearing as 2 pollen masses.

Two species endemic to Borneo.

55.40.1. Dimorphorchis rossii Fowlie, Orchid Digest 53: 14 (1989). Plate 45C. Type: MOUNT KINABALU, 500 m, *Fowlie & Ross 83P912* (holotype UCLA n.v.).

Epiphyte. Hill forest on ultramafic substrate. Elevation: 500–1200 m. Known only from the type.

Endemic to Mount Kinabalu.

55.41. DIPODIUM R. Br.

Prodr., 330 (1810).

Terrestrial or climbing herbs of diverse habit: either 1) erect-stemmed, sympodial and green-leaved, 2) climbing, monopodial and green-leaved, or 3) leafless, saprophytic (habit 1 and 2 in Borneo). *Leaves*, when present, distichous, imbricate, ensiform, plicate. *Inflorescence* lateral or terminal, racemose, ± erect, few- to many-flowered. *Flowers* medium-sized, resupinate; *sepals* and *petals* free, spreading, usually spotted and blotched on reverse; *lip* 3-lobed, close and parallel to column, side-lobes narrow, ± embracing column to form a tube, mid-lobe much larger, ovate, convex, pubescent; *column* short, thick, foot absent; *pollinia* 2, cleft, each with a rather long stipes.

About 20 species distributed from tropical Asia to New Guinea, Australia and the Pacific islands.

55.41.1. Dipodium pictum (Lindl.) Rchb. f., Xenia Orch. 2: 15 (1862). Plate 46A, 46B.

Wailesia picta Lindl., J. Hort. Soc. 4: 261 (1849).

Climber. Lower montane forest on ridge with *Agathis*. Elevation: 1100 m.

General distribution: Sabah, Brunei, Kalimantan, Sarawak, Peninsular Malaysia, Sumatra, Java, Philippines.

Collection. BAMBANGAN RIVER: 1100 m, *Lamb LKC 3158* (K).

55.41.2. Dipodium scandens (Blume) J. J. Sm., Orch. Java, 488 (1905).

Leopardanthus scandens Blume, Rumphia 4: 47 (1849).

Climber. Hill forest on ultramafic substrate. Elevation: 900 m.

General distribution: Sabah, Peninsular Malaysia, Java.

Collection. MOUNT KINABALU: 900 m, *Lamb SAN 93376* (K).

55.42. ENTOMOPHOBIA de Vogel

Blumea 30: 199 (1984).

De Vogel, E. F. (1986). The genus *Entomophobia*. Orch. Monogr. 1: 41-42.

Epiphytic or lithophytic herbs. *Rhizome* short, creeping. *Pseudobulbs* 2–5 cm long, all turned to one side of rhizome, 2-leaved. *Leaves* 16–65 × 1–1.9 cm, linear, acute, petiolate. *Inflorescence* proteranthous to synanthous, racemose, densely 15–42-flowered, erect; floral bracts persistent, ovate, acuminate to truncate, folded along midrib. *Flowers* resupinate, all turned to one side, not opening widely, glabrous; *sepals* 7–10 mm long, ovate-oblong, concave; *petals* spathulate; *lip* 5–8 mm long, with lateral margins adnate to basal half of column, base deeply saccate, separated from front part by a transverse, high, slightly bent, fleshy callus which more or less fits into the stigmatic cavity, apical margins undulate; *column* 4–5 mm long, laterally flattened, foot absent, apex 3-lobed; *pollinia* 4.

A monotypic genus endemic to Borneo.

55.42.1. Entomophobia kinabaluensis (Ames) de Vogel, Blumea 30: 199 (1984). Plate 46C.

Pholidota kinabaluensis Ames, Orch. 6: 68 (1920). Type: MARAI PARAI SPUR, *Clemens 279* (holotype AMES mf).

Epiphyte or lithophyte. Lower montane ridge forest on ultramafic substrate. Elevation: 1100–2100 m.

General distribution: Sabah, Brunei, Kalimantan, Sarawak.

Additional collections. MAHANDEI RIVER: 1100 m, *Carr 3112* (SING); MAMUT COPPER MINE: 1400–1500 m, *Beaman 10348* (K); MARAI PARAI: 2100 m, *Carr 3475, SFN 28059* (SING); PENIBUKAN: 1200–1500 m, *Clemens 30607* (K).

55.43. EPIGENEIUM Gagnep.

Bull. Mus. Hist. Nat. (Paris), ser. 2, 4: 592 (1932).

Seidenfaden, G. (1980). Orchid genera in Thailand IX. *Flickingeria* Hawkes & *Epigeneium* Gagnep. Dansk Bot. Ark. 34, 1: 1–104.

Epiphytic or terrestrial herbs. *Aerial shoots* terminating in either a 1-, 2- or 3-leaved pseudobulb consisting of 1 internode, the new shoot arising at the base of the pseudobulb, sometimes pendulous, covered with conspicuous imbricate brown sheaths when young. *Pseudobulbs* usually short and conical, or ovoid. *Leaves* oblong or obovate, coriaceous. *Inflorescence* slender, arising between or just below the leaves, racemose, 1- to several-flowered. *Flowers* medium-sized to large, resupinate, long lasting. *Sepals* and *petals* narrowly elliptic, subequal. *Dorsal sepal* enclosing column. *Lateral sepals* larger, adnate to column-foot forming a short mentum. *Petals* long-decurrent on mentum. *Lip* pandurate-oblong to 3-lobed; disc with a callus which is lobulate or ridged at base. *Column* short to rather long, with or without short stelidia, foot long. *Pollinia* 2, or 4 in 2 groups.

About 35 species distributed in mainland Asia, Taiwan, Malaysia, Indonesia and the Philippines.

55.43.1. **Epigeneium** aff. **geminatum** (Blume) Summerh., Kew Bull. 12: 262 (1957).

Desmotrichum geminatum Blume, Bijdr., 332 (1825).

Epiphyte. Hill forest?, on ultramafic substrate. Elevation: 900–1200 m.

Collection. PENATARAN RIVER: 900–1200 m, *Clemens 34323* (BM).

55.43.2. **Epigeneium kinabaluense** (Ridl.) Summerh., Kew Bull. 12: 262 (1957). Plate 47.

Dendrobium kinabaluense Ridl. in Stapf, Trans. Linn. Soc. Bot. 4: 234 (1894). Type: MOUNT KINABALU, 2100 m, *Haviland 1253* (holotype unlocated; isotype K!).

Sarcopodium kinabaluense (Ridl.) Rolfe, Orchid Rev. 18: 239 (1910).

Katherinea kinabaluense (Ridl.) A. D. Hawkes, Lloydia 19: 96 (1956).

Epiphytic or terrestrial. Lower montane forest, upper montane forest, mostly in scrub on ultramafic substrate. Elevation: 1500–3400 m.

General distribution: Sabah, Sarawak.

Additional collections. GURULAU SPUR: 2400–2700 m, *Clemens 50659* (BM); KEMBURONGOH: 2400 m, *Clemens s.n.* (BM), *197* (AMES, BM); LUMU-LUMU: 2100 m, *Clemens s.n.* (BM, BM, BM); MARAI PARAI: 1500 m, *Clemens 32208* (BM), 1500 m, *32746* (BM), 2400–2700 m, *33133* (BM, K), 1500 m, *Collenette A 17* (BM); MARAI PARAI SPUR: 1700 m, *Bailes & Cribb 816* (K), *Clemens 249* (AMES), *367* (AMES), *370A* (AMES), 1600 m, *Lamb AL 40/83* (K); MESILAU CAVE: 1900–2200 m, *Beaman 9567* (K); MESILAU SPUR: 2300 m, *Lamb LKC 3175* (K); MINIRINTEG CAVE: 2300 m, *Clemens 29214* (BM); PANAR LABAN: 3400 m, *Sidek S 50* (K, SING); PENIBUKAN: 1500 m, *Clemens 30943* (BM), 1800 m, *50292* (BM); PIG HILL: 1700–2000 m, *Sutton 5* (K), 1700–2000 m, *7* (K), *9* (K); SUMMIT AREA: 3200–3400 m, *Sands 3881* (K); TENOMPOK/LUMU-LUMU: 1800 m, *Clemens 27368* (BM).

55.43.3. Epigeneium longirepens (Ames & C. Schweinf.) Seidenf., Dansk Bot. Ark. 34: 18 (1980).Plate 48A, 48B.

Dendrobium longirepens Ames & C. Schweinf., Orch. 6: 105 (1920). Type: MARAI PARAI SPUR, *Clemens 245* (holotype AMES mf; isotypes BM!, K!, SING!).

Desmotrichum longirepens (Ames & C. Schweinf.) A. D. Hawkes, Lloydia 20: 126 (1957).

Ephemerantha longirepens (Ames & C. Schweinf.) P. F. Hunt & Summerh., Taxon 10: 105 (1961).

Flickingeria longirepens (Ames & C. Schweinf.) A. D. Hawkes, Orchid Weekly 2(46): 456 (1961).

Epiphyte. Lower montane forest, frequently on ultramafic substrate. Elevation: 900–1700 m.

Endemic to Mount Kinabalu.

Additional collections. GURULAU SPUR: 1500 m, *Clemens 50384* (BM, G, K), 1700 m, *50409* (BM, G, K); KIAU: 900 m, *Clemens 68A* (AMES); LUBANG: *Clemens 116A* (AMES); MARAI PARAI SPUR: *Clemens 372* (AMES); PENIBUKAN: 1200 m, *Clemens 30639* (BM, K), 1200 m, *32077* (BM), 1200 m, *40659* (BM), 1500 m, *50203* (BM, K), 1500 m, *50393* (BM, G, K); PINOSUK PLATEAU: 1400 m, *Beaman 10723* (K), 1500 m, *Lamb AL 20/82* (K); TENOMPOK: 1600 m, *Carr SFN 27196* (SING).

55.43.4. Epigeneium tricallosum (Ames & C. Schweinf.) J. J. Wood, Lindleyana 5: 99 (1990). Fig. 38, Plate 48C.

Dendrobium tricallosum Ames & C. Schweinf., Orch. 6: 114 (1920). Type: MOUNT KINABALU, *Clemens s.n.* (holotype AMES mf).

Epiphyte. Lower montane forest. Elevation: 800–1600 m.

General distribution: Sabah.

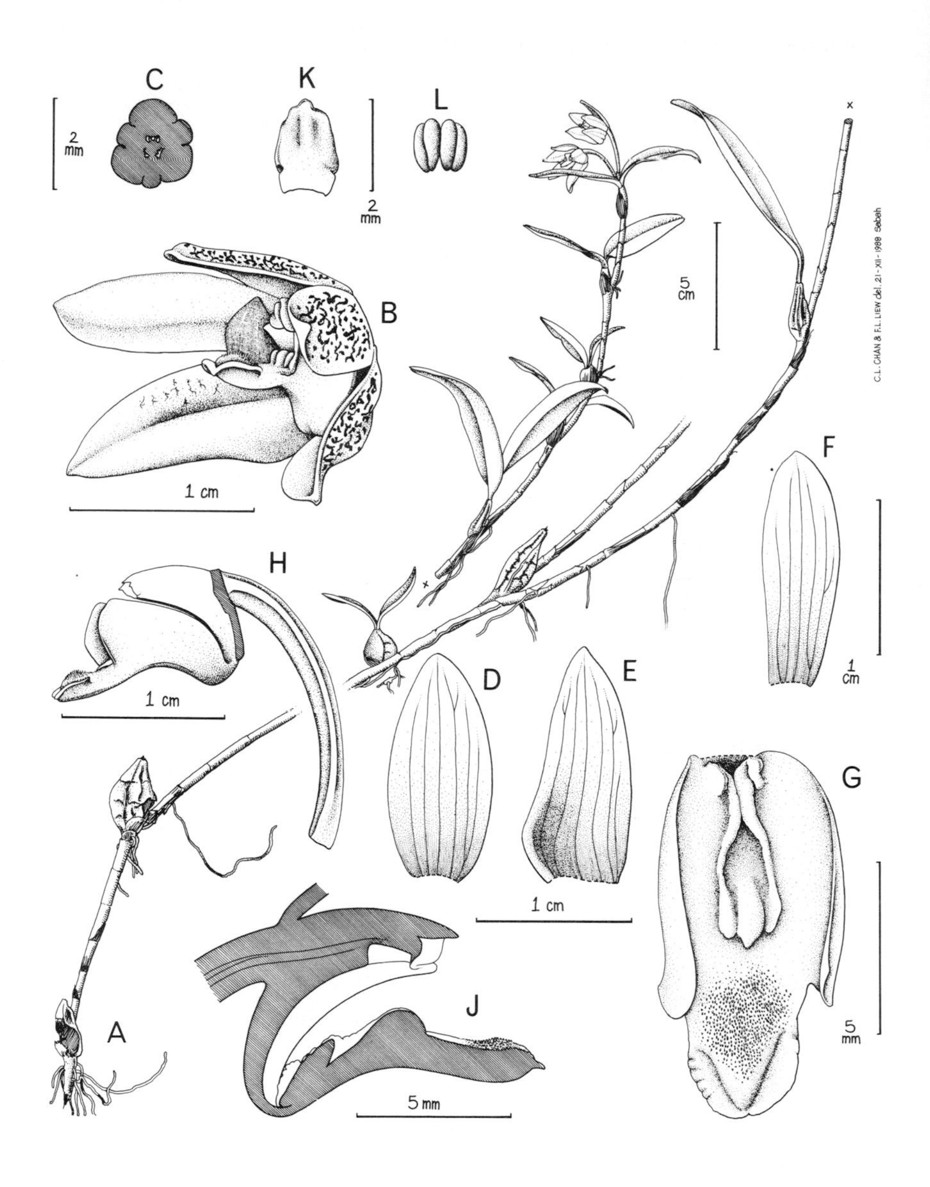

Fig. 38. Epigeneium tricallosum. **A**, habit; **B**, flower (oblique view); **C**, ovary (transverse section); **D**, dorsal sepal; **E**, lateral sepal; **F**, petal; **G**, lip (front view); **H**, pedicel-with-ovary, column and lip (side view); **J**, pedicel-with-ovary, column and lip (longitudinal section); **K**, anther-cap (back view); **L**, pollinia. All from *Chan & Gunsalam s.n.* Drawn by Chan Chew Lun and F. L. Liew.

PLATE 33.

A. Cymbidium dayanum. North Sumatra. Photo J. Dransfield.

B. Cymbidium elongatum. Sabah, Mt. Monkobo. Photo A. Lamb.

C. Cyrtosia javanica. Sabah, Danum Valley. Photo J. Dransfield.

PLATE 34.

A. **Dendrobium alabense.** Sabah, Mt. Alab. Photo J. B. Comber.

B. **Dendrobium bifarium.** Mount Kinabalu, Marai Parai Spur. Photo J. Dransfield.

C. **Dendrobium cinereum.** Mount Kinabalu, Hempuen Hill. Photo A. Lamb.

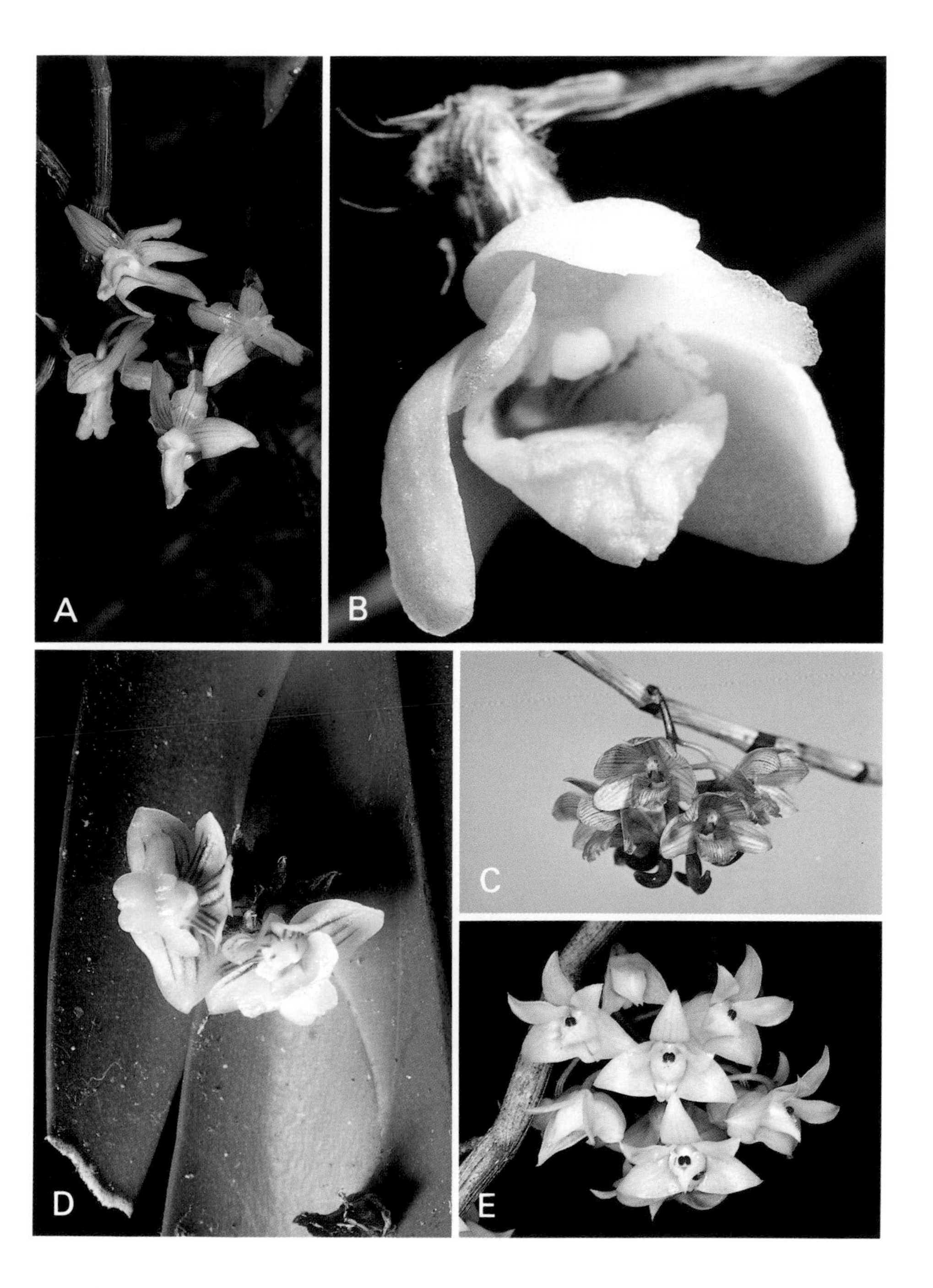

PLATE 35.

A. Dendrobium cumulatum. Mount Kinabalu, Marai Parai Spur. Photo S. Collenette.

B. Dendrobium cymbulipes. Sabah, Crocker Range. Photo J. B. Comber.

C. Dendrobium hamaticalcar. *Bacon 142* (holotype). Photo P. Woods.

D. Dendrobium kiauense. Sabah, Sipitang District. Photo J. B. Comber.

E. Dendrobium linguella. Cult. Poring Orchid Garden. Photo P. J. Cribb.

PLATE 36.

A. **Dendrobium microglaphys.** Mount Kinabalu. Photo A. Lamb.

B. **Dendrobium olivaceum.** Sabah, Crocker Range. Photo J. B. Comber.

C. **Dendrobium orbiculare.** Sabah, Nabawan. Photo J. B. Comber.

PLATE 37.

A. Dendrobium panduriferum. Mount Kinabalu, Lohan River. Photo A. Lamb.

B. Dendrobium parthenium. Cult. South Africa. Photo L. Vogelpoel.

C. Dendrobium patentilobum. Sabah, Mount Kinabalu, Marai Parai/Kiau Trail, *Bailes & Cribb 800.* Photo R. B. G. Kew.

D. Dendrobium piranha. Mount Kinabalu, Marai Parai Spur. Photo J. Dransfield.

PLATE 38.

A. **Dendrobium spectatissimum.** Mount Kinabalu. Photo J. Dransfield.

B. **Dendrobium spectatissimum.** Mount Kinabalu. Photo A. Lamb.

PLATE 39.

A. Dendrobium secundum. Cult. Bogor Botanic Gardens, Java. Photo J. B. Comber.

B. Dendrobium tetrachromum. Kalimantan Timur. Photo P. J. Cribb.

C. Dendrobium tridentatum. Mount Kinabalu, Marai Parai Spur. Photo A. Lamb.

PLATE 40.

A. Dendrobium uniflorum. Sabah, Mt. Alab. Photo P. J. Cribb.

B. Dendrobium ventripes. Sarawak, Mt. Mulu. Photo J. Dransfield.

C. Dendrochilum alatum. Mount Kinabalu, 2700–3000 m. Photo A. Lamb.

D. Dendrochilum alatum. Mount Kinabalu, 3150 m. Photo P. J. Cribb.

PLATE 41.

A. Dendrochilum alpinum. Mount Kinabalu, Panar Laban. Photo A. Lamb.

B. Dendrochilum crassum. Sabah, Crocker Range. Photo J. B. Comber.

C. Dendrochilum crassum. Sabah, Crocker Range. Photo J. B. Comber.

D. Dendrochilum dewindtianum. Mount Kinabalu, 2750 m. Photo G. Cubitt.

PLATE 42.

A. Dendrochilum gibbsiae. Mount Kinabalu, Park Headquarters, *Wood 621.* Photo J. B. Comber.

B. Dendrochilum grandiflorum. Mount Kinabalu, Summit Trail. Photo P. J. Cribb.

C. Dendrochilum haslamii. Mount Kinabalu. Photo P. J. Cribb.

D. Dendrochilum imbricatum. Mount Kinabalu. Photo P. J. Cribb.

E. Dendrochilum imbricatum. Mount Kinabalu, Park Headquarters, *Wood 618.* Photo J. B. Comber.

PLATE 43.

A. Dendrochilum stachyodes. Mount Kinabalu, Panar Laban. Photo A. Lamb.

B. Dendrochilum stachyodes. Mount Kinabalu, Panar Laban. Photo A. Lamb.

PLATE 44.

A. Dilochia cantleyi. Sabah. Photo A. Lamb.

B. Dilochia cantleyi. Peninsular Malaysia, Genting Highlands. Photo J. Dransfield.

PLATE 45.

A. **Dilochia parviflora.** Mount Kinabalu, Park Headquarters. Photo A. Lamb.

B. **Dilochia wallichii.** Sabah, Crocker Range. Photo J. B. Comber.

C. **Dimorphorchis rossii.** Sabah. Photo P. J. Cribb.

PLATE 46.

A. Dipodium pictum. Sabah, Mt. Lumaku. Photo J. B. Comber.

B. Dipodium pictum. Sabah, Mt. Lumaku. Photo J. B. Comber.

C. Entomophobia kinabaluensis. Sabah, Crocker Range. Photo A. Lamb.

PLATE 47.

A. Epigeneium kinabaluense. Mount Kinabalu, Mesilau Ridge. Photo A. Lamb.

B. Epigeneium kinabaluense. Mount Kinabalu, Marai Parai Spur. Photo J. Dransfield.

PLATE 48.

A. Epigeneium longirepens. Mount Kinabalu, 1500 m. Photo J. B. Comber.

B. Epigeneium longirepens. Mount Kinabalu, 1500 m. Photo J. B. Comber.

C. Epigeneium tricallosum. Mount Kinabalu, 1500 m. Photo J. B. Comber.

PLATE 49.

A. Epipogium roseum. Mount Kinabalu, Lohan River. Photo A. Lamb.

B. Eria brookesii. Sabah. Photo P. J. Cribb.

C. Eria brookesii. Sabah, cult. R. B. G. Kew. Photo J. B. Comber.

PLATE 50.

A. Eria cymbidifolia var. **cymbidifolia.** Mount Kinabalu, Golf Course Site, *Brentnall 133*. Photo M. Svanderlik.

B. Eria cymbidifolia var. **cymbidifolia.** Mount Kinabalu, Golf Course Site, *Brentnall 133*. Photo M. Svanderlik.

C. Eria grandis. Mount Kinabalu, Panar Laban, 3000–3300 m. Photo A. Lamb.

D. Eria grandis. Mount Kinabalu, Summit Trail, 3000 m. Photo P. J. Cribb.

PLATE 51.

A. Eria magnicallosa. Mount Kinabalu, East Mesilau/Menteki Rivers, *Wood 826.* Photo J. B. Comber.

B. Eria jenseniana. Mount Kinabalu, Langanan River, *Lohok 23.* Photo P. J. Cribb.

C. Eria nutans. Sabah, Nabawan. Photo S. Collenette.

PLATE 52.

A. Eria pannea. Sabah, Mt. Lumaku. Photo J. B. Comber.

B. Eria pellipes. Sarawak, Mt. Mulu. Photo J. Dransfield.

PLATE 53.

A. Eria robusta. Sabah. Photo J. B. Comber.

B. Eria pseudocymbiformis var. **pseudocymbiformis.** Sabah, Mt. Alab. Photo P. J. Cribb.

C. Erythrodes latifolia. Java, Mt. Halimun. Photo J. Dransfield.

PLATE 54.

A. **Eulophia zollingeri.** Peninsular Malaysia. Photo J. Dransfield.

B. **Gastrodia grandilabris.** Mount Kinabalu, Park Headquarters. Photo A. Lamb.

C. **Goodyera ustulata.** Sabah. Photo A. Lamb.

PLATE 55.

A. Habenaria setifolia. Mount Kinabalu, Haye-haye River. Photo A. Lamb.

B. Habenaria singapurensis. Peninsular Malaysia. Photo J. Dransfield.

PLATE 56.

A. Hylophila lanceolata. Mount Kinabalu, Park Headquarters. Photo A. Lamb.

B. Hylophila lanceolata. Mount Kinabalu, Park Headquarters, *Wood 842.* Photo J. B. Comber.

C. Kuhlhasseltia javanica. Mount Kinabalu, Kiau View Trail. Photo A. Lamb.

D. Liparis aurantiorbiculata. Mount Kinabalu. Photo P. J. Cribb.

PLATE 57.

A. Liparis compressa. Mount Kinabalu, Pinosuk Plateau. Photo A. Lamb.

B. Liparis kinabaluensis. Mount Kinabalu, West Mesilau River, *Brentnall 124.* Photo M. Svanderlik.

C. Liparis kinabaluensis. Mount Kinabalu, West Mesilau River, *Brentnall 124.* Photo M. Svanderlik.

D. Liparis lacerata. Sabah, cult. Tenom Orchid Centre. Photo P. J. Cribb.

E. Liparis latifolia. North Sumatra. Photo J. Dransfield.

F. Liparis lobongensis. Mount Kinabalu, 1500 m. Photo J. B. Comber.

PLATE 58.

A. Liparis pandurata. Sabah, Sipitang District. Photo J. B. Comber.

B. Liparis sp. An unidentified species from Mount Kinabalu and the Crocker Range. Photo A. Lamb.

C. Luisia curtisii. Cult. Poring Orchid Garden, *Lohok 18*. Photo P. J. Cribb.

PLATE 59.

A. Luisia volucris. Cult. Poring Orchid Garden, *Lohok 21.* Photo P. J. Cribb.

B. Macodes petola. Sabah. Photo A. Lamb.

C. Macodes lowii. Mount Kinabalu, Lohan River. Photo A. Lamb.

PLATE 60.

A. Malaxis kinabaluensis. Mount Kinabalu, East Mesilau/Menteki Rivers, *Wood 825*. Photo J. B. Comber.

B. Malaxis metallica. Mount Kinabalu, Hempuen Hill. Photo A. Lamb.

C. Malaxis punctata. Mount Kinabalu, Kiau/Tahubang River, *Sands 4011* (holotype). Photo R. B. G. Kew.

D. Malaxis punctata. Mount Kinabalu, Kiau/Tahubang River, *Sands 4011* (holotype). Photo R. B. G. Kew.

PLATE 61.

A. **Malaxis punctata,** purple-flowered form. Sabah, Sipitang District. Photo J. B. Comber.

B. **Malleola witteana.** North Sumatra. Photo J. Dransfield.

C. **Micropera callosa.** Mount Kinabalu, Hempuen Hill. Photo A. Lamb.

D. **Nabaluia angustifolia.** Cult. R. B. G. Kew. Photo W. Rossi.

PLATE 62.

A. Nabaluia clemensii. Mount Kinabalu, Summit Trail, 2000 m, *Bailes & Cribb 784.* Photo R. B. G. Kew.

B. Nephelaphyllum flabellatum. Sabah. Photo A. Lamb.

C. Nephelaphyllum pulchrum. Sabah. Photo A. Lamb.

D. Nephelaphyllum pulchrum. Peninsular Malaysia, Cameron Highlands. Photo J. Dransfield.

PLATE 63.

A. Neuwiedia borneensis. Mount Kinabalu, Hempuen Hill, *Beaman 7404.* Photo R. S. Beaman.

B. Neuwiedia borneensis. Mount Kinabalu, Hempuen Hill. Photo A. Lamb.

C. Neuwiedia zollingeri var. **javanica.** Mount Kinabalu, Hempuen Hill. Photo A. Lamb.

PLATE 64.

A. Oberonia kinabaluensis. Sabah, Sipitang District. Photo J. B. Comber.

B. Octarrhena parvula. Peninsular Malaysia, Pahang. Photo J. Dransfield.

C. Oeceoclades pulchra. Sabah. Photo A. Lamb.

D. Ornithochilus difformis var. **kinabaluensis.** Mount Kinabalu, Pinosuk Plateau. Photo A. Lamb.

Additional collections. DALLAS: 900 m, *Clemens s.n.* (BM); MESILAU TRAIL: *Madani SAN 76482* (K); MOUNT KINABALU: 1500 m, *Chan & Gunsalam s.n.* (K); MT. NUNGKEK: 800 m, *Clemens 32906* (BM); PENIBUKAN: 1200–1500 m, *Clemens s.n.* (BM), 1500 m, *30996* (BM), 1500 m, *50324* (BM); TENOMPOK: 1600 m, *Carr SFN 26955* (K, SING), 1500 m, *Clemens 27235* (BM, K), 1500 m, *28608* (BM), 1500 m, *29754* (BM, G, K), 1500 m, *29841* (BM).

55.43.5. **Epigeneium** indet.

Collections. KEMBURONGOH: 2300 m, *Carr 3124, SFN 26553A* (SING); MARAI PARAI: 1600 m, *Carr 3124, SFN 26553* (SING).

55.44. **EPIPOGIUM** J. G. Gmel. ex Borkh.

Tent. Disp. Pl. German., 139 (1792).

Leafless saprophytic herbs. *Rhizome* tuberous. *Stems* erect, fleshy. *Inflorescence* terminal, racemose, brittle, ephemeral. *Flowers* non-resupinate, nodding at first. *Sepals* and *petals* about equal, narrow, acute, free; *lip* sessile, concave, entire, with minutely papillose ridges, shortly spurred; *column* short; *pollinia* 2, each with a filiform caudicle.

Two species widely distributed in tropical Africa and from western Europe to Japan, SE Asia to Australia and the Pacific islands.

55.44.1. **Epipogium roseum** (D. Don) Lindl., J. Linn. Soc. Bot. 1: 177 (1857). Plate 49A.

Limodorum roseum D. Don, Prodr. Fl. Nep., 30 (1825).

Saprophyte. Hill forest in leaf litter on forest floor. Elevation: 800–1000 m.

General distribution: Sabah, widespread in Africa, mainland & SE Asia east to New Guinea, Australia, the Solomon Islands & New Caledonia.

Collections. SAYAP: 800–1000 m, *Beaman 9812* (K), 800–1000 m, *9813* (K).

55.45. **ERIA** Lindl.

Bot. Reg. 11: t. 904 (1825).

Seidenfaden, G. (1982). Orchid genera in Thailand X. *Trichotosia* Bl. & *Eria* Lindl. Opera Bot. 62: 1–157.

Epiphytic or, rarely, terrestrial polymorphic herbs. *Stems* pseudobulbous (short or long, slender or thick), or without pseudobulbs and cane-like, 2- to many-leaved, rarely 1-leaved. *Leaves* flat or terete, thin-textured to coriaceous. *Inflorescence* terminal or axillary, racemose or rarely 1-flowered, rachis often hirsute or woolly. *Flowers* mostly small to medium-sized, rarely showy. *Sepals* subequal, free, rarely connate, glabrous or hirsute. *Lateral sepals* adnate to column-foot forming a short to long and spur-like or saccate mentum. *Petals* similar to dorsal sepal. *Lip* simple or 3-lobed, with or without keels. *Column* short, broad, more or less 2-winged, with a prominent foot. *Pollinia* 8, attached in fours by narrow bases to viscidium.

About 370 species widespread in tropical Asia, extending east to New Guinea, Australia and the Pacific islands.

55.45.1. Eria angustifolia Ridl. in Stapf, Trans. Linn. Soc. Bot. 4: 237 (1894). Type: MOUNT KINABALU, 1800 m, *Haviland s.n.* (holotype unlocated).

Epiphyte. Lower montane forest? Elevation: 1800 m. Known only from the type.

Endemic to Mount Kinabalu.

55.45.2. Eria bancana J. J. Sm., Bull. Jard. Bot. Buit., ser. 3, 2: 53 (1920).

Epiphyte. Lower montane forest. Elevation: 1600 m.

General distribution: Sabah, Bangka.

Collections. LUMU-LUMU: 1600 m, *Carr SFN 27153* (SING), 1600 m, *SFN 27153A* (SING).

55.45.3. Eria biflora Griff., Not. 3: 302 (1851).

Epiphyte. Lower montane forest. Elevation: 1400–1700 m.

General distribution: Sabah, India, Burma, Peninsular Malaysia, Thailand, Laos, Vietnam, Sumatra, Java, Bali.

Collections. BUNDU TUHAN: 1400 m, *Darnton 194* (BM); GURULAU SPUR: 1500 m, *Clemens 50381* (BM); PINOSUK PLATEAU: 1500 m, *Lamb AL 16/82* (K); TENOMPOK: 1500 m, *Clemens 28831* (BM); WEST MESILAU RIVER: 1700 m, *Brentnall 165* (K).

55.45.4. Eria borneensis Rolfe in Gibbs, J. Linn. Soc. Bot. 42: 150 (1914). Type: KIAU, 900 m, *Gibbs 3955* (holotype K!; isotype BM!).

Epiphyte. Undergrowth in secondary hill forest. Elevation: 900 m. Known only from the type.

Endemic to Mount Kinabalu.

55.45.5. Eria brookesii Ridl., J. Straits Branch Roy. Asiat. Soc. 50: 136 (1908). Plate 49B, 49C.

Epiphyte. Lower montane forest, sometimes on ultramafic substrate. Elevation: 800–2400 m.

General distribution: Sabah, Sarawak.

Collections. MARAI PARAI: 1800–2400 m, *Clemens 35176* (BM); MARAI PARAI/KIAU TRAIL: 1600 m, *Bailes & Cribb 878* (K); MT. NUNGKEK: 800 m, *Clemens 32904* (BM); PINOSUK PLATEAU: 1600 m, *Bailes & Cribb 538* (K).

55.45.6. Eria aff. **brookesii** Ridl., J. Straits Branch Roy. Asiat. Soc. 50: 136 (1908).

Epiphyte. Lower montane forest. Elevation: 1500 m.

Collection. TENOMPOK: 1500 m, *Clemens 30153* (K).

55.45.7. Eria carnosissima Ames & C. Schweinf., Orch. 6: 120 (1920). Type: KIAU, *Clemens 314* (holotype AMES mf).

Epiphyte. Lower montane forest. Elevation: 900–1400 m.

Endemic to Mount Kinabalu.

Additional collections. BUNDU TUHAN: 1400 m, *Carr SFN 27043* (SING); KIAU: 900 m, *Clemens 181* (AMES).

55.45.8. Eria cepifolia Ridl., J. Linn. Soc. Bot. 31: 282 (1896).

Epiphyte. Lower montane forest. Elevation: 1200–1400 m.

General distribution: Sabah, Sarawak, Peninsular Malaysia, Thailand.

Collections. BUNDU TUHAN: 1200 m, *Carr SFN 27747* (K, SING); PINOSUK PLATEAU: 1400 m, *Beaman 10754* (K).

55.45.9. Eria cymbidifolia Ridl., J. Bot. 36: 212 (1898).

a. var. **cymbidifolia.** Plate 50A, 50B.

Epiphyte or lithophyte. Lower montane forest, sometimes on ultramafic substrate. Elevation: 900–2400 m.

General distribution: Sabah, Kalimantan, Sumatra.

Collections. EAST MESILAU RIVER: 1500–1600 m, *Bailes & Cribb 506* (K), 1500–1600 m, *508* (K); GOLF COURSE SITE: 1700–1800 m, *Brentnall 133* (K); GURULAU SPUR: 1500 m, *Gibbs 4009* (BM); KEMBURONGOH/LUMU-LUMU: 1800–2400 m, *Clemens 27160* (BM, K); KIAU: 900 m, *Clemens 147* (AMES, BM); LUMU-LUMU: 2100 m, *Clemens s.n.* (BM); MARAI PARAI SPUR: 2000 m, *Collenette A 62* (BM); MESILAU CAVE: 1900–2200 m, *Beaman 9571* (K); PENIBUKAN: 1200 m, *Clemens 30699* (BM); TENOMPOK: 1600 m, *Carr SFN 27809* (K), 1500 m, *Clemens 26842* (K), 1500 m, *Meijer SAN 20316* (K); TINEKUK FALLS: 2000 m, *Clemens 40911* (BM, K).

b. var. **pandanifolia** J. J. Wood, Lindleyana 5: 95 (1990).

Epiphyte. Lower montane forest, upper montane forest, mostly on ultramafic substrate. Elevation: 1200–2700 m.

General distribution: Sabah, Sarawak.

Collections. GURULAU SPUR: 2400–2700 m, *Clemens 50648* (BM); KINATEKI RIVER HEAD: 2700 m, *Clemens 35183* (BM), 2100 m, *35185* (BM); MARAI PARAI: 1800 m, *Clemens 35186* (BM); MESILAU BASIN: 2400 m, *Clemens 51627* (BM, K); PENIBUKAN: 1200 m, *Clemens 50143* (BM, K).

55.45.10. Eria discolor Lindl., J. Linn. Soc. Bot. 3: 51 (1859).

Epiphyte. Hill forest, lower montane forest. Elevation: 900–1500 m.

General distribution: Sabah, Sarawak, India, Burma, Peninsular Malaysia, Thailand, Laos, Vietnam, Sumatra, Java.

Collections. DALLAS: 900 m, *Clemens 26412* (BM); TENOMPOK: 1500 m, *Clemens s.n.* (BM).

55.45.11. Eria farinosa Ames & C. Schweinf., Orch. 6: 122 (1920). Type: MARAI PARAI SPUR, *Clemens 283* (holotype AMES mf).

Epiphyte. Lower montane forest, sometimes on ultramafic substrate. Elevation: 1200–2000 m.

General distribution: Sabah.

Additional collections. BUNDU TUHAN: 1200 m, *Clemens 28051* (BM); EAST MESILAU/MENTEKI RIVERS: 1700 m, *Beaman 9374* (K); GOLF COURSE SITE: 1700–1800 m, *Beaman 7225* (K); KINATEKI RIVER/MARAI PARAI: 1700 m, *Clemens 32784* (BM); MESILAU CAVE TRAIL: 1800 m, *Beaman 7480* (K); PENIBUKAN: 1200 m, *Carr 3048, SFN 26348* (SING), *Clemens s.n.* (BM), 1200 m, *30849* (BM), 1200 m, *31850* (BM), *35190* (BM); TENOMPOK: 1500 m, *Clemens s.n.* (BM, BM, BM), 1500 m, *28051* (K), 1500 m, *28419* (BM), 1500 m, *28614* (BM), 1500 m, *29849* (BM); TINEKUK FALLS: 2000 m, *Clemens 40917* (BM, K).

55.45.12. Eria aff. **farinosa** Ames & C. Schweinf., Orch. 6: 122 (1920).

Epiphyte. Lower montane forest. Elevation: 1600 m.

Collection. GOLF COURSE SITE: 1600 m, *Bailes & Cribb 736* (K).

55.45.13. Eria floribunda Lindl., Bot. Reg. 29: misc. 43, no. 56 (1843).

Epiphyte. Lower montane forest. Elevation: 1100–2400 m.

General distribution: Sabah, Kalimantan, Burma, Peninsular Malaysia, Thailand, Vietnam, Sumatra, Mentawai, Riau Archipelago, Philippines.

Collections. GOLF COURSE SITE: 1700–1800 m, *Beaman 7223* (K); KILEMBUN BASIN: 1500–1800 m, *Clemens 34443* (BM), 1500–1800 m, *34487* (BM); KILEMBUN RIVER HEAD: 1800–2400 m, *Clemens s.n.* (BM); KUNDASANG: 1400 m, *Kidman Cox 2525* (K); MAMUT RIVER: 1200 m, *RSNB 293* (K); MOUNT KINABALU: *Haslam s.n.* (AMES); MT. NUNGKEK: 1100 m, *Darnton 449* (BM); MURU-TURA RIDGE: 1200–1800 m, *Clemens 34363* (BM); TENOMPOK: 1500 m, *Beaman 10519* (K, MSC); TENOMPOK/RANAU: 1500 m, *Carr SFN 27039* (SING).

55.45.14. Eria cf. **floribunda** Lindl., Bot. Reg. 29: misc. 43, no. 56 (1843).

Epiphyte. Lower montane forest. Elevation: 1500 m.

Collection. TENOMPOK: 1500 m, *Beaman 10520* (K).

55.45.15. Eria gibbsiae Rolfe in Gibbs, J. Linn. Soc. Bot. 42: 151 (1914). Type: KIAU, 900 m, *Gibbs 3960* (holotype K!; isotype BM!).

Epiphyte. Secondary hill forest. Elevation: 900 m.

General distribution: Sabah, Sarawak?

Additional collection. DALLAS: 900 m, *Clemens 26413* (K).

55.45.16. Eria grandis Ridl. in Stapf, Trans. Linn. Soc. Bot. 4: 237 (1894). Fig. 39, Plate 50C, 50D. Type: MOUNT KINABALU, 3000–3700 m, *Haviland 1157* (holotype SING!; isotype K!).

Terrestrial. Upper montane forest and scrub, sometimes on ultramafic substrate. Elevation: 2100–3700 m.

Endemic to Mount Kinabalu.

Additional collections. EASTERN SHOULDER: 3000 m, *RSNB 733* (K); EASTERN SHOULDER, CAMP 4: 2900 m, *RSNB 1121* (K); GURULAU SPUR: 2400–2700 m, *Clemens 50647* (BM, K), 2400–2700 m, *50648* (K); KEMBURONGOH: 2700 m, *Clemens 27165* (BM), 2400 m, *27861* (BM, K); KILEMBUN RIVER HEAD: 2700–2900 m, *Clemens 33962* (BM); MARAI PARAI: 2100 m, *Carr SFN 27432* (SING); MESILAU BASIN: 2100 m, *Clemens 29254* (BM); MOUNT KINABALU: *Haslam s.n.* (AMES, BM, K); PAKA-PAKA CAVE: 3000 m, *Carr SFN 27534* (SING), *Clemens 112* (AMES, BM, K, SING), 3400 m, *27170* (BM), 2900–3200 m, *Gibbs 4268* (BM, K), 3000 m, *Meijer SAN 29293* (K); SUMMIT TRAIL: 3000 m, *Beaman 8307* (K).

55.45.17. Eria aff. **hyacinthoides** (Blume) Lindl., Gen. Sp. Orch., 66 (1830).

Dendrolirium hyacinthoides Blume, Bijdr., 346 (1825).

Epiphyte. Lower montane forest. Elevation: 1700–1900 m.

Collection. MESILAU CAVE TRAIL: 1700–1900 m, *Beaman 7995* (K).

55.45.18. Eria iridifolia Hook. f., Fl. Brit. Ind. 5: 790 (1890).

Epiphyte. Hill forest, lower montane forest, sometimes on ultramafic substrate. Elevation: 800–1600 m.

General distribution: Sabah, Peninsular Malaysia, Sumatra, Java.

Collections. GOLF COURSE SITE: 1600 m, *Bailes & Cribb 712* (K); LOHAN RIVER: 800–1000 m, *Beaman 9974* (K); UPPER KINABALU: *Clemens 29356* (K).

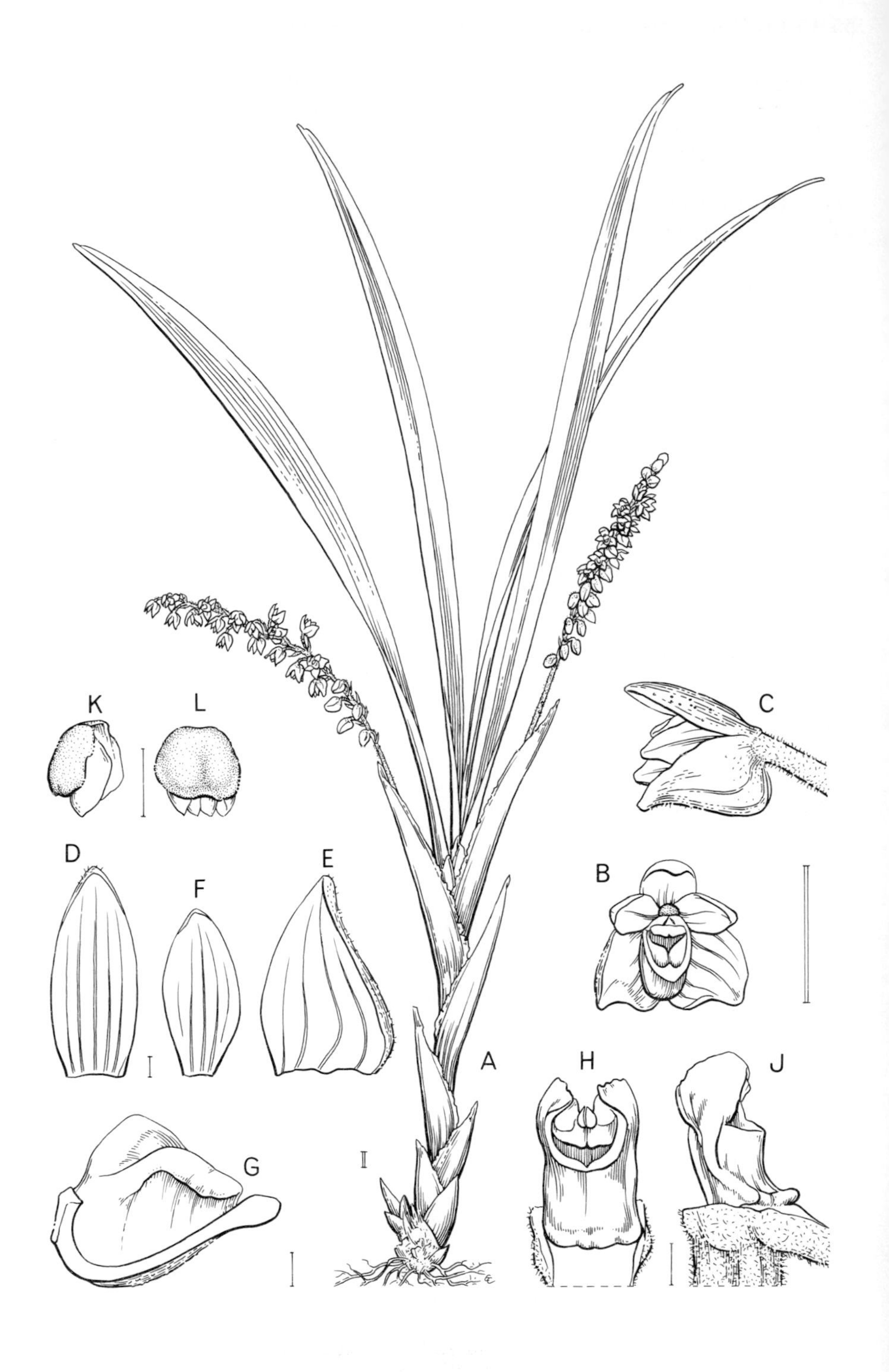
A
B
C
D
E
F
G
H
J
K
L

55.45.19. Eria javanica (Sw.) Blume, Rumphia 2: 23 (1836).

Dendrobium javanicum Sw., Kongl. Vetensk. Acad. Nya Handl. 21: 247 (1800).

Epiphyte. Lowlands, lower montane forest. Elevation: 400–2400 m.

General distribution: Sabah, Sarawak, widespread in S & SE Asia to New Guinea.

Collections. KEMBURONGOH: 2400 m, *Clemens 31680* (BM); MEKEDEU RIVER: 400–500 m, *Beaman 9646* (K); MENTEKI RIVER: 1600 m, *Beaman 10771* (K); MESILAU RIVER: 1500 m, *RSNB 4858* (K), 1500 m, *RSNB 7109* (K, SING); TAHUBANG RIVER: 1200–1500 m, *Clemens 31656* (BM).

55.45.20. Eria jenseniana J. J. Sm., Bull. Jard. Bot. Buit., ser. 3, 2: 50 (1920). Plate 51B.

Pendulous epiphyte. Hill forest on ultramafic substrate. Elevation: 600 m.

General distribution: Sabah, Sumatra.

Collection. HEMPUEN HILL, 600 m, *Lamb AL 1467/92* (K); LANGANAN RIVER: *Lohok 23* (K).

55.45.21. Eria aff. **jenseniana** J. J. Sm., Bull. Jard. Bot. Buit., ser. 3, 2: 50 (1920).

Terrestrial. Cultivated at Poring, of local origin.

Collection. PORING ORCHID GARDEN: *Lohok 4* (K).

55.45.22. Eria kinabaluensis Rolfe in Gibbs, J. Linn. Soc. Bot. 42: 151 (1914). Type: PAKA-PAKA CAVE, 2700 m, *Gibbs 4227* (syntype K!; isosyntype BM!), 3000 m, *4302* (syntype BM!).

Epiphyte. Lower montane forest, upper montane forest, sometimes on ultramafic substrate. Elevation: 1500–3000 m.

General distribution: Sabah.

Additional collections. BAMBANGAN RIVER: 1500 m, *RSNB 4554* (K); KEMBURONGOH: 2200 m, *Carr SFN 27484* (K, SING), 2400 m, *Clemens 27175* (BM, K); TENOMPOK: 1500 m, *Clemens 30155* (K); TENOMPOK/LUMU-LUMU: 1700 m, *Sinclair et al. 9226* (K, SING).

Fig. 39. Eria grandis. A, habit; **B,** flower (front view); **C,** flower (side view); **D,** dorsal sepal; **E,** lateral sepal; **F,** petal; **G,** lip (longitudinal section); **H,** column (front view); **J,** column (side view); **K,** anther-cap (side view); **L,** anther-cap (back view). **A** from *RSNB 1121*; **B–L** from *Beaman 8307.* Scale: single bar = 1 mm; double bar = 1 cm. Drawn by Eleanor Catherine.

55.45.23. Eria latiuscula Ames & C. Schweinf., Orch. 6: 125 (1020). Type: MOUNT KINABALU, *Clemens s.n.* (holotype AMES mf).

Epiphyte. Lower montane forest, sometimes on ultramafic substrate. Elevation: 1100–2100 m.

General distribution: Sabah.

Additional collections. GOLF COURSE SITE: 1700–1800 m, *Beaman 7224* (K); KINATEKI RIVER HEAD: 2100 m, *Clemens 31832* (BM); MAMUT COPPER MINE: 1600–1700 m, *Beaman 9952* (K); MESILAU BASIN: 2100 m, *Clemens 29292* (BM); MESILAU CAVE: 1800 m, *RSNB 4777* (K); MESILAU TRAIL: 1500 m, *Bailes & Cribb 519* (K); PARK HEADQUARTERS: 1700 m, *Kanis & Sinanggul SAN 51488* (K); PENATARAN RIVER: 1100 m, *Clemens 34058* (BM), 1800 m, *34330* (BM); PENIBUKAN: 1200 m, *Clemens 30828* (BM), 1200–1500 m, *31591* (BM); TENOMPOK: 1600 m, *Carr SFN 26935* (K, SING), 1500 m, *Clemens 29635* (BM), 1500 m, *29675* (BM, K), 1500 m, *29822* (K), 1500 m, *30152* (K).

55.45.24. Eria aff. **latiuscula** Ames & C. Schweinf., Orch. 6: 125 (1920).

Epiphyte. Hill forest, upper montane forest, sometimes on ultramafic substrate. Elevation: 900–2700 m.

Collections. DALLAS: 900 m, *Clemens 26128* (BM); DALLAS/TENOMPOK: 1200 m, *Clemens s.n.* (BM); KINATEKI RIVER HEAD: 2100 m, *Clemens 31733* (BM), 2700 m, *35177* (BM); MESILAU RIVER: 2100 m, *Clemens 29098* (BM).

55.45.25. Eria leiophylla Lindl., J. Linn. Soc. Bot. 3: 57 (1859).

Epiphyte. Lower montane forest, sometimes on ultramafic substrate. Elevation: 1200–2000 m.

General distribution: Sabah, Sarawak, Peninsular Malaysia, Sumatra, Sulawesi, Seram.

Collections. BUNDU TUHAN: 1200 m, *Carr SFN 27407A* (K); GOLF COURSE SITE: 1700–1800 m, *Beaman 10677a* (K); GURULAU SPUR: 1700 m, *Clemens 50418* (K); KILEMBUN RIVER HEAD: 1800 m, *Clemens 32524* (BM); LITTLE MAMUT RIVER: 1600 m, *Collenette 1054* (K); MARAI PARAI SPUR: *Clemens 250* (AMES); PARK HEADQUARTERS: 1700–2000 m, *Vermeulen & Chan 408* (K); PENIBUKAN: 1200 m, *Clemens 30853* (BM), 1200–1500 m, *31258* (BM), 1200 m, *40856* (BM, K), *50397* (BM); TENOMPOK: 1700 m, *Vermeulen 548* (K); TENOMPOK ORCHID GARDEN: 1500 m, *Clemens 50165* (BM), 1500 m, *50365* (BM); TENOMPOK/RANAU: 1400 m, *Carr SFN 27011* (SING).

55.45.26. Eria linearifolia Ames in Merr., Bibl. Enum. Born. Pl., 172 (1921).

Epiphyte. Hill forest, riverine forest, on ultramafic substrate. Elevation: 800–1000 m.

General distribution: Sabah, Kalimantan, Sarawak.

Collections. BAMBANGAN RIVER: *Lamb LKC 3160* (K); LOHAN RIVER: 800–1000 m, *Beaman 10007* (K).

55.45.27. Eria macrophylla Ames & C. Schweinf., Orch. 6: 127 (1920). Type: MOUNT KINABALU, *Haslam s.n.* (holotype AMES mf).

Epiphyte. Known only from the type.

Endemic to Mount Kinabalu.

55.45.28. Eria magnicallosa Ames & C. Schweinf., Orch. 6: 129, pl. 92 (1920). Plate 51A. Type: KIAU, 900 m, *Clemens 32* (holotype AMES mf).

Epiphytic or terrestrial. Hill forest, lower montane forest, sometimes on ultramafic substrate. Elevation: 900–2400 m.

General distribution: Sabah, Brunei.

Additional collections. BAMBANGAN RIVER: 1500 m, *RSNB 4402* (K), 1500 m, *RSNB 1304* (K, SING); DALLAS: 900 m, *Clemens 26313* (K), 900 m, *26723* (BM, K), 900 m, *26766* (BM, K), 900 m, *26782* (BM); EAST MESILAU/MENTEKI RIVERS: *Wood 826* (K); GOLF COURSE SITE: 1700–1800 m, *Beaman 7232* (K), 1700–1800 m, *10671* (K, MSC); KEMBURONGOH: 2400 m, *Clemens 28859* (BM); KIAU: 900 m, *Clemens 76* (AMES); KIBAMBANG LUBANG: 1400 m, *Clemens 32440* (BM); KILEMBUN BASIN: 2300 m, *Clemens 33726* (BM), 1500–1800 m, *35188* (BM); KUNDASANG: 1400 m, *Kidman Cox 2516* (K); LIWAGU RIVER: 1500 m, *Darnton 565* (BM); MAHANDEI RIVER: 1100 m, *Carr 3106, SFN 26471* (SING); MARAI PARAI/KILEMBUN RIVER: 1500 m, *Clemens 33171* (BM); MESILAU CAVE: 1800 m, *RSNB 4705* (SING); MESILAU RIVER: *Madani SAN 76464* (K); MOUNT KINABALU: *Clemens s.n.* (AMES), *Haslam s.n.* (AMES); PARK HEADQUARTERS/POWER STATION: 1700 m, *Cribb 89/35* (K); PENATARAN RIVER: 900–1200 m, *Clemens 34320* (BM); PENIBUKAN: 1500 m, *Clemens 30991* (BM); TENOMPOK: 1500 m, *Clemens 27246* (BM, K), 1500 m, *28687* (BM, K), 1500 m, *29678* (BM, K), 1500 m, *29843* (BM), 1500 m, *29903* (BM, K), 1500 m, *30156* (K); TINEKUK FALLS: 1800 m, *Clemens 40918* (BM, K).

55.45.29. Eria major Ridl. ex Stapf, Trans. Linn. Soc. Bot. 4: 237 (1894). Type: MOUNT KINABALU, 1800 m, *Haviland 1250* (holotype? K!).

Eria scortechinii Stapf, nomen delendum, non Hook. f., Trans. Linn. Soc. Bot. 4: 237 (1894).

Eria villosissima Rolfe in Gibbs, J. Linn. Soc. Bot. 42: 150 (1914). Type: MARAI PARAI SPUR, 2100 m, *Gibbs 4090* (holotype K!; isotype BM!).

Epiphytic, lithophytic or terrestrial. Lower and upper montane forest, xerophyllous scrub forest, frequently on ultramafic substrate. Elevation: 1100–2900 m.

General distribution: Sabah, Maluku.

Additional collections. BAMBANGAN RIVER: 1500 m, *RSNB 4563* (K), 1500 m, *RSNB 4622* (K); EAST MESILAU/MENTEKI RIVERS: 1700 m, *Beaman 8745* (K), 1700 m, *9392a* (K); GURULAU SPUR: 2100–2700 m, *Clemens 50770* (BM); KIBAMBANG LUBANG: 2100 m, *Clemens 32556* (BM); KILEMBUN BASIN: 1100 m, *Clemens s.n.* (BM); LUMU-LUMU: 1800 m, *Clemens 29844* (BM, K); MARAI PARAI: 1800 m, *Clemens 32618* (BM); MARAI PARAI SPUR: *Clemens 281* (AMES), 2400–2700 m, *376* (AMES), 2100 m, *Collenette A 67* (BM); MESILAU BASIN: 2100 m, *Clemens 29356* (BM); MESILAU CAVE: 1900–2200 m, *Beaman 9575a* (K); MESILAU CAVE/JANET'S HALT: 2000 m, *Collenette 21642* (K); MESILAU RIVER: 1500 m, *RSNB 4857* (K); MOUNT KINABALU: *Clemens s.n.* (AMES), *Haslam s.n.* (AMES); PARK HEADQUARTERS: 1400 m, *Price 125* (K); PENATARAN RIVER: 2400–2900 m, *Clemens 35189* (BM); PENIBUKAN: 1200–1500 m, *Clemens 31454* (K); PIG HILL: 2000–2300 m, *Beaman 9832* (K); TENOMPOK: 1500 m, *Carr SFN 27168* (SING).

55.45.30. Eria monophylla Schltr., Bull. Herb. Boissier, ser. 2, 6: 461 (1906).

Epiphyte. Hill forest on ultramafic substrate. Elevation: 800 m.

General distribution: Sabah, Kalimantan.

Collection. HEMPUEN HILL: 800 m, *Madani SAN 89513* (K).

55.45.31. Eria nutans Lindl., Bot. Reg. 26: misc. 83, no. 196 (1840). Plate 51C.

Epiphyte. Hill forest, lower montane forest, sometimes on ultramafic substrate. Elevation: 900–1500 m.

General distribution: Sabah, Sarawak, Peninsular Malaysia, Singapore, Thailand, Lingga Archipelago, Natuna Islands, Riau Archipelago.

Collections. DALLAS: 900 m, *Clemens 26516* (K); PENATARAN RIVER: 1400 m, *Clemens 40216* (BM); PENIBUKAN: *Clemens s.n.* (BM), 1200–1500 m, *31339* (BM), 1400 m, *40559* (BM, K).

55.45.32. Eria obliqua (Lindl.) Lindl., J. Linn. Soc. Bot. 3: 55 (1859).

Mycaranthes obliqua Lindl., Bot. Reg. 26: misc. 77, no. 184 (1840).

Epiphyte. Lower montane forest. Elevation: 1500 m.

General distribution: Sabah, Sarawak, Peninsular Malaysia, Singapore, Sumatra, Bangka.

Collection. TENOMPOK: 1500 m, *Clemens 29849* (K).

55.45.33. Eria oblitterata (Blume) Rchb. f., Bonplandia 5: 55 (1857).

Mycaranthes oblitterata Blume, Bijdr., 353 (1825).

Eria ridleyi Rolfe in Gibbs, J. Linn. Soc. Bot. 42: 150 (1914).

Epiphytic or terrestrial. Lower montane forest. Elevation: 1500–1600 m.

General distribution: Sabah, Kalimantan, Sarawak, Peninsular Malaysia, Thailand, Cambodia, Vietnam, Sumatra, Java, Bali.

Collections. EAST MESILAU RIVER: 1500–1600 m, *Bailes & Cribb 525* (K); MESILAU RIVER: 1500 m, *RSNB 4176* (K); TENOMPOK: 1500 m, *Carr SFN 27131* (K, SING).

55.45.34. Eria ornata (Blume) Lindl., Gen. Sp. Orch., 66 (1830).

Dendrolirium ornatum Blume, Bijdr., 345 (1825).

Epiphyte. Lowlands, hill forest. Elevation: 300–1500 m.

General distribution: Sabah, Kalimantan, Sarawak, Peninsular Malaysia, Thailand, Sumatra, Java, Sulawesi, Philippines.

Collections. DALLAS: 900 m, *Clemens 26502* (BM, K), 900 m, *29008* (K), 900 m, *30122* (K); DALLAS/TENOMPOK: *Clemens 29676* (BM, K); KADAMAIAN RIVER: 700 m, *Carr SFN 27038* (SING);

KAUNG: 300 m, *Collenette 3* (BM); KIAU: *Clemens 29008* (BM); TENOMPOK: 1500 m, *Clemens s.n.* (K).

55.45.35. Eria aff. **pachystachya** Lindl., J. Linn. Soc. Bot. 3: 60 (1859).

Epiphyte. Hill forest. Elevation: 900 m.

Collection. DALLAS: 900 m, *Clemens 27516* (BM).

55.45.36. Eria pannea Lindl., Bot. Reg. 28: misc. 64, no. 79 (1842). Plate 52A.

Epiphyte. Hill forest. Elevation: 1200 m.

General distribution: Sabah, Sarawak, Kalimantan, widespread from India & China (Yunnan) southeast to Sumatra & Bangka.

Collection. BUNDU TUHAN: 1200 m, *Carr SFN 27444* (SING).

55.45.37. Eria pellipes Hook. f., Fl. Brit. Ind. 5: 802 (1890). Plate 52B.

Epiphyte. Lower montane ridge forest on ultramafic substrate. Elevation: 1400–1500 m.

General distribution: Sabah, Sarawak, Peninsular Malaysia, Thailand.

Collections. MAMUT COPPER MINE: 1400–1500 m, *Beaman 10356* (K); PENIBUKAN: 1500 m, *Clemens 50391* (BM, K).

55.45.38. Eria pseudocymbiformis J. J. Wood, Kew Bull. 39: 84, f. 9 (1984).

a. var. **pseudocymbiformis.** Plate 53B.

Epiphyte. Lower montane forest, sometimes on ultramafic substrate. Elevation: 1200–2100 m.

General distribution: Sabah, Sarawak.

Collections. GOLF COURSE SITE: 1700–1800 m, *Beaman 10672* (K); KILEMBUN RIVER HEAD: 1800–2100 m, *Clemens 35180* (BM); LUMU-LUMU: 2100 m, *Clemens 29835* (BM); MAMUT COPPER MINE: 1600–1700 m, *Beaman 9951* (K); MARAI PARAI/NUNGKEK: 1400 m, *Clemens 32555* (BM); PENIBUKAN: 1200 m, *Clemens s.n.* (BM); TENOMPOK ORCHID GARDEN: 1500 m, *Clemens 50487* (BM).

b. var. **hirsuta** J. J. Wood, Lindleyana 5: 97 (1990). Type: KILEMBUN RIVER HEAD, 2100–2400 m, *Clemens 33938* (holotype BM!; isotype K!).

Epiphyte or lithophyte. Lower montane forest, upper montane mossy ridge and scrub forest, frequently on ultramafic substrate. Elevation: 900–2900 m.

General distribution: Sabah.

Additional collections. EASTERN SHOULDER: 2400 m, *RSNB 201* (K); GURULAU SPUR: 1800 m, *Clemens 50410* (BM); KIAU: 900 m, *Clemens 77* (AMES, BM, K, SING); KILEMBUN RIVER HEAD: 1500–1800 m, *Clemens 35181* (BM), 1500–1800 m, *35182* (BM); MARAI PARAI: 1800 m, *Clemens 32667* (BM); MESILAU CAMP: *Poore H 301* (K); MINETUHAN: 1800–2400 m, *Clemens 33787* (BM); MT. NUNGKEK: 1500 m, *Clemens s.n.* (BM); PARK HEADQUARTERS: 1700–2000 m, *Vermeulen & Chan 391* (K); PENATARAN RIDGE: 2400–2900 m, *Clemens s.n.* (BM); PENIBUKAN: 1200 m, *Clemens 30579* (BM), 1500 m, *35187* (BM), 1500 m, *50326* (BM, K), 1500 m, *50342* (BM, K); PIG HILL: 2000–2300 m, *Beaman 9894* (K), 1700–2000 m, *Sutton 2* (K); TAHUBANG RIVER HEAD: 2100 m, *Clemens 35184* (BM); TENOMPOK ORCHID GARDEN: 1500 m, *Clemens 50355* (BM); TENOMPOK/KEMBURONGOH: 2000 m, *Meijer SAN 20417* (K); TINEKUK FALLS: 2000 m, *Clemens 40912* (BM).

55.45.39. Eria pseudoleiophylla J. J. Wood, Orchid Rev. 89: 209 (1981). Type: MOUNT KINABALU, 800 m, *Puasa 154* (holotype K!; isotype SAN!).

Epiphyte. Hill forest, lower montane forest. Elevation: 800–1700 m.

General distribution: Sabah, Sulawesi, New Guinea.

Additional collections. DALLAS: 1200 m, *Clemens 26458* (BM, K); GURULAU SPUR: 1700 m, *Clemens 50418* (BM).

55.45.40. Eria rigida Blume, Mus. Bot. Lugd. 2: 183 (1856).

Epiphyte. Lower montane forest. Elevation: 1500 m.

General distribution: Sabah, Kalimantan, Sarawak, Peninsular Malaysia, Thailand, Sumatra, New Guinea.

Collection. TENOMPOK: 1500 m, *Clemens 28817* (BM).

55.45.41. Eria robusta (Blume) Lindl., Gen. Sp. Orch., 69 (1830). Plate 53A.

Dendrolirium robustum Blume, Bijdr., 347 (1825).

Epiphyte. Lowlands, hill forest, lower montane forest, upper montane forest, mossy forest, sometimes on ultramafic substrate. Elevation: 400–3000 m.

General distribution: Sabah, Brunei, Kalimantan, Sarawak, Peninsular Malaysia, Thailand, Sumatra, Java, New Caledonia, Samoa, Solomon Islands, Fiji.

Collections. DACHANG: 3000 m, *Clemens 29284* (BM, K); DAIRY ANNEX: 1600 m, *Beaman 10791* (K); DALLAS: 900 m, *Clemens 27752* (K); DALLAS/TENOMPOK: 900 m, *Clemens 27752* (BM); EAST MESILAU/MENTEKI RIVERS: *Wood 834* (K); GURULAU SPUR: *Clemens 50536* (BM), 2400–2700 m, *50646* (BM, K); JANET'S HALT/SHEILA'S PLATEAU: 2900 m, *Collenette 21536* (K); KAUNG: 400 m, *Darnton 399* (BM); KIBAMBANG LUBANG: 1400 m, *Clemens 32434* (BM); KILEMBUN RIVER: 1400–1800 m, *Clemens 34357* (K); LUMU-LUMU: 2100 m, *Clemens 27858* (BM); MARAI PARAI: 2000 m, *Clemens 32325* (BM), 1500 m, *Collenette A 25* (BM); MARAI PARAI SPUR: 1100 m, *Clemens 32777* (BM), 1700 m, *34314* (BM); MARAI PARAI/NUNGKEK: 1400 m, *Clemens 32549* (BM); MT. NUNGKEK: 1400 m, *Clemens 32778* (BM), 800 m, *32902* (BM); MURU-TURA RIDGE: 1500–1800 m, *Clemens 34357* (BM); PENIBUKAN: 1200 m, *Clemens 30775* (BM), 1400 m, *40227* (BM), 1100 m, *40444* (BM), *40860* (BM), 1500 m, *50209* (BM, K); PIG HILL: 2000–2300 m, *Beaman 9893* (K); SEDIKEN RIVER/MARAI PARAI: 1500 m, *Clemens 32309* (BM).

55.45.42. Eria aff. **saccifera** Hook. f., Fl. Brit. Ind. 5: 797 (1890).

Epiphyte. Lower montane forest. Elevation: 1600 m.

Collection. EAST MESILAU RIVER: 1600 m, *Bailes & Cribb 538* (K).

55.45.43. Eria xanthocheila Ridl., Mat. Fl. Malay. Penins. 1: 102 (1907).

Epiphyte. Hill forest, lower montane forest, sometimes on ultramafic substrate. Elevation: 700–1600 m.

General distribution: Sabah, Kalimantan, Burma, Peninsular Malaysia, Thailand, Sumatra, Java, Mentawai.

Collections. GOLF COURSE SITE: 1600 m, *Bailes & Cribb 737* (K); LOHAN RIVER: 700–900 m, *Beaman 9257* (K); PINOSUK PLATEAU: 1400–1500 m, *Lamb AL 27/82* (K).

55.45.44. Eria sp. 1, sect. Cylindrolobus

Epiphyte. Hill forest. Elevation: 900–1200 m.

Collections. DALLAS: 900 m, *Clemens 26514* (K), 900 m, *26603* (BM, K); KIAU: 900 m, *Clemens 30470* (BM); PENIBUKAN: 1200 m, *Clemens 30781* (BM).

55.45.45. Eria sp. 2, sect. Cylindrolobus

Epiphyte. Lower montane forest. Elevation: 1200–1500 m.

Collection. PENIBUKAN: 1200–1500 m, *Clemens 31019* (BM, K).

55.45.46. Eria sp. 3, sect. Mycaranthes

Epiphyte. Lower montane forest. Elevation: 1700–2000 m.

Collection. PARK HEADQUARTERS: 1700–2000 m, *Vermeulen & Chan 411* (K).

55.45.47. Eria sp. 4, sect. Mycaranthes

Epiphyte. Hill forest. Elevation: 900 m.

Collection. MINITINDUK GORGE: 900 m, *Carr 3166, SFN 26636* (K).

55.45.48. Eria sp. 5, sect. Mycaranthes

Epiphyte. Hill forest, lower montane forest. Elevation: 900–1500 m.

Collections. DALLAS: 900 m, *Clemens 26233* (K); TENOMPOK: 1500 m, *Clemens 29236* (K).

55.45.49. Eria sp. 6, sect. Cymboglossum

Epiphyte. Lower montane forest. Elevation: 2000 m.

Collection. TINEKUK FALLS: 2000 m, *Clemens 40913* (BM, K).

55.45.50. Eria sp. 7, sect. Cymboglossum

Epiphyte. Lower montane forest. Elevation: 1700–2100 m.

Collections. KEMBURONGOH: 2100 m, *Price 226* (K); TINEKUK FALLS: 1700 m, *Clemens 40825* (BM, K).

55.45.51. Eria sp. 8

Epiphyte. Lower montane forest. Elevation: 2100 m.

Collection. SUMMIT TRAIL: 2100 m, *Fuchs 21058* (K).

55.45.52. Eria sp. 9

Epiphyte. Lower montane forest. Elevation: 1500 m.

Collection. TENOMPOK: 1500 m, *Clemens 29945* (BM, K).

55.45.53. Eria sp. 10

Epiphyte. Hill forest. Elevation: 900 m.

Collection. DALLAS: 900 m, *Clemens 26457* (K).

55.45.54. Eria sp. 11

Epiphyte. Hill forest. Elevation: 900 m.

Collection. DALLAS: 900 m, *Clemens 29996* (K).

55.45.55. Eria sp. 12

Epiphyte. Lower montane forest. Elevation: 1500–1800 m.

Collection. PINOSUK PLATEAU: 1500–1800 m, *Chow & Leopold SAN 76430* (K).

55.45.56. Eria sp. 13

Epiphyte. Lower montane forest. Elevation: 1000–2100 m.

Collections. KEMBURONGOH: 2100 m, *Price 224* (K); PARK HEADQUARTERS: 1700 m, *Lamb KL 3020* (K); TAHUBANG RIVER: 1000 m, *Carr 3063, SFN 26378* (K, SING); TENOMPOK: 1500 m, *Clemens 26779* (BM, K), 1500 m, *26833* (BM, K), 1500 m, *Meijer SAN 20323* (K).

55.45.57. **Eria** indet.

Collections. BAMBANGAN RIVER: 1500 m, *RSNB 4454* (SING), 1500 m, *RSNB 4465* (K); BUNDU TUHAN: 800 m, *Carr 3589* (SING), 1200 m, *SFN 27064A* (SING), 1200 m, *SFN 27349* (SING), 1200 m, *SFN 27406* (SING), 1200 m, *SFN 27407A* (SING), 1200 m, *SFN 27415* (SING), 1200 m, *SFN 27765* (SING), 1200 m, *SFN 27791* (SING), 900 m, *SFN 28028* (SING), 1400 m, *Darnton 225* (BM), 1400 m, *227* (BM); DALLAS: 900 m, *Clemens s.n.* (BM), 900 m, *26143* (BM), 900 m, *27515* (K); GOLF COURSE SITE: 1700–1800 m, *Beaman 7220* (K); GURULAU SPUR: 1400 m, *Carr SFN 27064* (SING), 1400 m, *SFN 28000* (SING), 2100–2700 m, *Clemens s.n.* (K), 1500 m, *50371* (BM), 1500 m, *50413* (BM), 2400–2700 m, *50660* (BM); KAUNG: 500 m, *Carr SFN 27421* (SING), 400 m, *SFN 27766* (SING), 400 m, *SFN 27948* (SING), 400 m, *SFN 28032* (SING); KEMBURONGOH: 2200 m, *Carr SFN 27106* (SING), 2100 m, *SFN 27465* (SING), 2100 m, *SFN 27563* (SING), 2100 m, *SFN 27661* (SING), 2400 m, *Clemens s.n.* (BM); KIAU: 900 m, *Carr SFN 26282* (SING); KIBAMBANG LUBANG: 1400 m, *Clemens 32435* (BM); KILEMBUN BASIN: 1400 m, *Clemens s.n.* (BM); KINATEKI RIVER: 900 m, *Carr SFN 26831* (SING), 1100 m, *SFN 26842* (SING); KINUNUT VALLEY HEAD: 1400 m, *Carr SFN 27150* (SING), 1200 m, *SFN 27232* (SING), 1200 m, *SFN 27365* (SING); LUBANG: 1400 m, *Carr SFN 27094* (SING); LUMU-LUMU: 1600 m, *Carr SFN 27462* (SING), 1700 m, *SFN 27809A* (SING), 2000 m, *Clemens s.n.* (BM, BM, BM), 2000 m, *26838* (BM), 2000 m, *27359* (BM), 1800 m, *27367* (BM); MAHANDEI RIVER: 1100 m, *Carr 3135, SFN 26589* (SING), 1200 m, *SFN 27986* (SING); MAHANDEI RIVER HEAD: 1400 m, *Carr SFN 27780* (SING); MARAI PARAI: 2000 m, *Carr SFN 27406A* (SING), 2100 m, *SFN 27433* (SING), 1800 m, *Clemens 32443* (BM); MARAI PARAI/NUNGKEK: 1400 m, *Clemens s.n.* (BM), 1400 m, *32550* (BM); MINETUHAN: 2300 m, *Clemens 35179* (BM); MINITINDUK GORGE: 900 m, *Carr SFN 26636* (SING); NUMERUK RIDGE: 1500 m, *Clemens 40061* (BM); PAKA-PAKA CAVE: 3500 m, *Carr SFN 27534C* (SING); PENATARAN RIDGE: 1800 m, *Clemens 40129* (BM); PENIBUKAN: 1200 m, *Carr SFN 26343* (SING), 1200–1500 m, *Clemens s.n.* (BM), 1200 m, *30850* (BM), 1200–1500 m, *31451* (BM), 1200–1500 m, *31453* (BM), 1200 m, *32087* (BM), 1100 m, *40345* (BM), 1400 m, *40558* (BM), 1500 m, *40703* (BM), 1200 m, *40787* (BM), 1200 m, *50104* (BM), 1200 m, *50112* (BM), 1500 m, *50168* (BM), 1500 m, *50199* (BM), 1700 m, *50361* (BM); PINOSUK: 800 m, *Carr SFN 27039B* (SING); SUMMIT TRAIL: 2000 m, *Carr 3535* (SING), 1800 m, *SFN 27465A* (SING); TENOMPOK: 1500 m, *Carr 3301* (SING), 1500 m, *SFN 27039A* (SING), 1500 m, *Clemens s.n.* (BM, BM, BM, BM), 1500 m, *29899* (BM); TENOMPOK ORCHID GARDEN: 1500 m, *Clemens 50161* (BM), 1500 m, *50166* (BM); TENOMPOK/RANAU: 1600 m, *Carr SFN 26999* (SING), 1500 m, *SFN 27051* (SING); TIBABAR RIVER: 1800 m, *Carr SFN 27561* (SING); TINEKUK FALLS: 2000 m, *Clemens 40907* (BM); TINEKUK RIVER: 1100 m, *Carr SFN 27931* (SING), 1200 m, *Clemens 51464* (BM).

55.46. **ERYTHRODES** Blume

Bijdr., 410 (1825).

Terrestrial herbs. *Leaves* ovate, oblique, green, petioles sheathing. *Inflorescence* lax, racemose; *peduncle* long; *rachis* and *ovary* pubescent. *Flowers* resupinate; *dorsal sepal* and *petals* connivent, forming a hood; *lip* spurred, concave, apex reflexed, spur 2-lobed, projecting between lateral sepals, internal warts and hairs absent; *column* short; *stigma* hollow, at foot of rostellum; *pollinia* 2.

Between 60 and 100 species, primarily in tropical America, with about 15–20 represented in the E Asiatic-Pacific area.

55.46.1. **Erythrodes latifolia** Blume, Bijdr., 411, t. 72 (1825). Plate 53C.

Terrestrial. Hill forest. Elevation: 900 m.

General distribution: Sabah, Peninsular Malaysia, Java.

Collections. KIAU: 900 m, *Carr 3771* (SING); MOUNT KINABALU: *Clemens s.n.* (BM).

55.46.2. Erythrodes triloba Carr, Gardens' Bull. 8: 181 (1935). Type: TENOMPOK, 1400 m, *Carr 3564, SFN 27859* (holotype SING!; isotypes AMES mf, K!).

Terrestrial. Lower montane forest. Elevation: 1400–1500 m.

Endemic to Mount Kinabalu.

Additional collection. TENOMPOK: *Clemens 27689* (AMES, BM).

55.47. EULOPHIA R. Br. ex Lindl.

Bot. Reg. 8: t. 686 (1823).

Terrestrial, sometimes saprophytic, herbs. *Rhizome* short. *Stems* pseudobulbous, squat, with either 1) terminal leaves, 2) slender, leafy and sheathed in basal cataphylls, or 3) leafless. *Leaves*, when present, narrowly elliptic, plicate. *Inflorescence* lateral, arising from near base of stem, erect, racemose, few- to many-flowered. *Flowers* small to large, resupinate; *sepals* free, similar; *petals* smaller than sepals; *lip* entire or 3-lobed, spurred at base; *column* slender; *pollinia* 4.

A pantropical genus of about 200 species, particularly well represented in Africa.

55.47.1. Eulophia graminea Lindl., Gen. Sp. Orch., 182 (1833).

Terrestrial. Open grassy, sandy areas. Elevation: 500 m.

General distribution: Sabah, Brunei, widespread from Sri Lanka, India & Nepal southeast to Indonesia, north to China, Taiwan, Hong Kong, Ryukyu Islands.

Collection. LOHAN RIVER: 500 m, *Lamb AL 310/85* (K).

55.47.2. Eulophia spectabilis (Dennst.) Suresh, Regnum Veg. 119: 300 (1988).

Wolfia spectabilis Dennst., Schlüssel Hortus Malab., 11, 25, 38 (1818).

Eulophia nuda Lindl., Gen. Sp. Orch., 180 (1833).

Eulophia squalida Lindl., Bot. Reg. 27: misc. 77, no. 164 (1841).

Terrestrial. Open, grassy areas. Elevation: 400–900 m.

General distribution: Sabah, Sarawak, Kalimantan, widespread from mainland & SE Asia east to New Guinea & the Pacific islands.

Collections. BUNDU TUHAN: 800 m, *Gibbs 4319* (BM); DALLAS: 900 m, *Clemens 27629* (BM, K); KAUNG: 400 m, *Carr SFN 26263* (SING); KIAU: 900 m, *Clemens 149* (AMES), *345* (AMES), *371* (AMES); LOHAN/MAMUT COPPER MINE: 900 m, *Beaman 10365* (K); MAMUT ROAD: 900 m, *Lamb SAN 92339* (K); MOUNT KINABALU: *Clemens s.n.* (AMES), *Haslam s.n.* (AMES).

55.47.3. Eulophia zollingeri (Rchb. f.) J. J. Sm., Orch. Java, 228 (1905). Plate 54A.

Cyrtopera zollingeri Rchb. f., Bonplandia 5: 38 (1857).

Saprophyte. Primary and secondary hill forest, possibly lower montane forest. Elevation: 500–1500 m.

General distribution: Sabah, widespread from India, Malaysia, Indonesia to the Philippines, north to Taiwan & Japan, east to New Guinea & Australia.

Collections. BUNDU TUHAN: 900 m, *Lamb AL 13/82* (K); DALLAS: 900 m, *Clemens s.n.* (BM, BM); KULUNG HILL: 500 m, *Darnton 118* (BM); TENOMPOK: 1500 m, *Clemens 28829* (BM), 1500 m, *29782* (K).

55.48. FLICKINGERIA A. D. Hawkes

Orchid Weekly 2: 451 (1961).

Seidenfaden, G. (1980). Orchid genera in Thailand IX. *Flickingeria* Hawkes & *Epigeneium* Gagnep. Dansk Bot. Ark. 34, 1: 1–104.

Epiphytic herbs. *Aerial shoots* terminating in a 1-leaved pseudobulb consisting of 1 internode, erect and bushy or drooping and laxly branched, new branches usually arising at base of pseudobulb. *Leaves* narrowly- to oblong-elliptic, coriaceous. *Inflorescences* borne from chaffy bracts in front or behind the leaf base, 1–2-flowered. *Flowers* fragile, ephemeral, lasting less than a day. *Sepals* and *petals* acute. *Lateral sepals* adnate to column-foot forming a rather long mentum. *Petals* narrower than sepals. *Lip* 3-lobed, 2–3-keeled, side-lobes erect, mid-lobe variable in shape, straight, curved or very undulate-plicate and transversely bilobed, often narrow at base forming a mesochile. *Column* short, with a long foot. *Pollinia* 4, naked, i.e. without a stipes or caudicle.

55.48.1. Flickingeria bicarinata (Ames & C. Schweinf.) A. D. Hawkes, Orchid Weekly 2, 46: 452 (1961).

Dendrobium bicarinatum Ames & C. Schweinf., Orch. 6: 98 (1920). Type: KIAU, *Clemens 335* (holotype AMES mf).

Epiphyte. Hill forest. Elevation: 900 m.

General distribution: Sabah.

Additional collection. MINITINDUK RIVER: 900 m, *Carr 3189, SFN 26775* (SING).

55.48.2. Flickingeria dura (J. J. Sm.) A. D. Hawkes, Orchid Weekly 2, 46: 454 (1961).

Dendrobium durum J. J. Sm., Orch. Java, 320 (1905).

Terrestrial. Lower montane forest. Elevation: 1600 m.

General distribution: Sabah, Java.

Collection. LUMU-LUMU: 1600 m, *Carr SFN 27151* (SING).

55.48.3. Flickingeria fimbriata (Blume) A. D. Hawkes, Orchid Weekly 2, 46: 454 (1961).

Desmotrichum fimbriatum Blume, Bijdr., 329 (1825).

Epiphyte. Lowlands and hill forest on ultramafic substrate. Elevation: 500–800 m.

General distribution: Sabah, Sarawak, Kalimantan, widespread in SE Asia.

Collections. KAUNG: 500 m, *Carr SFN 27869* (SING); LOHAN RIVER: 800 m, *Beaman 8366* (K); PINAWANTAI: 500 m, *Shea & Aban s.n.* (K).

55.48.4. Flickingeria pseudoconvexa (Ames) A. D. Hawkes, Orchid Weekly 2, 46: 458 (1961).

Dendrobium pseudoconvexum Ames, Orch. 5: 135 (1915).

Epiphyte. Hill forest, lower montane forest. Elevation: 900–1800 m.

General distribution: Sabah, Philippines.

Collections. BUNDU TUHAN: 900 m, *Darnton 540* (BM); DALLAS: *Clemens s.n.* (BM, BM); EAST MESILAU/MENTEKI RIVERS: 1700 m, *Beaman 9370* (K); GOLF COURSE SITE: 1700–1800 m, *Beaman 10668* (K); KIAU: 900 m, *Clemens 68* (AMES); PENIBUKAN: 1200 m, *Clemens 40663* (BM, K); PINOSUK PLATEAU: 1500–1700 m, *Vermeulen 492* (K); TENOMPOK: 1500 m, *Beaman 10521* (K), 1500 m, *Clemens 29789* (BM, K), 1500 m, *29841* (K), 1500 m, *30166* (K), 1500 m, *30167* (K).

55.48.5. Flickingeria aff. **pseudoconvexa** (Ames) A. D. Hawkes, Orchid Weekly 2, 46: 458 (1961).

Epiphyte. Lower montane forest, on ultramafic substrate. Elevation: 1500 m.

Collection. MARAI PARAI: 1500 m, *Carr 3127* (SING).

55.48.6. Flickingeria aff. **scopa** (Lindl.) F. G. Brieger, Schltr. Orch. 1 (11–12): 742 (1981).

Epiphyte. Low-stature hill forest on ultramafic substrate. Elevation: 700–900 m.

Collection. LOHAN RIVER: 700–900 m, *Beaman 9255* (K).

55.48.7. Flickingeria aff. **xantholeuca** (Rchb. f.) A. D. Hawkes, Orchid Weekly 2, 46: 460 (1961).

Epiphyte. Hill forest, sometimes on ultramafic substrate. Elevation: 800–1200 m.

Collections. DALLAS: 900 m, *Clemens 26414* (K); HEMPUEN HILL: 800–1200 m, *Beaman 7688* (K).

55.48.8. Flickingeria sp. 1

Epiphyte. Lowlands. Elevation: 400 m.

Collections. KAUNG: 400 m, *Darnton 393* (BM); TAKUTAN: *Amin Kalantas SAN 93600* (K).

55.48.9. Flickingeria indet.

Collection. DALLAS: 900 m, *Clemens 29304* (BM).

55.49. GALEOLA Lour.

Fl. Cochinch. 2: 520 (1790).

Saprophytic herbs with only scale-leaves. *Stems* stout, long-climbing. *Inflorescences* terminal, and also lateral in the axils of the upper scale-leaves of climbing stems; *rachis* pubescent-furfuraceous. *Flowers* resupinate, fleshy, yellowish or brownish, pubescent-furfuraceous; *sepals* and *petals* about equal, free, usually hardly spreading; base of *lip* surrounding the column, the blade concave to saccate, with longitudinal ridges; *column* stout, rather short, arcuate, clavate; *pollinia* 2, cleft, granular; *fruit* a long dry capsule, dehiscent, opening with 2 unequal valves; *seeds* winged, rather large.

About 10 species distributed in Madagascar eastward and from India (Sikkim) to China, southeastward through Thailand to Malaysia, Indonesia, the Philippines and New Guinea.

55.49.1. Galeola nudifolia Lour., Fl. Cochinch. 2: 521 (1790).

Climbing saprophyte. Hill dipterocarp forest. Sight record of A. Lamb on the old track between Lohan and Poring; no specimens seen.

General distribution: Sabah, widespread in China, Burma, Thailand & Indochina, through Malaysia & Indonesia to the Philippines.

55.50. **GASTROCHILUS** D. Don

Prodr. Fl. Nepal, 32 (1825).

Small- to medium-sized monopodial epiphytes. *Stems* usually short, up to 30 cm in a few species. *Leaves* narrowly elliptic or strap-shaped, apex unequally bilobed, acute or, rarely, with 3 setae, rather leathery. *Inflorescences* short and densely many flowered, ± umbellate, usually sessile, a few species having a short peduncle. *Flowers* with greenish yellow to yellow sepals and petals, often spotted red and with a white lip, the hypochile often spotted red and the epichile with a central yellow patch, spotted red; *sepals* and *petals* narrowly obovate; *lip* adnate to the column, immobile, divided into a semi-globose saccate hypochile and a fan-shaped, often broadly triangular epichile, epichile often hairy or papillose, margin entire to fimbriate; *column* short and stout, foot absent, *rostellar projection* oblong; *stipes* strap-shaped, longer than 2 times the diameter of the pollinia; *pollinia* 2, porate, or 4, unequal.

Around 38 species distributed from India & Sri Lanka through E Asia and Japan south to Indonesia.

55.50.1. Gastrochilus sororius Schltr., Feddes Repert. 12: 315 (1913).

Epiphyte. Hill forest, lower montane forest. Elevation: 900–1500 m.

General distribution: Sabah, Sumatra, Java.

Collections. BUNDU TUHAN: 1200 m, *Carr SFN 27744* (K, SING); DALLAS: 900 m, *Clemens s.n.* (BM); TENOMPOK: 1500 m, *Clemens 29845* (BM, K).

55.51. **GASTRODIA** R. Br.

Prodr., 330 (1810).

Leafless saprophytic herbs. *Rhizome* tuberous, ± horizontal. *Stems* tall or short, simple, erect, often elongating after fertilisation. *Inflorescence* terminal, racemose, 1- to many-flowered. *Flowers* campanulate; *sepals* and *petals* connate to form a 5-lobed tube which may or may not be gibbous at the base, and which may be split between the lateral sepals; *petals* smaller than sepals; *lip* shorter than sepals, entire or 3-lobed, disc with 2 or 3 keels and sometimes glandular basal calli; *column* quite long, foot short; *stigma* at base of column; *pollinia* 2.

About 20 species widespread from E and SE Asia to Australia, New Zealand and the Pacific islands.

55.51.1. Gastrodia grandilabris Carr, Gardens' Bull. 8: 179 (1935). Plate 54B. Type: TENOMPOK/RANAU, 1500 m, *Carr 3264, SFN 26970* (holotype SING!; isotype K!).

Saprophyte. Lower montane forest. Elevation: 1500–1600 m.

General distribution: Sabah.

Additional collections. GURULAU SPUR: *Clemens 51711* (BM, K); KIAU VIEW TRAIL: 1600 m, *Patrick & Thomas SNP 248* (K).

55.52. GEODORUM Jacks.

Bot. Repos. 10: t. 626 (1811).

Terrestrial herbs. *Stems* short, later swelling at base to form almost round, subterranean pseudobulbs. *Leaves* terminal, few, plicate, the uppermost the largest, broad, petiolate. *Inflorescence* separate from the leaf-bearing stem, erect; *rachis* decurved, nodding, straightening prior to seed dispersal, densely several- to many-flowered. *Flowers* not opening widely, bell-shaped; *sepals* and *petals* similar; *lip* entire or obscurely trilobed, forming with the column-foot a short saccate base, not spurred; *column* short with distinct foot; *pollinia* 4.

Between 5 and 10 species distributed from tropical Asia eastward through SE Asia to New Guinea, Australia and the Pacific islands.

55.52.1. Geodorum densiflorum (Lam.) Schltr., Feddes Repert. Beih. 4: 259 (1919).

Limodorum densiflorum Lam., Encycl. 3: 516 (1792).

Terrestrial. Secondary hill forest on ultramafic substrate, probably also lower montane forest. Elevation: 500–1800 m.

General distribution: Sabah, widespread from mainland Asia east to New Guinea, Australia & the Pacific islands.

Collections. HEMPUEN HILL: 500 m, *Lamb AL 308/85* (K); KADAMAIAN RIVER: 1800 m, *Carr SFN 27592* (SING); PINOSUK: 800 m, *Carr SFN 27865* (SING).

55.53. GOODYERA R. Br.

In W. Aiton & W. T. Aiton, Hort. Kew., ed. 2, 5: 197 (1813).

Terrestrial, rarely epiphytic, herbs. *Leaves* ovate, often asymmetric, sometimes variegated, petioles sheathing stem. *Stem, peduncle* and *ovaries* finely pubescent. *Inflorescence* erect, 1- to many-flowered. *Flowers* usually not opening widely, resupinate, white, pale green, pink to purplish; *sepals* parallel to floral axis, or the lateral sepals spreading; *dorsal sepal* connivent with petals, forming a hood; *petals* sometimes joined near their tips; *lip* concave or saccate, glabrous or papillose, with bristly hairs inside, narrowed to an acute, entire, often reflexed tip; *column* short, without basal appendages; *rostellum* usually long, deeply cleft; *stigma* not divided, large, ventral; *pollinia* 2.

A cosmopolitan genus of about 50 species, mostly distributed in warmer climates.

55.53.1. Goodyera bifida (Blume) Blume, Fl. Javae, ser. 2, 1, Orch.: 40 (1858).

Neottia bifida Blume, Bijdr., 408 (1825).

Terrestrial. Lower montane forest. Elevation: 1500–2100 m.

General distribution: Sabah, Peninsular Malaysia, Sumatra, Java, Sulawesi.

Collection. MESILAU: 1500–2100 m, *Clemens 51302* (BM).

55.53.2. Goodyera hylophiloides Carr, Gardens' Bull. 8: 195 (1935). Type: DALLAS, 900 m, *Clemens s.n.* (syntype BM!); MESILAU BASIN AND LUMU-LUMU, 1800 m, *Clemens 29993* (syntype BM!); TENOMPOK, 1600 m, *Carr 3256, SFN 26939* (syntype SING!; isosyntypes AMES mf, K!).

Terrestrial. Lower montane forest. Elevation: 900–2300 m.

General distribution: Sabah, Sarawak.

Additional collections. SUMMIT TRAIL: 2100–2300 m, *Chan & Gunsalam 39* (K); TENOMPOK RIDGE: 1400–1500 m, *Beaman 8213* (K).

55.53.3. Goodyera kinabaluensis Rolfe in Gibbs, J. Linn. Soc. Bot. 42: 159 (1914). Type: GURULAU SPUR, 1500–1700 m, *Gibbs 3997* (syntype unlocated), 1500–1700 m, *4003* (syntype BM!).

Terrestrial. Hill forest, lower montane forest. Elevation: 800–1700 m.

General distribution: Sabah.

Additional collections. DALLAS: 900 m, *Clemens 27695* (BM); KAUNG/KELAWAT: 800 m, *Carr 3419, SFN 27337* (K, SING); MINITINDUK: 900 m, *Carr 3179, SFN 27337* (K, SING).

55.53.4. Goodyera procera (Ker Gawl.) Hook., Exot. Fl. 1: t. 39 (1823).

Neottia procera Ker Gawl., Bot. Reg. 8: t. 639 (1822).

Terrestrial. Upper montane forest. Elevation: 2200–2400 m.

General distribution: Sabah, widespread from India & Sri Lanka east to Indonesia, north to China, Taiwan, Japan & the Philippines.

Collections. KEMBURONGOH: 2200 m, *Carr 3509, SFN 27470* (SING), 2400 m, *Clemens s.n.* (BM); LUMU-LUMU: *Clemens s.n.* (BM).

55.53.5. Goodyera reticulata (Blume) Blume, Fl. Javae, ser. 2, 1, Orch.: 35, t. 9b, f. 1 (1858).

Neottia reticulata Blume, Bijdr., 408 (1825).

Terrestrial. Lower montane forest. Sight record of A. Lamb between Nabalu and Tenompok; no specimens seen.

General distribution: Sabah, Java.

55.53.6. Goodyera rostellata Ames & C. Schweinf., Orch. 6: 12 (1920). Type: KIAU, *Clemens 401* (holotype AMES mf).

Terrestrial. Lower montane forest. Elevation: 900–2700 m.

General distribution: Sabah.

Additional collections. KADAMAIAN RIVER: 1800 m, *Carr SFN 27592* (SING); KIAU: 900 m, *Clemens 96* (AMES), *322* (AMES); MESILAU RIVER: 2100–2700 m, *Clemens 51733* (BM); POWER STATION: 1800 m, *Lamb SAN 88569* (K); TENOMPOK: 1500 m, *Carr 3437* (K).

55.53.7. Goodyera rubicunda (Blume) Lindl., Bot. Reg. 25: misc. 61, no. 92 (1839).

Neottia rubicunda Blume, Bijdr., 408 (1825).

Terrestrial. Hill forest, lower montane forest. Elevation: 500–1500 m.

General distribution: Sabah, widespread from Peninsular Malaysia, Indonesia, Philippines, New Guinea & Australia north to Taiwan, Ryukyu Islands & the Pacific islands.

Collections. DALLAS: 900 m, *Carr 3757, SFN 28054* (K, SING), 900 m, *Clemens 26308* (BM, K, SING), 900 m, *26896* (BM, K), 900 m, *26896A* (K), 900 m, *26997* (BM); DALLAS/TENOMPOK: *Clemens 26832* (K); KAUNG/DALLAS: 500 m, *Collenette A 130* (BM); TENOMPOK: 1500 m, *Clemens 27689* (K).

55.53.8. Goodyera ustulata Carr, Gardens' Bull. 8: 194 (1935). Plate 54C. Type: DALLAS, 900 m, *Clemens 26690a* (syntype BM!), 900 m, *26911* (syntype SING!; isosyntypes BM!, K!); KEGIITAN AGAYE HEAD, 1100 m, *Carr 3236, SFN 27359* (syntype SING!).

Terrestrial. Hill forest, sometimes in secondary vegetation. Elevation: 500–1100 m.

General distribution: Sabah.

Additional collections. PENATARAN RIDGE: 500 m, *Lamb SAN 91587* (K); PORING ORCHID GARDEN: *Lohok 24* (K).

55.53.9. Goodyera viridiflora (Blume) Blume, Fl. Javae, ser. 2, 1, Orch.: 41, t. 9c, f. 2 (1858).

Neottia viridiflora Blume, Bijdr., 415 (1825).

Terrestrial. Lowlands. Elevation: 400 m.

General distribution: Sabah, widespread in mainland & SE Asia east to New Guinea, north to the Philippines, Taiwan, Ryukyu Islands & China.

Collection. KAUNG: 400 m, *Carr 3572, SFN 27977* (K, SING).

55.53.10. Goodyera indet.

Collection. GOLF COURSE SITE: 1700 m, *Beaman 8554* (K).

55.54. GRAMMATOPHYLLUM Blume

Bijdr., 377 (1825).

Large robust epiphytic, rarely lithophytic, herbs. *Roots* long, spreading and erect, acuminate. *Pseudobulbs* medium-sized to very large, ovoid to elongate, clustered, few- to many-leaved. *Leaves* distichous, plicate, long and narrow, flexible. *Inflorescence* lateral, simple, erect or pendulous, a many-flowered raceme. *Flowers* resupinate, medium-sized to large, often showy, yellow, marked with brown or maroon; *sepals* and *petals* similar, spreading, obtuse; *lip* much smaller, 3-lobed, fused at base to base of column, usually hairy, disc 2- or 3-ridged; *column* short, clavate, foot absent, but often with a concave basal outgrowth; *pollinia* 2, deeply cleft; *stipes* deeply bilobulate; *viscidium* fat.

About 12 species distributed in SE Asia eastward to New Guinea and the Solomon Islands.

55.54.1. Grammatophyllum kinabaluense Ames & C. Schweinf., Orch. 6: 210 (1920). Type: KIAU, 900 m, *Clemens 55* (holotype AMES mf; isotypes BM!, K!, SING!).

Epiphyte. Lowlands and hill forest. Elevation: 500–900 m.

General distribution: Sabah.

Additional collection. KAUNG: 500 m, *Carr SFN 26598* (SING).

55.54.2. Grammatophyllum speciosum Blume, Bijdr., 378, t. 20 (1825).

Epiphyte. Lowlands. No specimens seen. Noted by Lamb and Chan (1978: 236) as occurring at Poring.

General distribution: Sabah, Kalimantan, Sarawak, Burma, Peninsular Malaysia, Thailand, Laos, Vietnam, Sumatra, Java, Bangka, Riau Archipelago, Philippines.

55.55. **GROSOURDYA** Rchb. f.

Bot. Zeit. 22: 297 (1864).

Small monopodial epiphytes. *Stems* short. *Leaves* few, flat, to 10 × 2 cm. *Inflorescences* usually shorter than the leaves, often many borne simultaneously on a plant, peduncle longer than the rachis, both prickly-hairy, with 1 or 2 flowers open at a time. *Flowers* ephemeral, to 1.5 cm across, yellow marked with red, lip white; *sepals* and *petals* free, spreading; *lip* mobile, 3-lobed, side-lobes narrow, erect, recurved, mid-lobe with a distinct spur, blade apically bilobed with a small median tooth, giving a '4-lobed' appearance; *column* long, slender, bent forward at an obtuse angle at the base of the stigma, as long as or longer than the foot, *rostellar projection* elongated; *stipes* narrowly triangular; *viscidium* triangular; pollinia 2, entire.

About 10 species distributed in the Andaman Islands, Burma, Thailand and Indochina to Malaysia, Indonesia and the Philippines.

55.55.1. Grosourdya appendiculata (Blume) Rchb. f., Xenia Orch. 2: 123 (1867).

Dendrocolla appendiculata Blume, Bijdr., 289 (1825).

Epiphyte on small branches. Hill forest. Elevation: 500 m.

General distribution: Sabah, Kalimantan, Sarawak, Andaman Islands, Burma, Peninsular Malaysia, Thailand, Vietnam, Java, Lesser Sunda Islands?, Philippines.

Collection. PORING: 500 m, *Lamb AL 28/82* (K).

55.56. **HABENARIA** Willd.

Sp. Pl. 4: 44 (1805).

Terrestrial herbs arising from tubers. *Stem* erect, few- to many-leaved, sheathed below. *Leaves* thin, narrowly elliptic to orbicular, not jointed at base, uppermost bract-like. *Inflorescence* terminal, usually quite long, with many small to fairly large flowers, lax or dense. *Flowers* white, green or pink, sometimes fragrant; *dorsal sepal* and *petals* usually forming a hood over column; *lateral sepals* spreading or reflexed; *lip* spurred, blade usually 3-lobed, sometimes finely divided and fringed; *column* short, consisting mainly of the anther, usually with a small auricle on either side; *stigmas* 2, usually separate, on stigmatophores in front of the column, the caudicles enclosed in long or short, often prominent, tubes (canals or thecae) separated by a 3-lobed rostellum.

A genus of between 600 and 800 species with a cosmopolitan distribution, but particularly well represented in Africa.

55.56.1. Habenaria damaiensis J. J. Sm., Bull. Jard. Bot. Buit., ser. 3, 8: 35 (1926).

Terrestrial. Hill forest, lower montane forest. Elevation: 900–1500 m.

General distribution: Sabah, Kalimantan.

Collections. KUNDASANG: 1100 m, *Carr 3641* (SING), 900 m, *SFN 28056* (SING); PENIBUKAN: 1200 m, *Clemens 30801* (BM), 1200 m, *32159* (BM); TENOMPOK: 1500 m, *Clemens s.n.* (BM).

55.56.2. Habenaria setifolia Carr, Gardens' Bull. 8: 171 (1935). Plate 55A. Type: TENOMPOK, 1500 m, *Clemens 28323* (holotype BM!; isotype AMES mf).

Terrestrial. Hill forest, lower montane forest. Elevation: 1000–1500 m.

General distribution: Sabah.

Additional collections. HAYE-HAYE RIVER: 1000 m, *Lamb AL 50/83* (K), 1000 m, *AL 746/87* (K); KUNDASANG: 1200 m, *Clemens 51470* (BM); PENIBUKAN: 1200–1500 m, *Clemens 31625* (BM).

55.56.3. Habenaria singapurensis Ridl., J. Linn. Soc. Bot. 32: 410 (1896). Plate 55B.

Terrestrial. Lower montane forest. Elevation: 1200–1400 m.

General distribution: Sabah, Peninsular Malaysia, Singapore.

Collection. MESILAU TRAIL: 1200–1400 m, *RSNB 5790* (K).

55.57. HETAERIA Blume

Bijdr., 409 (1825).

Terrestrial herbs. *Rhizome* decumbent, rooting at nodes. *Plants* 25–60 cm tall when in flower. *Leaves* broad, ovate to elliptic, usually asymmetric, with sheathing petioles, green. *Inflorescence* a lax or dense, many-flowered raceme. *Flowers* small, not resupinate, lip uppermost; *dorsal sepal* and *petals* connivent, forming a hood; *lateral sepals* enclosing lip base; *lip* ventricose at base, convex, apex shallow and narrow, entire or with divaricate apical lobes, containing papillae or glands above base on both sides; *column* short, with 2 parallel, keel-like lamellae in front; *stigma* bilobed; *pollinia* 2.

Around 20 species distributed in the Old World tropics, extending from India to Fiji, the majority of species native to Malaysia.

55.57.1. Hetaeria angustifolia Carr, Gardens' Bull. 8: 190 (1935). Type: KAUNG, 800 m, *Carr 3770* (holotype SING n.v.; isotype K!).

Terrestrial. Hill forest. Elevation: 700–1000 m.

General distribution: Sabah.

Additional collections. KIPUNGIT HILL: 700–1000 m, *Beaman 8233* (K); LOHAN/MAMUT COPPER MINE: 1000 m, *Beaman 10646* (K); MELANGKAP KAPA: 700–1000 m, *Beaman 8779* (K).

55.57.2. Hetaeria biloba (Ridl.) Seidenf. & J. J. Wood, Orch. Penin. Malaysia and Singapore: 95 (1992). Fig. 40.

Zeuxine biloba Ridl., J. Fed. Malay States Mus. 4: 73 (1909).

Hetaeria grandiflora Ridl., J. Malayan Branch Roy. Asiat. Soc. 87: 98 (1923), **syn. nov.**

Hetaeria rotundiloba J. J. Sm., Svensk Bot. Tidskr. 20: 470 (1926).

Terrestrial. Lower montane forest, upper montane forest. Elevation: 1600–2100 m.

General distribution: Sabah, Peninsular Malaysia, Thailand, Sumatra, Sulawesi.

Collections. MESILAU CAVE: 2000–2100 m, *Beaman 8132* (K); MESILAU CAVE TRAIL: 1700–1900 m, *Beaman 7960* (K), 1700–1900 m, *9129* (K); MESILAU RIVER: 1800 m, *Lamb SAN 91536* (K); WEST MESILAU RIVER: 1600–1700 m, *Beaman 8702* (K), 1600 m, *Collenette 640* (K).

55.58. HYLOPHILA Lindl.

Bot. Reg. 19: sub t. 1618 (1833).

Terrestrial herbs. *Rhizomes* fleshy, creeping. *Stems* erect, leafy. *Leaves* ovate to elliptic, often asymmetric, with sheathing petioles. *Inflorescence* erect, many-flowered, dense. *Flowers* small, resupinate, usually not opening widely; *dorsal sepal* and *petals* connivent, forming a hood; *lateral sepals* oblique, enclosing whole of lip; *lip* saccate, hairy within near the narrow reflexed blade and sometimes with several small interior keels; *column* short; *anther* long, acute; *rostellum* long, deeply divided; *stigma* convex, ventral, sometimes with a short, horn-like appendage spreading on either side. *Pollinia* 2.

About 6 species distributed from Thailand through Malaysia, Indonesia and the Philippines eastward to New Guinea and the Solomon Islands.

55.58.1. Hylophila lanceolata (Blume) Miq., Fl. Ned. Ind. 3: 746 (1859). Plate 56A, 56B.

Dicerostylis lanceolata Blume, Fl. Javae, ser. 2, 1, Orch.: 116, t. 38, f. 1 (1859).

Dicerostylis kinabaluensis Carr, Gardens' Bull. 8: 192 (1935). Type: BUNDU TUHAN, 900 m, *Carr 3614, SFN 27861* (holotype SING!; isotypes AMES mf, K!).

Terrestrial. Hill forest, lower montane forest, in damp shady areas on mossy rocks. Elevation: 900–1500 m.

General distribution: Sabah, Kalimantan, Thailand, Sumatra, Java, Flores, Philippines.

Additional collections. BUNDU TUHAN: 1200 m, *Carr 3491* (SING); KILEMBUN BASIN: 1400 m, *Clemens 33973* (BM); PARK HEADQUARTERS: 1500 m, *Phillipps SNP 2467* (K); 1500 m, *Wood 842* (K); TENOMPOK: 1500 m, *Clemens 29877* (AMES, BM), 1500 m, *48115* (BM).

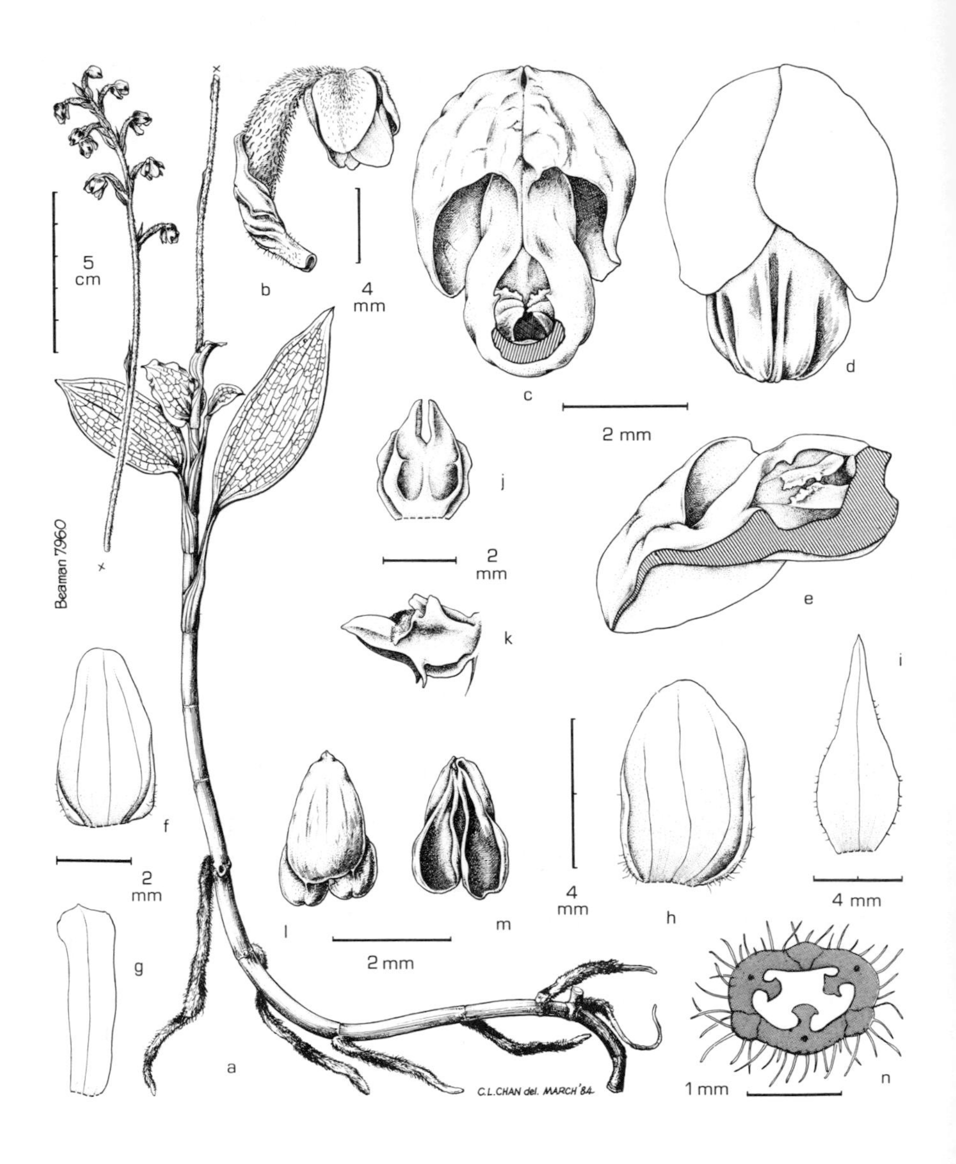

Fig. 40. Hetaeria biloba. **a**, habit; **b**, flower; **c**, lip (front view); **d**, lip (back view); **e**, lip (longitudinal section); **f**, dorsal sepal; **g**, petal; **h**, lateral sepal; **i**, floral bract; **j**, column (top view); **k**, column (side view); **l**, anther-cap (back view); **m**, anther-cap (front view); **n**, ovary (transverse section). All from *Beaman 7960.* Drawn by Chan Chew Lun.

55.59. KUHLHASSELTIA J. J. Sm.

Icon. Bogor. 4: 1 (1910).

Terrestrial herbs. *Rhizomes* decumbent, rooting at nodes. *Leaves* small, petioles sheathing, blade often purple on reverse, margin sometimes undulate. *Inflorescence* lax or dense, few- to several-flowered. *Flowers* small, resupinate; *dorsal sepal* and *petals* connivent, forming a hood; *lateral sepals* enclosing lip base; *lip* saccate at base and containing 2 glands, narrowed above into a claw with inflexed margins and a short spathulate apical blade; *column* without appendages; *stigma* entire; *pollinia* 2.

About 6 species distributed in Malaysia, Indonesia, the Philippines and New Guinea.

55.59.1. Kuhlhasseltia javanica J. J. Sm., Icon. Bogor. 4, t. 301 (1910). Plate 56C.

Kuhlhasseltia kinabaluensis Ames & C. Schweinf., Orch. 6: 14 (1920). Type: MARAI PARAI SPUR, *Clemens 398* (holotype AMES mf).

Terrestrial. Lower montane forest, in damp places in mossy forest, sometimes on ultramafic substrate. Elevation: 1400–1700 m.

General distribution: Sabah, Sarawak, Java.

Additional collections. KINATEKI RIVER: 1400 m, *Carr 3130, SFN 26552* (K); LIWAGU RIVER: 1400 m, *Darnton 573* (BM); LUBANG: 1600 m, *Carr 3130, SFN 26552A* (K); MARAI PARAI: 1500 m, *Clemens s.n.* (BM), *33005* (BM); MESILAU RIVER: 1700 m, *RSNB 7124* (K); MOUNT KINABALU: 1700 m, *Allen AK 66* (K).

55.60. LECANORCHIS Blume

Mus. Bot. Lugd. 2: 188 (1856).

Leafless saprophytic herbs. *Rhizomes* branched, fleshy. *Stems* simple or branched. *Inflorescence* terminal, erect, racemose, lax, few- to many-flowered. *Flowers* small, resupinate; *sepals* and *petals* free, subequal, usually spreading and encircled by a shallow, denticulate cup or calyculus; *lip* adnate to base of column, 3-lobed, rarely entire, not spurred, disc papillose or hirsute; *column* slender, elongate, clavate, partially adnate to claw of lip; *pollinia* 2.

About 20 species distributed in Thailand, Malaysia, Indonesia, north to Japan including the Ryukyu Islands, Taiwan and the Philippines eastward to New Guinea.

55.60.1. Lecanorchis multiflora J. J. Sm., Bull. Jard. Bot. Buit., ser. 2, 26: 8 (1918).

Saprophyte. Hill forest, lower montane forest. Elevation: 1100–1200 m.

General distribution: Sabah, Sarawak, Peninsular Malaysia, Thailand, Sumatra, Java.

Collections. DALLAS/TENOMPOK: 1200 m, *Clemens 27419* (BM, K); KINATEKI RIVER (NEAR MINITINDUK): 1100 m, *Carr 3206, SFN 26801* (SING); MESILAU: *Clemens 51628* (BM); MINITINDUK: *Carr s.n.* (K); NUMERUK RIDGE: 1100 m, *Clemens 40065* (BM); PENIBUKAN: 1100 m, *Clemens 40464* (BM, K).

55.61. LEPIDOGYNE Blume

Fl. Javae ser. 2, 1, Orch.: 93 (1859).

Tall terrestrial herbs. *Leaves* grouped at base, many, narrowly elliptic, acuminate, with sheathing base. *Inflorescence* densely many-flowered, long; *peduncle* up to 30 cm long; *rachis* up to 40 cm long, pubescent; *floral bracts* narrow, longer than flowers. *Flowers* reddish-brown; *dorsal sepal* and *petals* connivent, forming a hood; *lateral sepals* enclosing base of lip; *lip* with swollen concave base containing a transverse row of small calli, blade 3-lobed, side-lobes small, erect, mid-lobe long and narrow; *column* short; *anther* rostrate; *rostellum* very long, deeply cleft; *stigma* transverse, covered by a plate springing from its lower edge; *pollinia* 2, long, clavate, deeply divided.

Three species distributed in Malaysia, Indonesia, the Philippines and New Guinea.

55.61.1. Lepidogyne longifolia (Blume) Blume, Fl. Javae, ser. 2, 1, Orch.: 94, t. 25 (1859).

Neottia longifolia Blume, Bijdr., 406 (1825).

Terrestrial. Hill forest, lower montane forest. Elevation: 900–1500 m.

General distribution: Sabah, Peninsular Malaysia, east to the Philippines, New Guinea?

Collections. DALLAS: 900 m, *Clemens 27725* (BM, K); 900 m, *Carr 3137, SFN 27951* (K, SING); MARAI PARAI/KIAU TRAIL: 1200 m, *Bailes & Cribb 876* (K); PENIBUKAN: 1200 m, *Clemens 32190* (BM); PINOSUK PLATEAU: 1400–1500 m, *Beaman 9191* (K); TENOMPOK: 1500 m, *Clemens 29898* (BM, K).

55.62. LIPARIS Rich.

Mém. Mus. Hist. Nat. 4: 43, 52 (1818).

Seidenfaden, G. (1976). Orchid genera in Thailand IV. *Liparis* L. C. Rich. Dansk Bot. Ark. 31, 1: 1–105.

Epiphytes, lithophytes or terrestrials. *Pseudobulbs* ovoid to cylindrical, fleshy, 1- to several-leaved at apex. *Leaves* plicate or conduplicate, narrowly elliptic or ovate, often fleshy, membranaceous, sheathing at base. *Inflorescence* terminal, a several- to many-flowered raceme, erect, lax or dense. *Flowers* resupinate, small to medium-sized, sometimes self-fertilizing, green, purple, yellow-green or dull orange; *sepals* and *petals* free, often reflexed; *sepals* ovate to narrowly elliptic; *petals* usually narrow, linear; *lip* entire or 3-lobed, apex often deflexed, usually with a fleshy basal callus; *column* long, arcuate, sometimes swollen at base, narrowly winged above, foot absent; *pollinia* 4.

About 250 species with a cosmopolitan distribution, the greatest number being in Asia and New Guinea.

55.62.1. Liparis atrosanguinea Ridl., J. Straits Branch Roy. Asiat. Soc. 39: 71 (1903).

Terrestrial. Lower montane forest, mostly on ultramafic substrate. Elevation: 1200–1800 m.

General distribution: Sabah, Sarawak, Peninsular Malaysia, Thailand, Vietnam, Sumatra.

Collections. KINATEKI RIVER: 1200 m, *Clemens 31041* (BM); MARAI PARAI: 1500 m, *Clemens 32229* (BM), 1500 m, *32350* (BM), 1500 m, *Collenette A 102* (BM); MARAI PARAI SPUR: 1700 m, *Bailes & Cribb 850* (K); SUMMIT TRAIL: 1700 m, *Darnton 513* (BM); TENOMPOK: 1500 m, *Clemens s.n.* (BM); TINEKUK FALLS: 1800 m, *Clemens 40928* (BM).

55.62.2. Liparis aurantiorbiculata J. J. Wood & A. Lamb, **sp. nov.** Fig. 41, Plate 56D.

Liparidi grandi Ames et C. Schweinf., species etiam kinabaluensis, affinis, sed pseudobulbis multo longioribus, foliis angustioribus, pedicello cum ovario breviore clavato, sepalis latioribus triangulari-ovatis vel oblongo-ovatis, labello latiore late ovato vel orbiculari apice leviter retuso nunquam abrupte mucronato callo basali parvo carnoso bilobato distinguitor; etiam a *L. jarensi* Ames, species philippinensis, pseudobulbis longioribus, floribus aurantiacis, pedicello cum ovario breviore clavato, sepalis latioribus, labello grandiore distinguenda.

Type: EAST MALAYSIA, SABAH, MOUNT KINABALU, PINOSUK PLATEAU: 1700 m, 18 December 1982, *Lamb AL 22/82* (holotype K!, herbarium and spirit material).

Erect, epiphytic herb. *Rhizome* thick and tough, concealed by sheaths. *Pseudobulbs* 6–11 cm long, to 0.5 cm wide above, 1 cm wide at base, attenuate above, swollen below, enclosed by 2–3 green to pale brown acuminate sheaths 6–11 cm long, unifoliate. *Leaves* (28–)36–46 × 1.7–2.2 cm, jointed, ensiform, acute, conduplicate toward base, erect to spreading. *Inflorescence* a stiffly erect, laxly 12–15-flowered raceme, as long as or shorter than the leaves; *peduncle* 9–12 cm long, terete; *rachis* 11–16 cm long, 4-winged; *floral bracts* 0.7–1.2 × 0.3–0.4 cm, ovate, acute to acuminate. *Flowers* porrect to spreading, orange, with a yellowish green column. *Pedicel-with-ovary* 1.6–2 cm long, clavate. Dorsal sepal and petals strongly reflexed. *Dorsal sepal* 1–1.1 × 0.2–0.25 cm, narrowly triangular-ovate or narrowly oblong-ovate, acute. *Lateral sepals* 1.1 × 0.3 cm, oblong, twisted, apex carinate, subulate, hidden behind lip. *Petals* 1 cm long, 0.2–0.3 cm wide, linear, acute, sometimes obliquely linear. *Lip* 1.1–1.2 × 1.1–1.2 cm, broadly ovate to orbicular, apex

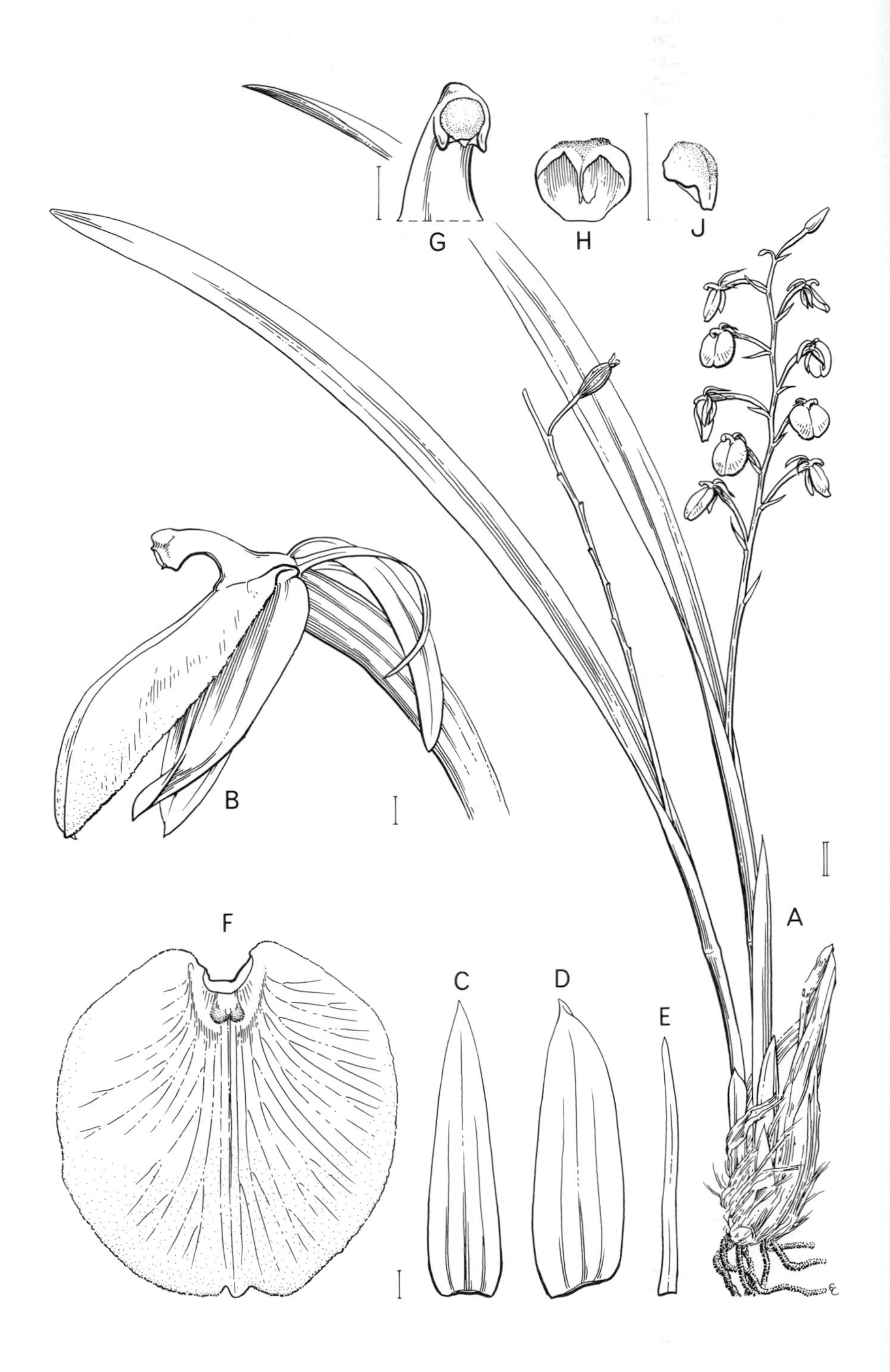
G
H
J
B
A
F
C
D
E

shallowly retuse, with a triangular, acute tooth in the sinus, clasping column at base, flat, margin minutely erose, upper and lower surface minutely papillose, particularly in apical half, disc with a small, fleshy bilobed basal callus. *Column* 4.5 mm long, thickened and 2 mm wide at base, narrow and curved in middle, apex with short, rather truncate wings. *Anther-cap* 0.5 × 0.6–0.7 mm, cucullate.

Liparis aurantiorbiculata belongs to section *Coriifoliae*, distinguished by having leaves that are jointed to the pseudobulbs. It is one of a large group within the section which includes *L. chlorantha* Schltr. from Papua New Guinea, *L. grandis* from Sabah, *L. jarensis* Ames from the Philippines and *L. orbiculata* L. O. Williams from Fiji and Vanuatu.

Liparis aurantiorbiculata is closely related to *L. grandis*, so far only known from Mount Kinabalu, but may be distinguished by its much longer pseudobulbs, narrower leaves, shorter clavate pedicel-with-ovary, broader triangular-ovate to oblong-ovate sepals and broader, broadly ovate to orbicular lip which has a shallowly retuse, never abruptly mucronate, apex and a small fleshy bilobed basal callus. It is also similar to *L. jarensis*, but this species lacks pseudobulbs and has green flowers with a longer, slender pedicel-with-ovary, narrower sepals and a smaller lip.

The specific epithet is derived from the Latin *aurantiacus*, orange, and *orbicularis*, circular, in reference to the lip.

Lower montane forest. Elevation: 1200–1700 m.

General distribution: Sabah.

Additional collections: CROCKER RANGE, KM 59 ON KOTA KINABALU-TAMBUNAN ROAD, 1400 m, *Beaman 7378* (K); MOUNT KINABALU: *Chan & Gunsalam s.n.* (K); PENIBUKAN: 1200 m, *Clemens 30698* (BM, mixed collection with *Liparis grandis*); TENOMPOK ORCHID GARDEN: 1500 m, *Clemens 50255* (BM).

55.62.3. Liparis bootanensis Griff., Itin. Pl. Khas. Mts., 98 (1848, descr.) & Not. 3, 278 (1851, name).

Epiphyte. Hill forest or lower montane forest on ultramafic substrate. Elevation: 1200 m.

General distribution: Sabah, Kalimantan, widespread from India, Bhutan, Burma, Peninsular Malaysia, Thailand, Vietnam, Java, Sumbawa, Philippines, north to China & Japan.

Collections. HEMPUEN HILL: 1200 m, *Madani SAN 89541* (K); KINATEKI RIVER: 1200 m, *Collenette A 107* (BM); MAMUT COPPER MINE: *Lamb SAN 93369* (K); PENIBUKAN: 1200 m, *Clemens 30599* (BM, K).

Fig. 41. Liparis aurantiorbiculata. A, habit; **B**, flower (side view); **C**, dorsal sepal; **D**, lateral sepal; **E**, petal; **F**, lip (front view); **G**, column (front view); **H**, anther-cap (front view); **J**, anther-cap (side view). **A** from *Lamb AL 22/82*; **B–J** from *Chan & Gunsalam s.n.* Scale: single bar = 1 mm; double bar = 1 cm. Drawn by Eleanor Catherine.

55.62.4. Liparis caespitosa (Thouars) Lindl., Bot. Reg. 11: sub t. 882 (1825).

Malaxis caespitosa Thouars, Orch. Iles Aust. Afr., t. 90 (1822).

Epiphyte. Hill forest, lower montane forest, sometimes on ultramafic substrate. Elevation: 400–1500 m.

General distribution: Sabah, Sarawak, Kalimantan, widespread from Africa through Asia to the Pacific islands.

Collections. DALLAS: 900 m, *Clemens 28143* (BM); GURULAU SPUR: 900 m, *Clemens 50380* (BM, K); HEMPUEN HILL: 800–1000 m, *Beaman 7411* (K); KAUNG: 400 m, *Carr SFN 26266* (SING), 400 m, *Darnton 318* (BM); PENIBUKAN: 1200–1500 m, *Clemens s.n.* (BM, BM), 1200 m, *30642* (BM); TAHUBANG RIVER: 1200–1500 m, *Clemens 31650* (BM); TENOMPOK: 1500 m, *Clemens s.n.* (BM), 900–1500 m, *27000* (BM, K).

55.62.5. Liparis compressa (Blume) Lindl., Gen. Sp. Orch., 32 (1830). Plate 57A.

Malaxis compressa Blume, Bijdr., 390, t. 54 (1825).

Terrestrial. Lower montane forest, primary forest in light shade, sometimes on ultramafic substrate. Elevation: 1500–1700 m.

General distribution: Sabah, Kalimantan, Sarawak, Peninsular Malaysia, Sumatra, Java, Sulawesi, Philippines.

Collections. MAMUT COPPER MINE: 1500 m, *Collenette 1049* (K); MARAI PARAI SPUR: 1500 m, *Clemens 33031* (BM); MESILAU TRAIL: 1700 m, *Chow & Leopold SAN 76439* (K).

55.62.6. Liparis condylobulbon Rchb. f., Hamburger Garten-Blumenzeitung 18: 34 (1862).

Epiphyte. Hill forest, lower montane forest. Elevation: 900–1500 m.

General distribution: Sabah, Sarawak, Kalimantan, widespread from Taiwan, Indonesia, Philippines east to New Guinea & the Pacific islands.

Collections. BUNDU TUHAN: 1400 m, *Darnton 200* (BM), 900 m, *537* (BM); KIAU: 900 m, *Clemens 30468* (BM); MOUNT KINABALU: 1500 m, *Clemens s.n.* (BM); TENOMPOK: 1500 m, *Clemens 28686* (K), 1500 m, *29679* (BM, K); UPPER KINABALU: *Clemens 26874* (K).

55.62.7. Liparis elegans Lindl., Gen. Sp. Orch., 30 (1830).

Epiphyte, rarely terrestrial. Lowlands, hill forest, lower montane forest, sometimes on ultramafic substrate. Elevation: 400–2600 m.

General distribution: Sabah, Kalimantan, Sarawak, Peninsular Malaysia, Sumatra, Anambas Islands, Mentawai, Riau Archipelago, Philippines.

Collections. DALLAS: 900 m, *Clemens 26831* (BM, K), 900 m, *26838* (BM), 900 m, *26894* (BM, K), 900 m, *26999* (BM, K); GURULAU SPUR: 1400 m, *Carr 3145* (SING), 1500 m, *Clemens 306* (AMES, BM, K); KAUNG: 400 m, *Darnton 268* (BM), 400 m, *303* (BM); KEMBURONGOH: 2600 m, *Clemens 27366* (BM); KIAU: 900 m, *Clemens 35A* (AMES), 900 m, *37* (AMES, BM, K, SING), 900 m, *46* (AMES, BM, K, SING), 900 m, *77* (AMES, BM, K), 900 m, *139* (AMES), 900 m, *141* (AMES, K,

SING), *Topping 1523* (K); LOHAN RIVER: 800 m, *Beaman 8363* (K); LUBANG: 1500 m, *Clemens 119* (AMES); MOUNT KINABALU: *Haslam s.n.* (AMES).

55.62.8. Liparis endertii J. J. Sm., Bull. Jard. Bot. Buit., ser. 3, 11: 122 (1931).

Epiphyte. Hill forest. Elevation: 1200 m.

General distribution: Sabah, Kalimantan.

Collection. BUNDU TUHAN: 1200 m, *Carr SFN 27872* (SING).

55.62.9. Liparis gibbosa Finet, Bull. Soc. Bot. France 55: 342, t. 11, f. 36–44 (1908).

Liparis disticha auct. non (Thouars) Lindl.

Epiphyte. Lower montane forest. Elevation: 1500–1900 m.

General distribution: Sabah, Sarawak, Kalimantan, widespread from Burma to New Guinea and the Pacific islands.

Collections. GURULAU SPUR: 1500 m, *Gibbs 4014* (BM); MESILAU TRAIL: 1900 m, *Collenette 630* (K).

55.62.10. Liparis grandis Ames & C. Schweinf., Orch. 6: 87 (1920). Type: MOUNT KINABALU, *Haslam s.n.* (holotype AMES mf; isotypes BM!, K!).

Epiphyte. Lower montane forest on ultramafic substrate. Elevation: 1200–1500 m.

Endemic to Mount Kinabalu.

Additional collections. NUMERUK RIDGE: 1400 m, *Clemens 40054* (BM); PENATARAN BASIN: 1400 m, *Clemens 34336* (BM); PENIBUKAN: 1200 m, *Carr 3043, SFN 26345* (SING), 1200 m, *Clemens 30698* (BM), 1200 m, *30782* (BM), 1200–1500 m, *31516* (BM, K).

55.62.11. Liparis kamborangensis Ames & C. Schweinf., Orch. 6: 89 (1920). Type: KEMBURONGOH, *Clemens 220* (holotype AMES mf).

Terrestrial, lithophyte. Lower montane forest, upper montane forest, on ultramafic substrate; wet mossy rock faces, mossy banks, ridge forest. Elevation: 1500–3200 m.

Endemic to Mount Kinabalu.

Additional collections. DACHANG: 3000 m, *Clemens 29219* (BM); GURULAU SPUR: 3200 m, *Clemens 50904* (BM, K); KINATEKI RIVER: 1500 m, *Chan & Lamb AL 313* (K); MARAI PARAI: 2400–3000 m, *Clemens 33172* (BM); MESILAU BASIN: 2700 m, *Clemens 51515* (BM); SUMMIT TRAIL: 2700 m, *Carr SFN 27598* (SING).

55.62.12. Liparis kinabaluensis J. J. Wood, Lindleyana 5: 84, f. 2 (1990). Fig. 42, Plate 57B, 57C. Type: PENIBUKAN, 1200 m, *Clemens 50099* (holotype K!).

Epiphyte or lithophyte. Hill forest on ultramafic substrate, lower montane forest. Elevation: 800–1800 m.

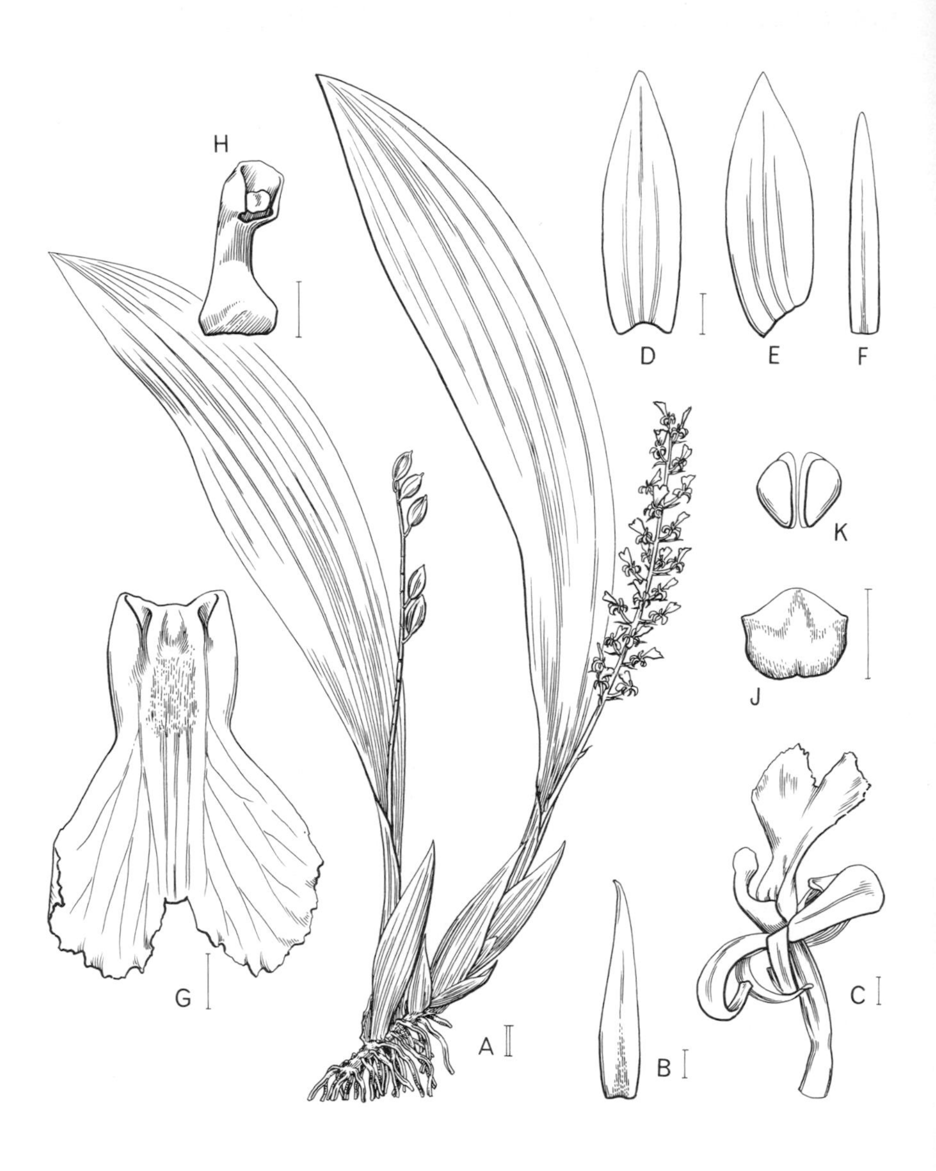

Fig. 42. Liparis kinabaluensis. **A**, habit; **B**, floral bract; **C**, flower; **D**, dorsal sepal; **E**, lateral sepal; **F**, petal; **G**, lip (front view); **H**, column (front view); **J**, anther-cap (back view); **K**, pollinia. **A** from *Clemens 30141, 50099* and *51462*; **B–K** from *Brentnall 124.* Scale: single bar = 1 mm; double bar = 1 cm. Drawn by Eleanor Catherine.

Endemic to Mount Kinabalu.

Additional collections. LOHAN RIVER: 800 m, *Bailes & Cribb 658* (K); TAHUBANG RIVER: 1100 m, *Clemens 51462* (BM, K); TENOMPOK: 1500 m, *Clemens 27245* (K), 1500 m, *29635* (BM), 1500 m, *30141* (K); TENOMPOK ORCHID GARDEN: 1500 m, *Clemens 51726* (BM); TINEKUK FALLS: 1800 m, *Clemens 40908* (BM, K); WEST MESILAU RIVER: *Brentnall 124* (K).

55.62.13. Liparis lacerata Ridl., J. Linn. Soc. Bot. 22: 284 (1886). Plate 57D.

Epiphyte. Hill forest, probably on ultramafic substrate. Elevation: 600 m.

General distribution: Sabah, Kalimantan, Sarawak, Burma, Peninsular Malaysia, Sumatra, Mentawai.

Collection. LOHAN RIVER: 600 m, *Lamb SAN 93351* (K).

55.62.14. Liparis latifolia (Blume) Lindl., Gen. Sp. Orch., 30 (1830). Plate 57E.

Malaxis latifolia Blume, Bijdr., 393 (1825).

Epiphyte. Lowlands, hill forest, lower montane forest. Elevation: 400–1500 m.

General distribution: Sabah, China (Hainan), Peninsular Malaysia, Thailand, Sumatra, Java, Timor, New Guinea.

Collections. BAMBANGAN RIVER: 1500 m, *RSNB 4529* (K); KAUNG: 500 m, *Carr SFN 27388* (SING), 400 m, *Darnton 385* (BM); KIAU: 900 m, *Clemens 180* (AMES); PENIBUKAN: 1200 m, *Clemens 40863* (BM); TAHUBANG RIVER: 1200 m, *Clemens 40376* (BM); TENOMPOK: 1500 m, *Clemens 26123* (BM, K), 1500 m, *29677* (BM).

55.62.15. Liparis lingulata Ames & C. Schweinf., Orch. 6: 90 (1920). Type: KIAU, *Clemens 324* (holotype AMES mf; isotypes BM!, K!, SING!).

Epiphyte. Hill forest, lower montane forest. Elevation: 800–1500 m.

General distribution: Sabah.

Additional collections. DALLAS: 800 m, *Clemens s.n.* (BM); PENIBUKAN: 1200 m, *Clemens 51730* (BM, K); TENOMPOK: 1500 m, *Carr SFN 27362* (K, SING), *Clemens 51741* (BM); TENOMPOK ORCHID GARDEN: *Clemens 51741* (K).

55.62.16. Liparis lobongensis Ames, Orch. 6: 92 (1920). Fig. 43, Plate 57F. Type: LUBANG, *Clemens 219* (holotype AMES mf).

Epiphyte. Lower montane forest, sometimes on ultramafic substrate. Elevation: 800–2300 m.

General distribution: Sabah.

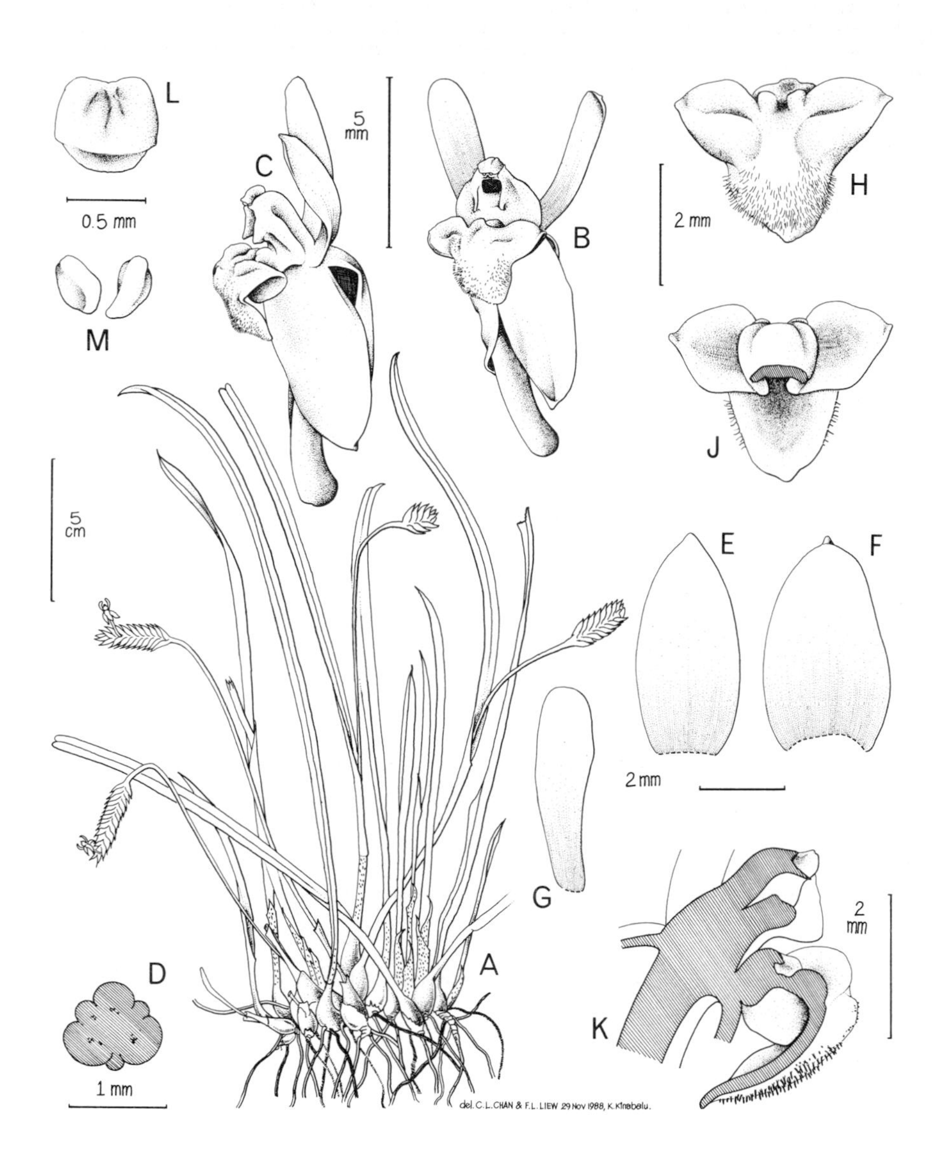

Fig. 43. Liparis lobongensis. **A**, habit; **B**, flower (oblique view); **C**, flower (side view); **D**, ovary (transverse section); **E**, dorsal sepal; **F**, lateral sepal; **G**, petal; **H**, lip (front view); **J**, lip (back view); **K**, pedicel-with-ovary, column and lip (longitudinal section); **L**, anther-cap (back view); **M**, pollinia. All from *Chan 121/89.* Drawn by Chan Chew Lun and F. L. Liew.

Additional collections. BAMBANGAN RIVER: 1500 m, *RSNB 4481* (K); BUNDU TUHAN: 1200 m, *Carr SFN 27420* (K); HEMPUEN HILL: 800–1000 m, *Cribb 89/30* (K); LUBANG: 1400 m, *Carr SFN 27343* (K, SING), *Clemens 103* (AMES, BM, K, SING); MARAI PARAI: 1500 m, *Clemens 33100* (BM); MESILAU CAVE/JANET'S HALT: 2300 m, *Collenette 21574* (K); PARK HEADQUARTERS: 1600 m, *Chan 121/89* (K); PENATARAN RIVER: 1400 m, *Clemens 32585* (BM); PENIBUKAN: 1200 m, *Clemens 40859* (BM, K); TENOMPOK: 1500 m, *Clemens 27260* (BM, K); TINEKUK FALLS: 1500–2100 m, *Clemens 40927* (BM, K).

55.62.17. Liparis aff. **lobongensis** Ames, Orch. 6: 92 (1920).

Epiphyte. Hill forest on ultramafic substrate. Elevation: 800–1200 m.

Collection. HEMPUEN HILL: 800–1200 m, *Beaman 7685* (K).

55.62.18. Liparis mucronata (Blume) Lindl., Gen. Sp. Orch., 32 (1830).

Malaxis mucronata Blume, Bijdr., 391 (1825).

Liparis divergens J. J. Sm., Icon. Bogor. 2: t. 109E, f. 1–4 (1903).

Epiphyte. Hill forest on ultramafic substrate, lower montane forest. Elevation: 800–1400 m.

General distribution: Sabah, Sumatra, Java, Bali.

Collections. HEMPUEN HILL: 800–1000 m, *Wood 840* (K); PINOSUK PLATEAU: 1400 m, *Beaman 10711* (K).

55.62.19. Liparis pandurata Ames, Orch. 6: 94 (1920). Fig. 44, Plate 58A. Type: LUBANG, *Clemens 117* (holotype AMES mf).

Epiphyte. Lower montane forest. Elevation: 1200–2600 m.

General distribution: Sabah.

Additional collections. BAMBANGAN RIVER: 1500 m, *RSNB 4486* (K); BUNDU TUHAN: 1200 m, *Carr 3041, SFN 27416* (K); GOLF COURSE SITE: 1700–1800 m, *Beaman 7226* (K), 1700–1800 m, *10675* (K); KILEMBUN RIVER: 2600 m, *Clemens 33790* (BM, K); LIWAGU RIVER: 1600 m, *Darnton 556* (BM); MESILAU BASIN: 2100 m, *Clemens 28992* (BM); MESILAU CAVE TRAIL: 1700–1900 m, *Beaman 8011* (K); MESILAU TRAIL: 1700 m, *Fuchs & Collenette 21498* (K); MINIRINTEG CAVE: 2100 m, *Clemens 29285* (BM, K); TENOMPOK: 1500 m, *Clemens 27642* (BM); WEST MESILAU RIVER: 1600–1700 m, *Beaman 8623* (K), 1600 m, *8998* (K).

55.62.20. Liparis parviflora (Blume) Lindl., Gen. Sp. Orch., 31 (1830).

Malaxis parviflora Blume, Bijdr., 392 (1825).

Epiphyte. Lowlands, hill forest, lower montane forest. Elevation: 500–2000 m.

General distribution: Sabah, Kalimantan, Sarawak, Peninsular Malaysia, Thailand, Sumatra, Java, Sulawesi, Bali, Krakatau, Philippines.

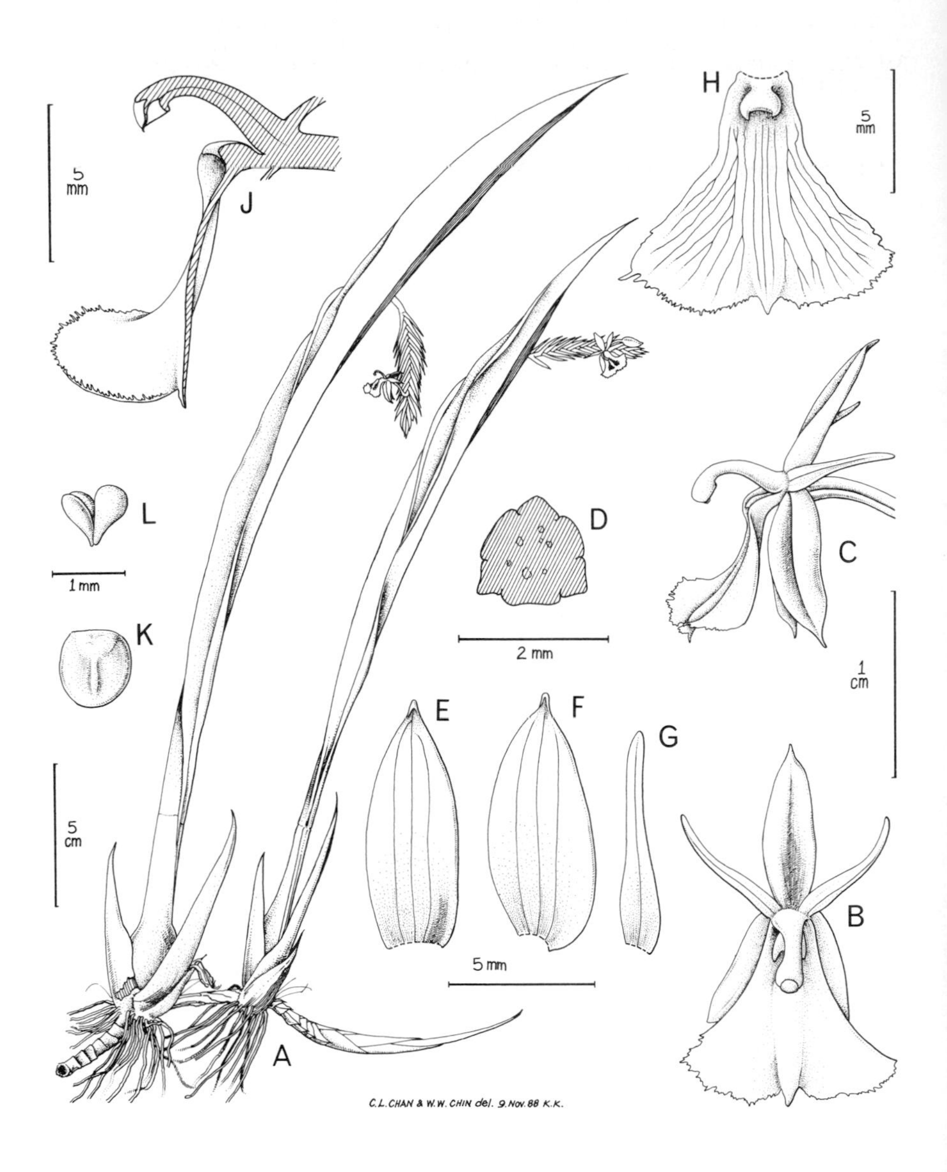

Fig. 44. Liparis pandurata. **A**, habit; **B**, flower (front view); **C**, flower (side view); **D**, ovary (transverse section); **E**, dorsal sepal; **F**, lateral sepal; **G**, petal; **H**, lip (front view); **J**, pedicel-with-ovary, column and lip (longitudinal section); **K**, anther-cap (back view); **L**, pollinia. All from *Chan s.n.*, Silau-silau Trail. Drawn by Chan Chew Lun and W. W. Chin.

Collections. KAUNG: 500 m, *Carr SFN 27306* (SING); KIAU: *Clemens 350* (AMES); LUMU-LUMU: 2000 m, *Clemens 26874* (BM).

55.62.21. Liparis rheedii (Blume) Lindl., Gen. Sp. Orch., 34 (1830).

Malaxis rheedii Blume, Bijdr., 389, t. 54 (1825).

Terrestrial. Hill forest. Elevation: 800 m.

General distribution: Sabah, Kalimantan, Peninsular Malaysia, Thailand, Sumatra, Java, Sulawesi, Sumbawa, New Guinea.

Collection. DALLAS: 800 m, *Clemens s.n.* (BM).

55.62.22. Liparis rhodochila Rolfe, Kew Bull. 1908: 412 (1908).

Epiphyte. Hill forest. Elevation: 1000 m.

General distribution: Sabah, Java.

Collection. BUNDU TUHAN: 1000 m, *Brentnall 107* (K).

55.62.23. Liparis viridiflora (Blume) Lindl., Gen. Sp. Orch., 31 (1830).

Malaxis viridiflora Blume, Bijdr., 392 (1825).

Epiphyte. Hill forest, lower montane forest. Elevation: 500–2000 m.

General distribution: Sabah, Sarawak, widespread from SE Asia east to Fiji & Samoa.

Collections. DALLAS: 900 m, *Clemens 26842* (BM, K), 900 m, *26895* (BM, K), 900 m, *27023* (BM, K), 900 m, *27207* (BM, K), 900 m, *27323* (BM, K), 900 m, *27727* (BM, K); GURULAU SPUR: 1200 m, *Carr SFN 27873* (SING), 1200 m, *Clemens 312* (AMES), *51722* (BM, K); KIAU: *Clemens s.n.* (AMES), 900 m, *35* (AMES), 900 m, *38* (AMES), 900 m, *49* (AMES, BM, SING), 900 m, *90* (AMES), 900 m, *175* (AMES), 800–900 m, *337* (AMES, BM); KINATEKI RIVER: 500 m, *Collenette A 126* (BM); KULUNG HILL: 500 m, *Darnton 129* (BM); LUMU-LUMU: 2000 m, *Clemens 26874* (BM); TENOMPOK: 1500 m, *Clemens s.n.* (BM).

55.62.24. Liparis wrayii Hook. f., Fl. Brit. Ind. 6: 181 (1890), sens. lat.

Terrestrial. Lower montane forest, in leaf litter in shade. Elevation: 1200–1400 m.

General distribution: Sabah, Brunei, Burma, Peninsular Malaysia, Thailand, Sumatra, Java, New Guinea.

Collection. TAWARAS TEKI RIVER: 1200–1400 m, *Lamb AL 304/85* (K).

55.62.25. Liparis sp. 1, sect. Distichae

Epiphyte. Lower and upper montane forest. Elevation: 1500–2700 m.

Collections. GURULAU SPUR: 2400–2700 m, *Clemens 50656* (BM, K); LUMU-LUMU: 2100 m, *Clemens 27178* (BM); MESILAU CAVE/JANET'S HALT: 2400 m, *Fuchs & Collenette 21407* (K); MESILAU RIVER: 2100 m, *Clemens 51469* (BM, K); TENOMPOK: 1500 m, *Clemens 27247* (BM, K), 1500 m, *51727* (BM, K).

55.62.26. Liparis indet.

Collections. BUNDU TUHAN: 1200 m, *Carr SFN 27416* (SING); GURULAU SPUR: 1400 m, *Carr SFN 26593* (SING); JANET'S HALT/SHEILA'S PLATEAU: 2500 m, *Collenette 21567* (K); KEMBURONGOH: 2100 m, *Carr SFN 27459* (SING); KINATEKI RIVER: 1400 m, *Carr SFN 26431* (SING), 1400 m, *SFN 26432* (SING), 1200 m, *Clemens 31044* (BM); MESILAU CAVE: 1800 m, *RSNB 4845* (K, SING); MINITINDUK GORGE: 800 m, *Carr SFN 27434* (SING); PENIBUKAN: 1200 m, *Carr SFN 27983* (SING), 1500 m, *Clemens 51317* (BM); TAHUBANG RIVER: 1100 m, *Carr SFN 26466* (SING); TENOMPOK: 1500 m, *Carr SFN 26879* (SING), 1500 m, *SFN 27820* (SING), 1500 m, *Clemens 51127* (K); TENOMPOK ORCHID GARDEN: 1500 m, *Clemens 50183* (BM).

55.63. LUISIA Gaudich.

In Freycinet, Voy. Uranie, 426 (1829 "1826").

Seidenfaden, G. (1971). Notes on the genus *Luisia.* Dansk Bot. Ark. 27: 1–101 (1971).

Erect or climbing epiphytes or lithophytes. *Stems* often branching at the base, forming a tufted habit, others with a single shoot up to 40 cm long. *Leaves* well spaced, terete, linear. *Inflorescences* dense, almost sessile, with fewer than 10 flowers; *peduncle* and *rachis* very short. *Flowers* usually small, about 1 cm across, occasionally up to 3 cm across, with yellow or yellowish green sepals and petals and a purple lip; *sepals* and *petals* free, spreading, petals often longer and narrower than sepals; *lateral sepals* often keeled; lip fleshy, immobile, usually divided into a ± concave hypochile and a ± wrinkled epichile; *column* short and stout, foot absent; *rostellar projection* and *stipes* short; *viscidium* short and broad; *pollinia* 2, porate.

About 40 species distributed from Sri Lanka and India to China south to Thailand, Indochina, east to Japan, New Guinea and the Pacific islands. The centre of distribution lies in Burma and Thailand.

55.63.1. Luisia curtisii Seidenf., Bot. Tidsskr. 68: 83 (1973). Plate 58C.

Epiphyte. Lower montane forest. Elevation: 1500 m.

General distribution: Sabah, Peninsular Malaysia, Thailand, Vietnam.

Collections. PARK HEADQUARTERS: 1500 m, *Lamb AL 3/82* (K); PORING ORCHID GARDEN: *Lohok 18* (K).

55.63.2. Luisia volucris Lindl., Fol. Orch., *Luisia*, 1 (1853). Plate 59A.

Epiphyte. Cultivated at Poring, of local origin.

General distribution: Sabah, India.

Collection. PORING ORCHID GARDEN: *Lohok 21* (K).

55.64. **MACODES** (Blume) Lindl.

Gen. Sp. Orch., 496 (1840).

Terrestrial herbs. *Rhizome* creeping, fleshy. *Stems* erect, short or long. *Leaves* rather fleshy, ovate to elliptic-subcircular, margins sometimes undulate, petioles sheathing, blade with coloured veins. *Inflorescence* laxly few- to many-flowered; *peduncle* and *floral bracts* pubescent. *Flowers* rather small, not resupinate, lip uppermost, asymmetric owing to twisting of lip and column; *sepals* pubescent on exterior; *lateral sepals* enclosing base of lip; *lip* bipartite or 3-lobed, *hypochile* saccate at base, sac containing 2 glands, *epichile* twisted to one side; *column* short, twisted, with 2 thin close parallel wings on the front, the wings descending down into the saccate hypochile; *anther* acute; *rostellum* cleft; *stigma* entire, large; *pollinia* 2.

About 10 species distributed from Malaysia to the Pacific islands.

55.64.1. Macodes lowii (E. Lowe) J. J. Wood, Orchid Digest 48: 155 (1984). Plate 59C.

Anoectochilus lowii E. Lowe, L'Illust. Hort. 30: 161, pl. 501 (1883).

Terrestrial. Hill forest, in humus under casuarina, among rocks on ultramafic substrate. Elevation: 700 m.

General distribution: Sabah.

Collection. LOHAN RIVER: 700 m, *Fowlie & Lamb s.n.* (K).

55.64.2. Macodes petola (Blume) Lindl., Gen. Sp. Orch., 497 (1840). Plate 59B.

Neottia petola Blume, Bijdr., 407, t. 2 (1825).

Terrestrial. Hill forest, lower montane forest, on ultramafic substrate. Elevation: 900–1500 m.

General distribution: Sabah, Peninsular Malaysia, Sumatra, Java, Philippines.

Collections. KILEMBUN RIVER: 900 m, *Clemens s.n.* (BM); MAHANDEI RIVER: 1100 m, *Carr 3334* (SING); PENIBUKAN: 1200–1500 m, *Clemens s.n.* (BM), 1400 m, *40513* (BM).

55.65. **MACROPODANTHUS** L. O. Williams

Bot. Mus. Leafl. Harvard Univ. 6: 103 (1938).

Medium-sized monopodial epiphytes. *Stems* to 10 cm long. *Leaves* narrow, strap-shaped, apex unequally bilobed. *Inflorescences* often clustered, as long as the leaves, with up to 15 flowers in a loose raceme, *rachis* sometimes keeled. *Flowers* showy, white or yellow, marked with red, lilac-pink or orange-brown; *dorsal sepal* free; *lateral sepals* and *petals* adnate to the column-foot; *lip* mobile, 3-lobed, lateral lobes linear or tooth-like, mid-lobe with a distinct saccate spur about 1 cm long; *column* short, slightly broadened at the truncate apex, foot very long; *rostellar projection* elongated; *stipes* very long and slender, to 5 mm long; *viscidium* obovate or oblong; *pollinia* 2, sulcate.

About 6 species distributed from the Andaman Islands, Thailand and Peninsular Malaysia east to Indonesia and the Philippines.

55.65.1. Macropodanthus sp. 1

Epiphyte. Lower montane forest. Elevation: 1500–1600 m.

Collections. GOLF COURSE SITE: 1600 m, *Bailes & Cribb 735* (K); TENOMPOK: 1500 m, *Clemens 29634* (BM, K).

55.66. MALAXIS Sol. ex Sw.

Prodr. 8: 119 (1788).

Seidenfaden, G. (1978). Orchid genera in Thailand VII. *Oberonia* Lindl. & *Malaxis* Sol. ex Sw. Dansk Bot. Ark. 33, 1: 1–94.

Terrestrial or, rarely, epiphytic herbs. *Rhizomes* creeping. *Stems* erect, leafy, basally extended into a small jointed pseudobulb. *Leaves* 1 to several, plicate or conduplicate, ovate or narrowly elliptic, acute, thin-textured or fleshy, ± petiolate and articulate to a tubular sheath. *Inflorescence* terminal, erect, racemose or subumbellate, laxly or densely many-flowered. *Flowers* small, non-resupinate; *sepals* and *petals* free, subequal, spreading, rarely reflexed; *lateral sepals* often connate at base; *petals* usually narrower than sepals, circinate; *lip* free from column, sessile, entire or lobed, apex often toothed, cordate or auriculate at base, enveloping base of column, often with a nectar-secreting hollow or fovea at base; *column* short, foot absent; *pollinia* 4.

About 300 species with a cosmopolitan distribution, most numerous from Asia to New Guinea.

55.66.1. Malaxis burbidgei (Rchb. f. ex Ridl.) Kuntze, Revis. Gen. Pl. 2: 673 (1891).

Microstylis burbidgei Rchb. f. ex Ridl., J. Linn. Soc. Bot. 24: 336 (1883).

Terrestrial or lithophytic. Lower montane forest, on mossy rocks. Elevation: 1600 m.

General distribution: Sabah, Kalimantan.

Collection. MESILAU CAVE: 1600 m, *Bailes & Cribb 689* (K).

55.66.2. Malaxis calophylla (Rchb. f.) Kuntze, Revis. Gen. Pl. 2: 673 (1891).

Microstylis calophylla Rchb. f., Gard. Chron. ser. 2, 12: 718 (1879).

Terrestrial. Hill forest on ultramafic substrate. Elevation: 1100 m.

General distribution: Sabah, India, Burma, Peninsular Malaysia, Thailand, Cambodia.

Collection. MAHANDEI RIVER: 1100 m, *Carr 3026, SFN 26308* (SING).

55.66.3. Malaxis commelinifolia (Zoll. & E. Morren) Kuntze, Revis. Gen. Pl. 2: 673 (1891).

Microstylis commelinifolia Zoll. & E. Morren, Natuur-Geneesk. Arch. Ned. Ind. 1: 402 (1844).

Terrestrial. Hill forest, lower montane forest, sometimes on ultramafic substrate. Elevation: 800–1400 m.

General distribution: Sabah, Brunei, Sarawak, Java.

Collections. DALLAS: 900 m, *Clemens s.n.* (K), 900 m, *26912* (BM); LOHAN RIVER: 800–1000 m, *Beaman 9065* (K); MAMUT RIVER: 1400 m, *Collenette 1039* (K); MELANGKAP TOMIS: 900–1000 m, *Beaman 8984* (K).

55.66.4. Malaxis graciliscapa Ames & C. Schweinf., Orch. 6: 73, pl. 88 (1920). Type: MARAI PARAI SPUR, *Clemens 258* (holotype AMES mf).

Terrestrial. Lower montane forest on ultramafic substrate. Elevation: 1200–2400 m.

General distribution: Sabah, Kalimantan.

Additional collections. MAHANDEI RIVER HEAD: 1400 m, *Carr SFN 27401* (SING); MARAI PARAI: 1400 m, *Clemens 32357* (BM); MESILAU BASIN: 2100–2400 m, *Clemens 29708* (BM); PENIBUKAN: 1200 m, *Clemens 30800* (BM); TENOMPOK: 1500 m, *Clemens 26411* (BM, K).

55.66.5. Malaxis kinabaluensis (Rolfe) Ames & C. Schweinf., Orch. 6: 75 (1920). Plate 60A.

Microstylis kinabaluensis Rolfe in Gibbs, J. Linn. Soc. Bot. 42: 146 (1914). Type: PENIBUKAN, 1800 m, *Gibbs 4065* (holotype BM!).

Terrestrial. Lower montane forest with *Drimys*, rattans. Elevation: 1200–2000 m.

General distribution: Sabah, Kalimantan.

Additional collections. EAST MESILAU/MENTEKI RIVERS: 1700–2000 m, *Beaman 9614* (K), 1700 m, *Wood 825* (K); EAST AND WEST MESILAU RIVERS: 1700 m, *Collenette 916* (K); PENATARAN BASIN: 1200 m, *Clemens 34446* (BM); PENIBUKAN: 1200 m, *Carr SFN 26344* (SING), 1200 m, *Clemens 30575* (BM), 1500 m, *31005* (BM), 1200–1500 m, *31568* (BM), 1500 m, *51719* (BM).

55.66.6. Malaxis aff. **kinabaluensis** (Rolfe) Ames & C. Schweinf., Orch. 6: 75 (1920).

Terrestrial. Hill forest. Elevation: 900 m.

Collection. MOUNT KINABALU: 900 m, *Lamb SAN 93469* (K).

55.66.7. Malaxis latifolia Sm. in Rees, Cycl., 22 (1812).

Terrestrial. Hill forest. Elevation: 900–1500 m.

General distribution: Sabah, Sarawak, Kalimantan, widespread from India, SE Asia & China to New Guinea & Australia.

Collections. DALLAS: 1000 m, *Carr SFN 27366* (SING), 900 m, *Clemens s.n.* (BM), 900 m, *26303* (BM), 900 m, *26481* (BM), 900 m, *26692* (BM), 900 m, *26722* (BM), 900 m, *27005* (BM, K); KIAU: 900 m, *Clemens 33894* (BM); MOUNT KINABALU: *Haslam s.n.* (AMES); TAHUBANG RIVER: 1200–1500 m, *Clemens 31340* (BM).

55.66.8. Malaxis lowii (E. Morren) Ames in Merr., Bibl. Enum. Born. Pl., 151 (1921).

Microstylis lowii E. Morren, Belgique Hort. 34: 281, t. 14, f. 2 (1884).

Terrestrial. Hill forest, lower montane forest, principally on ultramafic substrate. Elevation: 500–2100 m.

General distribution: Sabah, Sarawak.

Collections. DALLAS: 900 m, *Clemens s.n.* (BM), 900 m, *26130* (K), 900 m, *26602* (BM), 900 m, *26690* (BM); HEMPUEN HILL: 800–1000 m, *Cribb 89/28* (K); MESILAU RIVER: 2100 m, *Clemens 51148* (BM); PENATARAN RIVER: 500 m, *Beaman 8858* (K); TENOMPOK: 1500 m, *Clemens 27980* (BM).

55.66.9. Malaxis metallica (Rchb. f.) Kuntze, Revis. Gen. Pl. 2: 673 (1891). Plate 60B.

Microstylis metallica Rchb. f., Gard. Chron. ser. 2, 12: 750 (1879).

Terrestrial. Hill forest on ultramafic substrate. Elevation: 500–1000 m.

General distribution: Sabah.

Collections. HEMPUEN HILL: 500–600 m, *Clements 3375A* (K), 800–1000 m, *Cribb 89/29* (K), 900 m, *Lamb SAN 91557* (K), 500–600 m, *Wood 603* (K); LOHAN RIVER: 800 m, *Bailes & Cribb 657* (K), 800–900 m, *Beaman 9486* (K).

55.66.10. Malaxis multiflora Ames & C. Schweinf., Orch. 6: 75, pl. 88 (1920). Type: KIAU, 900 m, *Clemens 86* (holotype AMES mf).

Terrestrial. Hill forest. Elevation: 900–1200 m.

General distribution: Sabah.

Additional collection. BUNDU TUHAN: 1200 m, *Carr SFN 27425* (SING).

55.66.11. Malaxis perakensis (Ridl.) Holttum, Gardens' Bull. 11: 283 (1947).

Microstylis perakensis Ridl., J. Linn. Soc. Bot. 32: 222 (1896).

Terrestrial. Hill forest, in damp shady situations. Elevation: 800 m.

General distribution: Sabah, Sarawak, Peninsular Malaysia, Thailand, Java.

Collection. KINATEKI RIVER: 800 m, *Collenette A 120* (BM).

55.66.12. Malaxis punctata J. J. Wood, Kew Mag. 4: 76, pl. 78 (1987). Plate 60C, 60D, 61A. Type: KIAU/TAHUBANG RIVER, 1000 m, *Sands 4011* (holotype K!).

Terrestrial. Hill forest, lower montane forest, often on ultramafic substrate, in deep shade. Elevation: 800–1800 m.

General distribution: Sabah.

Additional collections. KILEMBUN RIVER: 900 m, *Clemens s.n.* (BM); KINATEKI RIVER: 800 m, *Collenette A 113* (BM); MESILAU CAVE TRAIL: 1800 m, *Wood 846* (K); PENATARAN BASIN: 1200 m, *Clemens 34059* (BM); PENIBUKAN: 1200 m, *Clemens 30577* (BM); TAHUBANG RIVER: 1400 m, *Clemens 40824* (BM); TINEKUK FALLS: 1800 m, *Clemens 40909* (BM, K).

55.66.13. Malaxis variabilis Ames & C. Schweinf., Orch. 6: 77, pl. 88 (1920). Type: KIAU, 900 m, *Clemens 75* (holotype AMES mf).

Terrestrial. Hill forest. Elevation: 900 m.

General distribution: Sabah, Sarawak.

Additional collections. KIAU: 900 m, *Clemens 134* (AMES), *156* (AMES); MINITINDUK GORGE: 900 m, *Carr 3164, SFN 26670* (SING); MOUNT KINABALU: *Haslam s.n.* (AMES).

55.66.14. Malaxis indet.

Collections. KIAU: 700 m, *Carr SFN 27964* (SING); KILEMBUN BASIN: 1400 m, *Clemens 33956* (BM), 1400 m, *34404* (BM), 1200 m, *34481* (BM); LUBANG: 1500 m, *Carr SFN 26344A* (SING); LUGAS HILL: 1300 m, *Beaman 10560* (K); MAHANDEI RIVER: 1200 m, *Carr SFN 26311* (SING); MAMUT COPPER MINE: 1400 m, *Collenette 1045* (K); MELANGKAP KAPA: 600–700 m, *Beaman 8569* (K); MESILAU BASIN: 2700 m, *Clemens 29287* (BM); MESILAU RIVER: 1500 m, *RSNB 4126* (K), 1700 m, *Lamb SAN 91537* (K), 1500 m, *SAN 91538* (K); PENATARAN BASIN: 1500 m, *Clemens 34097* (BM); PENIBUKAN: 1500 m, *Clemens 31012* (BM), 1200 m, *31110* (BM), 1200–1500 m, *31513* (BM), 1200 m, *33026* (BM), 1400 m, *40524* (K); PINOSUK PLATEAU: *Lamb AL 305/85* (K); TAHUBANG RIDGE: 1100 m, *Clemens 40349* (BM); TAHUBANG RIVER: 1100 m, *Carr SFN 26462* (SING); TENOMPOK: 1500 m, *Carr SFN 27412* (SING), 1500 m, *Clemens 29236* (BM), 1500 m, *30128* (K); TINEKUK FALLS: 2100 m, *Clemens 40926* (BM).

55.67. MALLEOLA J. J. Sm. & Schltr. ex Schltr.

Feddes Repert. Beih. 1: 979 (1913).

Small monopodial epiphytes. *Stems* short or elongate and pendulous, to 30 cm long. *Leaves* scattered along stem, often flushed with purple-red, 4–7 × 1–3.5 cm. *Inflorescences* many-flowered, mostly shorter than the leaves, usually pendent. *Flowers* small, yellow, flushed with purple, red or mauve; *sepals* and *petals* free, spreading; *dorsal sepal* to 4 mm long; *lip* 3-lobed, immobile, with a variably shaped cylindrical spur with interior ornaments, side-lobes short, broadly triangular, mid-lobe very small, conical or linear; *column* short and stout, hammer-shaped, foot absent; *anther* ± dorsal; *stipes* spathulate, very broad below the pollinia; *viscidium* very small; *pollinia* 2, entire.

Around 30 species distributed from Thailand and Vietnam to Malaysia and Indonesia, to the Philippines, New Guinea and the Pacific islands. The centre of distribution lies in the Malay Archipelago.

55.67.1. Malleola kinabaluensis Ames & C. Schweinf., Orch. 6: 225, pl. 96 (1920). Type: KIAU, *Clemens 330* (holotype AMES mf).

Epiphyte. Hill forest, lower montane forest. Elevation: 900–1500 m. Its relationship with *M. witteana* needs investigation.

General distribution: Sabah.

Additional collections. KIAU: 900 m, *Clemens 87* (AMES), 900 m, *163* (AMES), 900 m, *176* (AMES), *353* (AMES); LUBANG: 1500 m, *Clemens 111* (AMES), 1500 m, *133* (AMES); PENIBUKAN: 1100 m, *Clemens 40590* (BM, K); PINOSUK PLATEAU: 1200–1400 m, *Lamb AL 303/85* (K); TENOMPOK ORCHID GARDEN: 1500 m, *Clemens 50357* (BM, K).

55.67.2. Malleola witteana (Rchb. f.) J. J. Sm. & Schltr., Feddes Repert. Beih. 1: 981 (1913). Plate 61B.

Saccolabium witteanum Rchb. f., Gard. Chron. ser. 2, 20: 618 (1883).

Saccolabium kinabaluense Rolfe in Gibbs, J. Linn. Soc. Bot. 42: 158 (1914). Type: LUBANG, 1500 m, *Gibbs 4111* (holotype K!).

Epiphyte. Lower montane forest. Elevation: 1500 m.

General distribution: Sabah, Peninsular Malaysia, Sumatra, Java.

55.67.3. Malleola indet.

Collections. GURULAU SPUR: 1500 m, *Clemens 50447* (BM); HEMPUEN HILL: 600 m, *Lamb AL 868/87* (K); KAUNG: 400 m, *Carr SFN 27314* (SING); KEGITAN AGAYE HEAD: 1200 m, *Carr SFN 27411* (SING); LOHAN RIVER: 700–900 m, *Beaman 9261* (K), 500–600 m, *Clements 3350* (K); LUGAS HILL: 1300 m, *Beaman 8468* (K); MAHANDEI RIVER: 1100 m, *Carr SFN 26473* (SING); MINITINDUK: 900 m, *Carr 3169* (SING); PENIBUKAN: 1100 m, *Clemens 50064* (BM); TENOMPOK: 1500 m, *Carr SFN 27871* (SING), 1400 m, *SFN 27954* (K); TINEKUK FALLS: 2000 m, *Clemens 40923* (BM).

55.68. MICROPERA Lindl.

Bot. Reg. 18: sub t. 1522 (1832).

Small climbing monopodial epiphytes. *Stems* with internodes to about 2 cm, bearing many support roots at intervals. *Leaves* linear, to 17 × 2 cm but usually much less. *Inflorescences* simple, to about 15 cm long. *Flowers* small, not resupinate, fleshy, usually yellow with purple markings, or pink; *lateral sepals* connate for a short distance at the base; *lip* sac-like, with a back-wall callus, a longitudinal septum and usually a 2-lobed front-wall callus, apex 3-lobed, mid-lobe much smaller than the sac or spur; *column* variable, usually without a foot; *rostellar projection* elongated, slender, usually twisted to one side, recalling *Ludisia*; *stipes* linear; *viscidium* small, ovate; *pollinia* 4, appearing as 2 unequal masses.

About 14 or 15 species distributed from India (Sikkim) and Indochina east to Australia and the Solomon Islands.

55.68.1. Micropera callosa (Blume) Garay, Bot. Mus. Leafl. Harvard Univ. 23: 186 (1972). Plate 61C.

Cleisostoma callosum Blume, Bijdr., 364 (1825).

Epiphyte. Hill forest on ultramafic substrate.

General distribution: Sabah, Kalimantan, Sarawak, Java.

Collections. HEMPUEN HILL: *Lamb s.n.* (K photo); PORING ORCHID GARDEN: *Lohok 12* (K).

55.69. MICROSACCUS Blume

Bijdr. 367 (1825).

Small monopodial epiphytes. *Stems* up to 20 cm long, often curved, rooting at the base. *Leaves* fleshy, laterally flattened, arranged in 2 rows, usually imbricate at the base. *Inflorescences* very short, arising from leaf axils, few-flowered, usually 2 borne opposite one another. *Flowers* white, somewhat fleshy; *sepals* free; *lateral sepals* sometimes decurrent along the spur; *lip* spurred, immobile, entire, emarginate or slightly 2-lobed; *column* short, stout, without a foot, *stipes* linear-clavate, 1.5–3 times the diameter of the pollinia; *viscidium* narrowly elliptic; *pollinia* 4, equal.

About 12 or 13 species distributed from Burma to Indonesia and the Philippines.

55.69.1. Microsaccus griffithii (Parish & Rchb. f.) Seidenf., Opera Bot. 95: 25 (1988).

Saccolabium griffithii Parish & Rchb. f., Trans. Linn. Soc. Bot. 30: 145 (1874).

Epiphyte. Lower montane forest. Elevation: 1700–2000 m.

General distribution: Sabah, Kalimantan, Burma, Peninsular Malaysia, Singapore, Thailand, Cambodia, Sumatra, Java, Philippines.

Collection. PIG HILL: 1700–2000 m, *Sutton 15* (K).

55.69.2. Microsaccus longicalcaratus Ames & C. Schweinf., Orch. 6: 232 (1920). Type: KIAU, *Clemens 342* (holotype AMES mf; isotypes BM!, K!, SING!).

Epiphyte. Hill forest, lower montane forest, sometimes on ultramafic substrate. Elevation: 1100–1500 m.

General distribution: Sabah.

Additional collections. MAHANDEI RIVER: 1100 m, *Carr SFN 26508* (SING); PENIBUKAN: 1200 m, *Clemens 50106* (BM, K); TENOMPOK: 1500 m, *Clemens 30151* (K).

55.69.3. Microsaccus indet.

Collections. LUMU-LUMU: 1700 m, *Carr SFN 27960* (SING); MAHANDEI RIVER: 1100 m, *Carr 3110, SFN 26496* (SING); MARAI PARAI: 1400 m, *Carr SFN 26971* (SING); TENOMPOK: 1500 m, *Carr SFN 26941* (SING), 1500 m, *SFN 27814* (SING).

55.70. MICROTATORCHIS Schltr. in K. Schum. & Lauterb.

Nachtr. Fl. Schutzgeb. Südsee, 224 (1905).

Small monopodial epiphytes similar in habit to *Taeniophyllum*. *Stems* very short, with or without leaves. *Inflorescences* lateral, racemose, elongating gradually; rachis angled; floral bracts persistent, alternate, distichous. *Flowers* pale yellowish-green, not opening widely. *Sepals* and *petals* similar, fused at base forming a short tube. *Lip* entire or 3-lobed, with a bristle or tooth inside near apex, shortly spurred. *Column* short, foot absent. *Pollinia* 2, entire.

About 47 species distributed in Java, Borneo, Sulawesi and the Philippines eastward through New Guinea to the Solomon Islands, Vanuatu, New Caledonia, Fiji and other Pacific islands. The centre of distribution lies in New Guinea.

55.70.1. Microtatorchis javanica J. J. Sm.

Epiphyte. Lower montane forest. Elevation: 1500 m.

General distribution: Sabah, Java.

Collection. TENOMPOK: 1500 m, *Carr SFN 27200* (SING).

55.70.2. Microtatorchis sp. 1

Epiphyte. Hill forest on ultramafic substrate. Elevation: 1100 m.

Collection. MAHANDEI RIVER: 1100 m, *Carr 3100, SFN 26469* (SING).

55.71. MISCHOBULBUM Schltr.

Feddes Repert. Beih. 1: 98 (1911).

Turner, H. (1992). A revision of the orchid genera *Ania* Lindley, *Hancockia* Rolfe, *Mischobulbum* Schltr. and *Tainia* Blume. Orchid Monogr. 6: 43–100.

Terrestrial herbs. *Pseudobulbs* slender, unifoliate. *Leaf* apical, broadly ovate. *Inflorescence* arising from the base of a recently matured pseudobulb, with few to many, well-spaced, rather large flowers. *Flowers* resupinate; *sepals* and *petals* free, elliptic to lanceolate; *lateral sepals* joined to the column-foot to form a mentum; *lip* entire, with prominent keels on the upper surface; *column* with a foot to which the lip and lateral lobes are joined; *pollinia* 8.

Eight species distributed from China and India to Peninsular Malaysia east to New Guinea and the Solomon Islands.

55.71.1. Mischobulbum scapigerum (Hook. f.) Schltr., Feddes Repert. Beih. 1: 98 (1911).

Nephelaphyllum scapigerum Hook. f., Bot. Mag. 89: t. 5390 (1863).

Terrestrial. Hill forest. Elevation: 800–900 m.

General distribution: Sabah.

Collections. DALLAS: 900 m, *Clemens s.n.* (BM); KAUNG/KELAWAT: 800 m, *Carr 3420* (SING).

55.72. MYRMECHIS (Lindl.) Blume

Fl. Javae ser. 2, 1, Orch.: 76 (1859).

Small terrestrial herbs. *Rhizomes* decumbent. *Leaves* ovate, spread out along stem and with several clustered at base of inflorescence. *Inflorescence* 1–3-flowered. *Flowers* resupinate, white; *dorsal sepal* and *petals* connivent, forming a hood; *lateral sepals* with an oblique, concave base; *lip* long-clawed, spathulate, base saccate, containing 2 bilobed glands, apex shallowly bilobed; *column* short; *rostellum* short, bifid; *stigmas* 2, on short processes; *pollinia* 2.

Five or six species distributed in India, China and Japan, south to Malaysia, Indonesia and the Philippines.

55.72.1. Myrmechis kinabaluensis Carr, Gardens' Bull. 8: 188 (1935). Type: KEMBURONGOH, 2700 m, *Carr 3539, SFN 27595* (holotype SING!; isotype AMES mf).

Terrestrial. Upper montane forest, wet mossy stunted forest, mossy rock faces. Elevation: 2100–2800 m.

Endemic to Mount Kinabalu.

Additional collections. JANET'S HALT/SHEILA'S PLATEAU: 2800 m, *Collenette 623* (K); MESILAU: 2100 m, *Clemens 29217* (BM).

55.73. NABALUIA Ames

Orch. 6: 70 (1920).

De Vogel, E. F. (1986). Revisions in Coelogyninae (Orchidaceae) II. The genera *Bracisepalum, Chelonistele, Entomophobia, Geesinkorchis* & *Nabaluia*. Orch. Monogr. 1: 47–51.

Epiphytic or lithophytic herbs. *Rhizome* short, creeping. *Pseudobulbs* all turned to one side of rhizome, more or less flattened, smooth, 2-leaved. *Leaves* ovate-oblong to linear, petiole sulcate. *Inflorescences* synanthous, racemose, erect, rigid; rachis fractiflex or more or less straight; floral bracts distichous, caducous, folded along mid-rib. *Flowers* resupinate, alternating in 2 rows, many open together; *sepals* ovate to oblong, margins more or less recurved; *petals* linear, usually rolled back; *lip* divided into a short, saccate hypochile and a long, more or less flat, spathulate, retuse epichile; hypochile 2- or 4-lobed, 2- to 5-keeled; disc with a horseshoe-shaped callus between hypochile and epichile; *column* with a 3-lobed hood, foot absent; *pollinia* 4, rather flattened.

Three species endemic to Borneo.

55.73.1. Nabaluia angustifolia de Vogel, Blumea 30: 202, pl. 2d, e (1984). Plate 61D. Type: KEMBURONGOH, 1800–2700 m, *Clemens 27159* (holotype AMES mf; isotypes BM!, K!, SING!).

Epiphyte or lithophyte. Lower montane forest, upper montane forest, sometimes on ultramafic substrate. Elevation: 1400–2700 m.

General distribution: Sabah.

Additional collections. GOLF COURSE SITE: 1600 m, *Bailes & Cribb 727* (K), 1600 m, *740* (K); KEMBURONGOH: 2700 m, *Clemens 27171* (BM, SING), 2700 m, *27184* (BM); MARAI PARAI SPUR: *Clemens 268* (AMES); MESILAU TRAIL: 1400 m, *Fuchs & Collenette 21381* (L); MOUNT KINABALU: *Lamb s.n.* (K); PENIBUKAN: 1500 m, *Clemens 51320* (BM); PIG HILL: 1700–2000 m, *Sutton 1* (K), 1700–2000 m, *3* (K); WEST MESILAU RIVER: 1700 m, *Brentnall 154* (K), 1700 m, *160* (K).

55.73.2. Nabaluia clemensii Ames, Orch. 6: 71, pl. 87 (1920). Plate 62A. Type: KEMBURONGOH, *Clemens 210* (holotype AMES mf; isotypes BM!, K!, SING!).

Chelonistele keithiana W. W. Sm., Notes Roy. Bot. Gard. Edinburgh 13: 188 (1921). Type: MOUNT KINABALU, 2200 m, *Moulton 103* (holotype E n.v.; isotype K!).

Chelonistele clemensii (Ames) Carr, Gardens' Bull. 8: 220 (1935).

Epiphyte or lithophyte. Lower montane forest, upper montane forest, sometimes on ultramafic substrate. Elevation: 1600–2900 m.

General distribution: Sabah.

Additional collections. KEMBURONGOH: *Clemens 111* (AMES), *210* (SING); MARAI PARAI: 2100 m, *Carr 3474* (K, SING); MARAI PARAI SPUR: *Clemens 395* (AMES); MENTEKI RIVER: 1600 m, *Beaman 10770* (K); MESILAU SPUR: 1800 m, *Lamb SAN 91534* (K); MINETUHAN SPUR: 2900 m, *Clemens 33791* (BM); MOUNT KINABALU: *Carr s.n.* (L); SUMMIT TRAIL: 2000 m, *Bailes & Cribb 784* (K), 2000 m, *Beaman 6769* (K), 1700 m, *Cribb 89/33* (K).

55.74. NEOCLEMENSIA Carr

Gardens' Bull. 8: 180 (1935).

Leafless saprophytic herb. *Rhizome* cylindrical, villose. *Stem* simple. *Inflorescence* erect, terminal, laxly 2–3-flowered; floral bracts appressed to pedicel. *Flowers* medium-sized, resupinate, campanulate, white with orange petals and greenish-brown lip; *sepals* adnate, forming a tube, free, recurved and papillose at apex; *petals* adnate at base to lateral sepals, much shorter than sepals, linear-spathulate, fimbriate; *lip* adnate to base of column, entire, clawed, claw oblong or subquadrate and bearing 2 subglobose, papillose-vesiculose apical calli, disc transversely rugulose, with a short median bilobed keel; *column* stout, with acute stelidia; *pollinia* 2.

A monotypic genus endemic to Borneo.

55.74.1. Neoclemensia spathulata Carr, Gardens' Bull. 8: 180 (1935). Fig. 45. Type: PENIBUKAN, 1100 m, *Clemens s.n.* (holotype BM? n.v.; isotype K!).

Saprophyte. Hill forest. Elevation: 1100 m. Known only from the type.

Endemic to Mount Kinabalu.

55.75. NEPHELAPHYLLUM Blume

Bijdr., 372 (1825).

Terrestrial herbs. *Rhizome* creeping, rooting at nodes. *Pseudobulbs* slender, fusiform, rarely distinct from the short petioles. *Leaves* slightly fleshy, ± cordate, often marbled light and dark green and flushed with purple. *Inflorescence* terminal, lax or dense, few-flowered. *Flowers* small to medium-sized, non-resupinate; *sepals* and *petals* subequal, acute, often reflexed; *lip* entire or slightly 3-lobed, shortly spurred, disc keeled; *column* usually rather short, slightly winged, without a foot; *anther* bicornute; *pollinia* 8.

Some 8–12 species distributed from Indochina through Thailand and Peninsular Malaysia east through Indonesia and north to the Philippines.

55.75.1. Nephelaphyllum flabellatum Ames & C. Schweinf., Orch. 6: 19 (1920). Plate 62B. Type: MARAI PARAI SPUR, *Clemens s.n.* (holotype AMES mf).

Tainia flabellata (Ames & C. Schweinf.) Gagnep., Bull. Mus. Natl. Hist. Nat., ser. 2, 4: 709 (1932).

Terrestrial. Lower montane forest on ultramafic substrate.

General distribution: Sabah.

Additional collection. MARAI PARAI SPUR: *Clemens 263A* (AMES).

55.75.2. Nephelaphyllum pulchrum Blume, Bijdr., 373, t. 36, f. 22 (1825). Plate 62C, 62D.

Nephelaphyllum latilabre Ridl. in Stapf, Trans. Linn. Soc. Bot. 4: 238 (1894). Type: Mount Kinabalu, 2000 m, *Haviland 1165* (holotype unlocated). *Tainia latilabra* (Ridl.) Gagnep., Bull. Mus. Natl. Hist. Nat., ser. 2, 4: 706 (1932).

Terrestrial. Lower montane forest, sometimes on ultramafic substrate; in leaf litter. Elevation: 1500–2000 m.

General distribution: Sabah, widespread from India and Bhutan to the Philippines.

Additional collections. GOLF COURSE SITE: 1700–1800 m, *Beaman 7233* (K, MSC); KEMBURONGOH/LUMU-LUMU: 1800 m, *Sinclair et al. 9027* (SING); KIAU VIEW TRAIL: 1500 m, *Lamb & Chan AL 6/82* (K); LUMU-LUMU: 1800 m, *Clemens s.n.* (BM), 1700 m, *27806* (BM), 1800 m, *29119* (BM); MARAI PARAI SPUR: 1600 m, *Bailes & Cribb 838* (K); MEMPENING TRAIL: 1600 m, *Chan 20* (K); MOUNT KINABALU: 1800 m, *Allen AK 66* (K); TENOMPOK: 1600 m, *Carr 3254, SFN 26938* (BM); TENOMPOK/LUMU-LUMU: 1700 m, *Clemens s.n.* (BM).

55.75.3. Nephelaphyllum verruculosum Carr, Gardens' Bull. 8: 200 (1935). Type: MINITINDUK/KINATEKI DIVIDE, 900 m, *Carr 3160, SFN 26607* (holotype SING!).

Fig. 45. Neoclemensia spathulata. A, habit (lower part of flower damaged); **B**, perianth tube cut and flattened showing sepals and spathulate, ciliate petals; **C**, lip (front view); **D**, column and pedicel-with-ovary (front view). **Tropidia saprophytica. E,** habit; **F**, flower (side view); **G,** dorsal sepal; **H**, lateral sepal; **J**, petal; **K**, lip (front view); **L**, pedicel-with-ovary, lip and column (side view); **M**, anther-cap (front view); **N**, pollinarium. **A** from *Clemens s.n.;* **B–D** after a copy of a sketch of *Clemens s.n.* by C. E. Carr; **E** from *Argent & Coppins 1135* from Mt. Mulu, Sarawak; **F–N** from *Phillipps & Lamb SNP 3019.* Scale: single bar = 1 mm; double bar = 1 cm. Drawn by Eleanor Catherine.

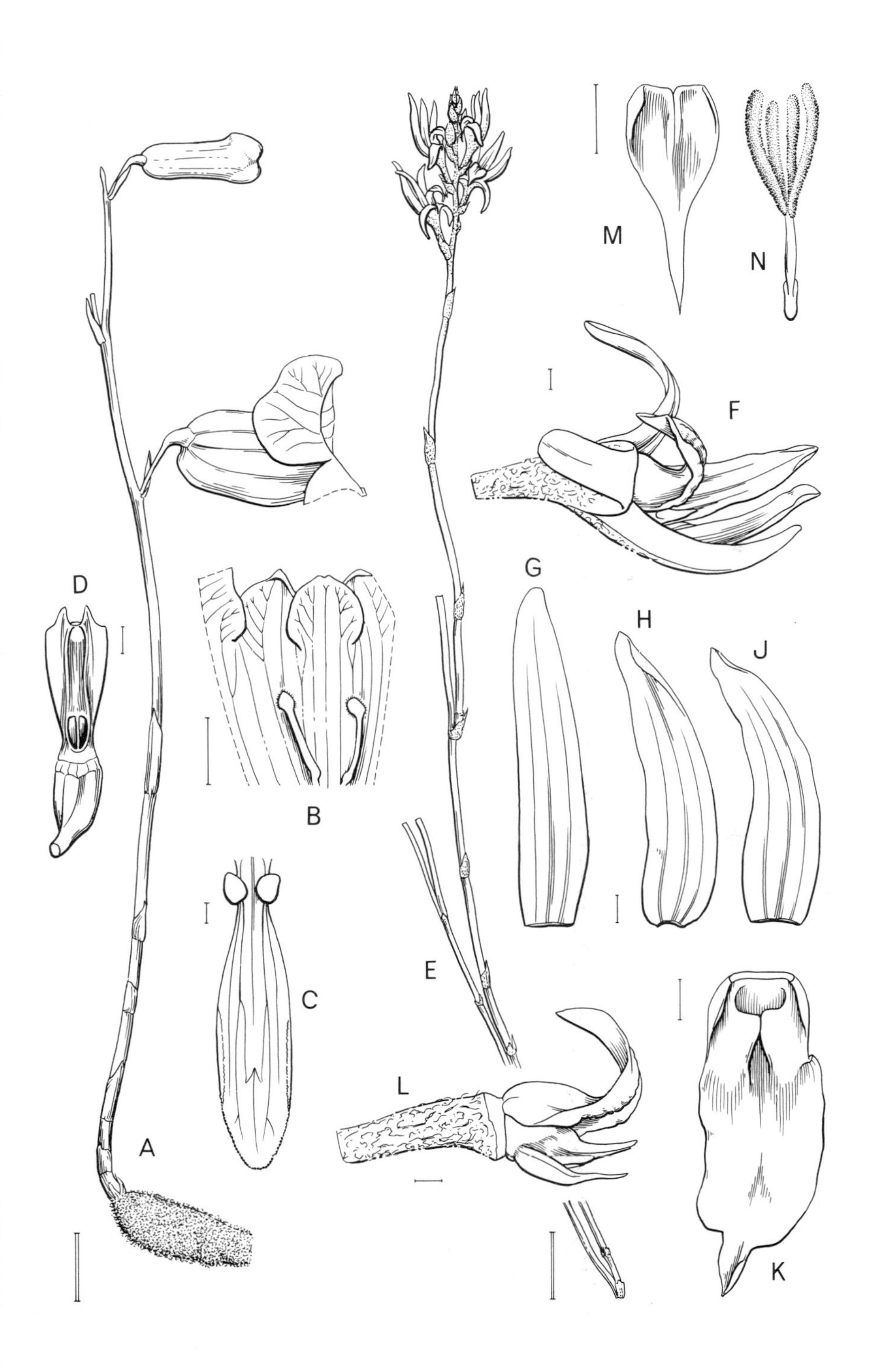
A
B
C
D
E
F
G
H
J
K
L
M
N

Terrestrial. Hill forest. Elevation: 900 m. Known only from the type.

Endemic to Mount Kinabalu.

55.76. NERVILIA Comm. ex Gaudich.

In Freycinet, Voy. Uranie, 421 (1829 "1826").

Terrestrial herbs arising from a rounded, reduced corm bearing short roots. *Leaf* solitary, appearing after flowers mature, on a separate stalk, erect to horizontal, plicate, often reniform or ovate to cordate, rarely lobed, glabrous or hirsute. *Inflorescence* erect, produced before the leaf appears, elongating after fertilization, terminating in a 1- to few-flowered raceme. *Flowers* small to large, often pendulous; *sepals* and *petals* similar, free, long and narrow; *lip* entire or 3-lobed, usually without a spur, base embracing column; *column* long, apex widened, without wings or foot; *pollinia* 2.

About 80 species distributed in the tropics and subtropics of Africa, Asia, New Guinea, Australia and the Pacific islands.

55.76.1. Nervilia punctata (Blume) Makino, Bot. Mag. Tokyo 16: 199 (1902).

Pogonia punctata Blume, Mus. Bot. Lugd. 1: 32 (1849).

Hill forest. Elevation: 1100 m.

General distribution: Sabah, Peninsular Malaysia, Thailand, Sumatra, Java.

Collection. KEGHITAN AGAYE HEAD: 1100 m, *Carr 3198, SFN 26796* (SING).

55.77. NEUWIEDIA Blume

Tijdschr. Natuurl. Gesch. Physiol. 1: 140 (1834).

De Vogel, E. F. (1969). Monograph of the tribe Apostasieae (Orchidaceae). Blumea 17: 313–350.

Erect or ascending, glabrous to hairy herbs, without rhizomes. *Stems* usually simple, leafy. *Leaves* narrowly elliptic or linear, acute, plicate. *Inflorescences* terminal, racemose, mostly unbranched, many-flowered; *floral bracts* conspicuous. *Flowers* yellow or white; *sepals* and *petals* free, slightly convex, not spreading; *lip* similar to petals but larger; *column* straight; *stamens* 3, filaments connate at base, partly adnate to style, *anthers* not clasping style, free from each other, dorsifixed; *ovary* ellipsoid, strongly contracted at apex, ± triangular to rounded.

Eight species distributed throughout SE Asia east to New Guinea, the Solomon Islands and Vanuatu.

55.77.1. Neuwiedia borneensis de Vogel, Blumea 17: 325 (1969). Plate 63A, 63B.

Terrestrial. Hill forest on ultramafic substrate. Elevation: 600–1000 m.

General distribution: Sabah, Brunei, Sarawak.

Collections. HEMPUEN HILL: 800–1000 m, *Beaman 7404* (MSC); 600–900 m, *Lohok in Lamb AL 355/85* (K); KULUNG HILL: 700–800 m, *Beaman 8355* (K).

55.77.2. Neuwiedia veratrifolia Blume, Tijdschr. Natuurl. Gesch. Physiol. 1: 142 (1834).

Terrestrial. Hill forest, sometimes on ultramafic substrate. Elevation: 500–1200 m.

General distribution: Sabah, Sarawak, Brunei, Kalimantan, widespread from Peninsular Malaysia east to the Philippines, New Guinea & the Solomon Islands.

Collections. DALLAS: 1200 m, *Carr 3387, SFN 27823* (K, SING), 900 m, *Clemens 26883* (BM, K), 900 m, *27696* (BM, K); HEMPUEN HILL: 800–900 m, *Lamb s.n.* (K), 800–900 m, *AL 155/83* (K); KIAU: 900 m, *Clemens 101* (BM); PORING: 500–600 m, *Lamb SAN 92328* (K).

55.77.3. Neuwiedia zollingeri Rchb. f.

a. var. **javanica** (J. J. Sm.) de Vogel, Blumea 17: 329 (1969). Plate 63C.

Neuwiedia javanica J. J. Sm., Bull. Jard. Bot. Buit., ser. 2, 14: 5 (1914).

Terrestrial. Hill forest, sometimes on ultramafic substrate. Elevation: 700–1000 m.

General distribution: Sabah, Sumatra, Java, Bali.

Collections. HEMPUEN HILL: 800–1000 m, *Beaman 7416* (K), *Madani SAN 89375* (K); MELANGKAP KAPA: 700–1000 m, *Beaman 8782* (K).

55.77.4. Neuwiedia indet.

Collection. HEMPUEN HILL: 600 m, *Wood & Charrington SAN 16370* (K).

55.78. OBERONIA Lindl.

Gen. Sp. Orch., 15 (1830).

Seidenfaden, G. (1978). Orchid genera in Thailand VII. *Oberonia* & *Malaxis* Sol. ex Sw. Dansk Bot. Ark. 33, 1: 1–94.

Epiphytic herbs. *Stems* short or long, leafy, often pendulous, lacking pseudobulbs. *Leaves* equitant, distichous, bilaterally flattened, sometimes jointed at base of blade. *Inflorescence* terminal, densely many-flowered, the flowers arranged in whorls, the lowermost usually opening last. *Flowers* minute, 1–2 mm long, non-resupinate, white, pale green, yellow, orange, red or brown; *sepals* and *petals* free, subequal, erect or reflexed; *sepals* broader than petals; *petals* often erose; *lip* sessile, entire or 3-lobed and erose, with 2 basal auricles that may or may not embrace base of column; *column* very short, without a foot; *pollinia* 4.

About 330 species distributed from tropical Africa to the Pacific islands and most numerous in mainland Asia.

55.78.1. Oberonia affinis Ames & C. Schweinf., Orch. 6: 79, pl. 89 (1920). Type: LUBANG, 1500 m, *Clemens 102* (holotype AMES mf; isotypes K!, SING!).

Epiphyte. Lower montane forest, sometimes on ultramafic substrate. Elevation: 1200–2100 m.

General distribution: Sabah.

Additional collections. GOLF COURSE SITE: 1700–1800 m, *Beaman 7230* (K); KIAU: *Clemens 380* (AMES); LUBANG: 1500 m, *Clemens 102* (BM); MARAI PARAI SPUR: *Clemens 275* (AMES); PENATARAN BASIN: *Clemens 34327* (BM); PENIBUKAN: 1200 m, *Clemens 40861* (BM); TAHUBANG FALLS: 1200 m, *Clemens 40336* (BM); TENOMPOK: 1500 m, *Clemens 50240* (BM, K); TINEKUK FALLS: 1400 m, *Clemens 40822* (BM, K), 1500–2100 m, *40924* (BM, K).

55.78.2. Oberonia djongkongensis J. J. Sm., Mitt. Inst. Allg. Bot. Hamburg 7: 43, t. 6, f. 33 (1927).

Epiphyte. Lower montane forest. Elevation: 1500 m.

General distribution: Sabah, Kalimantan.

Collection. TENOMPOK/RANAU: 1500 m, *Carr SFN 27346* (SING).

55.78.3. Oberonia aff. **equitans** (Forst. f.) Mutel, Mém. Soc. Roy. Centr. Agr. Sci. Arts Dépt. Nord: 84 (1837).

Epiphyte. Lowlands. Elevation: 400 m.

Collections. KAUNG: 400 m, *Carr SFN 26267* (SING), 400 m, *Darnton 355* (BM).

55.78.4. Oberonia kinabaluensis Ames & C. Schweinf., Orch. 6: 81, pl. 89 (1920). Plate 64A. Type: KIAU, *Clemens 329* (holotype AMES mf).

Epiphyte. Hill forest, lower montane forest, sometimes on ultramafic substrate. Elevation: 900–1500 m.

General distribution: Sabah.

Additional collections. BUNDU TUHAN: 1400 m, *Darnton 203* (BM); DALLAS: 900 m, *Clemens 26306* (BM), 900 m, *27006* (BM, K); KIAU: *Clemens 328* (AMES); KINATEKI RIVER: 1100 m, *Carr 3219* (K); LUBANG: 1500 m, *Clemens 102A* (AMES), 1500 m, *104A* (AMES); MAHANDEI RIVER: 1100 m, *Carr 3118, SFN 26536* (SING); MT. NUNGKEK: 1500 m, *Clemens 32881* (BM); PENATARAN BASIN: 900–1200 m, *Clemens 35171* (BM); PENATARAN RIVER: 900 m, *Clemens 32586* (BM);

TAHUBANG RIVER: 1100 m, *Clemens 50147* (BM); TENOMPOK: 1500 m, *Clemens 28414* (BM, K), 1500 m, *30138* (K).

55.78.5. Oberonia patentifolia Ames & C. Schweinf., Orch. 6: 83, pl. 90 (1920). Type: LUBANG, 1500 m, *Clemens 104* (holotype AMES mf).

Epiphyte. Hill forest, lower montane forest, sometimes on ultramafic substrate. Elevation: 900–1800 m.

General distribution: Sabah.

Additional collections. DALLAS: 900 m, *Clemens 26778* (BM), 900 m, *26783* (BM); DALLAS/TENOMPOK: *Clemens 26783* (K); GURULAU SPUR: 1700 m, *Clemens 50378* (BM); HEMPUEN HILL: *Madani SAN 89490* (K); KIAU: *Clemens 27* (AMES); LUMU-LUMU: 1800 m, *Clemens 27360* (BM); TENOMPOK: 1500 m, *Clemens 27003* (BM, SING), 1500 m, *27767* (BM, K), 1500 m, *50163* (BM); TENOMPOK ORCHID GARDEN: 1500 m, *Clemens 51743* (BM); TINEKUK FALLS: 1800 m, *Clemens 40922* (BM).

55.78.6. Oberonia rubra Ridl., J. Straits Branch Roy. Asiat. Soc. 50: 127 (1908).

Epiphyte. Lower montane forest. Elevation: 1400 m.

General distribution: Sabah, Sarawak.

Collection. PINOSUK PLATEAU: 1400 m, *Beaman 10718* (K).

55.78.7. Oberonia aff. **rubra** Ridl., J. Straits Branch Roy. Asiat. Soc. 50: 127 (1908).

Epiphyte. Hill forest, lower montane forest, sometimes on ultramafic substrate. Elevation: 900–1700 m.

Collections. DALLAS: 900–1200 m, *Clemens s.n.* (BM), 900 m, *27236* (K); MARAI PARAI: 1700 m, *Clemens 31887* (BM), 1500 m, *32449* (BM); NUMERUK RIDGE: 1500 m, *Clemens 40064* (BM); TENOMPOK ORCHID GARDEN: 1500 m, *Clemens 50354* (BM).

55.78.8. Oberonia triangularis Ames & C. Schweinf., Orch. 6: 85, pl. 90, f. 2 (1920). Type: LUBANG, 1500 m, *Clemens 104B* (holotype AMES mf).

Epiphyte. Lower montane forest. Elevation: 1200–1500 m.

General distribution: Sabah.

Additional collections. BUNDU TUHAN: 1200 m, *Carr SFN 27992* (SING); MARAI PARAI SPUR: *Clemens 275A* (AMES).

55.78.9. Oberonia sp. 1

Epiphyte. Low stature hill forest on ultramafic substrate. Elevation: 700–900 m.

Collection. LOHAN RIVER: 700–900 m, *Beaman 9263* (K).

55.78.10. Oberonia sp. 2

Epiphyte. Lower montane forest. Elevation: 1500 m.

Collections. TENOMPOK: 1500 m, *Carr SFN 27782* (K), 1500 m, *Clemens 28780* (BM, K), 1500 m, *30132* (K), 1500 m, *30133* (K).

55.78.11. Oberonia indet.

Collections. BUNDU TUHAN: 1400 m, *Carr SFN 27982* (SING), 900 m, *SFN 27997* (SING), 1200 m, *SFN 28022* (SING), 1200 m, *SFN 28023* (SING); GURULAU SPUR: 1400 m, *Carr SFN 26591* (SING); KAUNG: 400 m, *Carr SFN 27312* (SING); KEMBURONGOH: 2200 m, *Carr SFN 27790* (SING); KILEMBUN BASIN: 1400 m, *Clemens 34401* (BM); KILEMBUN RIVER: 1400 m, *Clemens 32584* (BM); KINATEKI RIVER: 1100 m, *Carr SFN 26853* (SING); LUBANG: 1400 m, *Carr SFN 26777* (SING); MARAI PARAI: 2000 m, *Clemens 35172* (BM); MEKEDEU RIVER: 400–500 m, *Beaman 9648* (K); MINITINDUK RIVER: 900 m, *Carr SFN 26940* (SING); PALUAN RIVER: 500 m, *Carr SFN 27392* (SING); PENATARAN BASIN: 900–1200 m, *Clemens 34325* (BM); PENIBUKAN: 1200 m, *Clemens 30574* (BM); TAHUBANG RIVER: 1100 m, *Carr SFN 26371* (SING); TENOMPOK: 1500 m, *Carr SFN 28025* (SING); TENOMPOK ORCHID GARDEN: 1500 m, *Clemens 50358* (BM).

55.79. OCTARRHENA Thwaites

Enum. Pl. Zeyl., 305 (1861).

Epiphytic herbs. *Stems* elongate, leafy toward apex, branching from the rooting base. *Leaves* distichous, terete or laterally compressed, jointed on their equitant sheaths. *Inflorescence* lateral, short, spiciform, racemose, secund, densely many-flowered. *Flowers* minute, resupinate, greenish-yellow. *Sepals* and *petals* free, spreading. *Petals* smaller than sepals. *Lip* entire, sessile at base of column, concave, sometimes uncinate. *Column* very short, without a foot. *Pollinia* 8.

About 20 species distributed from Sri Lanka to the Pacific islands; the majority are endemic to New Guinea.

55.79.1. Octarrhena parvula Thwaites, Enum. Pl. Zeyl., 305 (1861). Plate 64B.

Epiphyte. Hill forest, lower montane forest. Elevation: 900–1800 m.

General distribution: Sabah, Sri Lanka, Peninsular Malaysia, Sumatra, Java, Philippines.

Collections. GOLF COURSE SITE: 1700–1800 m, *Beaman 10676* (K); KINUNUT VALLEY: 1100 m, *Carr SFN 27190* (SING); KINUNUT VALLEY HEAD: 1200 m, *Carr SFN 27195* (SING); LOHAN/MAMUT COPPER MINE: 900 m, *Beaman 10596* (K); MINITINDUK GORGE: 900 m, *Carr SFN 26669* (SING).

55.79.2. Octarrhena indet.

Collection. TENOMPOK: 1600 m, *Carr SFN 26975* (SING).

55.80. **OECEOCLADES** Lindl.

Bot. Reg. 18: sub t. 1522 (1832).

Terrestrial herbs. *Pseudobulbs* short and one-noded to long, slender and few-noded. *Leaves* 1–3, terminal, conduplicate or plicate. *Inflorescence* lateral, arising from near the base of the pseudobulb, erect, racemose, many-flowered. *Flowers* with narrow, acute sepals; *petals* usually shorter and broader than sepals; *lip* 3-lobed or 4-lobed, with a short spur; *column* slender; *pollinia* 4.

About 20 species in the Old World tropics, mainly in Africa and Madagascar, extending east to the Pacific islands.

55.80.1. Oeceoclades pulchra (Thouars) P. J. Cribb & M. A. Clements, Austral. Orchid Res. 1: 99 (1989). Plate 64C.

Limodorum pulchrum Thouars, Orch. Iles Austr. Afr., t. 43 & 44 (1822).

Terrestrial. Hill forest. Elevation: 800–1100 m.

General distribution: Sabah, Kalimantan, widespread from Africa, Madagascar & Mascarene Islands east to India, Sri Lanka & SE Asia to New Guinea & Australia.

Collections. BUNDU TUHAN: 800 m, *Gibbs 4318* (BM); KEGHTAN AGAYE: 1100 m, *Carr SFN 27025* (SING).

55.81. **ORNITHOCHILUS** (Lindl.) Wall. ex Benth.

J. Linn. Soc. Bot. 18: 335 (1881).

Small monopodial epiphytes. *Stems* to 40 cm long. *Leaves* fleshy, to 12 × 4 cm. *Inflorescences* branching, many-flowered, as long as or longer than the leaves. *Flowers* to ca 1.3 cm across; *petals* narrower than sepals; *lip* with a cylindrical spur extending from the distal part of the hypochile, epichile 3-lobed, apex ciliate or crenulate; *column* to 4 mm long, foot absent; *rostellar projection* elongated, fleshy, obtuse; *anther-cap* truncate; *stipes* obovate-cuneiform, more than twice the diameter of the pollinia; *viscidium* obtriangular; *pollinia* 4, appearing as 2 pollen masses.

Three species distributed from India and Nepal to S China, south to Indonesia.

55.81.1. Ornithochilus difformis (Wall. ex Lindl.) Schltr.

a. var. **kinabaluensis** J. J. Wood, A. Lamb & Shim, ined. Plate 64D. Type: PINOSUK PLATEAU, 1400–1500 m, *Jukian & Lamb AL 4/82* (holotype K!).

Epiphyte. Lower montane forest. Elevation: 1400–1500 m. Known only from the type.

Endemic to Mount Kinabalu.

55.82. PANTLINGIA Prain

J. Asiat. Soc. Bengal, Pt. 2, Nat. Hist. 65(2): 107 (1896).

Delicate small tuberous terrestrial herbs. *Stems* erect, 1-leaved. *Leaf* positioned halfway up stem well above ground level, cordate, sessile, green. *Inflorescences* terminal, few-flowered, lax, racemose; floral bracts leafy. *Flowers* resupinate, green with purple callus on lip. *Sepals* and *petals* linear-ligulate, acute to acuminate. *Lip* orbicular, slightly retuse, not embracing column, spur absent, callus basal, simple to complicated, often with lateral processes. *Column* slender, curved, with a tooth-like process below, sometimes with a second process beneath stigma. *Pollinia* 2.

About 10 species distributed in the northern parts of eastern Asia, ranging from the Himalayas to Japan, extending southeast to New Guinea and New Caledonia. One species each in Java, Borneo and Sulawesi.

55.82.1. Pantlingia lamrii J. J. Wood & C. L. Chan, sp. nov. Fig. 46, Plate 65.

Pantlingiae celebicae (Schltr.) Rauschert, species celebica, affinis, sed callo basali labii parte supero porrecto rigido concavo cuspidato aculeiformi 1 mm longo apice deflexo ac appendiculo laterali filiformi utrinque prope basin instructo et parte infero erecto ligulato subacuto etiam 1 mm longo composito differt.

Type: EAST MALAYSIA, SABAH, MOUNT KINABALU, PINOSUK PLATEAU, MENTEKI RIVER, 1700–2000 m, 4 May 1984, *Beaman 9615* (holotype K!, spirit material only).

Tuberous terrestrial herb 4–10 cm high. *Stem* 1.5–4.3 cm, angular, weak, with a solitary leaf at apex. *Leaf* 0.6–0.9 × 0.9–1.2 cm, cordate, apiculate, margin irregular, thin-textured. *Inflorescence* erect, 2–7 flowered; *peduncle* 1.8–3 cm long, naked; *rachis* 0.5–2 cm long, angular, fractiflex; *floral bracts* 2–5 × 2–8 mm, ovate, mucronate, spreading. *Pedicel-with-ovary* 5–9 mm long, narrow, green. *Sepals* and *petals* green. *Dorsal sepal* 8–8.5 mm long, c. 1 mm wide when flattened, narrowly ligulate, acute, concave, erect to curving forward. *Lateral sepals* 3.5–4 × 0.2 mm, linear-subulate, directed downward behind lip. *Petals* 6–7 × 1 mm, ligulate, subfalcate, acuminate, curving forward. *Lip* 5 × 6 mm, orbicular, slightly retuse, margin minutely papillose and irregular, transparent green with a brownish purple central patch; basal callus dark purple, upper portion porrect, rigid, concave, cuspidate, thorn-like, c. 1 mm long, its tip deflexed, its base fused to the base of the column and having a non-rigid filiform lateral appendage 1.5 mm long borne on each side near the base, lower portion ligulate, subacute, ± erect, 1 mm long. *Column* 5 mm long, narrow, green, with an obtuse porrect projection 0.8–0.9 mm long borne just below the middle, column apex strongly curving

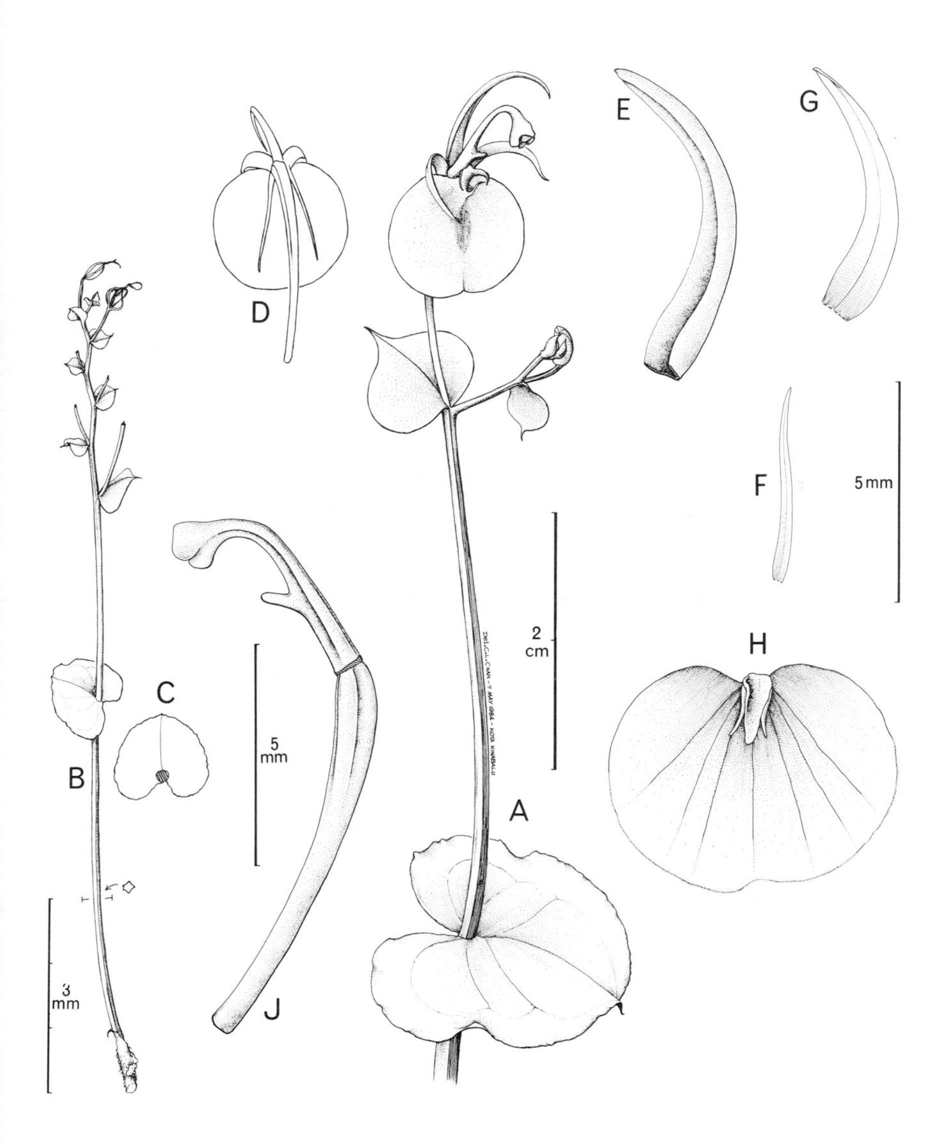

Fig. 46. Pantlingia lamrii. **A**, flowering plant; **B**, fruiting plant; **C**, leaf (back view); **D**, flower (back view); **E**, dorsal sepal; **F**, lateral sepal; **G**, petal; **H**, lip (front view); **J**, pedicel-with-ovary and column (side view). All from *Beaman 9615*. Drawn by Chan Chew Lun.

forward, with 2 wings on each side, the lower oblong, obtuse, the upper broader, rounded. *Anther-cap* lost, reported to be purple.

This curious little plant is related to *P. celebica* (Schltr.) Rauschert from Sulawesi. It differs in having a much more complicated basal callus on the lip, which is composed of a porrect, rigid, concave, cuspidate, thorn-like upper structure measuring 1 mm long. This has a deflexed tip and a filiform lateral appendage borne each side near the base. In addition, there is an erect, strap-shaped lower structure also measuring 1 mm long.

The specific epithet honours Datuk Lamri Ali, Director of Sabah Parks.

Lower montane forest. Elevation: 1500–2000 m.

Endemic to Mount Kinabalu.

Additional collection: PARK HEADQUARTERS: 1500 m, *Phillipps SNP 2123* (K, spirit material only).

55.83. PAPHIOPEDILUM Pfitzer

Morph. Stud. über Orchideenbl., 11 (1886).

Cribb, P. (1987). The Genus *Paphiopedilum*. The Royal Botanic Gardens, Kew in association with Collingridge.

Terrestrial or, rarely, epiphytic herbs. *Rhizomes* very short. *Stem* usually short, sometimes elongated, ascending and leafy. *Leaves* arranged in a fan, linear-ligulate to elliptic oblong, fleshy or leathery, often tessellated above, sometimes purple stained below. *Inflorescence* erect or arcuate, usually hirsute, 1- to many-flowered; *floral bracts* lanceolate and acute to elliptic and obtuse. *Flowers* large, usually showy, usually unscented, long-lived, often waxy-textured; *lateral sepals* united to form a synsepalum; *petals* deflexed and/or reflexed and often twisted; *lip* slipper- or helmet-shaped, deeply saccate; *column* bearing 2 ventral anthers and an apical, often shield-shaped staminode.

About 65 species distributed from India, southern China and SE Asia to the Philippines, the Malay Archipelago, New Guinea and the Solomon Islands.

Two or three named hybrids in *Paphiopedilum* may have occurred naturally on Mount Kinabalu (Cribb 1987). These are *P.* × *burbidgei* (Rchb. f.) Pfitzer, a hybrid of *P. hookerae* and *P. javanicum* var. *virens; P.* × *kimballianum* (Linden) Rolfe, a hybrid of *P. rothschildianum* and *P. dayanum*; and *P.* × *shipwayae* Rolfe, a hybrid of *P. dayanum* and *P. hookerae.* Cribb casts some doubt on the likelihood of *P.* × *shipwayae* occurring naturally, because the putative parents have not been found growing near each other. We have not seen wild-collected specimens of these hybrids and therefore have not included them in the Enumeration.

55.83.1. Paphiopedilum dayanum (Lindl.) Stein, Orchideenbuch, 464 (1892). Plate 66A.

Cypripedium spectabile Rchb. f. (sphalm. for *superbiens*) var. *dayanum* Lindl., Gard. Chron. 1860: 693 (1860). Type: MOUNT KINABALU, *Cult. Hort. Soc. s.n.* (holotype K!).

Cypripedium petri Rchb. f., Gard. Chron. ser. 2, 13: 680 (1880). Type: MOUNT KINABALU, *Hort. Veitch s.n.* (holotype W n.v.).

Terrestrial. Lowlands to lower montane forest, sometimes on ultramafic substrate, on steep ridges. Elevation: 300–1400 m.

General distribution: Sabah.

55.83.2. Paphiopedilum hookerae (Rchb. f.) Stein

a. var. **volonteanum** (Sander ex Rolfe) Kerch., Livre Orchid., 456 (1894). Plate 67A.

Cypripedium hookerae Rchb. f. var. *volonteanum* Sander ex Rolfe, Gard. Chron. ser. 3, 8: 66 (1890).

Terrestrial. Hill forest, lower montane forest, on ultramafic substrate, in the shade of bushes and casuarina. Elevation: 900–2000 m.

General distribution: Sabah.

55.83.3. Paphiopedilum javanicum (Reinw. ex Lindl.) Pfitzer

a. var. **virens** (Rchb. f.) Stein, Orchideenbuch, 471 (1892). Plate 66B.

Cypripedium virens Rchb. f., Bot. Zeitung 21: 128 (1863).

Paphiopedilum purpurascens Fowlie, Orchid Digest 38: 155 (1974).

Terrestrial. Hill forest, lower montane forest, on steep boulder-strewn slopes, often above rivers. Elevation: 900–1700 m. The type of *P. purpurascens* was cultivated, but originated from Kinabalu.

General distribution: Sabah.

55.83.4. Paphiopedilum lowii (Lindl.) Stein, Orchideenbuch, 476 (1892).

Cypripedium lowii Lindl., Gard. Chron. 1847: 765 (1847). Type: MOUNT KINABALU, *Low s.n.* (holotype K!).

Epiphyte. Lower montane forest. Elevation: 1400–1700 m.

General distribution: Sabah, Sarawak, Peninsular Malaysia, Sumatra, Java, Sulawesi.

55.83.5. Paphiopedilum rothschildianum (Rchb. f.) Stein, Orchideenbuch, 482 (1892). Plate 68.

Cypripedium rothschildianum Rchb. f., Gard. Chron. ser. 3, 3: 457 (1888). Type: MOUNT KINABALU, *Hort. Sander s.n.* (holotype W n.v.).

Terrestrial. Hill forest on ultramafic substrate on open and shaded ledges, steep slopes and cliffs. Elevation: 500–1800 m.

Endemic to Mount Kinabalu.

55.84. PARAPHALAENOPSIS A. D. Hawkes

Orquídea (Rio de Janeiro) 25: 212 (1964 "1963").

Large monopodial epiphytes. *Stems* short. *Leaves* few, terete, canaliculate, 20–165 cm long. *Inflorescences* few-flowered; *peduncle* distinct; *rachis* congested. *Flowers* showy; *dorsal sepal* free; *lateral sepals* spreading, decurrent on the column-foot; *lip* immobile, 3-lobed, side-lobes erect, mid-lobe narrow, with a broader forked or bilobed apex, callus conduplicate, plate-like, situated at the junction of the lobes, disc excavate behind the callus into a small sac-like nectary; *column* extended into a 3-fingered foot; *pollinia* 2, sulcate.

Four species endemic to Borneo.

55.84.1. Paraphalaenopsis labukensis Shim, A. Lamb & C. L. Chan, Orchid Digest 45: 139 (1981). Plate 69A, 69B.

Epiphyte. Hill forest on steep slopes, on ultramafic substrate, in shade. Elevation: 500–1000 m.

General distribution: Sabah.

55.85. PENNILABIUM J. J. Sm.

Bull. Jard. Bot. Buit., ser. 2, 13: 47 (1914).

Small monopodial epiphytes. *Stems* very short. *Leaves* few, clustered, rather fleshy, often twisted at the base, to 11 × 3 cm. *Inflorescences* racemose, 3–8 cm long; rachis somewhat thickened and complanate, with 1 or 2 flowers open at a time, placed in 2 rows. *Flowers* lasting for a day or two, white, cream, yellow or orange; *sepals* and *petals* 1–2 cm long; *petals* sometimes slightly toothed; *lip* 3-lobed, immobile, spurred, internal callosities absent, side-lobes either well developed and truncate or reduced to small ear-like lobes, when present often fimbriate or toothed, mid-lobe large, fleshy and solid or reduced to a small fleshy lobe; *column* short, more or less compressed dorsally, foot absent, *rostellar projection* elongate, prominent; *stipes* long, much broadened below the pollinia, often spathulate, 3–5 times the diameter of the pollinia; *viscidium* very small; *pollinia* 2, entire.

Some 10–12 species distributed from India (Assam) through Thailand and Malaysia to Indonesia and the Philippines.

55.85.1. Pennilabium angraecoides (Schltr.) J. J. Sm., Bull. Jard. Bot. Buit., ser. 2, 13: 47 (1914).

Saccolabium angraecoides Schltr., Bull. Herb. Boissier, ser. 2, 6: 472 (1906).

Epiphyte. Lower montane forest. Elevation: 1400 m.

General distribution: Sabah, Kalimantan.

Collection. LUBANG: 1400 m, *Carr SFN 27447* (SING).

55.86. PERISTYLUS Blume

Bijdr., 404 (1825).

Terrestrial herbs arising from tubers. *Stem* erect, few- to several-leaved, either basal, spaced along stem or grouped near the centre. *Leaves* thin, linear to broadly elliptic, not jointed at base, uppermost often bract-like. *Inflorescence* terminal, few- to many-flowered, lax or dense. *Flowers* small, white or green; pedicel-with-ovary porrect, close to rachis; *dorsal sepal* and *petals* forming a hood over column; *petals* usually a little broader than lateral sepals; *lip* 3-lobed, spur shorter than ovary, often reduced to a globular sac shorter than sepals; *column* short; *stigmas* 2, sessile, convex, cushion-like, often adnate to lip edge, but in some species appearing flat and clavate, protruding and adnate to hypochile; *pollinia* 2, the caudicles very short, without tubes (canals or thecae), attached to short rostellar side-lobules, rostellar mid-lobule small, ± hidden between the anthers.

Some 135 species are listed in *Index Kewensis*, although half this figure is probably more correct. They are distributed from India and Sri Lanka to China, Taiwan and the Ryukyu Islands, south through Thailand, Indochina, Malaysia and Indonesia east to New Guinea, Australia and the Pacific islands.

55.86.1. Peristylus brevicalcar Carr, Gardens' Bull. 8: 167 (1935). Type: BUNDU TUHAN, 900 m, *Carr 3353* (holotype SING!).

Terrestrial. Hill forest. Elevation: 900 m. Known only from the type.

Endemic to Mount Kinabalu.

55.86.2. Peristylus candidus J. J. Sm., Orch. Java, 36, f. 18 (1905).

Terrestrial. Hill forest. Elevation: 900 m.

General distribution: Sabah, Sarawak, Peninsular Malaysia, Cambodia, Vietnam, Sumatra, Java, Sulawesi, Ambon, Buru, Lingga Archipelago.

Collection. BUNDU TUHAN: 900 m, *Carr 3320, SFN 27811* (BM, K, SING).

55.86.3. Peristylus ciliatus Carr, Gardens' Bull. 8: 169 (1935). Type: KOLOPIS RIVER HEAD, 1300 m, *Carr 3368, SFN 27111* (syntype SING!; isosyntype AMES mf); TENOMPOK, 1500 m, *Clemens 29816* (syntype BM!; isosyntype K!).

Terrestrial. Lower montane forest. Elevation: 1300–1500 m.

Endemic to Mount Kinabalu.

55.86.4. Peristylus goodyeroides (D. Don) Lindl., Gen. Sp. Orch., 299 (1835).

Habenaria goodyeroides D. Don, Prodr. Fl. Nep., 25 (1825).

Terrestrial. Hill forest. Elevation: 900 m.

General distribution: Sabah, widespread from NW Himalayas to China & the Philippines, through Malaysia & Indonesia to New Guinea.

Collections. BUNDU TUHAN: 900 m, *Carr 3575, SFN 27779* (BM, SING); DALLAS: 900 m, *Clemens 26314* (BM), 900 m, *26482* (BM); KIAU: 900 m, *Haslam s.n.* (AMES).

55.86.5. Peristylus gracilis Blume, Bijdr., 404 (1825).

Terrestrial. Hill forest, lower montane forest, sometimes on ultramafic substrate. Elevation: 900 m.

General distribution: Sabah, India, Peninsular Malaysia, Thailand, Sumatra, Java, Krakatau.

Collections. KIAU: 900 m, *Clemens 193* (AMES); MARAI PARAI SPUR: *Clemens 397* (AMES); MOUNT KINABALU: *Clemens s.n.* (BM).

55.86.6. Peristylus grandis Blume, Bijdr., 405, t. 30 (1825).

Terrestrial. Hill forest. Elevation: 900 m.

General distribution: Sabah, Peninsular Malaysia, Indonesia, New Guinea.

Collections. DALLAS: 900 m, *Clemens s.n.* (BM), 900 m, *26770* (BM); KIAU: 900 m, *Clemens 45* (AMES); MINITINDUK: 900 m, *Carr 3196, SFN 27149* (SING).

55.86.7. Peristylus hallieri J. J. Sm., Bull. Dép. Agric. Indes Néerl. 22: 1 (1909). Plate 69C.

Terrestrial. Lower montane forest, in open grassy areas, rough mossy ground, landslides. Elevation: 800–1600 m.

General distribution: Sabah, Kalimantan, Sarawak.

Collections. BUNDU TUHAN: 1200 m, *Carr 3489, SFN 27455* (BM, K, SING); MT. NUNGKEK: 800 m, *Clemens 32775* (BM); PARK HEADQUARTERS: 1500 m, *Lamb AL 24/82* (K); PARK HEADQUARTERS/POWER STATION: 1600 m, *Cribb 89/36* (K).

55.86.8. Peristylus kinabaluensis Carr, Gardens' Bull. 8: 170 (1935). Type: DALLAS/TENOMPOK, 1200 m, *Carr 3389, SFN 27308* (holotype SING!).

Terrestrial. Hill forest or lower montane forest, by stream in young secondary forest. Elevation: 1200 m. Known only from the type.

Endemic to Mount Kinabalu.

55.87. PHAIUS Lour.

Fl. Cochinch. 2: 517, 529 (1790).

Terrestrial herbs. *Stems* clustered, short, long, cylindrical or pseudobulbous, few-leaved. *Pseudobulbs*, when present, conical or ovoid. *Leaves* medium to large, petiolate, narrowly obovate or narrowly elliptic, acuminate, plicate. *Inflorescence* lateral from base of pseudobulb, or axillary halfway up stem, lax or rather dense, racemose. *Flowers* usually large, often showy, resupinate, turning blue-black when old or damaged; *sepals* and *petals* similar, free, rather fleshy, spreading or reflexed; *lip* entire or obscurely 3-lobed, erect, sessile and partially adnate to and embracing column to form a tube, gibbous or shortly spurred; *column* long or short, fleshy, curved, with an inflexed foot; *pollinia* 8.

About 30 species distributed in the Old World tropics of Africa, Madagascar, S and SE Asia, Australia, New Guinea and the Pacific islands.

55.87.1. Phaius baconii J. J. Wood & Shim, ined. Type: PENIBUKAN, 1500 m, *Lamb AL 39/83* (holotype K!).

Terrestrial. Lower montane forest, mostly on ultramafic substrate, sometimes in bamboo thickets. Elevation: 1200–1500 m.

Endemic to Mount Kinabalu.

Additional collections. PENIBUKAN: 1200 m, *Carr SFN 26534* (SING), 1200 m, *Clemens 30848* (BM), 1500 m, *30995* (BM), 1200–1500 m, *31366* (BM, K), 1200 m, *40253* (BM), 1500 m, *51716* (K); PINOSUK PLATEAU: 1400–1500 m, *Beaman 9192* (K), 1400 m, *10712* (K).

55.87.2. Phaius borneensis J. J. Sm., Icon. Bogor. 2: 61, t. 3, f. c (1903).

Terrestrial. Hill forest and lower montane forest, probably on ultramafic substrate. Elevation: 500–1500 m.

General distribution: Sabah, Kalimantan.

Collections. MESILAU RIVER: 1500 m, *RSNB 7030* (K); PENATARAN RIVER: 500–600 m, *Lamb AL 551/86* (K).

55.87.3. Phaius pauciflorus (Blume) Blume

a. subsp. **sabahensis** J. J. Wood & A. Lamb, **subsp. nov.** Fig. 47.

A subsp. *paucifloro* (Blume) Blume, subspecies malayana, sumatrana atque javanica, inflorescentia perbrevi, labio non in medio incrassato, calcari recto differt.

Type: EAST MALAYSIA, SABAH, RANAU DISTRICT, MOUNT KINABALU, HEMPUEN HILL: 540–600 m, January 1991, *Lamb & Surat in Lamb AL 1320/91* (holotype K!, herbarium and spirit material).

Terrestrial herb. *Stems* 4–5 per plant, up to 60 × 0.5–1 cm, internodes 5–8 cm long, dark green. *Leaves* 4–5, borne at stem apex, blade 15–33 × 6.5–10.5 cm, elliptic, acuminate, plicate, glabrous, abruptly narrowed to a sheathing base 5–6 cm long which embraces the stem. *Inflorescences* usually 2-flowered, emerging from the nodes along the middle portion of the stem; *peduncle* 1 × 0.35 cm, green or greenish white, enclosed by two 8 mm long ovate, acute, white bracts; *rachis* 0.6 × 0.25 cm, green; *floral bracts* 0.6–1 cm long, ovate, acuminate, white. *Flowers* not opening widely. *Pedicel-with-ovary* 1.8–2 cm long, clavate, curved, greenish white. *Sepals* and *petals* off-white to cream. *Dorsal sepal* 2.5–2.9 × 0.8–0.9 cm, narrowly ovate-elliptic, acute. *Lateral sepals* 2.6–2.9 × 0.8–0.9 cm, obliquely narrowly ovate-elliptic, acuminate. *Petals* 2.4–2.6 × 0.7 cm, narrowly elliptic, acute, apiculate. *Lip* 2.2 cm long, 0.9–1(–1.2) cm at its widest point, oblong, acute, apiculate, entire, margin irregular, white to cream, often pale yellow near apex, with reddish lateral streaks on lower portion and white hairs at spur entrance, disc on lower half of blade with 2 obscure, low dark yellow ridges extending from near the base to about 4–5 mm below apex, and a third obscure, low shorter apical ridge, spur 1.5 cm long, narrowly conical, acute, straight, white. *Column* 1.1–1.3 cm long, with very small rounded apical wings, white to cream. *Anther-cap* 3 × 2 mm, ovate, apex attenuated and truncate.

Subspecies *sabahensis* differs from subsp. *pauciflorus,* from Peninsular Malaysia, Sumatra and Java, by its shorter inflorescence and lip which is not thickened in the middle and has a straight spur.

Fig. 47. Phaius pauciflorus subsp. **sabahensis. A**, habit; **B**, flower (side view); **C**, dorsal sepal; **E**, petal; **F**, lip (front view); **G**, column (front view); **H**, column (side view); **J**, anther-cap (front view); **K**, anther-cap (side view). All from *Lamb AL 1320/91.* Scale: single bar = 1 mm; double bar = 1 cm. Drawn by Eleanor Catherine.

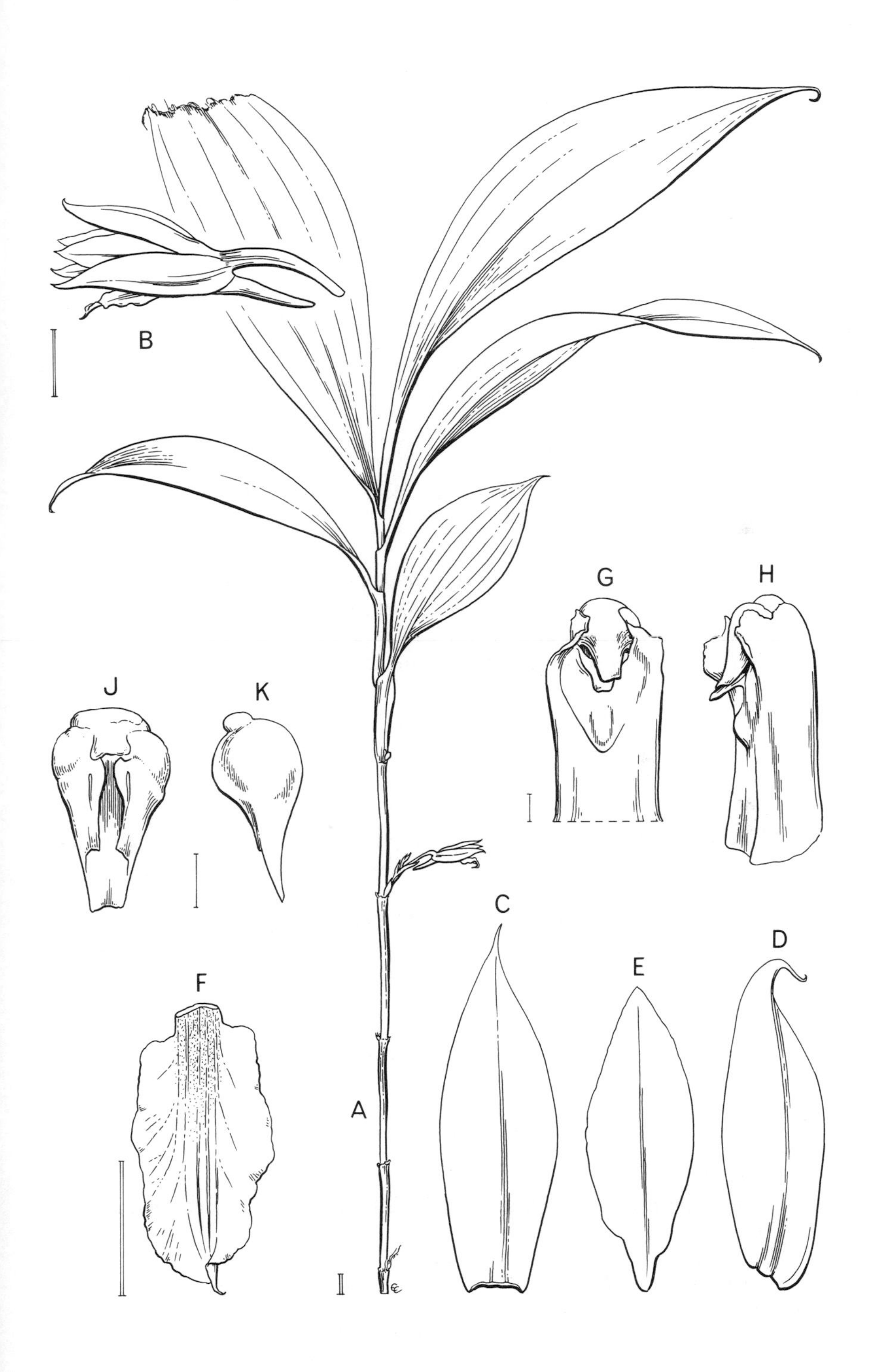
B
G
H
J
K
C
E
D
F
A

Flower colour and spur length vary a great deal throughout the range of subsp. *pauciflorus*. In Peninsular Malaysia, for example, the species is represented by var. *pallidus* (Ridl.) Holttum, which has flowers with pale yellow sepals, white violet-spotted petals, an orange-yellow lip spotted with red, with a 2 cm long spur. In Java, plants have pure white sepals and petals, a white and yellow lip sometimes striated with red, and a shorter spur.

The lack of a central thickening on the lip would seem to be the most distinctive feature of the Bornean subspecies. A fuller knowledge of variation throughout the range of the species needs to be gained before the status of the Bornean populations can be finally ascertained.

The infraspecific epithet refers to the East Malaysian state of Sabah.

Hill forest, lower montane forest, sometimes on ultramafic substrate. Elevation: 500–1500 m.

Endemic to Mount Kinabalu.

Additional collections: MARAI PARAI/KIAU TRAIL: 1300 m, *Bailes & Cribb 812* (K); TENOMPOK: 1500 m, *Clemens 28830* (BM).

55.87.4. Phaius reflexipetalus J. J. Wood & Shim, ined. Plate 70A. Type: PENATARAN RIDGE, 1100 m, *Lamb AL 25/82* (holotype K!).

Terrestrial. Hill forest on ultramafic substrate. Elevation: 1100 m.

Endemic to Mount Kinabalu.

Additional collections. LOHAN RIVER: *Clements 3398A* (K), *Lamb AL 164/83* (K).

55.87.5. Phaius subtrilobus Ames & C. Schweinf., Orch. 6: 157 (1920). Plate 70B. Type: MOUNT KINABALU, *Haslam s.n.* (holotype AMES mf).

Terrestrial. Lower montane forest. Elevation: 1300–1500 m.

General distribution: Sabah.

Additional collections. LUBANG: 1300 m, *Carr SFN 27409* (AMES, SING); MESILAU RIVER: 1500 m, *RSNB 4031* (K); MESILAU TRAIL: 1500 m, *Collenette 539* (K); PARK HEADQUARTERS: 1500 m, *Lamb SAN 87134* (K); PENATARAN RIVER: 1500 m, *Clemens 32528* (K).

55.87.6. Phaius tankervilleae (Banks ex L'Hérit.) Blume, Mus. Bot. Lugd. 2: 177 (1856).

Limodorum tankervilleae (Banks ex L'Hérit.) Blume, Sert. Angl., 28 (1789).

Terrestrial. Lowlands to lower montane forest. Elevation: 400–1500 m.

General distribution: Sabah, widespread from India through S & SE Asia east to Australia & the Pacific islands.

Collections. BUNDU TUHAN: 1200 m, *Carr SFN 26262A* (SING); DALLAS: 900 m, *Clemens 26768* (BM); KADAMAIAN RIVER: 600 m, *Darnton 598* (BM); KAUNG: 400 m, *Carr SFN 26262* (SING), *Clemens 40460* (BM, K), 500 m, *51581* (BM); KIAU: 900 m, *Clemens 130* (AMES), 900 m, *184* (AMES), *277A* (AMES), *346* (AMES), *378* (AMES); MOUNT KINABALU: *Clemens s.n.* (AMES); PENATARAN RIVER: 1500 m, *Clemens 32528* (BM); TENOMPOK: 1500 m, *Clemens 29637* (BM, K).

55.88. **PHALAENOPSIS** Blume

Bijdr., 294 (1825).

Sweet, H. (1980). The genus *Phalaenopsis*. The Orchid Digest Inc.

Medium-sized to large monopodial epiphytes or lithophytes. *Stems* very short. *Leaves* few, narrowly oblong to broadly elliptic, normally widest in apical half, often fleshy, sometimes transversely striped, rather shiny, sometimes deciduous. *Inflorescences* 1- to many-flowered, racemose or paniculate; *peduncle* long; *rachis* sometimes flattened. *Flowers* small to large, long lasting, waxy in texture, white, pink, mauve or violet to yellow with reddish-brown markings; *sepals* and *petals* spreading, free; *dorsal sepal* 1.5–4 cm long; *petals* sometimes much broader than the sepals and with a clawed base; *lip* immobile, 3-lobed, side-lobes erect, mid-lobe porrect, with a bifid or complex basal callus, apex sometimes extended into 2 recurved appendages or filiform 'antennae'; *column* erect, often expanded at apex, with a short foot; *rostellar projection* and *stipes* long and slender, pointing toward the base of the column; *viscidium* elliptic, shorter than the stipes; *pollinia* usually 2, sulcate (in Bornean species), rarely 4.

Between 40 and 45 species distributed from India to S China, Thailand, Indochina, Malaysia and Indonesia to the Philippines and New Guinea. The majority of species occur in Indonesia and the Philippines.

55.88.1. **Phalaenopsis amabilis** (L.) Blume, Bijdr., 294 (1825).

Epidendrum amabile L., Sp. Pl., 953 (1753).

Epiphyte. Hill forest on ultramafic substrate, lower montane forest. Elevation: 700–1500 m.

General distribution: Sabah, Sarawak, Kalimantan, widespread from Sumatra, Java & Sulawesi east to New Guinea & Australia, north to the Philippines.

55.88.2. **Phalaenopsis maculata** Rchb. f., Gard. Chron. ser. 2, 16: 134 (1881).

Epiphyte or lithophyte. Hill forest, on rocks and rocky banks with thick moss and leaf litter. Elevation: 800–900 m.

General distribution: Sabah, Kalimantan, Sarawak, Peninsular Malaysia.

55.88.3. Phalaenopsis modesta J. J. Sm., Icon. Bogor. 3: 47, t. 218 (1906).

Epiphyte. Hill forest, sometimes on ultramafic substrate. Elevation: 600–900 m.

General distribution: Sabah, Kalimantan.

55.89. PHOLIDOTA Lindl. ex Hook.

Exot. Fl. 2: t. 138 (1825).

De Vogel, E. (1988). Revisions in Coelogyninae (Orchidaceae) III. The genus *Pholidota*. Orch. Monogr. 3: 1–118.

Pendulous or erect epiphytic or lithophytic herbs. *Pseudobulbs* closely- or well-spaced, slender or swollen, 1- or 2-leaved. *Leaves* narrowly elliptic, ovate or oblong, coriaceous. *Inflorescence* terminal, emerging from apex of pseudobulb, slender, racemose, laxly or densely many-flowered, distichous, pendent; floral bracts large, concave, persistent. *Flowers* small, fleshy, resupinate, white. *Dorsal sepal* concave or convex, broadly ovate to elliptic; *lateral sepals* concave or convex, ovate to ovate-oblong, often carinate. *Petals* ovate to linear. *Lip* sessile, with a saccate basal hypochile and subentire or 3- to 4-lobed, deflexed epichile. *Column* short, with a broad hooded apex. *Pollinia* 4.

A genus of 29 species distributed in mainland and SE Asia, New Guinea, Australia and the Pacific islands.

55.89.1. Pholidota carnea (Blume) Lindl., Gen. Sp. Orch., 37 (1830).

a. var. **carnea.** Plate 71B.

Crinonia carnea Blume, Bijdr., 339 (1825).

Epiphyte. Lower montane forest. Elevation: 1200–2100 m.

General distribution: Sabah, Sarawak, Peninsular Malaysia, Thailand, Sumatra, Java, Sulawesi, Bali, Flores, Lombok, Philippines, New Guinea.

Collections. EAST MESILAU RIVER: 1500–1600 m, *Bailes & Cribb 522* (K); GURULAU SPUR: 1500 m, *Clemens 50349* (BM, K), 1800 m, *50430* (BM), 1700 m, *50700* (K); KIAU: *Clemens 331* (AMES); KINUNUT VALLEY HEAD: 1200 m, *Carr 3336, SFN 27398* (SING); LUBANG: *Clemens 107* (AMES), 1500 m, *127A* (AMES); MESILAU CAVE TRAIL: 1700–1900 m, *Beaman 7999* (K); PENIBUKAN: 2100 m, *Clemens 31209A* (BM), 1200 m, *50100* (BM, K), 1500 m, *50346* (BM), 1700 m, *50363* (BM), 1700 m, *50396* (BM); TENOMPOK: 1500 m, *Clemens 27239* (BM), 1500 m, *27398* (BM, K).

55.89.2. Pholidota clemensii Ames, Orch. 6: 66 (1920). Type: MARAI PARAI SPUR, *Clemens 390* (holotype AMES mf).

Epiphyte, occasionally terrestrial. Lower montane forest, rarely upper montane forest? Elevation: 1200–3400 m.

General distribution: Sabah, Kalimantan, Sarawak.

Additional collections. BAMBANGAN RIVER: 1500 m, *RSNB 4483* (K), 1500 m, *RSNB 4624* (K); BUNDU TUHAN: 1200 m, *Clemens s.n.* (BM); EAST MESILAU/MENTEKI RIVERS: 1700 m, *Beaman 8756* (K), 1700 m, *9391* (K), 1700–2000 m, *9602* (K); GOLF COURSE SITE: 1700 m, *Beaman 9046* (K); LUMU-LUMU: 1800 m, *Carr 3563, SFN 27781* (BM, K, SING), 2100 m, *Clemens 29992* (BM); MESILAU RIVER: 2100 m, *Clemens 51144* (BM); PAKA-PAKA CAVE: 3400 m, *Clemens 27805* (BM, K); PENIBUKAN: *Clemens s.n.* (BM), 1200 m, *30589* (BM), 1500 m, *30608* (BM), 1200 m, *30847* (BM), 1500 m, *31655* (BM); TENOMPOK: 1500 m, *Clemens s.n.* (BM), 1500 m, *29992* (K).

55.89.3. Pholidota gibbosa (Blume) De Vriese, Ill. Orch. Ind. Orient., t. 5, f. 1 (1854).

Chelonanthera gibbosa Blume, Bijdr., 383 (1825).

Epiphyte. Hill forest, lower montane forest. Elevation: 900–1700 m.

General distribution: Sabah, Brunei, Kalimantan, Sarawak, Peninsular Malaysia, Sumatra, Java, Sulawesi, Bali, Solomon Islands.

Collections. KIAU: 900 m, *Clemens 24* (AMES), 900 m, *83* (AMES), 900 m, *92* (AMES); LUBANG: 1500 m, *Clemens 127* (AMES, BM, K, SING); PENIBUKAN: 1200 m, *Clemens 40527* (BM), 1200 m, *40662* (BM), 1200 m, *40786* (BM); TAHUBANG RIVER: 1200 m, *Clemens 30706A* (BM); TENOMPOK: 1500 m, *Carr 3758* (SING), 1500 m, *Clemens 26835* (BM, K); TENOMPOK ORCHID GARDEN: 1500 m, *Clemens 50248* (BM, K), 1500 m, *51724* (BM); TINEKUK FALLS: 1700 m, *Clemens 40819* (BM, K).

55.89.4. Pholidota imbricata Hook., Exot. Fl. 2: t. 138 (1825).

Epiphyte, sometimes terrestrial. Lowlands, hill forest. Elevation: 400–1100 m.

General distribution: Sabah, Sarawak, Kalimantan, widespread in mainland & SE Asia east to New Guinea, Australia & the Pacific islands.

Collections. BUNDU TUHAN: 1100 m, *Brentnall 100* (K); DALLAS: 900 m, *Clemens s.n.* (BM), 900 m, *26301* (BM, K), 900 m, *26773* (BM, K), 900 m, *26781* (BM), 900 m, *27631* (BM, K); KAUNG: 400 m, *Carr 3012* (SING), 500 m, *Clemens 40458* (BM, K), 400 m, *Darnton 379* (BM).

55.89.5. Pholidota pectinata Ames, Orch. 6: 69, pl. 86 (1920). Type: MARAI PARAI SPUR, *Clemens 273* (holotype AMES mf; isotype BM!).

Epiphyte. Lower montane forest, probably on ultramafic substrate.

General distribution: Sabah, Kalimantan, Sarawak.

55.89.6. Pholidota ventricosa (Blume) Rchb. f., Bonplandia 5: 43 (1857).

Chelonanthera ventricosa Blume, Bijdr., 383 (1825).

Epiphyte, sometimes terrestrial. Lower montane forest, rarely lowlands. Elevation: 500–1700 m.

General distribution: Sabah, Kalimantan, Sarawak, Peninsular Malaysia, Sumatra, Java, Sulawesi, Philippines, New Guinea.

Collections. GURULAU SPUR: 1500 m, *Clemens 50450* (BM); KAUNG: 500 m, *Clemens 26022* (BM); MESILAU RIVER: 1500 m, *RSNB 4877* (K); PENATARAN BASIN: 1200 m, *Clemens 34445* (BM); PENIBUKAN: 1700 m, *Clemens 50290* (BM), 1700 m, *50364* (BM, K).

55.90. PHREATIA Lindl.

Gen. Sp. Orch., 63 (1830).

Epiphytic herbs. *Stems* pseudobulbous or caulescent, the former 1- to 3-leaved, the latter with up to 12 leaves, very short or elongate. *Leaves* distichous or arranged in a fan, erect to spreading, terminal, sometimes fleshy, jointed on equitant sheaths. *Inflorescences* lateral or arising from base of pseudobulb, racemose, densely many-flowered. *Flowers* minute, resupinate, pale green or white, often self-pollinating; *sepals* and *petals* similar, spreading; *lateral sepals* adnate to column-foot to form a mentum; *lip* entire or obscurely 3-lobed, usually without a spur, concave; *column* short, with a distinct foot; *pollinia* 8.

About 150 species distributed from mainland Asia, through SE Asia east to Australia, New Guinea and the Pacific islands, most species occurring in Indonesia and New Guinea.

55.90.1. Phreatia amesii Kraenzl., Pflanzenr. IV. 50. II. B. 23: 16 (1911).

Epiphyte. Lower montane forest. Elevation: 1200–1500 m.

General distribution: Sabah, Philippines.

Collections. MOUNT KINABALU: *Haslam s.n.* (AMES); TENOMPOK: 1200 m, *Clemens s.n.* (BM); TENOMPOK ORCHID GARDEN: 1500 m, *Clemens 50241* (BM).

55.90.2. Phreatia densiflora (Blume) Lindl., Gen. Sp. Orch., 64 (1830).

Dendrolirium densiflorum Blume, Bijdr., 350 (1825).

Epiphyte. Lower montane forest, sometimes on ultramafic substrate. Elevation: 1200–2100 m.

General distribution: Sabah, Sarawak, Peninsular Malaysia, Thailand, Sumatra, Java, Maluku, Philippines.

Collections. KILEMBUN BASIN: 1400 m, *Clemens 34238* (BM); KINATEKI RIVER: 2100 m, *Clemens 31731* (BM); LUBANG: 1800 m, *Gibbs 4123* (BM, K); MESILAU CAVE: 1800 m, *RSNB 4775* (K, SING); MURU-TURA RIDGE: 1700 m, *Clemens 34106* (BM); PENIBUKAN: 1200–1500 m, *Clemens 31451* (BM); TAHUBANG RIVER HEAD: 2100 m, *Clemens 32975* (BM, K, SING).

55.90.3. Phreatia listrophora Ridl., J. Linn. Soc. Bot. 32: 307 (1896).

Epiphyte. Lower montane forest. Elevation: 1200–1800 m.

General distribution: Sabah, Peninsular Malaysia, Thailand, Sulawesi, Seram.

Collections. EAST MESILAU/MENTEKI RIVERS: 1700 m, *Beaman 9390* (K); MARAI PARAI: 1700 m, *Clemens 31886* (BM), 1800 m, *32443* (BM); PENIBUKAN: 1200–1500 m, *Clemens 31382* (BM); TENOMPOK: 1500 m, *Clemens s.n.* (BM), 1500 m, *29464* (BM, K), 1500 m, *30139* (K); WEST MESILAU RIVER: 1600 m, *Bailes & Cribb 541* (K).

55.90.4. Phreatia monticola Rolfe in Gibbs, J. Linn. Soc. Bot. 42: 152 (1914). Type: KIAU, 900 m, *Gibbs 3959* (holotype BM!; isotype K!).

Epiphyte. Secondary hill forest, lower montane forest. Elevation: 900–1500 m.

General distribution: Sabah, Kalimantan.

Additional collections. BUNDU TUHAN: 1400 m, *Darnton 202* (BM); DALLAS: 900 m, *Clemens 30120* (K); KINUNUT VALLEY HEAD: 1200 m, *Carr SFN 27192* (SING); TENOMPOK: 1500 m, *Clemens 30140* (K).

55.90.5. Phreatia secunda (Blume) Lindl., Gen. Sp. Orch., 64 (1830). Plate 71C.

Dendrolirium secundum Blume, Bijdr., 350 (1825).

Phreatia minutiflora Lindl., J. Linn. Soc. Bot. 3: 62 (1859).

Epiphyte or lithophyte. Lower montane forest, sometimes on ultramafic substrate. Elevation: 900–2000 m.

General distribution: Sabah, Kalimantan, Sarawak, Peninsular Malaysia, Thailand, Vietnam, Sumatra, Java, Sulawesi, Mentawai, Seram, Philippines.

Collections. BUNDU TUHAN: 1400 m, *Darnton 201* (BM); DALLAS: 900 m, *Clemens 27634* (BM, K); EAST MESILAU RIVER: 2000 m, *Collenette 897* (K); GURULAU SPUR: *Clemens s.n.* (BM); KILEMBUN BASIN: 1400 m, *Clemens 34398* (BM); MARAI PARAI: 1500 m, *Clemens 33006* (BM); MOUNT KINABALU: *Haslam s.n.* (AMES); PARK HEADQUARTERS: 1700 m, *Bailes & Cribb 755* (K); TENOMPOK: 1500 m, *Clemens 27685* (BM, K), 1500 m, *30140* (K).

55.90.6. Phreatia sulcata (Blume) J. J. Sm., Orch. Java, 505 (1905).

Dendrolirium sulcatum Blume, Bijdr., 347 (1825).

Epiphyte. Lower montane forest. Elevation: 1100–1800 m.

General distribution: Sabah, Kalimantan, Peninsular Malaysia, Thailand, Sumatra, Java, Bali, Maluku, Seram, Philippines.

Collections. BUNDU TUHAN: 1400 m, *Darnton 213* (BM); LUMU-LUMU: 1800 m, *Clemens 29124* (BM); MARAI PARAI: 1200 m, *Clemens 32782* (BM); TAHUBANG RIDGE: 1100 m, *Clemens 40957*

(BM); TENOMPOK: 1500 m, *Clemens s.n.* (BM), 1500 m, *28897* (BM), 1500 m, *29234* (BM, K), 1500 m, *30137* (K), 1500 m, *30156* (K); WEST MESILAU RIVER: 1600 m, *Bailes & Cribb 540* (K).

55.90.7. Phreatia sp. 1

Epiphyte. Lower montane forest. Elevation: 1800 m.

Collection. MESILAU RIVER: 1800 m, *Clemens 51149* (BM, K).

55.90.8. Phreatia sp. 2

Epiphyte. Lower montane forest. Elevation: 1200–1500 m.

Collection. PENIBUKAN: *Clemens 31456* (BM, K).

55.90.9. Phreatia indet.

Collections. BUNDU TUHAN: 1200 m, *Carr 3324* (SING), 1400 m, *SFN 26973* (SING), 1200 m, *SFN 27192A* (SING), 600 m, *SFN 27898* (SING), 900 m, *Darnton 523* (BM); KADAMAIAN RIVER: 1800 m, *Carr SFN 27593* (SING); KILEMBUN RIVER: 900 m, *Clemens 35201* (BM); KINUNUT VALLEY HEAD: 1200 m, *Carr SFN 27206* (SING), 1400 m, *SFN 27347* (SING); MAHANDEI RIVER: 1100 m, *Carr SFN 26828* (SING); MAHANDEI RIVER HEAD: 1400 m, *Carr SFN 26396* (SING); MARAI PARAI: 1800 m, *Clemens 32442* (BM); MINITINDUK RIVER: 900 m, *Carr SFN 26635* (SING); MT. NUNGKEK: 1500 m, *Clemens 32877* (BM); PENATARAN BASIN: 1400 m, *Clemens 34054* (BM); PENIBUKAN: 1200 m, *Clemens 30779* (BM); TENOMPOK: 1500 m, *Carr 3570, SFN 27749A* (SING), 1500 m, *SFN 26892* (SING), 1600 m, *SFN 27019* (SING), 1500 m, *SFN 27749* (SING), 1600 m, *SFN 27791* (SING), 1500 m, *Clemens s.n.* (BM); TENOMPOK ORCHID GARDEN: 1500 m, *Clemens 50180* (BM).

55.91. PILOPHYLLUM Schltr.

Orchideen, 131 (1914).

Terrestrial herb. *Rhizomes* short, creeping. *Pseudobulbs* alternately 1–4 bearing a solitary leaf and 1 bearing an inflorescence, clearly articulated at junction with petiole or scape. *Leaf* large, convolute, erect, ovate, acuminate, brown-hairy. *Inflorescence* erect, racemose, laxly many-flowered; peduncle, rachis and bracts brown-hairy. *Flowers* non-resupinate, few open at a time, glabrous, tepals yellow with dark red stripes, lip white spotted with purple; *sepals* and *petals* free, spreading; *lateral sepals* and *petals* falcate; *lip* 3-lobed, side-lobes oblong, spreading, mid-lobe narrowed from a broad base and expanded abruptly into a reniform, apiculate epichile; *column* with lateral appendages on its front edge, with a distinct foot; *pollinia* 2.

A monotypic genus from Peninsular Malaysia, Java, Borneo, the Philippines, New Guinea and the Solomon Islands.

55.91.1. Pilophyllum villosum (Blume) Schltr., Orchideen, 131 (1914). Plate 72A.

Chrysoglossum villosum Blume, Bijdr., 338 (1825).

Terrestrial. Lower montane forest. Elevation: 1200–1500 m.

General distribution: Sabah, Peninsular Malaysia, Java, Seram, Philippines, New Guinea, Solomon Islands.

Collections. GURULAU SPUR: 1500 m, *Clemens 50377* (BM), 1500 m, *51023* (BM, K); MT. NUNGKEK: 1200 m, *Darnton 482* (BM); PENIBUKAN: 1200 m, *Clemens 30696* (BM), 1500 m, *50240* (BM).

55.92. PLATANTHERA Rich.

Mém. Mus. Hist. Nat. 4: 48 (1818).

Terrestrial herbs arising from tubers or a short rhizome. *Stem* erect, bearing a few leaves, often near the ground, sheathed at the base. *Leaves* thin, usually broad, elliptic, not jointed at base, uppermost bract-like. *Inflorescence* terminal, usually many-flowered, lax or dense. *Flowers* white or green, sometimes fragrant; *dorsal sepal* and *petals* usually forming a hood over column; *lateral sepals* spreading or reflexed; *lip* spurred, blade simple, horizontal to pendent; *column* short; *stigmas* 2, flat, joined across below rostellum, not freely extending in front of column; *pollinia* 2, separate, the caudicles enclosed in tubes (canals or thecae) separated by a rostellum.

About 85 species in Europe, temperate and tropical Asia, New Guinea, North Africa, North and Central America. A couple of species are found inside the Arctic Circle.

55.92.1. **Platanthera angustata** (Blume) Lindl., Gen. Sp. Orch., 290 (1835).

Mecosa angustata Blume, Bijdr., 404, t. 1 (1825).

Terrestrial. Upper montane forest, in moss under trees. Elevation: 2200 m.

General distribution: Sabah, Peninsular Malaysia & Thailand east to Indonesia, the Philippines & New Guinea north to Taiwan, Hong Kong & the Ryukyu Islands.

Collections. KEMBURONGOH: 2200 m, *Carr 3515, SFN 27521* (K, SING).

55.92.2. **Platanthera borneensis** (Ridl.) J. J. Wood, **comb. nov.**

Habenaria borneensis Ridl. in Stapf, Trans. Linn. Soc. Bot. 4: 240 (1894). Type: MOUNT KINABALU, 3000 m, *Haviland s.n.* (holotype unlocated).

Terrestrial. Upper montane forest. Elevation: 3000 m. Known only from the type.

Endemic to Mount Kinabalu.

55.92.3. Platanthera crassinervia (Ames & C. Schweinf.) J. J. Sm., Mitt. Inst. Allg. Bot. Hamburg 7: 12 (1927).

Habenaria crassinervia Ames & C. Schweinf., Orch. 6: 6 (1920). Type: PAKA-PAKA CAVE, *Clemens 221* (holotype AMES mf).

Terrestrial. Upper montane forest. Known only from the type.

Endemic to Mount Kinabalu.

55.92.4. Platanthera gibbsiae Rolfe in Gibbs, J. Linn. Soc. Bot. 42: 160 (1914). Type: PAKA-PAKA CAVE, 2700 m, *Gibbs 4258* (holotype K!).

Habenaria gibbsiae (Rolfe) Ames & C. Schweinf., Orch. 6: 8 (1920).

Terrestrial. Upper montane forest in open scrub, sometimes on ultramafic substrate. Elevation: 2700–3400 m.

Endemic to Mount Kinabalu.

Additional collections. LAYANG-LAYANG: 2900 m, *Lamb AL 586/86* (K); SHANGRI LA VALLEY: 3400 m, *Collenette 21503* (K).

55.92.5. Platanthera kinabaluensis Kraenzl. ex Rolfe in Gibbs, J. Linn. Soc. Bot. 42: 160 (1914). Plate 72B. Type: MOUNT KINABALU, *Haviland s.n.* (holotype K!).

Habenaria kinabaluensis (Kraenzl.) Ames & C. Schweinf., Orch. 6: 9 (1920).

Terrestrial. Upper montane forest, sometimes on ultramafic substrate. Elevation: 1500–4000 m.

Endemic to Mount Kinabalu.

Additional collections. GURULAU SPUR: 3700–4000 m, *Clemens 51194* (BM); MARAI PARAI: *Clemens 35199* (BM); MARAI PARAI SPUR: *Clemens 261* (AMES), 1600 m, *Lamb AL 49/83* (K); PAKA-PAKA CAVE: 2900 m, *Carr 3523, SFN 27533* (K, SING), 3100 m, *3549, SFN 27641C* (SING); SAYAT-SAYAT: 3400 m, *Carr 3523, SFN 27533A* (SING), 3200 m, *3549, SFN 27641* (K, SING); SUMMIT TRAIL: 3000 m, *Carr 3549, SFN 27641A* (SING), 2900 m, *3549, SFN 27641B* (SING); TENOMPOK: 1500 m, *Clemens 29782* (BM).

55.92.6. Platanthera saprophytica J. J. Sm., Mitt. Inst. Allg. Bot. Hamburg 7: 12, t. 1, f. 2 (1927).

Saprophyte. Lower montane forest on ultramafic substrate. Elevation: 1100–1700 m.

General distribution: Sabah, Kalimantan.

Collections. MARAI PARAI SPUR: 1500 m, *Clemens 33036* (BM); PENATARAN BASIN: 1700 m, *Clemens 40132* (BM); PENIBUKAN: 1100 m, *Carr 3114, SFN 26511* (SING).

55.92.7. Platanthera stapfii Kraenzl. ex Rolfe in Gibbs, J. Linn. Soc. Bot. 42: 160 (1914). Fig. 48, Plate 72C. Type: MARAI PARAI, 1500 m, *Haviland 1158* (holotype K!).

Habenaria stapfii (Kraenzl. ex Rolfe) Ames & C. Schweinf., Orch. 6: 9 (1920).

Terrestrial. Lower montane forest on ultramafic substrate, in damp open areas or between boulders. Elevation: 1500–1700 m.

Endemic to Mount Kinabalu.

Additional collections. MARAI PARAI: 1700 m, *Clemens 30808* (BM), 1500 m, *32228* (BM, K), 1500 m, *Collenette A 30* (BM); MARAI PARAI SPUR: 1700 m, *Bailes & Cribb 837* (K), 1500 m, *Carr 3126, SFN 26560* (K, SING); MESILAU CAVE TRAIL: *Wood 849* (K).

55.93. PLOCOGLOTTIS Blume

Bijdr., 380 (1825).

Terrestrial, rarely epiphytic, herbs. *Stems* slenderly pseudobulbous, 1-leaved, or elongated into a leafy stem. *Leaves* plicate, narrowly elliptic, sheathing at base, sometimes mottled yellow or, rarely, variegated with silver lines. *Inflorescences* lateral, from base of pseudobulb, erect; peduncle long; rachis bearing a succession of many flowers, a few open at a time; floral bracts small to long-acuminate, persistent. *Flowers* small to medium-sized, resupinate; *sepals* and *petals* similar, free, spreading, sepals pubescent on outer surface; *lateral sepals* adnate to column-foot forming a very short mentum; *lip* short, fleshy, entire to obscurely lobed, usually convex, adnate to sides and apex of column-foot to form a sac and usually with an elastic hinge that springs when lip is touched, apex often narrow and strongly reflexed; *column* rather short; *pollinia* 4.

About 35 species distributed from Peninsular Malaysia eastward to New Guinea and the Solomon Islands.

55.93.1. Plocoglottis acuminata Blume, Mus. Bot. Lugd. 1: 46 (1849). Plate 73A.

Terrestrial. Lowland dipterocarp forest. Elevation: 400–500 m.

General distribution: Sabah, Brunei, Kalimantan, Sarawak, Sumatra, Java, Philippines.

Collection. MEKEDEU RIVER: 400–500 m, *Beaman 9647* (K).

55.93.2. Plocoglottis borneensis Ridl., J. Straits Branch Roy. Asiat. Soc. 49: 33 (1908).

Terrestrial. Hill forest. Elevation: 500–600 m.

General distribution: Sabah, Brunei, Sarawak.

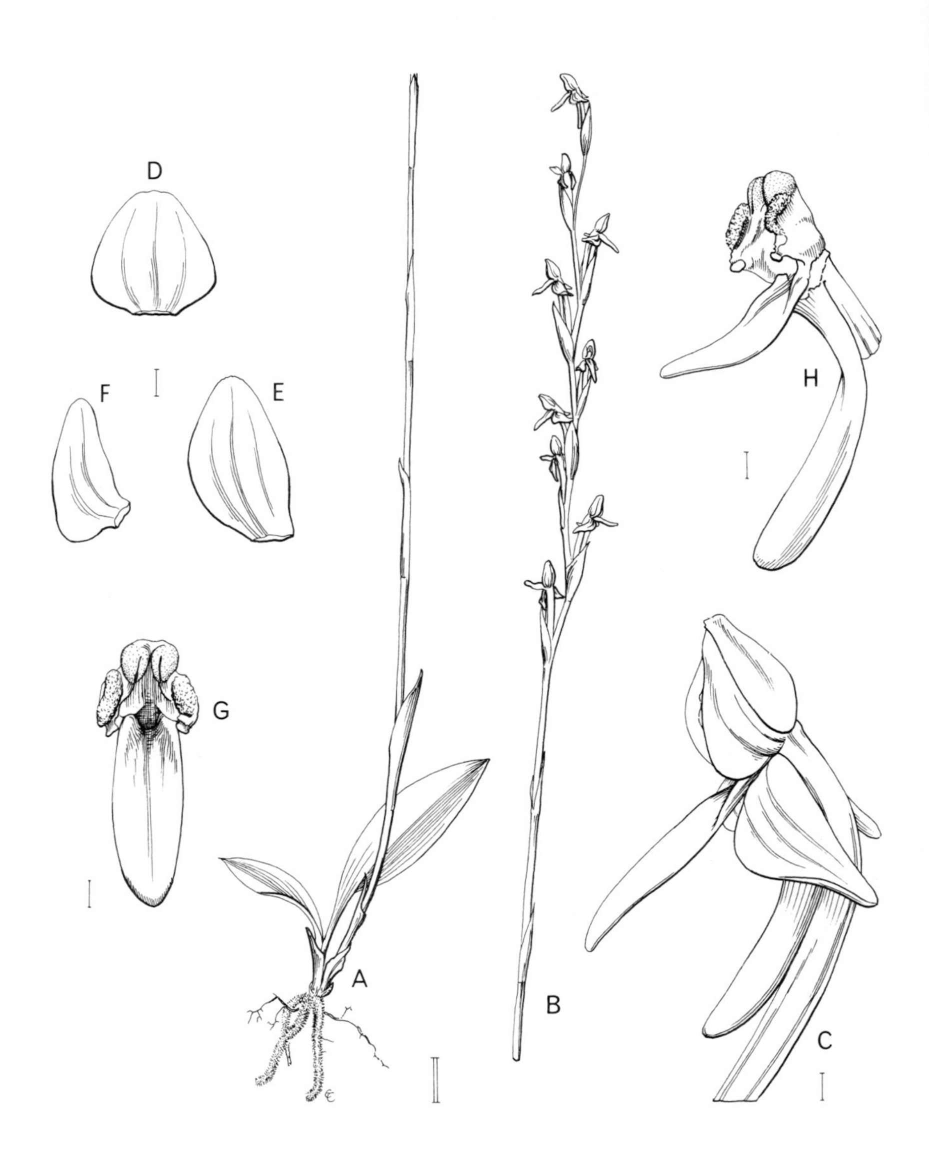

Fig. 48. Platanthera stapfii. **A**, habit; **B**, inflorescence; **C**, flower (side view); **D**, dorsal sepal; **E**, lateral sepal; **F**, petal; **G**, column and lip (front view); **H**, pedicel-with-ovary, lip, spur and column (side view). **A** from *Bailes & Cribb 837*; **B** from *Wood 849*; **C–H** from *Bailes & Cribb 837.* Scale: single bar = 1 mm; double bar = 1 cm. Drawn by Eleanor Catherine.

Collections. BUNDU TUHAN: 600 m, *Carr SFN 27901* (K, SING); PALUAN RIVER: 500 m, *Carr SFN 27391* (K).

55.93.3. Plocoglottis gigantea (Hook. f.) J. J. Sm., Feddes Repert. 32: 228 (1933). Plate 73B.

Calanthe gigantea Hook. f., Fl. Brit. Ind. 5: 856 (1890).

Terrestrial. Hill forest. Elevation: 900 m.

General distribution: Sabah, Peninsular Malaysia, Sumatra, Mentawai.

Collections. DALLAS: 900 m, *Clemens s.n.* (BM), 900 m, *26519* (BM); LANGANAN RIVER: *Lohok 10* (K).

55.93.4. Plocoglottis lowii Rchb. f., Gard. Chron. 1865: 434 (1865).

Terrestrial. Hill forest, mostly on ultramafic substrate. Elevation: 900–1000 m.

General distribution: Sabah, Sarawak, Andaman Islands, Peninsular Malaysia, Thailand, Sumatra, Maluku.

Collections. MELANGKAP TOMIS: 900–1000 m, *Beaman 8981* (K); MOUNT KINABALU: *Lamb SAN 92333* (K).

55.94. POAEPHYLLUM Ridl.

Mat. Fl. Malay. Penins. 1: 108 (1907).

Epiphytic, rarely terrestrial herbs. *Stems* simple, leafy, covered in sheathing leaf-bases and resembling *Appendicula. Leaves* narrow, distichous. *Inflorescences* lateral, axillary, usually short. *Flowers* minute to medium-sized, resupinate or non-resupinate; *sepals* acute, laterals adnate to column-foot forming a short mentum; *petals* often smaller and narrower than sepals; *lip* entire to 3-lobed, lacking a basal appendage and always adnate to the sides of the column-foot to form a sac; *column* with a short foot; *pollinia* 8.

Ten or more species distributed from the Nicobar Islands to Peninsular Malaysia, Sumatra, Java, Borneo, Sulawesi, the Philippines and New Guinea.

55.94.1. Poaephyllum pauciflorum (Hook. f.) Ridl., Mat. Fl. Malay. Penins. 1: 109 (1907).

Agrostophyllum pauciflorum Hook. f., Fl. Brit. Ind. 5: 824 (1890).

Epiphyte. Lowlands. Elevation: 400–600 m.

General distribution: Sabah, Sarawak, Peninsular Malaysia, Sumatra, Java, Philippines.

Collections. KAUNG: 400 m, *Carr SFN 27307* (SING), 400 m, *Darnton 359* (BM); NEAR RANAU: 600 m, *Lamb AL 442/85* (K).

55.94.2. Poaephyllum cf. **podochiloides** Ridl., Trans. Linn. Soc. Bot. 9: 192 (1916).

Epiphyte or lithophyte. Hill forest on ultramafic substrate. Elevation: 700–1000 m.

Collections. LOHAN RIVER: 700–900 m, *Beaman 9262* (K); MELANGKAP TOMIS: 900–1000 m, *Beaman 8983* (K).

55.94.3. Poaephyllum sp. 1

Epiphyte? Hill forest. Elevation: 1100 m.

Collection. PENIBUKAN: 1100 m, *Clemens 40955* (BM, K).

55.94.4. Poaephyllum sp. 2

Epiphyte. Lower montane forest on ultramafic substrate. Elevation: 1400 m.

Collection. PINOSUK PLATEAU: 1400 m, *Beaman 10710* (K).

55.94.5. Poaephyllum indet.

Collection. PENIBUKAN: 1100 m, *Clemens s.n.* (BM).

55.95. PODOCHILUS Blume

Bijdr., 295 (1825).

Small, rather delicate epiphytes or lithophytes, often forming dense mats. *Stems* slender, tufted, leafy, erect to spreading. *Leaves* short, distichous, lying in one plane by the twisting of the sheathing bases, not jointed on the sheaths. *Inflorescences* terminal, lateral or both, rarely exceeding 2 cm long, few- to many-flowered. *Flowers* minute or small, resupinate, white or green, often with purple markings; *sepals* and *petals* adnate at base, or free. *Lateral sepals* adnate to column-foot forming an often spur-like mentum; *petals* smaller than sepals; *lip* narrow, entire or obscurely 3-lobed, with a simple or bilobed basal appendage; *column* short, foot long and often curved upward; *pollinia* 4.

About 60 species distributed from India and Sri Lanka to China, south and east through Indonesia and New Guinea to the Pacific islands.

55.95.1. Podochilus lucescens Blume, Bijdr., 295, t. 12 (1825). Plate 73C.

Epiphyte. Hill forest, lower montane forest. Elevation: 900–1500 m.

General distribution: Sabah, Brunei, Kalimantan, Sarawak, Burma, Peninsular Malaysia, Thailand, Sumatra, Java, Sulawesi, Anambas Islands, Lingga Archipelago, Natuna Islands, Philippines.

Collections. KILEMBUN RIVER: 900 m, *Clemens 35167* (BM); PENIBUKAN: 1200–1500 m, *Clemens s.n.* (K), 1500 m, *30583* (BM, K), 1200 m, *30585* (BM), 1500 m, *50323* (K), 1500 m, *51715* (BM); TAHUBANG RIVER HEAD: 1100 m, *Carr 3095, SFN 26467* (SING).

55.95.2. Podochilus aff. **lucescens** Blume, Bijdr., 295, t. 12 (1825).

Epiphyte. Lower montane forest on ultramafic substrate. Elevation: 1400–1700 m.

Collections. MAMUT COPPER MINE: 1600–1700 m, *Beaman 9950* (K); PINOSUK PLATEAU: 1400 m, *Beaman 10715* (K).

55.95.3. Podochilus microphyllus Lindl., Gen. Sp. Orch., 234 (1833).

Epiphyte or lithophyte. Lower montane forest. Elevation: 900–1400 m.

General distribution: Sabah, Brunei, Kalimantan, Sarawak, Burma, Peninsular Malaysia, Thailand, Cambodia, Vietnam, Sumatra, Java, Lingga Archipelago.

Collections. KILEMBUN RIVER: 900 m, *Clemens 35167* (BM); LITTLE MAMUT RIVER: 1400 m, *Collenette 1010* (K); MOUNT KINABALU: *Clemens s.n.* (BM); PENIBUKAN: 1400 m, *Clemens 40522* (BM).

55.95.4. Podochilus sciuroides Rchb. f., Bonplandia 5: 41 (1857).

Epiphyte. Hill forest, lower montane forest, mostly on ultramafic substrate. Elevation: 800–1500 m.

General distribution: Sabah, Kalimantan, Sarawak, Sumatra, Riau Archipelago, Sumbawa.

Collections. LOHAN RIVER: 800–1000 m, *Beaman 10013* (K); MARAI PARAI SPUR: *Clemens 225A* (AMES), *253* (AMES); PENIBUKAN: 1200 m, *Carr 3047, SFN 26347* (SING); TAHUBANG RIVER HEAD: 1100 m, *Carr 3096, SFN 26468* (SING); TENOMPOK: 1500 m, *Carr SFN 26347A* (SING).

55.95.5. Podochilus serpyllifolius (Blume) Lindl., J. Linn. Soc. Bot. 3: 37 (1859).

Cryptoglottis serpyllifolia Blume, Bijdr., 297 (1825).

Epiphyte. Lower montane mossy forest. Elevation: 1500–1800 m.

General distribution: Sabah, Brunei, Sarawak, Sumatra, Java.

Collection. DAPATAN/PENIBUKAN RIDGES: 1500–1800 m, *Gibbs 4060* (BM).

55.95.6. **Podochilus tenuis** (Blume) Lindl., Gen. Sp. Orch., 235 (1833).

Apista tenuis Blume, Bijdr., 296 (1825).

Epiphyte or lithophyte. Lower montane forest, rarely hill forest, occasionally on ultramafic substrate, on tree trunks with moss. Elevation: 800–1900 m.

General distribution: Sabah, Kalimantan, Sarawak, Peninsular Malaysia, Sumatra, Java.

Collections. GURULAU SPUR: 1500 m, *Clemens 50382* (BM, K); KIAU VIEW TRAIL: 1500–1600 m, *Lamb AL 2/82* (K); KILEMBUN RIVER: 1500–1800 m, *Clemens 34486* (BM, K); LOHAN RIVER: 800–1000 m, *Beaman 10012* (K); MARAI PARAI SPUR: *Clemens 225* (AMES, BM, K, SING), *295* (AMES, BM, K, SING), *405* (AMES), *Topping 1485* (BM, K); MESILAU CAVE TRAIL: 1700–1900 m, *Beaman 8004* (K, MSC); MOUNT KINABALU: 1800 m, *Kidman Cox 2535* (K); MT. LENAU (BETWEEN TENOMPOK AND KEMBURONGOH): 1600 m, *Sinclair et al. 9009* (K, SING); PENIBUKAN: 1200 m, *Clemens 30588* (BM), 1200 m, *30777* (BM), 1500 m, *31008* (BM, K); TENOMPOK: 1500 m, *Clemens 26844* (BM, K), 1500 m, *27363* (BM, K), 1600 m, *Darnton 575* (BM); ULAR HILL TRAIL: *Collenette 3214* (K); WEST MESILAU RIVER: 1600–1700 m, *Beaman 8668* (K), 1600 m, *9034* (K).

55.95.7. **Podochilus sp. 1**

Lithophyte. Hill forest on ultramafic substrate. Elevation: 1200 m.

Collection. HEMPUEN HILL: 1200 m, *Madani SAN 89544* (K).

55.96. **POLYSTACHYA** Hook.

Exot. Fl. 2: t. 103 (1824).

Epiphytic or lithophytic herbs. *Stems* erect, often pseudobulbous or fusiform, simple or superposed, leafy. *Leaves* conduplicate, solitary to several. *Inflorescences* terminal, few- to many-flowered, racemose or paniculate; peduncle often enclosed in scarious sheaths. *Flowers* non-resupinate, mostly small; *sepals* connivent or spreading, free; *lateral sepals* adnate to column-foot; *petals* smaller, usually linear; *lip* uppermost, entire or 3-lobed, articulate with column-foot, disc often farinaceous; *column* short, with a distinct foot; *pollinia* 4.

About 200 species centred in Africa, with a couple of species in tropical America. One pantropical species is native to tropical Asia from Sri Lanka and India eastward to Sulawesi and the Philippines.

55.96.1. **Polystachya concreta** (Jacq.) Garay & Sweet, Orquideologia 9, 3: 206 (1974).

Epidendrum concretum Jacq., Enum. Syst. Pl., 30 (1760).

Epiphyte. Hill forest. Elevation: 900 m.

General distribution: Sabah; pantropical.

Collection. PINOSUK: 900 m, *Carr SFN 27197* (SING).

55.97. POMATOCALPA Breda

Gen. Sp. Orchid. Asclep., t. 15 (1829).

Small- to medium-sized monopodial epiphytes or terrestrials. *Stems* up to 40 cm. *Leaves* strap-shaped, to 20 × 5 cm. *Inflorescences* branched, often longer than the rest of the plant, on a long peduncle, many-flowered. *Flowers* small, non-resupinate, usually yellow marked with red; *sepals* and *petals* free, spreading; *lip* fleshy, 3-lobed, spurred or saccate, with a tongue-like, often bifurcate callus projecting from the back-wall; *column* short and stout; *rostellar projection* hammer-shaped; *pollinia* 4, appearing as 2 unequal masses.

Some 35–40 species distributed from India throughout tropical Asia to New Guinea, Australia and the Pacific islands.

55.97.1. Pomatocalpa sp. 1

Epiphyte. Hill dipterocarp forest. Elevation: 900 m.

Collection. LOHAN/MAMUT COPPER MINE: 900 m, *Beaman 10589* (K).

55.98. PORPAX Lindl.

Bot. Reg. 31: misc. 62, no. 66 (1845).

Tiny clump-forming epiphytic or lithophytic herbs. *Pseudobulbs* crowded together, flattened, covered by a sheath which disintegrates into a fine fibrous network or into radiating fibres. *Leaves* 2–3, often appearing after the flowers, oblong to elliptic, obtuse or acute, sometimes minutely hairy along margins. *Inflorescence* 1-flowered, borne either from base of pseudobulb, breaking through the sheath, or from apex of a developed pseudobulb; *peduncle* and *pedicel* very short, the flower appearing sessile at edge or centre of pseudobulb; floral bract conspicuous, enclosing lower part of flower. *Flowers* orange-red to deep dull red, sometimes flushed greenish-yellow; *dorsal sepal* connate with lateral sepals at least at its base; *lateral sepals* connate nearly, or completely, to apex, forming a tube and forming a mentum with column-foot; *petals* narrow, shorter than sepals; *lip* minute, much shorter than petals, usually recurved, obscurely 3-lobed, disc with a basal callus; *column* minute, with a large broad rostellum which more or less covers the stigma entrance; *pollinia* 8.

Eleven species distributed on mainland Asia from India through Thailand and Indochina to Peninsular Malaysia, and one outlying species in Borneo.

55.98.1. Porpax borneensis J. J. Wood & A. Lamb, **sp. nov.** Fig. 49, Plate 74A.

Porpaci ustulatae (Parish & Rchb. f.) Rolfe, species burmanica atque siamensis, affinis, sed sepalis minute papillosis nunquam hirsutis, labio margine irregulariter serrulato papillis pluribus basin instructo differt.

Type: EAST MALAYSIA, SABAH, SANDAKAN DISTRICT, GUNUNG TAWAI (TAVAI), SOUTH OF TELUPID, 600 m, August 1989, *Lamb AL 1164/89* (holotype K!, spirit material only).

Dwarf epiphytic or lithophytic herb. *Rhizome* creeping. *Pseudobulbs* 1–2-leaved, 0.8 × 6–1.4 × 1 cm, 3–4 mm thick, flattened, ovoid, often bilobed, when young covered by a sheath which disintegrates with age leaving the pseudobulb naked. *Leaves* 0.9 × 0.8–1.5 × 1(–4.5 × 1) cm, oblong or oblong-ovate, apex minutely retuse or asymmetric, glabrous. *Inflorescences* proteranthous, 1-flowered, borne from base of pseudobulbs, consisting of a very short scape 4 mm long, bearing 2 leaves above and 3 imbricate, ovate, acute, carinate sheaths 2–3 mm long below; *floral bract* 3 mm long, 3 mm wide at base, ovate, acute, carinate, apex reflexed and minutely serrate, translucent creamy-white. *Flower* resupinate, dark cardinal-red, sepals with pale yellow tips. *Sepals* 9–10 × 3 mm, connate at base, narrowly ovate-elliptic, acute, concave, minutely papillose on exterior. *Mentum* obtuse. *Petals* 1.5 × 8 mm, free, narrowly elliptic, acute, somewhat concave. *Lip* 4 mm long, 1.8–2 mm wide when flattened, entire, oblong-ligulate, obtuse, conduplicate, curved, margin irregularly serrulate, surface minutely papillose, especially along centre of undersurface, disc with several papillae of varying sizes toward base. *Column* minute, foot c. 2 mm long. *Anther-cap* ovate, cucullate, minutely papillose. *Pollinia* 8.

Porpax borneensis appears to be most closely related to *P. ustulata* (Parish & Rchb. f.) Rolfe from Burma and Thailand. It differs in having minutely papillose, never hirsute, sepals and an entire lip with an irregularly serrulate margin and several basal papillae.

Seidenfaden (1977: 1–14) recognised ten species, all confined to the Asian mainland, east as far as Indochina and south to Peninsular Malaysia. Six of these are represented in Thailand. *Porpax borneensis* brings the total number of species to eleven and also extends the range eastward. These plants are rarely collected on account of their tiny size; it seems likely that they may also be present in Sumatra and Java.

The specific epithet refers to the island of Borneo.

Low primary hill forest on ultramafic substrate, adpressed to rocks. Elevation: 600–1000 m.

General distribution: Sabah.

Additional collections: MT. TAWAI (TAVAI): 900 m, *Lamb AL 651/86* (K, sketch only); MELANGKAP TOMIS: 900–1000 m, *Beaman 8994a* (K).

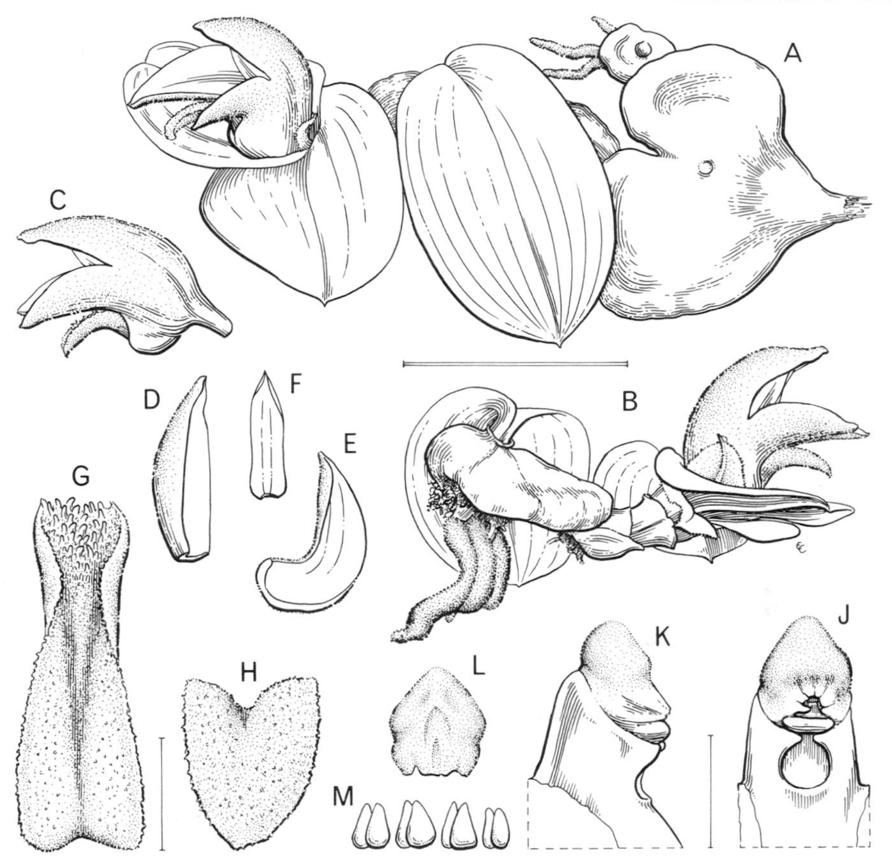

Fig. 49. Porpax borneensis. A and **B**, habit; **C**, flower (side view); **D**, dorsal sepal; **E**, lateral sepal; **F**, petal; **G**, lip (front view); **H**, lip apex; **J**, column and anther-cap (front view); **K**, column and anther-cap (side view); **L**, anther-cap (back view); **M**, pollinia. All from *Lamb AL 1164/89.* Scale: single bar = 1 mm; double bar = 1 cm. Drawn by Eleanor Catherine.

55.99. PORRORHACHIS Garay

Bot. Mus. Leafl. Harvard Univ. 23: 191 (1972).

Small monopodial epiphytes. *Stems* short or long. *Leaves* narrow, rigid, 2.5 × 0.6–15 × 1 cm. *Inflorescences* simple, stiff, perpendicular to the stem, sometimes longer than the leaves, with up to 10 lax flowers. *Flowers* very small, greenish or yellow; *sepals* 2.5–7 mm long; *dorsal sepal* incurved over the column; *lateral sepals* adpressed to the lip; *lip* to 3 mm long, immobile, fleshy, narrow, laterally compressed, with a spur-like tubular cavity; *column* short and stout, foot absent; *rostellar projection* very short; *stipes* broadly rhomboid, about as long as the pollinia; *viscidium* small; *pollinia* 2, entire.

Two species only, viz. *P. galbina* from Java and Borneo, and *P. macrosepala* (Schltr.) Garay from Sulawesi.

55.99.1. Porrorhachis galbina (J. J. Sm.) Garay, Bot. Mus. Leafl. Harvard Univ. 23, 4: 191 (1972). Fig. 50, Plate 74B, 74C.

Saccolabium galbinum J. J. Sm., Bull. Jard. Bot. Buit., ser. 2, 26: 97 (1918).

Epiphyte. Lower montane forest. Elevation: 1400–1500 m.

General distribution: Sabah, Java.

Collections. BUNDU TUHAN: 1400 m, *Darnton 199* (BM); TENOMPOK: 1500 m, *Clemens 29580* (BM), 1500 m, *30162* (K).

55.100. PRISTIGLOTTIS Cretz. & J. J. Sm.

Acta Fauna Fl. Universali, ser. 2, Bot. 1, no. 14: 4 (1934).

Terrestrial herbs. *Leaves* rather small, green or slightly coloured. *Inflorescence* short, with a few rather large flowers. *Flowers* resupinate, white; *dorsal sepal* and *petals* connivent, forming a hood; *lateral sepals* spreading; *lip* with a short saccate base enclosed by lateral sepals, containing 2 sessile glands, narrowed to a long channelled claw which expands into a bilobed blade; *rostellum* and *anther* long; *stigma* entire, ventral, below which are 2 narrow wings which project into base of lip; *pollinia* 2.

About 15 species distributed from India and China southeastward through Malaysia and Indonesia to the Pacific islands.

55.100.1. Pristiglottis hasseltii (Blume) Cretz. & J. J. Sm., Acta Fauna Fl. Universali, ser. 2, Bot. 1, no. 14: 4 (1934). Plate 75A.

Cystopus hasseltii Blume, Fl. Javae, ser. 2, 1, Orch.: 86, t. 30, f. 4, t. 36B (1859).

Terrestrial. Lower montane forest on mossy banks in shady mossy forest. Elevation: 1500 m.

General distribution: Sabah, Sarawak, Java.

Collections. KILEMBUN BASIN: *Clemens 32583* (BM); TENOMPOK: 1500 m, *Carr 3736, SFN 28055* (K, SING).

55.101. PTEROCERAS Hasselt ex Hassk.

Flora 25, Beibl.: 6 (1842).

Pedersen, H. A. (1992). The genus *Pteroceras* Hassk. (Orchidaceae) – a taxonomic revision. Cand. Scient. Thesis, Aarhus Univ.

Small, monopodial epiphytes. *Stems* short. *Leaves* few to about 10, apex usually unequally bilobed, to 10 × 1.5 cm. *Inflorescences* often longer, few- to

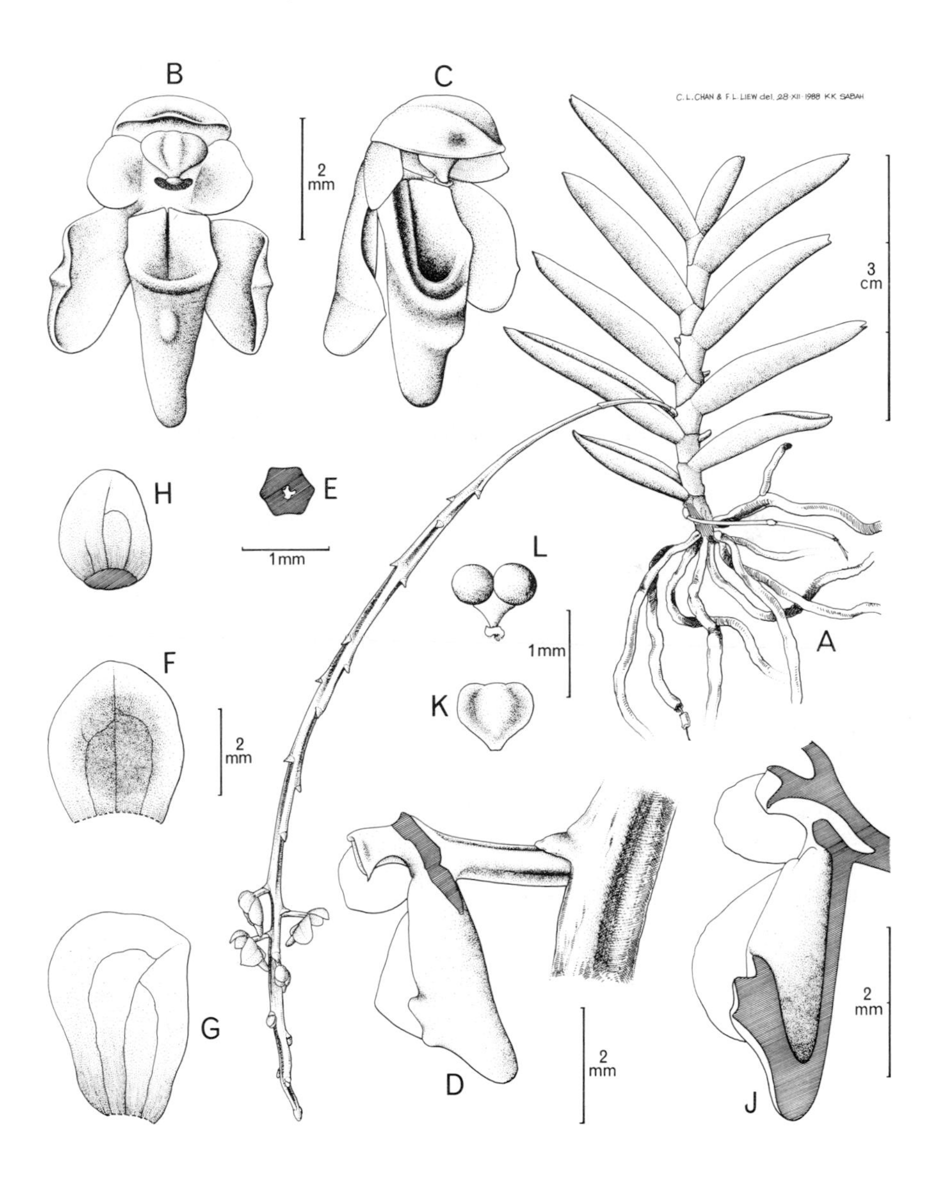

Fig. 50. Porrorhachis galbina. **A**, habit; **B**, flower (front view); **C**, flower (oblique view); **D**, rachis and flower with dorsal sepal, one lateral sepal and one petal removed; **E**, ovary (transverse section); **F**, dorsal sepal; **G**, lateral sepal; **H**, petal; **J**, flower (longitudinal section with dorsal sepal, one lateral sepal and one petal removed); **K**, anther-cap (back view); **L**, pollinarium. Crocker Range. Drawn by Chan Chew Lun and F. L. Liew.

many-flowered. *Flowers* small, to about 1.5 cm across, lasting one day; *sepals* and *petals* free, spreading; *lateral sepals* may be distinctly broader than petals, not adnate to the column-foot; *lip* 3-lobed, fleshy, mobile, hinged to the column-foot, with a spur or sac usually pointing forward in line with the foot, without interior ornaments, although the front wall may be quite fleshy, side-lobes usually large, erect, mid-lobe very short; *column* short and stout, foot long; *rostellar projection* oblong; *stipes* oblong; *viscidium* ± triangular; *pollinia* 2, sulcate.

Nineteen species distributed from NE India to Maluku. The centre of distribution is Borneo.

55.101.1. Pteroceras fragrans (Ridl.) Garay, Bot. Mus. Leafl. Harvard Univ. 23(4): 193 (1972). Plate 75B, 75C.

Sarcochilus fragrans Ridl., J. Straits Branch Roy. Asiat. Soc. 49: 38 (1907).

Epiphyte. Hill forest on ultramafic substrate. Elevation: 800 m.

General distribution: Sabah, Kalimantan, Sarawak.

Collections. LOHAN RIVER: 800 m, *Bailes & Cribb 676* (K), *Clements 3359* (K), *3385* (K).

55.101.2. Pteroceras leopardinum (Parish & Rchb. f.) Seidenf. & Smitin., Orch. Thailand 4: 535 (1963).

Thrixspermum leopardinum Parish & Rchb. f., Trans. Linn. Soc. London 30: 145 (1874).

Epiphyte. Hill forest, lower montane forest.

General distribution: Sabah, Burma, Thailand, Vietnam, Philippines.

Collection. LOHAN RIVER: *Clements 3367* (K).

55.101.3. Pteroceras spathibrachiatum (J. J. Sm.) Garay, Bot. Mus. Leafl. Harvard Univ. 23(4): 194 (1972).

Sarcochilus spathibrachiatus J. J. Sm., Blumea 5: 310 (1943). Type: KINATEKI RIVER HEAD, 1700 m, *Clemens s.n.* (holotype BO n.v.).

Epiphyte. Lower montane forest. Elevation: 1500–1700 m.

General distribution: Sabah.

Additional collections. BUNDU TUHAN: *Carr SFN 27804* (AMES); TENOMPOK/RANAU: 1500 m, *Carr SFN 27358* (AMES, SING).

55.101.4. Pteroceras teres (Blume) Holttum, Kew Bull. 14: 271 (1960). Plate 75D.

Dendrocolla teres Blume, Bijdr., 289 (1825).

Epiphyte. Lowlands, lower montane forest. Elevation: 500–1500 m.

General distribution: Sabah, Sarawak, Nepal, India, Burma, Thailand, Cambodia, Laos, Vietnam, Sumatra, Java, Sulawesi, Flores, Maluku, Sumbawa, Philippines.

Collections. KAUNG: 500 m, *Carr SFN 27981* (SING); KUNDASANG: *Kidman Cox 902* (L); LANGANAN RIVER: *Lohok 13* (K); PORING ORCHID GARDEN: *Lohok 20* (K); TENOMPOK: 1500 m, *Clemens 29761* (BM, K).

55.101.5. Pteroceras indet.

Collections. BUNDU TUHAN: 1200 m, *Carr SFN 27757* (SING), 800 m, *SFN 27863* (SING), 1200 m, *SFN 27988* (SING); DALLAS: 800 m, *Clemens s.n.* (SING); KAUNG: 400 m, *Carr SFN 27333* (SING); MAHANDEI RIVER: 1100 m, *Carr SFN 27778* (SING); MINITINDUK RIVER: 900 m, *Carr SFN 27189* (SING); TENOMPOK: 1500 m, *Carr SFN 27207* (SING), 1500 m, *SFN 27812* (SING).

55.102. RENANTHERA Lour.

Fl. Cochinch., 516, 521 (1790).

Robust monopodial epiphytes or, rarely, terrestrials with long, climbing stems, often up to several metres. *Leaves* oblong, branched, up to 80 cm long, many-flowered. *Flowers* usually red, sometimes orange or yellow; *lateral sepals* usually broader than the dorsal, edges undulate; *lip* much smaller than the sepals and petals, immobile, 3-lobed, saccate or spurred, mid-lobe small, often recurved, with basal calli, side-lobes erect; *column* short and stout, to 7 mm long; *rostellar projection* short; *viscidium* transverse, fleshy; *pollinia* 4, appearing as 2 pollen masses.

About 14 species distributed from E India (Assam) through China to the Philippines and south to Malaysia, Indonesia, New Guinea and the Solomon Islands.

55.102.1. Renanthera bella J. J. Wood, Orchid Rev. 89: 116, f. 97–99 (1981). Plate 76A, 76B. Type: MOUNT KINABALU, 1100 m, *Lamb SAN 89640* (holotype K!; isotype SAN!).

Epiphyte. Hill forest on ultramafic substrate, low on branches and trunks. Elevation: 800–1100 m.

General distribution: Sabah.

55.102.2. Renanthera matutina (Blume) Lindl., Gen. Sp. Orch., 218 (1833).

Aerides matutina Blume, Bijdr., 366 (1825).

Epiphyte. Lowland dipterocarp forest. Reported as *R. isosepala* Holttum by Lamb & Chan (1978: 236); no specimens seen.

General distribution: Sabah, Peninsular Malaysia, Sumatra, Java, Philippines.

55.103. ROBIQUETIA Gaudich.

In Freycinet, Voy. Uranie, 426 (1829).

Medium-sized monopodial epiphytes. *Stems* pendent, to 50 cm long. *Leaves* oblong to narrowly elliptic, apex unequally bilobed, to 20 × 5 cm. *Inflorescences* simple or branched, densely many-flowered, pendent, to 25 cm long. *Flowers* small, yellow marked with red or brownish, purple-red or pink; *sepals* and *petals* spreading; *lip* immobile, 3-lobed, with an often apically inflated spur which occasionally has some callosities or scales inside on either the back or front, or both, side-lobes small, sometimes fleshy; *column* short and stout, without a foot; *stipes* long, linear, spathulate, often uncinnate, rarely hamate; *viscidium* usually small; *pollinia* 2, sulcate.

About 40 species distributed from the Himalayan region to Australia and the Pacific islands, with most species in Indonesia.

55.103.1. Robiquetia crockerensis J. J. Wood & A. Lamb, **sp. nov.** Fig. 51, Plate 76C.

Robiquetiae pachyphyllae, species burmanica atque siamensis, affinis, sed foliis non crasse carnosis neque floribus translucentibus albido-viridibus, flavidisve nec purpureo-maculatis; differt etiam lobis lateralibus labelli promentioribus, calcari breviore ad pedicellum-cum-ovario parallelo, stipitatibus spathulatis.

Type: East Malaysia, Sabah, Mount Kinabalu, Mahandei River, 1050 m, 10 March 1933, *Carr 3107, SFN 26473* (holotype K!, herbarium material only; label indicates that duplicates were distributed to A, C, L, LAE, SING).

Epiphytic herb. *Stems* (2–)6–20 cm long, rooting at the base, spreading to pendulous. *Leaves* (4–)10–18 × 1.2–2.3 cm, ligulate, margins undulate, apex obtusely unequally bilobed, purple on the reverse, sheaths 1–2 cm long. *Inflorescence* simple and racemose, or with one branch, densely many-flowered; *peduncle* 1.5–4.5 cm long; *sterile bracts* 3–6 mm long, ovate, obtuse, amplexicaul; *rachis* 3–10 cm long, branch 2–9 cm long; *floral bracts* 3–4 mm long, triangular, acuminate. *Pedicel-with-ovary* 5–6 mm long, slender, straight, slightly ramentaceous, purple-green. *Flowers* unscented. *Sepals* and *petals*

Fig. 51. Robiquetia crockerensis. A, habit; **B**, small specimen; **C**, flower (back view); **D**, flower (oblique view); **E**, dorsal sepal; **F**, lateral sepal; **G**, petal; **H**, column and lip (side view); **J**, column and lip (longitudinal section); **K**, anther-cap (side view); **L**, stipes and viscidium (front and side views); **M**, pollinia. **A** and **B** from *Carr SFN 26473*; **C–M** from *Lamb AL 31/82*. Scale: single bar = 1 mm; double bar = 1 cm. Drawn by Eleanor Catherine.

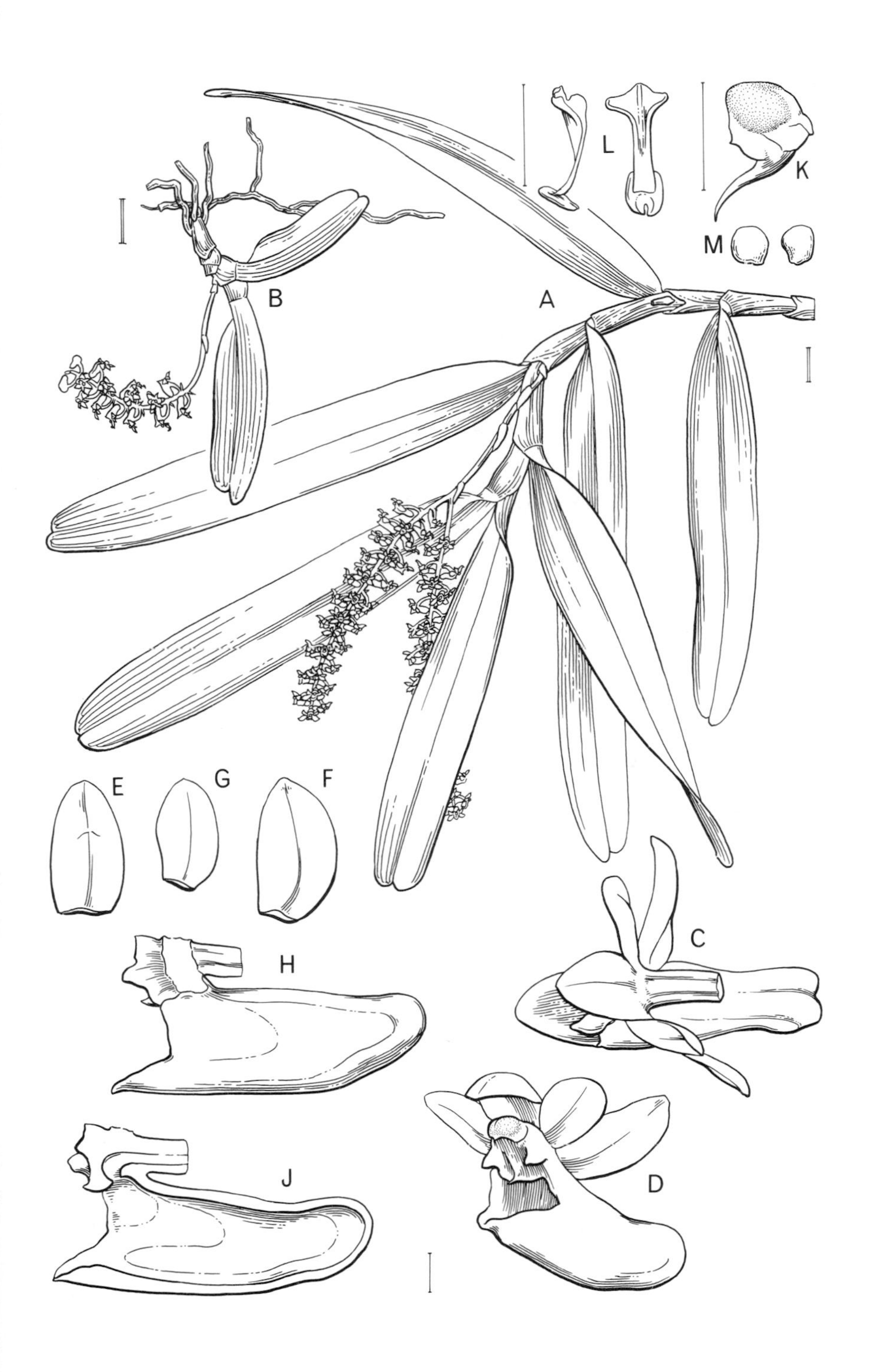

A
B
C
D
E
F
G
H
J
K
L
M

translucent greenish white or yellow, spotted with purple, lip and spur white with some purple spots, slightly ramentaceous at base. *Dorsal sepal* 3 × 2 mm, ovate, obtuse, strongly concave, curving over column. *Lateral sepals* 4 × 2 mm, slightly obliquely oblong-elliptic, obtuse, spreading. *Petals* 2–2.5 × 1–1.5 mm, oblong, obtuse, spreading. *Lip* fleshy, side-lobes 1.2–1.5 mm wide, irregularly triangular, acute, erect, clasping base of column, mid-lobe 1.5–1.6 mm long, 1.4–1.5 mm wide at base, oblong-ovate, obtuse, fleshier than side-lobes, flat, spur saccate, apex obtuse, slightly retuse, 4–5 mm long, 2–2.1 mm wide, parallel to, and with the apex sometimes touching the pedicel-with-ovary, interior unadorned. *Column* 1.1–1.5 mm long. *Anther-cap* extinctoriform, tail flat. *Pollinia* 2, cleft, pubescent. *Stipes* spathulate. *Viscidium* small.

Robiquetia crockerensis is related to *R. pachyphylla* (Rchb. f.) Garay from Burma and Thailand, but is at once distinguished by its thinner leaves, which are never thick and fleshy, and translucent greenish white or yellow flowers, which are spotted with purple. The lip has more prominent side-lobes and a shorter spur lying parallel to the pedicel-with-ovary, and the stipes is spathulate.

The specific epithet refers to the Crocker Range, which extends down the entire length of western Sabah and of which Mount Kinabalu is a part.

Hill forest, lower montane forest, sometimes on ultramafic substrate. Elevation: 800–1900 m.

General distribution: Sabah.

Additional collections: CROCKER RANGE, MT. ALAB, 1100 m, *Lamb AL 31/82* (K); GOLF COURSE SITE: 1600 m, *Bailes & Cribb 707* (K); HEMPUEN HILL: 800–1200 m, *Beaman 7717* (K); MARAI PARAI SPUR: 1900 m, *Collenette A 65* (BM); PENIBUKAN: 1200 m, *Clemens 32219* (BM), 1200 m, *35198* (BM); TENOMPOK: *Clemens 28554* (BM).

55.103.2. Robiquetia pinosukensis J. J. Wood & A. Lamb, sp. nov. Fig. 52.

Ab omnibus aliis speciebus huius generis apicibus foliorum perinaequaliter bilobatis, in apice quoque lobulo hoc usque ad 4 cm longo altero modo 0.2–0.4 cm longo dentiformi, vaginis foliorum purpureo-tinctis vel maculatis, calcari labelli late conoideo usque ad 3.5 mm longo, operculo anthera extinctoriformi, cauda elongata ad apicem sursum versa instructo distinguitor.

Type: EAST MALAYSIA, SABAH, MOUNT KINABALU, MESILAU TRAIL, 1950 m, 23 September 1972, *Chow & Leopold SAN 74504* (holotype K!, herbarium material only; label indicates that duplicates were distributed to A, L, SAN, SAR, SING).

Fig. 52. Robiquetia pinosukensis. **A**, habit; **B**, leaf apex; **C**, flower (oblique view); **D**, dorsal sepal; **E**, lateral sepal; **F**, petal; **G**, pedicel-with-ovary, column and lip (side view); **H**, pedicel-with-ovary, column and lip (side view, longitudinal section); **J**, anther-cap (side view); **K**, stipes and viscidium (side view); **L**, pollinia. **A** from *Chow & Leopold SAN 74504*; **B** from *Beaman 7227*; **C–L** from *Brentnall 156*. Scale: single bar = 1 mm; double bar = 1 cm. Drawn by Eleanor Catherine.

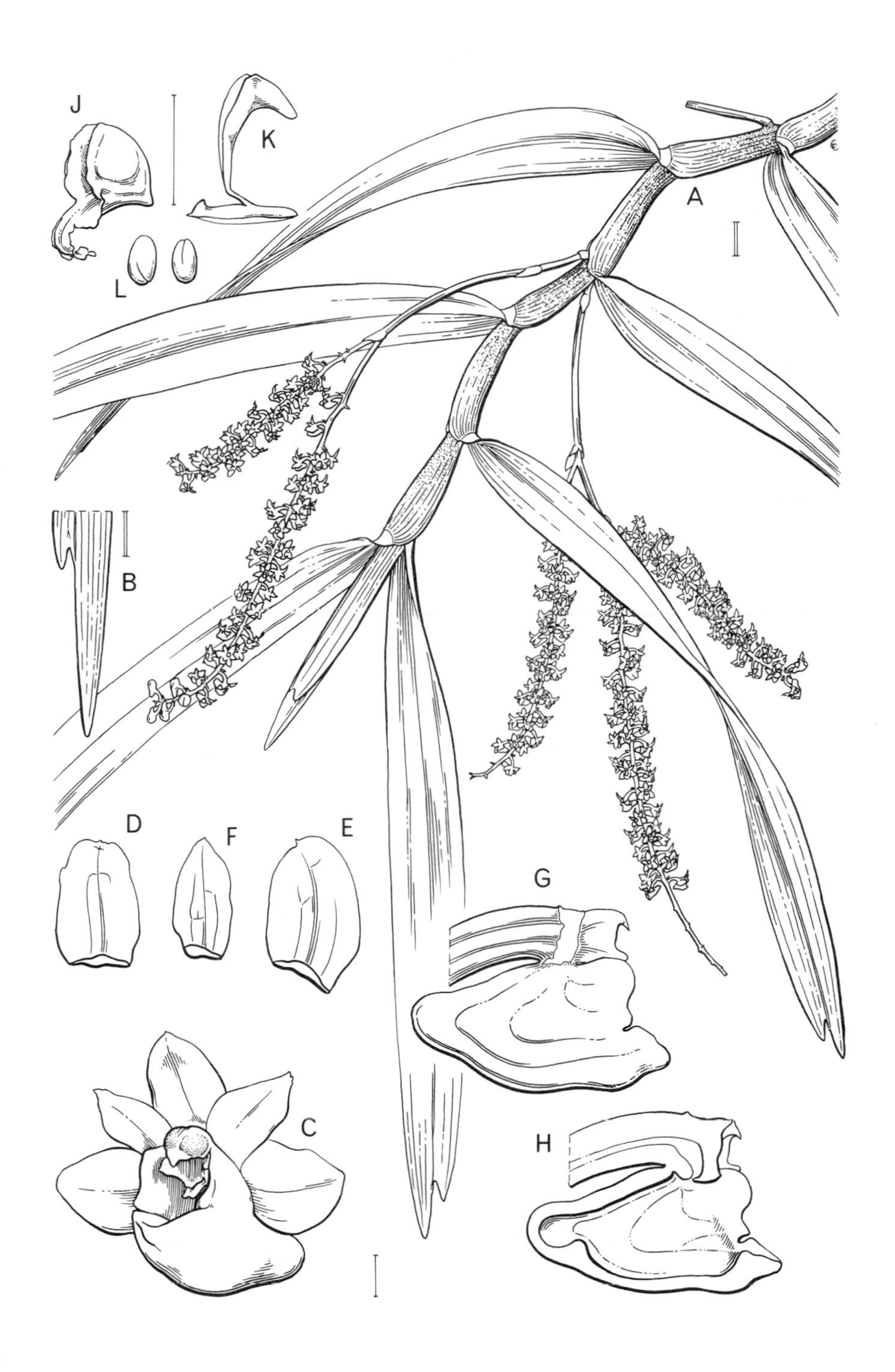
J
K
A
L
B
D
F
E
G
C
H

Epiphytic herb. *Stems* 25–40 cm long, producing long verruculose roots toward the base, spreading to pendulous. *Leaves* 13–28 × (1–)1.5–2.1 cm, ligulate, coriaceous, apex very unequally bilobed, one lobule up to 4 cm long, the other only 0.2–0.4 cm long and tooth-like, sheaths 2–3.5 cm long, usually stained or speckled purple. *Inflorescence* usually paniculate, with up to 3 branches, sometimes simple and racemose, densely many-flowered; *peduncle* 5–11.5 cm long; *sterile bracts* 2–3, remote, 0.4–1 cm long, tubular, obtuse, amplexicaul; *rachis* 8–15 cm long, branches 4.5–9 cm long; *floral bracts* 2–3 × 1.5 mm, ovate, aristate. *Flowers* unscented or smelling faintly of cucumber, variously described as having orange-yellow spotted red sepals and petals, lip pale yellow, speckled pale purple apically, spur apex greenish yellow, or greenish yellow spotted purple, or cream, spotted red, or orange-white, or lemon, spotted purple. *Pedicel-with-ovary* 3 mm long, slender, greatly curved. *Sepals* and *petals* spreading in mature flowers. *Dorsal sepal* 2–3 × 2 mm, ovate or oblong, obtuse, concave. *Lateral sepals* 4 × 2 mm, oblong, slightly oblique, apex obtuse and minutely erose. *Petals* 2–3 × 1 mm, oblong to elliptic, apex obtuse and minutely erose. *Lip* fleshy, lip wall swollen and callus-like at junction between side-lobes and mid-lobe; *side-lobes* very low, shallowly unequally bilobed, 1.5–2 mm wide, clasping base of column; *mid-lobe* 1–1.2 mm long, 1–1.2 mm wide at base when unflattened, triangular-ovate, subacute, concave and V-shaped in cross section; *spur* 3–3.5 mm long, 3 mm wide at base, saccate, broadly conical, ventral surface slightly constricted below middle, apex obtuse, not retuse, interior unadorned, borne at an acute angle to the pedicel-with-ovary. *Column* 0.5 mm long. *Anther-cap* extinctoriform, pointed at the top, elongated into a little tail which is upturned at the apex. *Pollinia* 2, cleft, glabrous. *Stipes* uncinate. *Viscidium* large.

Robiquetia pinosukensis is distinguished from all other species by its very unequally bilobed leaf tips with one lobule up to 4 cm long, the other only 0.2–0.4 cm long and tooth-like, and purple-stained or speckled leaf sheaths. The lip has a broadly conical spur to 3.5 mm long and the extinctoriform, i.e. candle snuffer-shaped, anther-cap has an elongated 'tail', which is upturned at the apex.

The specific epithet refers to the Pinosuk Plateau area of Mount Kinabalu, from where several specimens have been collected.

Lower montane forest. Elevation: 1500–2000 m.

Endemic to Mount Kinabalu.

Additional collections: EAST MESILAU RIVER: 1600–1700 m, *Bailes & Cribb 743* (K); GOLF COURSE SITE: 1700–1800 m, *Beaman 7227* (K), 1700 m, *Brentnall 156* (K); MESILAU: 1500 m, *Cockburn SAN 70120* (K); TENOMPOK: 1500 m, *Clemens 28141* (BM), 1500 m, *29252* (BM).

55.103.3. Robiquetia transversisaccata (Ames & C. Schweinf.) J. J. Wood, **comb. nov.** Fig. 53.

Malleola transversisaccata Ames & C. Schweinf., Orch. 6: 228 (1920). Type: KIAU, 900 m, *Clemens 166* (holotype AMES mf).

Epiphyte. Hill forest, lower montane forest, sometimes on ultramafic substrate. Elevation: 900–1500 m.

General distribution: Sabah.

Additional collections. KIAU: 900 m, *Clemens 67* (AMES); MARAI PARAI SPUR: *Clemens 246* (AMES); PENIBUKAN: 1500 m, *Clemens 50340* (BM).

55.103.4. Robiquetia indet.

Collections. BUNDU TUHAN: 1200 m, *Carr SFN 28026* (SING); MAHANDEI RIVER: 1100 m, *Carr SFN 27243* (SING).

55.104. SCHOENORCHIS Reinw.

In Hornschuh, Syll. Pl. Nov. 2: 4 (1825).

Small monopodial epiphytes. *Stems* with condensed or elongated internodes, up to 30 cm long. *Leaves* fleshy, either flat or semiterete, to 13 cm long. *Inflorescences* simple or branched, pendulous or horizontal, with many small white or red-purple flowers. *Flowers* usually not opening widely; *sepals* and *petals* free; *sepals* often dorsally keeled; *dorsal sepal* less than 2 mm long; *lip* spurred, very fleshy, longer than the sepals, 3-lobed; *column* very short, without a foot; *anther* and *rostellar projection* long, pointed and geniculate; *viscidium* narrowly elliptic to ovate; *pollinia* 4, appcaring as 2 unequal masses.

Christenson (1985) estimated some 24 species, others say around 15. These are distributed from the Himalayan region, S India and Sri Lanka to Hainan and the Philippines, Thailand, south to Indonesia and east to New Guinea, Australia and the Pacific islands.

55.104.1. Schoenorchis endertii (J. J. Sm.) Christenson & J. J. Wood, Lindleyana 5: 101 (1990). Plate 77A.

Robiquetia endertii J. J. Sm., Bull. Jard. Bot. Buit., ser. 3, 11: 153 (1931).

Epiphyte. Low-stature hill forest on ultramafic substrate. Elevation: 500–600 m.

General distribution: Sabah, Kalimantan.

Collections. LOHAN RIVER: 500–600 m, *Clements 3392* (K), *Lamb AL 433/85* (K).

55.104.2. Schoenorchis juncifolia Blume, Bijdr., 361 (1825).

Epiphyte. Known on Kinabalu only from a collection in Carr's Tenompok orchid garden. Elevation: 1500 m.

General distribution: Sabah, Sumatra, Java.

Collection. TENOMPOK ORCHID GARDEN: 1500 m, *Clemens 51057* (BM, K).

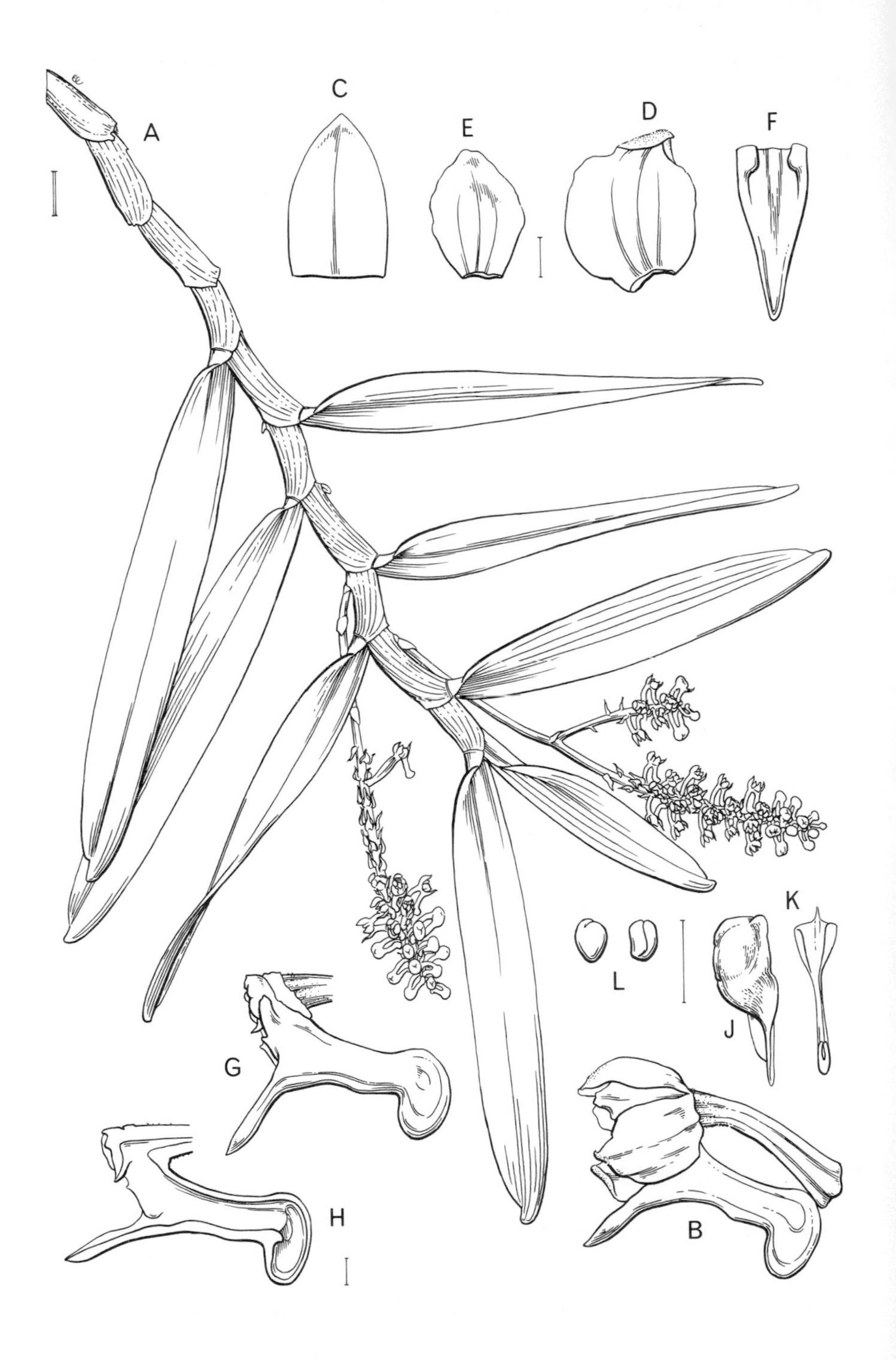
A
C
E
D
F
K
L
J
G
H
B

55.104.3. Schoenorchis micrantha Blume, Bijdr., 362 (1825).

Epiphyte. Hill forest, lower montane forest. Elevation: 800–1500 m.

General distribution: Sabah, Kalimantan, Sarawak, Peninsular Malaysia, Thailand, Vietnam, Sumatra, Java, Philippines, New Guinea east to Fiji.

Collections. BUNDU TUHAN: 800 m, *Carr SFN 27969* (SING); KIAU: 900 m, *Clemens 58* (AMES, BM, K); LUBANG: 1500 m, *Clemens 113* (AMES, BM, K, SING); MOUNT KINABALU: *Clemens s.n.* (AMES).

55.104.4. Schoenorchis aff. **micrantha** Blume, Bijdr., 362 (1825).

Epiphyte. Lower montane forest. Elevation: 1400–1500 m.

Collections. BUNDU TUHAN: 1400 m, *Darnton 212* (BM); TENOMPOK: 1500 m, *Clemens 28135* (BM, K), 1500 m, *28276* (BM, K), 1500 m, *29580* (BM, K), 1500 m, *30160* (K).

55.104.5. Schoenorchis paniculata Blume, Bijdr., 362 (1825).

Epiphyte. Hill forest on ultramafic substrate, on tree branches. Elevation: 500 m.

General distribution: Sabah, Sumatra, Java, Bali.

Collections. HEMPUEN HILL: *Madani SAN 89459* (K); LOHAN RIVER: 500 m, *Lamb AL 386/85* (K).

55.104.6. Schoenorchis indet.

Collection. BUNDU TUHAN: 900 m, *Carr SFN 27912* (SING).

55.105. SPATHOGLOTTIS Blume

Bijdr., 400 (1825).

Terrestrial herbs. *Pseudobulbs* conical to ovoid, sometimes depressed, covered with scarious sheaths. *Leaves* 1–2, lanceolate, plicate, sheathing at base. *Inflorescence* lateral, erect, racemose, many-flowered above, peduncle tall

---

Fig. 53. Robiquetia transversisaccata. A, habit; **B**, flower (side view); **C**, dorsal sepal; **D**, lateral sepal; **E**, petal; **F**, lip (with spur removed, front view); **G**, pedicel-with-ovary, column, lip and spur (side view); **H**, pedicel-with-ovary, column, lip and spur (side view, longitudinal section); **J**, anther-cap (side view); **K**, stipes and viscidium (front view); **L**, pollinia. **A** from *Clemens 50340*; **B–L** from *Lamb AL 196/84*, Crocker Range, Sabah. Scale: single bar = 1 mm; double bar = 1 cm. Drawn by Eleanor Catherine.

and slender. *Flowers* pedicellate, showy, resupinate; *sepals* and *petals* similar, free, erect to spreading; *petals* broader than sepals; *lip* strongly 3-lobed, side-lobes narrow, oblong, curving upward, mid-lobe spathulate, often emarginate, with a very narrow claw at the base of which are 2 small ovoid, often pubescent calli and 2 small laterally spreading teeth; *column* erect, incurved, slender, dilated above, without a foot; *pollinia* 8, in 2 groups of 4.

About 40 species distributed from tropical Asia to Australia and the Pacific islands.

55.105.1. Spathoglottis gracilis Rolfe ex Hook. f., Bot. Mag.: t. 7366 (1894). Plate 77B. Type: BORNEO, *Forstermann s.n.* (holotype K!).

Terrestrial. Hill forest, lower montane forest, on ultramafic substrate. Elevation: 800–1700 m.

General distribution: Sabah, Peninsular Malaysia.

Additional collections. BAMBANGAN RIVER: 1700 m, *RSNB 1323* (K); DAPATAN/MARAI PARAI SPURS: 1500 m, *Gibbs 4076* (BM); KINATEKI RIVER: 1200 m, *Haviland 1291* (K); KINATEKI RIVER/MARAI PARAI: 900–1500 m, *Collenette A 2* (BM); LOHAN RIVER: 800–1000 m, *Beaman 9070a* (K), 800–1000 m, *9340* (K); MAHANDEI RIVER HEAD: 1400 m, *Carr SFN 26397* (SING); MAMUT COPPER MINE: 1400–1500 m, *Beaman 10341* (K); MARAI PARAI: 1700 m, *Clemens 30809* (BM), 1400 m, *32361* (BM); MARAI PARAI SPUR: 1700 m, *Bailes & Cribb 827* (K); MOUNT KINABALU: 1400 m, *Lamb SAN 93371* (K); PENIBUKAN: 1500 m, *Clemens 31010* (BM), 1200–1500 m, *31514* (BM), 1500 m, *51718* (BM), 1400 m, *Lamb AL 45/83* (K).

55.105.2. Spathoglottis kimballiana Hook. f., Bot. Mag.: t. 7443 (1895).

Terrestrial. Hill forest, lower montane forest, on ultramafic substrate. Elevation: 600–1500 m.

General distribution: Sabah, Sarawak.

Collections. MAMUT COPPER MINE: 1400–1500 m, *Lamb SAN 93367* (K); PENATARAN RIDGE: 600 m, *Lamb SAN 93370* (K).

55.105.3. Spathoglottis microchilina Kraenzl., Bot. Jahrb. Syst. 17: 484 (1893). Fig. 54, Plate 77C.

Terrestrial. Hill forest, lower montane forest, on roadside banks, margins of secondary vegetation. Elevation: 900–1700 m.

General distribution: Sabah, Kalimantan, Sarawak, Peninsular Malaysia, Sumatra.

Collections. DALLAS: 900 m, *Clemens s.n.* (BM, BM), 900 m, *26767* (BM), 900 m, *28086* (BM), 900 m, *30097* (K); KIAU: 900 m, *Clemens 33* (AMES), 900 m, *150* (AMES), 1200 m, *Darnton 599* (BM), 900 m, *Gibbs 3957* (BM, K); MARAI PARAI: 1600 m, *Carr SFN 26554* (SING), 1500 m, *Clemens 32227* (BM), 1500 m, *Collenette A 24* (BM); MARAI PARAI SPUR: 1700 m, *Bailes & Cribb 828* (K), *Clemens 228* (AMES); MOUNT KINABALU: *Clemens s.n.* (AMES), *Haslam s.n.* (AMES); PARK HEADQUARTERS: 1500 m, *Tan & Phillipps SNP 579* (K); TENOMPOK: 1500 m, *Clemens s.n.* (BM), 1500 m, *28291* (BM, K).

Fig. 54. Spathoglottis microchilina. **A**, habit; **B**, basal part of plant and inflorescence; **C**, flower (oblique view); **D**, ovary (transverse section); **E**, dorsal sepal; **F**, lateral sepal; **G**, petal; **H**, lip, flattened (front view); **J**, pedicel-with-ovary, column and lip (longitudinal section); **K**, column (front view); **L**, anther-cap (back view); **M**, pollinia. All from *Chan 127*, Crocker Range, Sabah. Drawn by Chan Chew Lun and F. L. Liew.

55.105.4. Spathoglottis plicata Blume, Bijdr., 401, t. 76 (1825).

Terrestrial. Hill forest, roadside banks, open grassy areas, secondary vegetation. Elevation: 800–1200 m.

General distribution: Sabah, widespread from India to the Pacific islands.

Collections. DALLAS: 900 m, *Clemens s.n.* (BM), 900 m, *27617* (BM); DALLAS/TENOMPOK: *Clemens 26836* (BM); KADAMAIAN RIVER: 800 m, *Carr SFN 27093* (SING); KIAU: *Clemens 344* (AMES); LOHAN/MAMUT COPPER MINE: 1200 m, *Beaman 10659* (K).

55.105.5. Spathoglottis sp. 1

Terrestrial. Upper montane forest?

Collection. UPPER KINABALU: *Clemens 30021* (K).

55.105.6. Spathoglottis indet.

Collection. PARK HEADQUARTERS/POWER STATION: 1800 m, *Saikeh SAN 82763* (K).

55.106. SPIRANTHES Rich.

Mém. Mus. Hist. Nat. 4: 50 (1818).

Terrestrial herbs. *Roots* fleshy, fasciculate. *Stems* erect, rather short, leafy. *Leaves* narrow, conduplicate, fleshy. *Inflorescence* ± densely spicate, flowers arranged spirally on rachis. *Flowers* small, tubular, white or pink; *sepals* free; *dorsal sepal* porrect, connivent with petals, forming a hood; *lateral sepals* oblique, ± spreading; *lip* concave at base, containing 2 appendages, obscurely 3-lobed, undulate at apex; *column* short; *stigma* convex; *pollinia* 2.

A cosmopolitan genus of over 50 species, particularly well represented in the Americas.

55.106.1. Spiranthes sinensis (Pers.) Ames, Orch. 2: 53 (1908).

Neottia sinensis Pers., Syn. Pl. 2: 511 (1807).

Terrestrial. Hill forest, lower montane forest, in grassy areas, roadsides, open sunny places. Elevation: 900–1500 m.

General distribution: Sabah, probably all of Borneo, Asia east to Japan, southeast to Australia, New Zealand & the Pacific islands.

Collections. BUNDU TUHAN: 1400 m, *Carr 3315, SFN 27059* (BM, SING); DALLAS: 900 m, *Clemens 27117* (BM), 900 m, *28144* (BM, K, SING); KIAU: 900 m, *Clemens 140* (AMES), 900 m, *190* (AMES); MARAI PARAI SPUR: *Clemens 399A* (AMES); PARK HEADQUARTERS: 1500 m, *Lamb AL 1/82* (K); TENOMPOK: 1500 m, *Clemens 28258* (BM), 1500 m, *50821* (BM).

55.107. STEREOSANDRA Blume

Mus. Bot. Ludg. 2: 179 (1856).

Leafless saprophytic herbs. *Rhizome* tuberous. *Stems* erect. *Inflorescence* terminal, few- to many-flowered. *Flowers* resupinate; *sepals* and *petals* about equal, narrow; *lip* narrow, entire, with 2 basal glands; *column* short; *anther* erect, on a broad filament rising from the back of the column; *stigma* forming an erect bilobed structure with the rostellum on front of column; *pollinia* 2, powdery, with a caudicle.

About 5 species distributed from SE Asia to New Guinea and the Solomon Islands.

55.107.1. Stereosandra javanica Blume, Mus. Bot. 2: 176 (1856).

Saprophyte. Lowland dipterocarp forest. Sight record of A. Lamb (pers. comm.) near Kipungit Falls; no specimens seen.

General distribution: Sabah, Sarawak, Taiwan, Ryukyu Islands, Peninsular Malaysia, Thailand, Java, Philippines, New Guinea.

55.108. TAENIOPHYLLUM Blume

Bijdr., 355 (1825).

Small monopodial epiphytes or lithophytes with short stems bearing many long, spreading, flattened or terete greyish-green roots which are usually appressed to the substrate. *Leaves* reduced to tiny brown scales. *Inflorescences* lateral, with a short peduncle; *rachis* slowly elongating, bearing flowers in succession, 1 or 2 at a time; *floral bracts* alternate, in 2 ranks. *Flowers* small; *sepals* and *petals* free and widely spreading or connate at the base and not opening widely; *lip* fixed to the base of the column, spurred or saccate, entire or 3-lobed, often with an apical tooth or bristle; *column* short and stout, foot absent; *rostellar projection* very variable; *viscidium* usually narrowly elliptic or ovoid; *pollinia* 4, equal.

The function of the leaves is carried out by the numerous green roots.

A genus of 120–180 species distributed from tropical Africa (1 sp.) through tropical Asia to Australia and the Pacific islands. The centre of distribution is New Guinea, which has about 100 species.

55.108.1. Taeniophyllum esetiferum J. J. Sm., Bull. Jard. Bot. Buit., ser. 3, 11: 157 (1931).

Epiphyte. Hill forest. Elevation: 900 m.

General distribution: Sabah, Kalimantan, Peninsular Malaysia, Thailand.

Collection. BUNDU TUHAN: 900 m, *Carr SFN 27769* (AMES, SING).

55.108.2. Taeniophyllum gracillimum Schltr., Bull. Herb. Boissier, ser. 2, 6: 466 (1906).

Epiphyte. Lower montane forest. Elevation: 1200 m.

General distribution: Sabah, Peninsular Malaysia, Thailand.

Collection. BUNDU TUHAN: 1200 m, *Carr SFN 27419* (SING).

55.108.3. Taeniophyllum hirtum Blume, Bijdr., 356 (1825).

Epiphyte. Lower montane forest. Elevation: 1500 m.

General distribution: Sabah, Java.

Collections. TENOMPOK: 1500 m, *Clemens s.n.* (BM), 1500 m, *29376* (BM).

55.108.4. Taeniophyllum obtusum Blume, Bijdr., 357 (1825).

Epiphyte. Lowlands. Elevation: 400 m.

General distribution: Sabah, Sarawak, Peninsular Malaysia, Singapore, Thailand, Cambodia, Sumatra, Java.

Collection. KAUNG: 400 m, *Carr SFN 27313* (SING).

55.108.5. Taeniophyllum proliferum J. J. Sm., Bull. Jard. Bot. Buit., ser. 2, 26: 121 (1918).

Epiphyte. Hill forest. Elevation: 900 m.

General distribution: Sabah, Sumatra, Java.

Collection. MINITINDUK GORGE: 900 m, *Carr SFN 26798* (AMES, SING).

55.108.6. Taeniophyllum rubrum Ridl., J. Linn. Soc. Bot. 32: 363 (1896).

Epiphyte. Hill forest. Elevation: 800 m.

General distribution: Sabah, Peninsular Malaysia.

Collection. BUNDU TUHAN: 800 m, *Carr SFN 27810* (SING).

55.108.7. Taeniophyllum stella Carr, Gardens' Bull. 7: 69, t. 8B (1932).

Epiphyte. Hill forest on ultramafic substrate. Elevation: 1100 m.

General distribution: Sabah, Peninsular Malaysia.

Collections. KINATEKI RIVER: 1100 m, *Carr SFN 26802* (SING); LOHAN RIVER: *Lamb s.n.* (K photo).

55.108.8. Taeniophyllum sp. 1

Epiphyte on dead twigs. Low-stature hill forest, lower montane forest on ultramafic substrate. Elevation: 800–1500 m.

Collections. LOHAN RIVER: 800–1000 m, *Beaman 10015* (K); MAMUT COPPER MINE: 1400–1500 m, *Beaman 10352* (K).

55.108.9. Taeniophyllum indet.

Collections. BUNDU TUHAN: 1200 m, *Carr SFN 27408* (SING); PAKA-PAKA CAVE: 3100 m, *Carr SFN 27634* (SING), 2700 m, *SFN 27646* (SING); TENOMPOK: 1600 m, *Carr SFN 26985* (SING).

55.109. TAINIA Blume

Bijdr., 354 (1825).

Turner, H. (1992). A revision of the orchid genera *Ania* Lindley, *Hancockia* Rolfe, *Mischobulbum* Schltr. and *Tainia* Blume. Orchid Monogr. 6: 43–100 (1992).

Terrestrial herbs. *Shoots* arising from basal part of last pseudobulb; sterile shoots with 1 terminal leaf. *Pseudobulb* consisting of only 1 internode, erect, cylindrical to slightly ovoid. *Leaf* petiolate, elliptic to ovate, acute to acuminate. *Inflorescence* terminal, rarely lateral, erect, arising from basal part of terminal internode, rarely from subterminal internode of previous shoot. *Flowers* resupinate, glabrous; *sepals* and *petals* free, triangular to ovate, or elliptic to obovate; *lateral sepals* slightly decurrent along column-foot; *lip* without a spur, rarely obscurely saccate; *column* winged; *pollina* 8, rarely 6, in 4 pairs.

Fourteen species distributed in Sri Lanka, India, north to China, Hong Kong and Japan, south from Burma to New Guinea, Australia and the Pacific islands.

55.109.1. Tainia ovalifolia (Ames & C. Schweinf.) Garay & W. Kittr., Bot. Mus. Leafl. Harvard Univ. 30(3): 59 (1986).

Eulophia ovalifolia Ames & C. Schweinf., Orch. 6: 208 (1920). Type: KIAU, 900 m, *Clemens 93* (holotype AMES mf).

Terrestrial. Hill forest. Elevation: 900 m. Excluded from *Tainia* by Turner (1992) but not assigned elsewhere. Known only from the type.

Endemic to Mount Kinabalu.

55.109.2. Tainia paucifolia (Breda) J. J. Sm., Bull. Jard. Bot. Buit., ser. 2, 8: 5 (1912). Plate 77D.

Octomeria paucifolia Breda, Gen. Sp. Orchid. Asclep. t. 11 (1829).

Terrestrial. Hill forest. Elevation: 800 m.

General distribution: Sabah, Java.

Collection. MINITINDUK GORGE: 800 m, *Carr 3158* (SING).

55.109.3. Tainia purpureifolia Carr, Gardens' Bull. 8: 199 (1935). Type: GURULAU SPUR, 1400 m, *Carr 3150, SFN 26597* (holotype SING!; isotypes AMES mf, K!).

Terrestrial. Lower montane forest, in moss and humus in shade. Elevation: 1400 m.

General distribution: Sabah.

55.109.4. Tainia indet.

Collection. MESILAU RIVER: 2400 m, *Clemens 51625* (BM).

55.110. THECOPUS Seidenf.

Opera Bot. 72: 101 (1983).

Epiphytic herbs. *Pseudobulbs* of one node, one-leaved. *Leaves* oblong-elliptic, petiolate. *Inflorescences* lateral, arising from near base of pseudobulbs, racemose, several- to many-flowered, pendulous. *Flowers* large, resupinate; *sepals* and *petals* free; *petals* narrower than sepals; *lip* 3-lobed, joined at its base with an outgrowth from the column and with column-foot to form a tube at right angles to base of column, side-lobes erect, tips curving inward, mid-lobe acute, convex, hirsute, disc with or without keels; *column* arcuate, with spreading apical arms, foot hollow, with an entrance near the articulate lip base; *anther-cap* conical; *pollinia* 4, unequal, in pairs on long narrow stipes; *viscidium* obscure.

Two species distributed in Peninsular Malaysia, Thailand and Borneo.

55.110.1. Thecopus maingayi (Hook. f.) Seidenf., Opera Bot. 72: 101 (1983).

Thecostele maingayi Hook. f., Fl. Brit. Ind. 6: 20 (1890).

Epiphytic or terrestrial.

General distribution: Sabah, Peninsular Malaysia, Thailand.

Collections. MOUNT KINABALU: *Clemens s.n.* (AMES); PORING ORCHID GARDEN: *Cribb 89/61* (K).

55.111. **THECOSTELE** Rchb. f.

Bonplandia 5: 37 (1857).

Epiphytic herbs. *Pseudobulbs* of one node, one-leaved. *Leaves* oblong-elliptic or narrowly elliptic, shortly petiolate. *Inflorescences* lateral, arising from near base of pseudobulbs, racemose, many-flowered, pendulous. *Flowers* rather small, resupinate; *sepals* and *petals* free, spreading; *petals* narrower than sepals; *lip* 3-lobed, joined to the column in the same manner as in *Thecopus*, mid-lobe broad, convex, decurved, finely hairy, disc with 2 short keels between side-lobes; *column* curved into an S-shape, with a narrow nectary opening at base of lip on upper side of foot; *anther-cap* semiglobular; *pollinia* 2, cleft; *stipes* small; *viscidium* semicircular.

A monotypic genus distributed in Burma, Malaysia, Thailand and Indonesia to the Philippines.

55.111.1. Thecostele alata (Roxb.) Parish & Rchb. f., Trans. Linn. Soc. 30: 135 & 144, t. 29 (1874). Plate 78A.

Cymbidium alatum Roxb., Fl. Ind. 2, ed. 3: 459 (1832).

Epiphyte. Lowlands. Elevation: 400–500 m.

General distribution: Sabah, Sarawak, Kalimantan, India, Burma & Thailand east to the Philippines.

Collections. KAUNG: 400 m, *Carr 3009, SFN 26265* (SING), 400 m, *Darnton 306* (BM), 400 m, *307* (BM); MOUNT KINABALU: *Clemens s.n.* (AMES); RANAU: 500 m, *Darnton 176* (BM).

55.112. **THELASIS** Blume

Bijdr., 385 (1825).

Epiphytic herbs. *Stems* either pseudobulbous, 1- or 2-leaved, with sheaths and sometimes additional smaller leaves at base, or short and unthickened with several leaves in 2 close opposite ranks, laterally compressed and overlapping at base. *Leaves* narrow, rather thin. *Inflorescence* lateral, many-flowered. *Flowers* very small, greenish yellow or white, resupinate. *Sepals* and *petals* similar, only spreading at the apex; *lip* usually broadest and concave at the base, narrowed at apex; *column* short, without a foot; *pollinia* 8, in two groups of 4.

About 20 species distributed from India eastward to New Guinea and the Pacific islands.

55.112.1. Thelasis capitata Blume, Bijdr., 386 (1825).

Epiphyte on dipterocarps. Hill forest on ultramafic substrate. Elevation: 600 m.

General distribution: Sabah, Kalimantan, Peninsular Malaysia, Sumatra, Java, Seram, Philippines, Christmas Island.

Collection. LOHAN RIVER: 600 m, *Lamb SAN 91515* (K).

55.112.2. Thelasis carinata Blume, Bijdr., 386 (1825).

Epiphyte. Lowlands, hill forest. Elevation: 400–900 m.

General distribution: Sabah, Peninsular Malaysia, Sumatra, Java, Sulawesi, Bali, Buru, Philippines, New Guinea.

Collections. DALLAS: 900 m, *Clemens 27726* (BM); KAUNG: 400 m, *Carr SFN 27318* (SING), 500 m, *Clemens 26021* (BM).

55.112.3. Thelasis carnosa Ames & C. Schweinf., Orch. 6: 204 (1920). Type: KIAU, 900 m, *Clemens 88* (holotype AMES mf).

Epiphyte, one record on a dead tree. Hill forest, lower montane forest. Elevation: 900 m.

General distribution: Sabah.

Additional collections. MESILAU TRAIL: *Chow & Leopold SAN 74539* (K); MINITINDUK RIVER: 900 m, *Carr 3167, SFN 26638* (SING).

55.112.4. Thelasis micrantha (Brongn.) J. J. Sm., Orch. Java, 495, f. 374 (1905).

Oxyanthera micrantha Brongn. in Duperrey, Voy. Monde Phan.: 198, t. 37B (1834).

Epiphyte. Hill forest on ultramafic substrate, lower montane forest. Elevation: 500–1600 m.

General distribution: Sabah, Brunei, Kalimantan, Sarawak, Burma, Peninsular Malaysia, Thailand, Vietnam, Sumatra, Java, Philippines, New Guinea.

Collections. GOLF COURSE SITE: 1600 m, *Bailes & Cribb 721* (K); PENATARAN RIVER: 500 m, *Beaman 8857* (K).

55.112.5. Thelasis variabilis Ames & C. Schweinf., Orch. 6: 205 (1920). Fig. 55. Type: KIAU, 900 m, *Clemens 84* (holotype AMES mf).

Epiphyte. Hill forest, lower montane forest. Elevation: 800–1600 m. Its relationship with *T. macrobulbon* Ridl. from Peninsular Malaysia and Sarawak needs investigation.

General distribution: Sabah.

Additional collections. DALLAS: 800 m, *Clemens s.n.* (BM, BM), 900 m, *26309* (BM); LUBANG: 1500 m, *Clemens 109* (AMES); MT. NUNGKEK: 800 m, *Clemens 32907* (BM); PARK HEADQUARTERS: 1600 m, *Lamb AL 159/83* (K); PENIBUKAN: 1200 m, *Clemens 30641* (BM), 1200 m, *32060* (BM); TENOMPOK: 1500 m, *Clemens s.n.* (BM, BM).

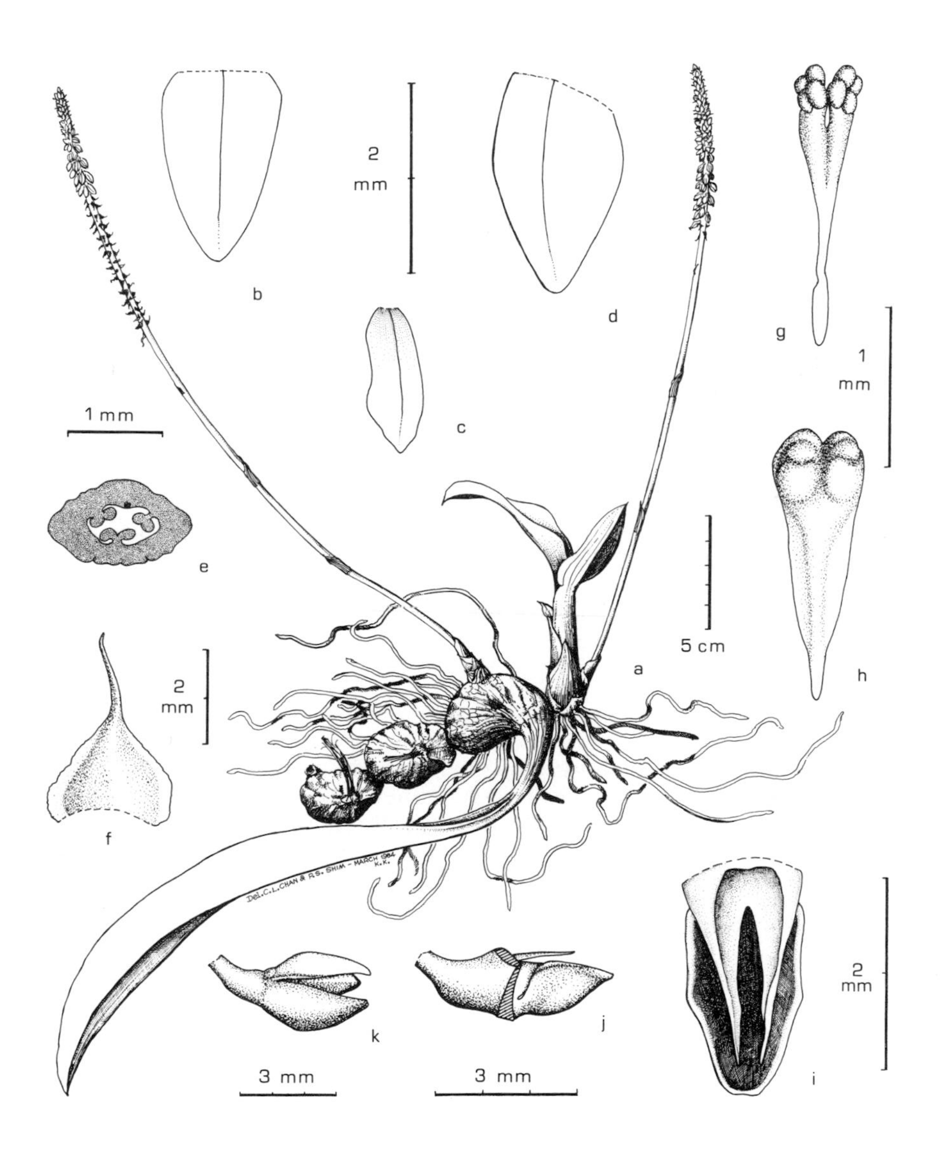

Fig. 55. Thelasis variabilis. **a**, habit; **b**, dorsal sepal; **c**, petal; **d**, lateral sepal; **e**, ovary (transverse section); **f**, floral bract; **g**, pollinarium; **h**, anther-cap (back view); **i**, lip (front view); **j**, pedicel-with-ovary, column and lip (side view); **k**, flower (side view). All from *Shim s.n.*, Mt. Alab, Sabah. Drawn by Chan Chew Lun and P. S. Shim.

55.112.6. Thelasis indet.

Collections. MAHANDEI RIVER: 1100 m, *Carr SFN 26472* (SING); PALUAN RIVER: 500 m, *Carr SFN 27393* (SING); PINOSUK: 800 m, *Carr 3403, SFN 27393A* (SING); TENOMPOK/RANAU: 1500 m, *Carr SFN 27130* (SING).

55.113. THRIXSPERMUM Lour.

Fl. Cochinch., 516, 519 (1790).

Medium-sized monopodial epiphytes or, more rarely, terrestrials. *Stems* varying from a few cm up to 1 m in length. *Leaves* usually well spaced, less often a few close together, flattened, sometimes fleshy, never terete or laterally compressed as in *Cordiglottis*. *Inflorescences* short or long, sometimes dense, a few flowers opening at a time, the flowering of many lowland species initiated by a sudden afternoon rainstorm. *Flowers* ephemeral, often fully open for only half a day, very variable, from a few mm to several cm across; *tepals* ± equal; *lip* immobile, 3-lobed, saccate but not truly spurred, usually with a partly hairy or papillose front-wall callus, side-lobes ± erect, mid-lobe usually fleshy; *column* short and stout, sometimes winged, with a long foot; *stipes* very short and broad, shorter than the diameter of the pollinia; *pollinia* 4, appearing as 2 unequal masses; *fruits* long and slender.

Around 100 species distributed from Sri Lanka and the Himalayan region east to the Pacific islands. The centre of distribution appears to be in Sumatra.

Three sections are recognised: 1. *Thrixspermum* having inflorescences with a flattened rachis bearing often spidery flowers on the edges; 2. *Dendrocolla* having short stems with flowers facing in all directions; 3. *Katocolla* having long limply pendent stems, closely set, rather short leaves and few-flowered inflorescences much shorter than the leaves, the flowers again arranged in all directions.

55.113.1. Thrixspermum amplexicaule (Blume) Rchb. f., Xenia Orch. 2: 121 (1867).

Dendrocolla amplexicaulis Blume, Bijdr., 288 (1825).

Epiphyte. Lowlands, lower montane forest. Elevation: 400–1500 m.

General distribution: Sabah, Sarawak, Kalimantan, widespread from the Andaman Islands, Thailand & Vietnam to Malaysia & Indonesia, the Philippines, New Guinea & the Solomon Islands.

Collections. BUNDU TUHAN: 1200 m, *Clemens 28040* (BM); KAUNG: 400 m, *Carr SFN 27367* (SING), 400 m, *Darnton 396* (BM); TENOMPOK: 1500 m, *Clemens 28418* (BM, K), 1500 m, *28462* (BM).

55.113.2. Thrixspermum centipeda Lour., Fl. Cochinch. 2: 520 (1790).

Epiphyte. Lowlands, lower montane forest. Elevation: 400–2100 m.

General distribution: Sabah, Sarawak, Brunei, Kalimantan, widespread from the Himalayas, India & China through Burma east to the Philippines.

Collections. KAUNG: 400 m, *Carr SFN 27373* (SING); KEMBURONGOH: 2100 m, *Price 220* (K).

55.113.3. Thrixspermum crescentiforme Ames & C. Schweinf., Orch. 6: 215 (1920). Type: MARAI PARAI SPUR, *Clemens 238* (holotype AMES mf).

Epiphyte. Lower montane forest. Elevation: 1200–1500 m.

General distribution: Sabah.

Additional collections. BUNDU TUHAN: 1400 m, *Darnton 207* (BM); GURULAU SPUR: 1200–1500 m, *Clemens 51472* (BM, K); KINUNUT VALLEY HEAD: 1200 m, *Carr SFN 27181* (K); TENOMPOK: 1500 m, *Clemens 28356* (BM, K), 1500 m, *29634* (BM), 1500 m, *29792* (BM, K).

55.113.4. Thrixspermum longicauda Ridl., J. Linn. Soc. Bot. 31: 299 (1896).

Epiphyte. Lower montane forest. Elevation: 1500 m.

General distribution: Sabah, Sarawak.

Collections. TENOMPOK: 1500 m, *Clemens 27978* (BM); TENOMPOK ORCHID GARDEN: *Clemens 50181* (BM, K).

55.113.5. Thrixspermum pardale (Ridl.) Schltr., Orchis 5, 4: 56 (1911).

Sarcochilus pardalis Ridl., Trans. Linn. Soc. Bot., ser. 3: 371 (1893).

Epiphyte. Hill forest. Elevation: 900 m.

General distribution: Sabah, Peninsular Malaysia, Sumatra.

Collection. BUNDU TUHAN: 900 m, *Carr SFN 27788* (SING).

55.113.6. Thrixspermum pensile Schltr., Bot. Jahrb. Syst. 45, Beibl. 104: 59 (1911). Plate 78B.

Epiphyte. Hill forest on ultramafic substrate. Sight record and photograph of A. Lamb on the Lohan River; no specimens seen.

General distribution: Sabah, Thailand, Sumatra, Java.

55.113.7. Thrixspermum triangulare Ames & C. Schweinf., Orch. 6: 217 (1920). Plate 78C. Type: KEMBURONGOH, *Clemens 201* (holotype AMES mf; isotypes BM!, K!, SING!).

Epiphyte or lithophyte. Lower montane forest, upper montane mossy forest, on tree roots, branches, rocks, in ridge scrub, often on ultramafics. Elevation: 1200–3400 m.

Endemic to Mount Kinabalu.

Additional collections. GURULAU SPUR: 2400–2700 m, *Clemens 50657* (BM, K), 2400–2700 m, *50663* (BM, K), 2700 m, *51009* (BM, K), 2400–3400 m, *51193* (BM, K); KEMBURONGOH: *Carr SFN 27503* (K); KILEMBUN BASIN: 2300 m, *Clemens 32889* (BM); KINATEKI RIVER HEAD: 1500 m, *Clemens s.n.* (BM); LAYANG-LAYANG: 3000 m, *Collenette 888* (K); LUMU-LUMU: 2100 m, *Clemens 27173* (BM); MARAI PARAI: 2100–2700 m, *Clemens 33145* (BM); MESILAU CAVE: 1700–2000 m, *Beaman 10687* (K), 1800 m, *RSNB 4787* (K); MESILAU CAVE TRAIL: 1700–1900 m, *Beaman 9146* (K); MOUNT KINABALU: *Haslam s.n.* (AMES, BM, K); PAKA-PAKA CAVE: 3000 m, *Clemens 27869* (BM, K); PENIBUKAN: 1200–1500 m, *Clemens s.n.* (BM); SUMMIT TRAIL: 2700–3400 m, *Gunsalam 2* (K), *Wood 611* (K); WEST MESILAU RIVER: 1600 m, *Beaman 8999* (K).

55.113.8. Thrixspermum aff. **triangulare** Ames & C. Schweinf., Orch. 6: 217 (1920).

Epiphyte. Lower montane forest on ultramafic substrate. Elevation: 1200–1800 m.

Collections. KILEMBUN BASIN: 1200 m, *Clemens 34484* (BM); KILEMBUN RIVER HEAD: 1400 m, *Clemens 32485* (BM); MARAI PARAI: 1800 m, *Clemens 32444* (BM); MARAI PARAI SPUR: 1800 m, *Collenette A 88* (BM); MESILAU CAVE: 1600 m, *Bailes & Cribb 694* (K).

55.113.9. Thrixspermum trichoglottis (Hook. f.) Kuntze, Revis. Gen. Pl. 2: 682 (1891).

Sarcochilus trichoglottis Hook. f., Fl. Brit. Ind. 6: 39 (1890).

Epiphyte. Lowlands, low stature lower montane ridge forest on ultramafic substrate. Elevation: 400–1500 m.

General distribution: Sabah, Sarawak, India & Andaman Islands east to Java.

Collections. KAUNG: 400 m, *Darnton 356* (BM); MAMUT COPPER MINE: 1400–1500 m, *Beaman 10353* (K).

55.113.10. Thrixspermum sp. 1

Epiphyte. Hill forest. Elevation: 900 m.

Collection. BUNDU TUHAN: 900 m, *Darnton 525* (BM).

55.113.11. Thrixspermum sp. 2

Epiphyte. Lower montane forest. Elevation: 1400 m.

Collection. BUNDU TUHAN: 1400 m, *Darnton 224* (BM).

55.113.12. Thrixspermum indet.

Collections. BUNDU TUHAN: 1200 m, *Carr SFN 27787* (SING), 900 m, *SFN 27796* (SING), 800 m, *SFN 27914* (SING); EAST MESILAU RIVER: 1600–1700 m, *Bailes & Cribb 745* (K); KAUNG: 400 m, *Carr SFN 27369* (SING), 400 m, *Darnton 403* (BM); KEMBURONGOH/PAKA-PAKA CAVE: 2800 m,

Carr SFN 27530 (SING); KINUNUT VALLEY HEAD: 1200 m, *Carr SFN 27363* (SING); KOLOPIS RIVER: 1100 m, *Carr SFN 27179* (SING); MAHANDEI RIVER: 1100 m, *Carr SFN 26470* (SING); MARAI PARAI: 1500 m, *Carr SFN 27832* (SING); MINITINDUK: 900 m, *Carr SFN 27219* (SING); PAKA-PAKA CAVE: 3200 m, *Carr SFN 27623* (SING); PALUAN RIVER: 500 m, *Carr SFN 27985* (SING); PENIBUKAN: 1200 m, *Carr SFN 27784* (SING), 1200 m, *SFN 27784A* (SING), 1200–1500 m, *Clemens s.n.* (BM); PINOSUK: 900 m, *Carr SFN 28009* (SING), 800 m, *SFN 28030* (SING); TAHUBANG RIVER: 1100 m, *Clemens 35201* (BM), 900 m, *40459* (BM); TENOMPOK: 1500 m, *Carr SFN 26878* (SING); TENOMPOK ORCHID GARDEN: 1500 m, *Clemens 50251* (SING).

55.114. TRICHOGLOTTIS Blume

Bijdr., 359 (1825).

Climbing monopodial epiphytes. *Stems* short or long, straggling, up to 80 cm long, with elongated internodes. *Leaves* linear to elliptic, apex usually unequally bilobed. *Inflorescences* with a short peduncle, 1–4-flowered, but often more than one per node. *Flowers* rather small, opening widely, lasting about a week, resupinate, usually yellowish with light brown or purple markings; *sepals* and *petals* free; *dorsal sepal* up to 1.5 cm long; *lip* spurred or saccate at base, 3-lobed, immobile, with a small, usually hairy basal tongue which emerges from the back-wall, disc often hairy; *column* short and stout, without a foot, often with small roughly hairy stelidia; *stipes* linear-oblong; *viscidium* small, ovate or elliptic; *pollinia* 4, appearing as 2 unequal masses.

About 55–60 species are known, distributed from Sri Lanka and the Nicobar Islands east to New Guinea, Australia and the Solomon Islands, north to Thailand. The centre of distribution lies in Indonesia and the Philippines.

55.114.1. **Trichoglottis bipenicillata** J. J. Sm., Icon. Bogor. 2: 125, t. 125A (1903). Plate 78D, 79.

Epiphyte. Hill forest on ultramafic substrate. Elevation: 600 m.

General distribution: Sabah, Kalimantan.

Collection. PENATARAN RIDGE: 600 m, *Lamb AL 185/84* (K).

55.114.2. **Trichoglottis collenetteae** J. J. Wood, C. L. Chan & A. Lamb, **sp. nov.** Fig. 56, Plate 80A, 80B.

Ab omnibus speciebus generis aliis habitu ad *Acampen papillosam* referenti, foliis crassis coriaceis ligulatis vaginis rugulosis nigro-maculosis-lineatisque instructis, sepalis prorsum curvatis, marginibus petalorum decurvatis distinguitur. Labellum lateraliter compressum, lingua in pariete posteriore ligulata truncata, lobis lateralibus bidentatis, lobo medio tridentato caliis basalibus duobus carnosis acutis praedito instructum, intra labellum quoque est structura quae sursum versus linguam iam commemoratam eminet.

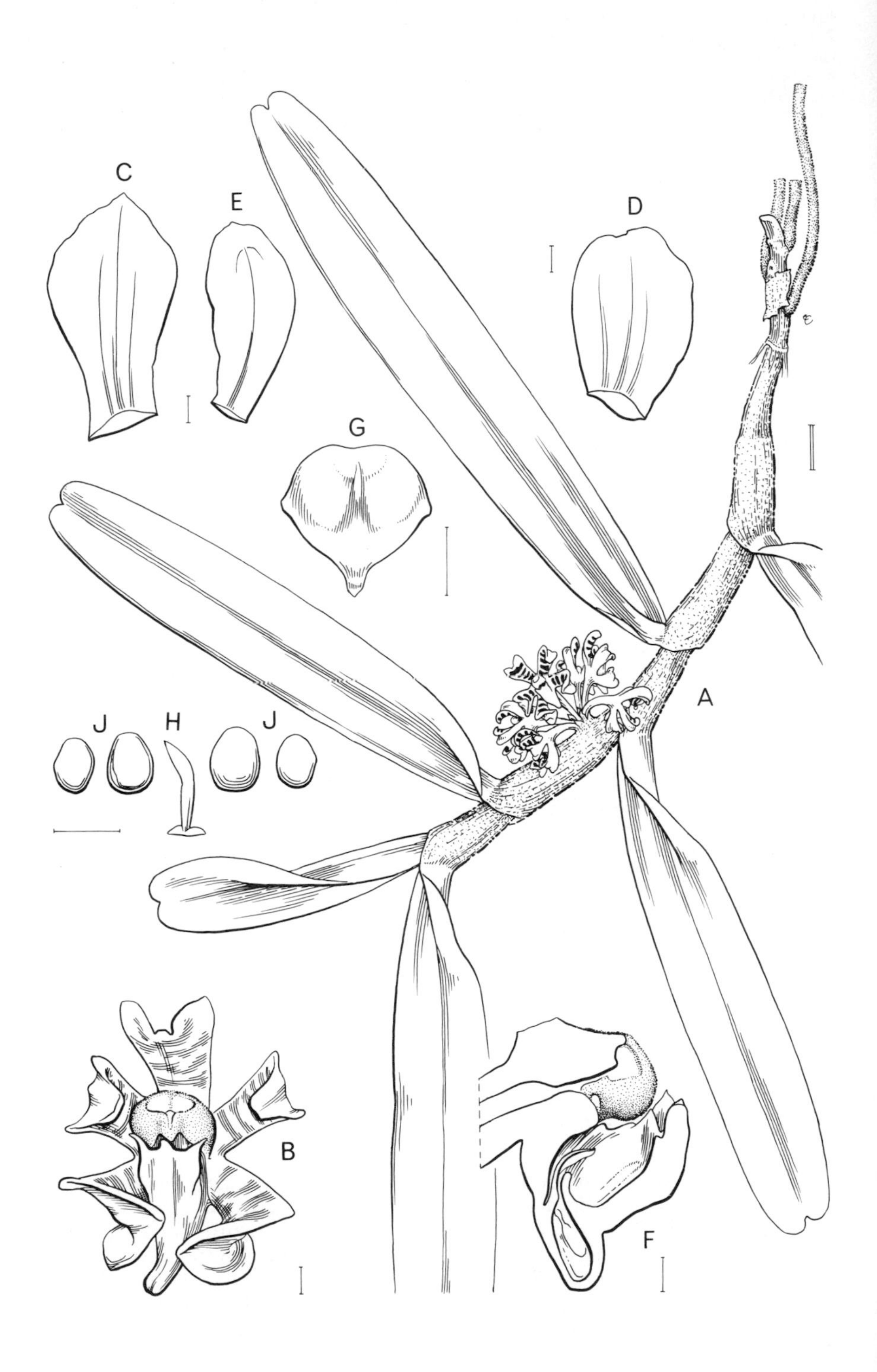
C
E
D
G
A
J
H
J
B
F

Type: EAST MALAYSIA, SABAH, CROCKER RANGE, SINSURON ROAD, BETWEEN KOTA KINABALU AND TAMBUNAN, 1440 m, 27 February 1981, *Collenette 2295* (holotype K, herbarium and spirit material).

Pendulous to spreading epiphyte, in habit recalling *Acampe papillosa* (Lindl.) Lindl. *Stems* occasionally branching, 16–30 cm long, 1 cm thick, with numerous thick rugulose roots at base, internodes 2.5–3.5 cm long. *Leaves* 11–16 × 2–2.5(–3) cm, ligulate, thick and leathery, apex retuse to unequally bilobed, sheaths 2.5–3.5 cm long, rugulose, with black spots and streaks. *Inflorescence* sessile, emerging from near base of leaf sheath, up to 6-flowered; rachis 6–7 × 5–6 mm, thick and fleshy; *floral bracts* reduced and scale-like. *Flowers* resupinate, pedicel-with-ovary white, sepals and petals cream with transverse brownish orange or chestnut-brown markings, lip white, faintly flushed with lilac or purple, column white, flushed with brownish orange or mustard-yellow. *Pedicel-with-ovary* 1 cm long. *Sepals* and *petals* curving forward, their margins deflexed, obtuse. *Dorsal sepal* 0.8–1 × 0.4 cm, spathulate. *Lateral sepals* 0.7–1 × 0.5–0.6 cm, spathulate. *Petals* 0.7–0.9 × 0.3 cm, spathulate. *Lip* immobile, 5–6 mm long, 2.5 mm wide across apex, with an obtuse spur 1–1.5 mm long; back-wall tongue ligulate, truncate, minutely papillose-hairy on under-surface; body of lip laterally compressed, side-lobes bidentate, the upper tooth-like lobes broader, subacute, the lower tooth-like lobes narrowly triangular, acute; mid-lobe tridentate, the outer tooth-like lobes spreading, broadly triangular, acute, 0.8 mm long, central tooth much shorter, raised, keel-like, base of mid-lobe with 2 fleshy, pointed calli, interior of lip with a minutely papillose-hairy, fleshy, conical structure. *Column* 3 × 3 mm, truncate, sides pubescent. *Anther-cap* 2 × 2 mm, ovate, apex attenuate, reflexed. *Pollinia* 4, unequal; *stipes* ligulate, cucullate above; *viscidium* small, elliptic.

Trichoglottis collenetteae is distinguished from all other species of *Trichoglottis* by its habit, which resembles *Acampe papillosa,* thick leathery ligulate leaves with black-spotted and -streaked rugulose sheaths, and forward curving sepals and petals with deflexed margins. The laterally compressed lip has a ligulate, truncate back-wall tongue, bidentate side-lobes and a tridentate mid-lobe with 2 fleshy, pointed calli at the base. Inside the lip is a minutely papillose-hairy conical structure projecting up toward the back-wall tongue.

The specific epithet honours Mrs. Sheila Collenette (née Darnton), who collected the type. Her numerous well-documented collections have added greatly to our knowledge of the Kinabalu flora.

Fig. 56. Trichoglottis collenetteae. A, habit; **B**, flower (front view); **C**, dorsal sepal; **D**, lateral sepal; **E**, petal; **F**, pedicel-with-ovary, lip and column (longitudinal section); **G**, anther-cap; **H**, stipes and viscidium; **J**, pollinia. **A**, from *Chan 52/86;* **B–J** from *Collenette 2295*. Scale: single bar = 1 mm; double bar = 1 cm. Drawn by Eleanor Catherine.

Lower montane forest. Elevation: 1100–1500 m.

General distribution: Sabah.

Additional collections: CROCKER RANGE, MT. ALAB, 1050 m, *Lamb SAN 92260, LMC 1395* (K sketch); LIWAGU RIVER TRAIL: 1600 m, *Chan 52/86* (K); MT. NUNGKEK: 1500 m, *Clemens s.n.* (BM); PENIBUKAN: 1500 m, *Clemens 30990* (BM); TENOM DISTRICT, NEAR SAPONG ESTATE, 800 m, *H. F. Comber* colour slide.

55.114.3. Trichoglottis kinabaluensis Rolfe in Gibbs, J. Linn. Soc. Bot. 42: 157 (1914). Type: GURULAU SPUR, 1700 m, *Gibbs 3993* (holotype K!).

Epiphyte. Hill forest, lower montane forest. Elevation: 900–1700 m.

General distribution: Sabah.

Additional collections. BUNDU TUHAN: 1400 m, *Darnton 226* (BM); GURULAU SPUR: *Clemens 315* (AMES); KIAU: 900 m, *Clemens 31* (AMES); MINITINDUK: 900 m, *Carr SFN 27040* (SING); PENIBUKAN: 1200–1500 m, *Clemens 31116* (BM), 1200 m, *32138* (BM); TENOMPOK: 1200 m, *Clemens 26126* (BM, K), 1500 m, *28137* (BM).

55.114.4. Trichoglottis lanceolaria Blume, Bijdr., 360 (1825).

Pendulous epiphyte. Hill forest. Elevation: 900 m. Known from Kinabalu only from a colour slide by A. Lamb.

General distribution: Sabah, Peninsular Malaysia, Thailand, Vietnam, Sumatra, Java, Natuna Islands.

Collection. PORING HOT SPRINGS/LANGANAN WATER FALLS: 900 m, *Lamb s.n.* (K photo).

55.114.5. Trichoglottis aff. **lanceolaria** Blume, Bijdr., 360 (1825).

Pendulous epiphyte. Hill forest. Known only from a colour slide by A. Lamb.

Collection. HEMPUEN HILL: *Lamb s.n.* (K photo).

55.114.6. Trichoglottis magnicallosa Ames & C. Schweinf., Orch. 6: 221 (1920). Type: MOUNT KINABALU, *Haslam s.n.* (holotype AMES mf; isotypes BM!, K!).

Epiphyte. Lowlands, hill forest. Elevation: 500–900 m.

General distribution: Sabah.

Additional collections. DALLAS: 900 m, *Clemens 27163* (BM, K); KAUNG: 500 m, *Carr SFN 27321* (SING).

55.114.7. Trichoglottis smithii Carr, Gardens' Bull. 8: 125 (1935). Plate 80C, 80D.

Epiphyte or lithophyte. Low-stature hill forest on ultramafic substrate; on loose rocks of steep hillside. Elevation: 700–1000 m.

General distribution: Sabah, Sarawak, Sumatra?

Collections. HEMPUEN HILL: 800–1000 m, *Cribb 89/26* (K); LOHAN RIVER: 700–900 m, *Beaman 9254* (K), 800–1000 m, *10006* (K).

55.114.8. Trichoglottis tenuis Ames & C. Schweinf., Orch. 6: 223 (1920). Type: KIAU, 900 m, *Clemens 60* (holotype AMES mf).

Epiphyte. Hill forest. Elevation: 900 m.

Endemic to Mount Kinabalu.

Additional collection. DALLAS: 900 m, *Clemens 27417* (BM).

55.114.9. Trichoglottis vandiflora J. J. Sm., Bull. Dép. Agric. Indes Néerl. 22: 49 (1909).

Epiphyte. Hill forest on ultramafic substrate. Elevation: 500 m.

General distribution: Sabah, Kalimantan.

Collection. HEMPUEN HILL: 500 m, *Lamb AL 864/87* (K).

55.114.10. Trichoglottis winkleri J. J. Sm.

a. var. **minor** J. J. Sm., Bull. Jard. Bot. Buit., ser. 2, 26: 102 (1918). Plate 81A.

Epiphyte. Hill forest on ultramafic substrate. Known from Kinabalu only from cultivated material; from the Lohan River fide A. Lamb (pers. comm.).

General distribution: Sabah, Peninsular Malaysia, Java.

Collection. PORING ORCHID GARDEN: *Cribb 89/60* (K).

55.114.11. Trichoglottis sp. 1

Epiphyte. Elevation: 600–800 m.

Collection. MOUNT KINABALU: 600–800 m, *Lamb SAN 93375* (K).

55.114.12. Trichoglottis sp. 2

Epiphyte. Lower montane forest. Elevation: 1700–1800 m.

Collection. GOLF COURSE SITE: 1700–1800 m, *Beaman 7229* (K).

55.114.13. Trichoglottis sp. 3

Epiphyte. Lower montane forest. Elevation: 1200–1500 m.

Collections. BUNDU TUHAN: 1400 m, *Collenette 597* (K); PENIBUKAN: 1200–1500 m, *Clemens s.n.* (BM).

55.114.14. Trichoglottis sp. 4

Epiphyte. Lower montane forest. Elevation: 1200–1500 m.

Collections. KIBAMBANG RIVER: 1200–1500 m, *Clemens 34311* (BM); TAHUBANG RIVER: 1200–1500 m, *Clemens s.n.* (BM).

55.114.15. Trichoglottis indet.

Collections. KINATEKI RIVER: 1200–1500 m, *Clemens 31113* (BM); KINUNUT VALLEY HEAD: 1100 m, *Carr SFN 27193* (SING); PENIBUKAN: 1700 m, *Clemens 51725* (BM); TENOMPOK: 1600 m, *Carr SFN 27003* (SING).

55.115. TRICHOTOSIA Blume

Bijdr., 342 (1825).

Seidenfaden, G. (1982). Orchid genera in Thailand X. *Trichotosia* & *Eria* Lindl. Opera Bot. 62: 1–157.

Epiphytic, rarely terrestrial, herbs. *Stems* long or short, leafy throughout except at base, usually covered throughout with reddish brown, rarely white, hispid hairs, sometimes hairs restricted to leaf-sheaths and inflorescences. *Inflorescences* lateral, from any node, piercing the leaf-sheath, short and few-flowered, or long, pendulous and many-flowered; floral bracts hairy, at right angles to the rachis, large, concave. *Flowers* small to medium sized, not opening widely, resupinate; *sepals* red-hairy on outside, laterals adnate to column-foot forming a mentum; *lip* entire to obscurely 3-lobed, disc with or without keels, sometimes papillose; *column* with a foot; *pollinia* 8.

About 50 species distributed from mainland Asia, through SE Asia to New Guinea and the Pacific islands.

55.115.1. Trichotosia aporina (Hook. f.) Kraenzl., Pflanzenr. IV. 50. II. B. 21: 150 (1911).

Eria aporina Hook. f., Fl. Brit. Ind. 5: 808 (1890).

Epiphyte. Lower montane forest. Elevation: 1700 m.

General distribution: Sabah, Sarawak, Peninsular Malaysia.

Collection. GURULAU SPUR: 1700 m, *Clemens 50431* (BM, K).

55.115.2. Trichotosia aff. **aporina** (Hook. f.) Kraenzl., Pflanzenr. IV. 50. II. B. 21: 150 (1911).

Epiphyte. Hill forest, lower montane forest. Elevation: 900–1500 m.

Collections. DALLAS: 900 m, *Clemens 26305* (BM), 900 m, *26601* (BM, K); PENIBUKAN: 1200 m, *Clemens 40552* (BM); TENOMPOK: 1500 m, *Clemens 28665* (BM, K), 1500 m, *29461* (BM, K).

55.115.3. Trichotosia aurea (Ridl.) Carr, Gardens' Bull. 8: 99 (1935).

Eria aurea Ridl., J. Straits Branch Roy. Asiat. Soc. 49: 31 (1908).

Epiphyte. Lower montane forest. Elevation: 1400–2000 m.

General distribution: Sabah, Brunei, Kalimantan, Sarawak.

Collections. GOLF COURSE SITE: 1700–1800 m, *Beaman 7228* (K); LITTLE MAMUT RIVER: 1400 m, *Collenette 1021* (K); PARK HEADQUARTERS: 1700 m, *Kanis & Sinanggul SAN 51496* (K); PENIBUKAN: 1700 m, *Clemens 50395* (BM, K); SUMMIT TRAIL: 2000 m, *Bailes & Cribb 775* (K); TENOMPOK: 1500 m, *Clemens 28715* (BM, K); TENOMPOK/RANAU: 1500 m, *Carr SFN 26969* (K).

55.115.4. Trichotosia aff. **aurea** (Ridl.) Carr, Gardens' Bull. 8: 99 (1935).

Epiphyte. Lower montane forest, frequently on ultramafic substrate. Elevation: 1100–2000 m.

Collections. BAMBANGAN RIVER: 1500 m, *RSNB 4463* (K); GURULAU SPUR: 1500 m, *Clemens 50345* (BM, K), 1800 m, *50417* (BM); KILEMBUN BASIN: 1700 m, *Clemens 40052* (K); KILEMBUN RIVER HEAD: 1800 m, *Clemens 32526* (BM, K); MARAI PARAI SPUR: 1100 m, *Clemens 32780* (BM); MURU-TURA RIDGE: 1700 m, *Clemens 34105* (BM), 1500 m, *40053* (BM); NUMERUK RIDGE: 1700 m, *Clemens 40052* (BM); PENIBUKAN: 1400 m, *Clemens 40604* (BM, K); SUMMIT TRAIL: 2000 m, *Beaman 6763* (K); TENOMPOK: 1500 m, *Clemens s.n.* (BM); TENOMPOK/RANAU: 1500 m, *Carr SFN 27345* (SING).

55.115.5. Trichotosia brevipedunculata (Ames & C. Schweinf.) J. J. Wood, **comb. nov.** Fig. 57, Plate 81B.

Eria brevipedunculata Ames & C. Schweinf., Orch. 6: 118 (1920). Type: MARAI PARAI SPUR, *Clemens 255* (holotype AMES mf).

Epiphytic or terrestrial. Lower montane forest, sometimes on ultramafic substrate. Elevation: 1200–2100 m.

General distribution: Sabah.

Additional collections. GOLF COURSE SITE: 1700–1800 m, *Beaman 7235* (K); LUMU-LUMU: 2100 m, *Clemens 27187* (BM); MARAI PARAI SPUR: *Clemens 291* (AMES), *373* (AMES); MESILAU CAVE TRAIL: 1800 m, *Beaman 7491* (K); MESILAU RIVER: 1500 m, *RSNB 4099* (K), 1500 m, *RSNB 4258* (K); PARK HEADQUARTERS: 1500 m, *Wood 619* (K); PENIBUKAN: 1200 m, *Clemens 30851* (K), 1400 m, *40603* (BM), 1700 m, *50285* (BM); SEDIKEN RIVER/MARAI PARAI: 1500 m, *Clemens 32452* (BM); TENOMPOK: 1500 m, *Clemens 26775* (BM, K), 1500 m, *26837* (BM, K); TENOMPOK ORCHID GARDEN: 1500 m, *Clemens 50173* (BM); TINEKUK FALLS: 2000 m, *Clemens 40916* (BM, K).

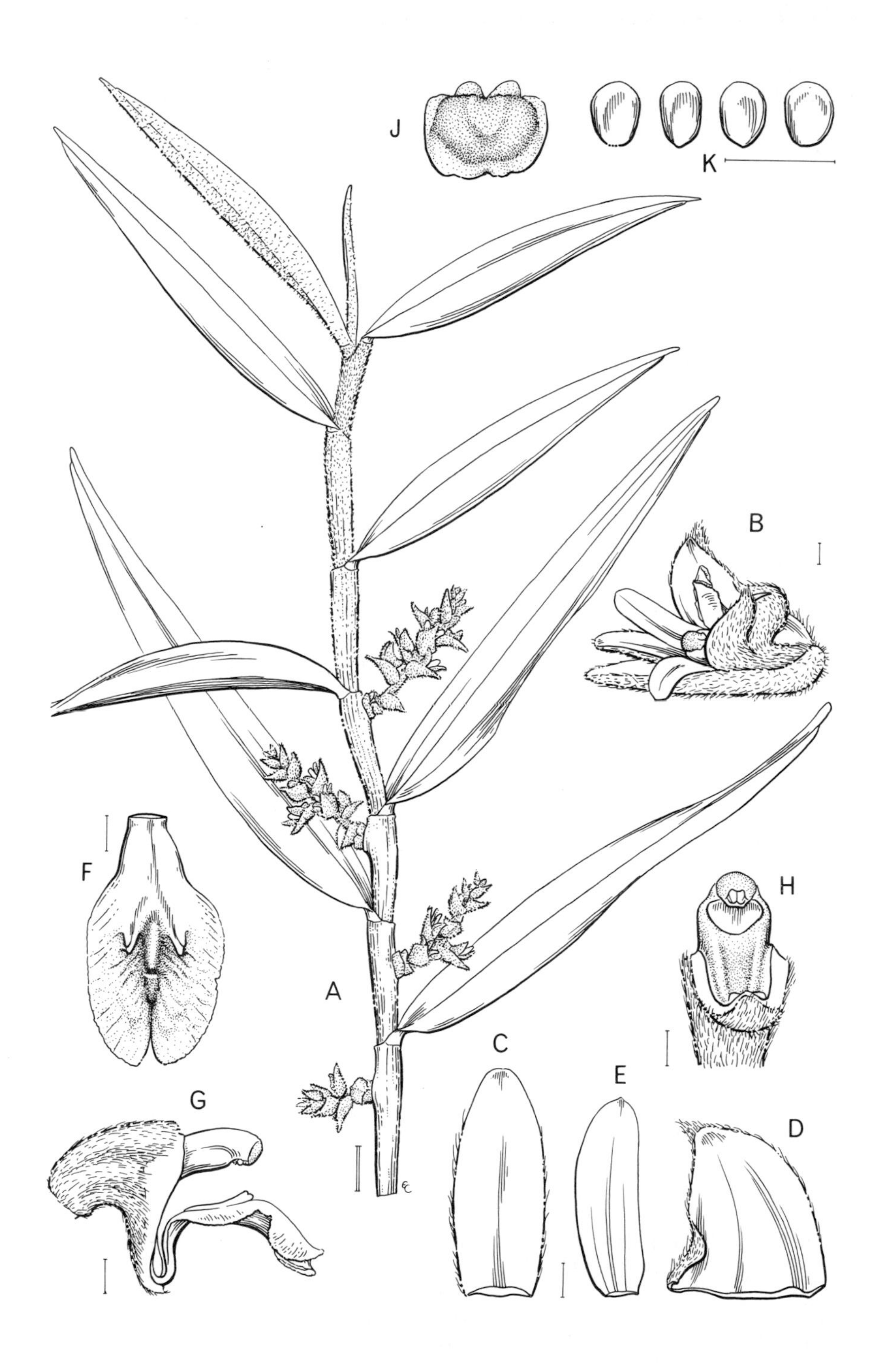
J
K
B
A
F
H
C
E
G
D

55.115.6. Trichotosia ferox Blume, Bijdr., 342 (1825). Plate 81C.

Epiphyte. Upper montane forest, on tree trunks among moss. Elevation: 1500–3000 m.

General distribution: Sabah, Sarawak, Peninsular Malaysia, Thailand, Sumatra, Java, Lombok.

Collections. DACHANG: 2700 m, *Clemens 29286* (BM); EASTERN SHOULDER: 2000 m, *RSNB 179* (K, SING); GOLF COURSE SITE: 1600 m, *Bailes & Cribb 713* (K); LUMU-LUMU: 2100 m, *Clemens 27176* (BM); MOUNT KINABALU: 1900 m, *Tiong & Dewol SAN 85738* (K); PAKA-PAKA CAVE: 2700–3000 m, *Gibbs 4279* (BM); TENOMPOK: 1500 m, *Clemens 29953* (BM).

55.115.7. Trichotosia aff. **ferox** Blume, Bijdr., 342 (1825).

Epiphyte. Hill forest, lower and upper montane forest, sometimes on ultramafic substrate. Elevation: 900–2700 m.

Collections. DALLAS: 900 m, *Clemens 30123* (K); GURULAU SPUR: 2100–2700 m, *Clemens 51732* (BM); KEMBURONGOH: 2200 m, *Carr SFN 27468* (K, SING), 2300 m, *Mikil SAN 38455* (K), 2400 m, *Sinanggul SAN 38313* (K); MARAI PARAI: 2100–2700 m, *Clemens 33134* (BM, K); MESILAU RIVER: 1500 m, *Lajangah SAN 36159* (K); MINETUHAN SPUR: 2700 m, *Clemens 33793* (BM); MT. TEMBUYUKEN: 2400 m, *Aban SAN 55422* (K); PENIBUKAN: 1200–1500 m, *Clemens 30469* (BM, K), 1200 m, *30586* (BM), 1200 m, *31050* (BM, K), 1200 m, *40661* (BM, K), 1400 m, *40827* (BM, K), 1200 m, *40855* (BM), 1200 m, *50327* (BM), 1200 m, *50496* (BM, K); TENOMPOK: 1200 m, *Carr SFN 27742* (K).

55.115.8. Trichotosia microphylla Blume, Bijdr., 343 (1825). Plate 82A.

Epiphyte. Lower montane forest, on mossy branches and trunks. Elevation: 1100–1500 m.

General distribution: Sabah, Kalimantan, Sarawak, Peninsular Malaysia, Thailand, Vietnam, Sumatra, Java.

Collections. BUNDU TUHAN: 1400 m, *Darnton 217* (BM), *519* (BM); MAHANDEI RIVER: 1100 m, *Carr 3103, SFN 26494* (SING); PARK HEADQUARTERS: 1500 m, *Wood 620* (K); PENIBUKAN: 1200 m, *Clemens 31048* (BM, K), 1100 m, *50151* (BM).

55.115.9. Trichotosia mollicaulis (Ames & C. Schweinf.) J. J. Wood, **comb. nov.** Plate 82B.

Eria mollicaulis Ames & C. Schweinf., Orch. 6: 131 (1920). Type: KIAU, 900 m, *Clemens 66* (holotype AMES mf; isotype SING!).

Figure 57. Trichotosia brevipedunculata. A, habit; **B**, floral bract and flower (side view); **C**, dorsal sepal; **D**, lateral sepal; **E**, petal; **F**, lip (front view, flattened); **G**, pedicel-with-ovary, column and lip (side view); **H**, column with anther-cap (front view); **J**, anther-cap (back view), **K**, pollinia. **A** from *Clemens 26837*; **B–K** from *Wood 619.* Scale: single bar = 1 mm; double bar = 1 cm. Drawn by Eleanor Catherine.

Epiphyte. Hill forest, lower montane forest. Elevation: 900–1700 m.

General distribution: Sabah.

Additional collections. DALLAS: 900 m, *Clemens 26604* (K); GURULAU SPUR: 1200 m, *Carr SFN 27035* (SING), *Clemens 303* (AMES), 1700 m, *50556* (BM, K); KIAU: 900 m, *Clemens 66* (BM), 900 m, *160* (AMES); MOUNT KINABALU: *Haslam s.n.* (AMES, BM); PENIBUKAN: 1200–1500 m, *Clemens 30851* (BM), 1200 m, *32086* (BM), 1200 m, *40657* (BM, K), 1500 m, *50207* (BM, K), 1500 m, *50328* (BM), 1200 m, *50511* (BM); SUMMIT TRAIL: *Wood 610* (K); TAHUBANG RIVER: 1200 m, *Clemens 30707* (BM).

55.115.10. Trichotosia pilosissima (Rolfe) J. J. Wood, comb. nov.

Eria pilosissima Rolfe in Gibbs, J. Linn. Soc. Bot. 42: 152 (1914). Type: LUBANG, 1800 m, *Gibbs 4117* (holotype K!; isotype BM!).

Epiphyte. Lower montane forest, sometimes on ultramafic substrate. Elevation: 1500–2400 m.

General distribution: Sabah, Brunei, Sarawak.

Additional collections. EAST MESILAU/MENTEKI RIVERS: 1700 m, *Beaman 9388* (K); KEMBURONGOH: 2400 m, *Clemens 29129* (BM); LUMU-LUMU: 2300 m, *Clemens 27180* (BM); MARAI PARAI SPUR: *Clemens 290* (AMES); MURU-TURA RIDGE: 1500–1800 m, *Clemens 34482* (BM); TENOMPOK: 1500 m, *Clemens 28779* (BM, K); TINEKUK FALLS: 1800 m, *Clemens 40334* (BM).

55.115.11. Trichotosia rubiginosa (Blume) Kraenzl., Pflanzenr. IV. 50. II. B. 21: 155 (1911).

Eria rubiginosa Blume, Mus. Bot. Lugd. 2: 184 (1856).

Epiphytic or terrestrial. Lower montane forest. Elevation: 1700 m.

General distribution: Sabah, Kalimantan, Sarawak.

Collection. GURULAU SPUR: 1700 m, *Gibbs 4016* (BM, K).

55.115.12. Trichotosia indet.

Collections. KINATEKI RIVER HEAD: 2700 m, *Clemens 31931* (BM); LOHAN RIVER: 500–600 m, *Clements 3293* (K); MAHANDEI RIVER HEAD: 1200 m, *Carr 3079, SFN 26409* (SING); MAMUT COPPER MINE: 1400–1500 m, *Beaman 10351* (K); PENIBUKAN: 1500 m, *Clemens 30992* (BM), 1200–1500 m, *31381* (BM); SUMMIT TRAIL: 2000 m, *Fuchs 21054* (K); TAHUBANG RIVER HEAD: 2100 m, *Clemens 32976* (BM); TENOMPOK: 1600 m, *Carr SFN 26936* (SING), 1500 m, *SFN 27828* (SING), 1500 m, *SFN 27972* (SING); TENOMPOK ORCHID GARDEN: 1500 m, *Clemens 50373* (BM).

55.116. TROPIDIA Lindl.

Bot. Reg. 19: sub t. 1618 (1833).

Terrestrial herbs. *Stems* erect, leafy. *Leaves* distichous, tough, plicate, sheathing at the base. *Inflorescences* terminal, simple, with 1 or 2 flowers opening in succession; *floral bracts* large, imbricate, sheathing. *Flowers* resupinate or non-resupinate, often opening only partially; *sepals* and *petals* free, or lateral sepals sometimes connate, together enclosing base of lip; *lip* entire, concave, saccate or spurred; *column* short, *rostellum* long, erect. *Pollinia* 2.

A pantropical genus consisting of about 20 species, two of which are saprophytic.

55.116.1. Tropidia curculigoides Lindl., Gen. Sp. Orch., 497 (1840).

Terrestrial. Hill forest, sometimes on ultramafic substrate. Elevation: 800–1300 m.

General distribution: Sabah, Sarawak, China (Hainan), Taiwan, India, Burma, Peninsular Malaysia, Thailand, Cambodia, Vietnam, Java, Timor.

Collections. KULUNG HILL: 800 m, *Beaman 7785* (K); LUGAS HILL: 1300 m, *Beaman 10551* (K).

55.116.2. Tropidia cf. **curculigoides** Lindl., Gen. Sp. Orch., 497 (1840).

Terrestrial. Hill forest? Elevation: 1200 m.

Collection. PENIBUKAN: 1200 m, *Lamb AL 851/87* (K).

55.116.3. Tropidia pedunculata Blume, Fl. Javae, ser. 2, 1, Orch., 122, t. 40 (1859).

Terrestrial. Hill forest, sometimes on ultramafic substrate; interface between old secondary and primary forest. Elevation: 700–1000 m.

General distribution: Sabah, Sarawak, Peninsular Malaysia, Thailand, Laos, Sumatra, Java, Tanimbar, Timor.

Collections. DALLAS: 900 m, *Carr 3764* (K, SING); HEMPUEN HILL: 800–1000 m, *Beaman 7381* (K); MELANGKAP KAPA: 700–1000 m, *Beaman 8796* (K); SAYAP: 800–1000 m, *Beaman 9789* (K).

55.116.4. Tropidia saprophytica J. J. Sm., Mitt. Inst. Allg. Bot. Hamburg 7: 27, t. 3, f. 16 (1927). Fig. 45, Plate 83A.

Saprophyte. Lower montane forest, on alluvial soils near river. Elevation: 1900 m.

General distribution: Sabah, Kalimantan, Sarawak.

Collection. HAYE-HAYE/TAHUBANG RIVERS: 1900 m, *Phillipps & Lamb SNP 3019* (K).

55.117. TUBEROLABIUM Yamam.

Bot. Mag. Tokyo 38: 209 (1924).

Wood, J. J. (1990). Notes on *Trachoma, Tuberolabium* and *Parapteroceras* (Orchidaceae). Nordic J. Bot. 10: 481–486.

Small monopodial epiphytes. *Stems* short. *Leaves* few, linear-falcate or strap-shaped, 7–14 × 2–3 cm. *Inflorescences* few- to many-flowered, a few flowers open at once or all open together; *peduncle* short; *rachis* fleshy, sulcate, terete, sometimes clavate. *Flowers* rather short-lived or lasting for about a week, up to 9 mm across, white, yellowish or greenish with various purple, brownish purple or red markings; *sepals* and *petals* free, spreading; *dorsal sepal* 4.5 × 2.5 mm; *petals* narrower; *lip* 3-lobed, adnate to the base of the column, immobile, very fleshy, side-lobes very small, tooth-like, mid-lobe very fleshy, laterally compressed, with incurved margins, 3 × 1.5 mm, with a conical spur as long as the blade, spur ornaments absent; *column* short and stout, foot absent; *rostellar projection* short; *stipes* linear, 1–2 times the diameter of the pollinia, *viscidium* small, ovate; *pollinia* 2, entire.

Eleven species distributed from India, Thailand and Peninsular Malaysia, north to Taiwan and the Philippines, south to Indonesia, east to New Guinea, Australia and the Pacific islands.

55.117.1. Tuberolabium rhopalorrhachis (Rchb. f.) J. J. Wood, Nordic J. Bot. 10: 482 (1990).

Dendrocolla rhopalorrhachis Rchb. f., Xenia Orch. 1: 214, t. 86 (1856).

Epiphyte. Low-stature hill forest on ultramafic substrate, lower montane forest. Elevation: 600–1500 m.

General distribution: Sabah, Peninsular Malaysia, Thailand, Sumatra, Java, Maluku.

Collections. HEMPUEN HILL: *Lohok 8* (K); LOHAN RIVER: 800–1000 m, *Beaman 10011* (K), 600 m, *Lamb AL 361/85* (K), 800–1000 m, *AL 443/85* (K); TENOMPOK: 1500 m, *Clemens s.n.* (BM).

55.118. VANDA Jones ex R. Br.

Bot. Reg. 6: t. 506 (1820).

Medium-sized to large monopodial epiphytes or lithophytes. *Stems* usually stiffly erect, 10–100 cm long. *Leaves* linear, strap-shaped, apex praemorse, rigid, V-shaped in cross-section, arranged in 2 rows at an acute angle to the stem, usually decurved. *Inflorescences* large, usually simple, with rather few well-spaced flowers. *Flowers* showy, up to 5 cm across; *sepals* and *petals* free, elliptic-obovate, narrowed at the base, margins often reflexed, twisted or undulate, variously coloured, often tessellated; *dorsal sepal* 2–4 cm long; *lip* immobile, 3-lobed, usually divided into a hypochile and epichile, often shorter than sepals and petals, with a short spur, no adornments within the spur; *column* short and stout, without a distinct foot; *rostellar projection* broad, shelf-like; *stipes* and *viscidium* short and broad; *pollinia* 2, sulcate.

Between 40 and 50 species distributed from Sri Lanka and India north to S China, south to Indonesia, eastward to Australia, New Guinea and the Solomon Islands.

55.118.1. Vanda hastifera Rchb. f.

a. var. **gibbsiae** (Rolfe) P. J. Cribb, **comb. et stat. nov.** Plate 83B.

Vanda gibbsiae Rolfe in Gibbs, J. Linn. Soc. Bot. 42: 158 (1914). Type: KIAU/KAUNG, 800 m, *Gibbs 3970* (holotype K!; isotype BM!).

Epiphyte. Hill forest. Elevation: 800–900 m.

General distribution: Sabah.

Additional collections. DALLAS: 900 m, *Clemens 26307* (BM, K); KIAU: 900 m, *Carr 3188, SFN 26760* (SING); LUBANG: *Clemens 311* (AMES); MINITINDUK GORGE: 800 m, *Carr SFN 26692* (K, SING); MOUNT KINABALU: *Haslam s.n.* (AMES, BM, K).

55.118.2. Vanda helvola Blume, Rumphia 4: 49 (1849). Plate 84A.

Epiphyte. Hill forest, lower montane forest. Elevation: 800–1500 m.

General distribution: Sabah, Sumatra, Java.

Collections. BUNDU TUHAN: 1000 m, *Brentnall 103* (K); KUNDASANG: *Chan 45* (K), 800 m, *Jukian & Lamb AL 33* (K); TENOMPOK: 1500 m, *Clemens 29009* (BM, K).

55.119. VANILLA Mill.

Gard. Dict. abr. ed. 4 (1754).

Scandent or climbing plants, bearing a leaf and a root at each node. *Stems* long, branched, green. *Leaves* large, fleshy, sessile or shortly petiolate, or replaced by small scales. *Inflorescences* lateral, arising from leaf axils, short, few- to many-flowered. *Flowers* rather large, fugaceous; *sepals* and *petals* about equal, free, spreading; *lip* tubular, entire or obscurely lobed, its claw adnate to column, usually with hairy appendages inside; *column* long, foot absent; *anther* pointing downward on the front of the column; *rostellum* broad; *pollinia* granular; *fruit* a long, fleshy, cylindric 'pod', often fragrant when mature; *seeds* relatively large, unwinged.

About 100 species widely distributed in the Old and New World tropics.

55.119.1. Vanilla albida Blume, Catalogus, 100 (1823).

Climber. Hill forest. Elevation: 900 m.

General distribution: Sabah, Peninsular Malaysia, Thailand, Sumatra, Java.

Collections. KIAU: 900 m, *Clemens 148* (AMES), 900 m, *308* (AMES).

55.119.2. Vanilla kinabaluensis Carr, Gardens' Bull. 8: 176 (1935). Type: DALLAS, 900 m, *Clemens 26300* (syntype BM!), 900 m, *26725* (syntype BM!; isosyntype K!); KADAMAIAN RIVER NEAR MINITINDUK GORGE, 800 m, *Carr 3157* (syntype SING n.v.; isosyntype K!).

Climber. Hill forest; climbing up trees in damp forest. Elevation: 800–900 m.

General distribution: Sabah, Peninsular Malaysia.

55.119.3. Vanilla pilifera Holttum, Gardens' Bull. 13: 253 (1951). Plate 84B.

Climber. Hill forest. Known from Kinabalu only from a colour slide by A. Lamb.

General distribution: Sabah, Peninsular Malaysia, Thailand.

Collection. HEMPUEN HILL: *Lamb s.n.* (K photo).

55.119.4. Vanilla aff. **pilifera** Holttum, Gardens' Bull. 13: 253 (1951).

Climber. Hill forest on ultramafic substrate. Elevation: 800–1000 m.

Collection. HEMPUEN HILL: 800–1000 m, *Beaman 7408* (K).

55.119.5. Vanilla sumatrana J. J. Sm., Bull. Jard. Bot. Buit., ser. 3, 2: 22 (1920).

Climber. Lowlands. Elevation: 400 m.

General distribution: Sabah, Sumatra.

Collection. KAUNG: 400 m, *Carr 3417, SFN 27334* (K, SING).

55.119.6. Vanilla sp. 1

Climber. Hill forest on ultramafic substrate. Elevation: 800 m.

Collection. HEMPUEN HILL: 800 m, *Madani SAN 89511* (K).

55.119.7. Vanilla sp. 2

Climber. Lower montane forest. Elevation: 1500 m.

Collection. PINOSUK PLATEAU: 1500 m, *Collenette 643* (K).

PLATE 65.

Pantlingia lamrii. Mount Kinabalu. Photo J. Dransfield.

PLATE 66.

A. Paphiopedilum dayanum. Cult. R. B. G. Kew. Photo R. B. G. Kew.

B. Paphiopedilum javanicum var. **virens.** Mount Kinabalu. Photo R. S. Beaman.

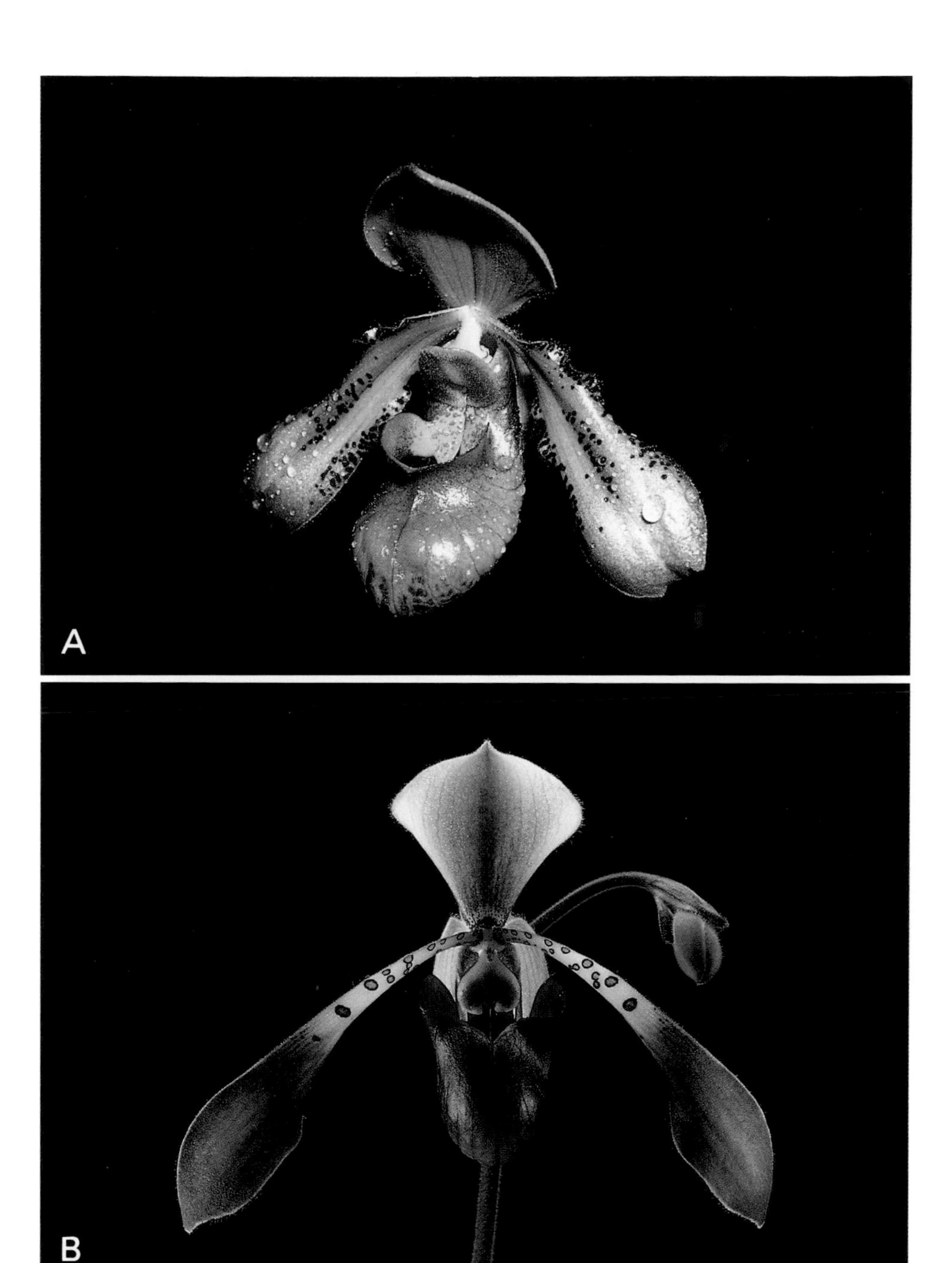

PLATE 67.

A. Paphiopedilum hookerae var. **volonteanum.** Mount Kinabalu, *Beaman 8990.* Photo R. S. Beaman.

B. Paphiopedilum lowii. Cult. R. B. G. Kew. Photo R. B. G. Kew.

PLATE 68.

A. Paphiopedilum rothschildianum. Cult. R. B. G. Kew. Photo R. B. G. Kew.

B. Paphiopedilum rothschildianum. Cult. R. B. G. Kew. Photo R. B. G. Kew.

C. Paphiopedilum rothschildianum. Cult. R. B. G. Kew. Photo R. B. G. Kew.

PLATE 69.

A. **Paraphalaenopsis labukensis.** Cult. Singapore Botanic Gardens. Photo J. B. Comber.

B. **Paraphalaenopsis labukensis.** Cult. Singapore Botanic Gardens. Photo J. B. Comber.

C. **Peristylus hallieri.** Sabah, Sipitang District. Photo J. B. Comber.

PLATE 70.

A. Phaius reflexipetalus. Mount Kinabalu, Lohan River. Photo S. Collenette.

B. Phaius subtrilobus. Mount Kinabalu, Park Headquarters. Photo A. Lamb.

PLATE 71.

A. **Phalaenopsis fuscata.** Sabah. Photo A. Lamb.

B. **Pholidota carnea** var. **carnea.** North Sumatra. Photo J. Dransfield.

C. **Phreatia secunda.** Sabah, Sipitang District. Photo J. B. Comber.

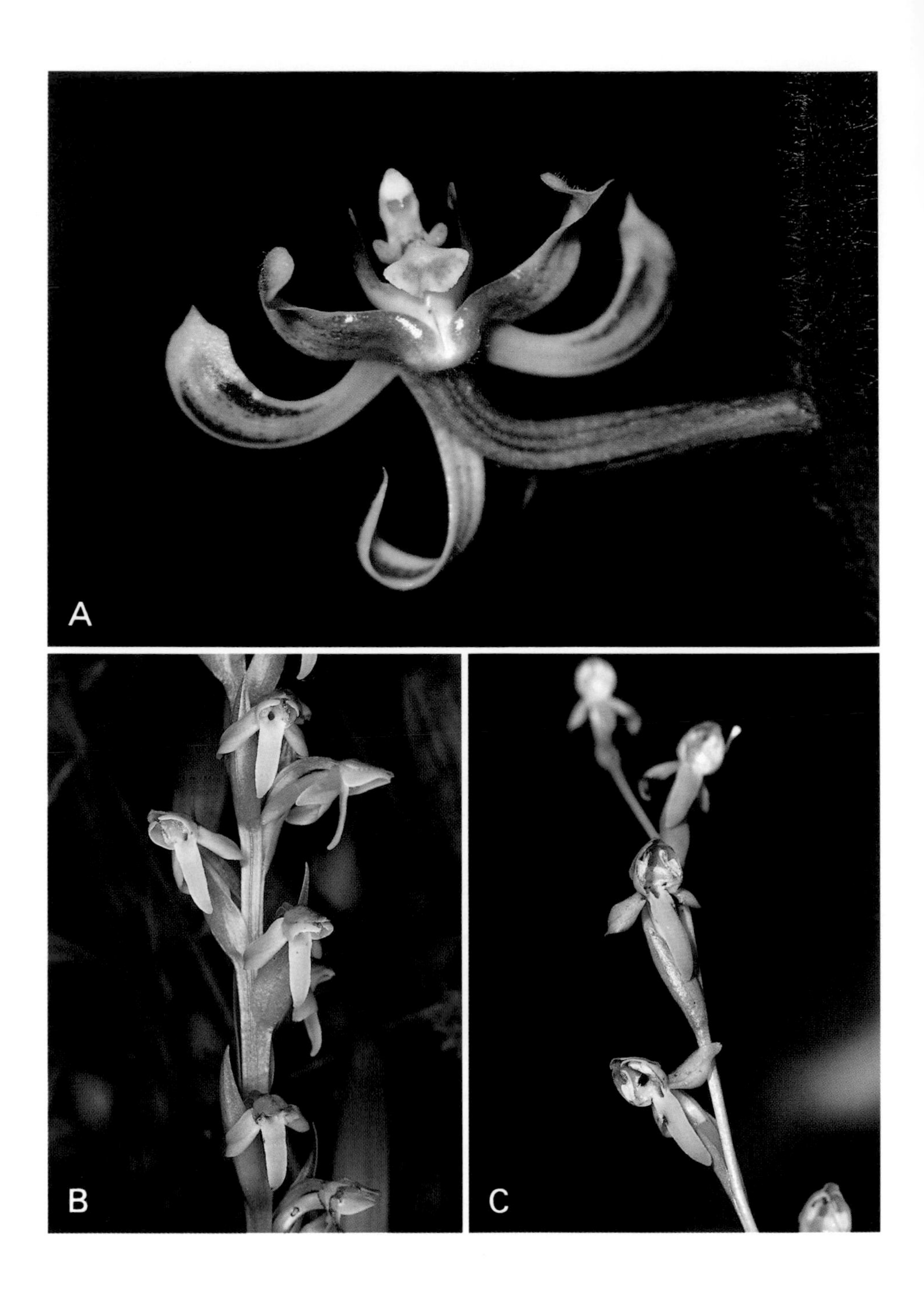

PLATE 72.

A. **Pilophyllum villosum.** Mount Kinabalu, Tahubang River. Photo A. Lamb.

B. **Platanthera kinabaluensis.** Mount Kinabalu. Photo P. J. Cribb.

C. **Platanthera stapfii.** Mount Kinabalu. Photo A. Lamb.

PLATE 73.

A. **Plocoglottis acuminata.** Cult. Poring Orchid Garden. Photo P. J. Cribb.

B. **Plocoglottis gigantea.** Mount Kinabalu, Langanan River, *Lohok 10.* Photo P. J. Cribb.

C. **Podochilus lucescens.** Sabah, Mt. Lumaku. Photo J. B. Comber.

PLATE 74.

A. Porpax borneensis. Mount Kinabalu, Penataran Ridge. Photo A. Lamb.

B. Porrorhachis galbina. Sabah, Crocker Range. Photo A. Lamb.

C. Porrorhachis galbina. Sabah, Crocker Range. Photo A. Lamb.

PLATE 75.

A. Pristiglottis hasseltii. Sabah. Photo A. Lamb.

B. Pteroceras fragrans. Kalimantan, cult. R. B. G. Kew. Photo J. B. Comber.

C. Pteroceras fragrans. Kalimantan, cult. R. B. G. Kew. Photo J. B. Comber.

D. Pteroceras teres. Cult. Poring Orchid Garden, *Lohok 20.* Photo P. J. Cribb.

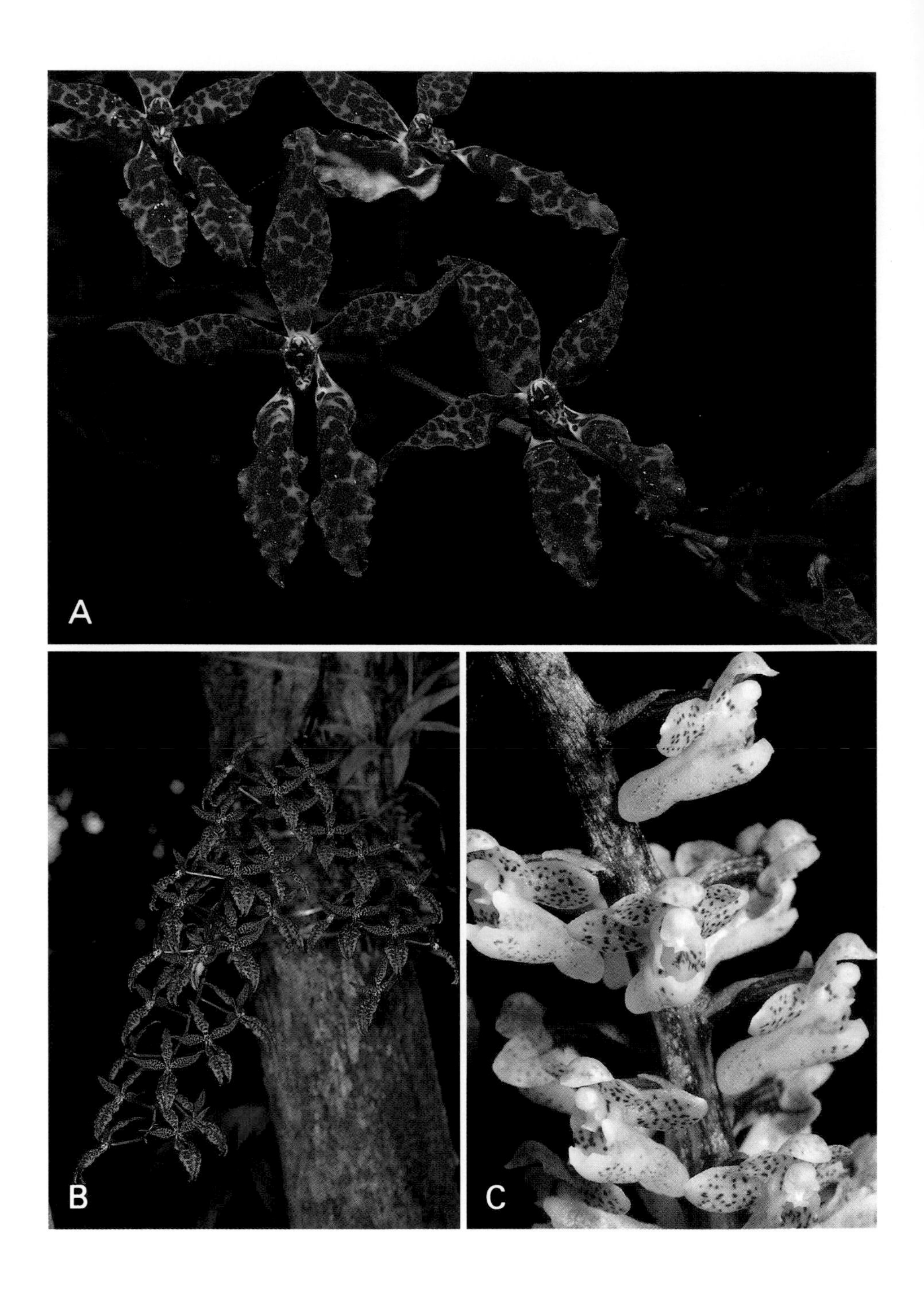

PLATE 76.

A. Renanthera bella. Cult. Tenom Orchid Centre. Photo J. B. Comber.

B. Renanthera bella. Cult. Poring Orchid Garden. Photo P. J. Cribb.

C. Robiquetia crockerensis. Cult. R. B. G. Kew. Photo J. B. Comber.

PLATE 77.

A. **Schoenorchis endertii.** Mount Kinabalu, Lohan River. Photo A. Lamb.

B. **Spathoglottis gracilis.** Mount Kinabalu, Penibukan. Photo A. Lamb.

C. **Spathoglottis microchilina.** Mount Kinabalu. Photo A. Lamb.

D. **Tainia paucifolia.** Sabah, Tenom District. Photo A. Lamb.

PLATE 78.

A. Thecostele alata. Sabah. Photo A. Lamb.

B. Thrixspermum pensile. Java. Photo J. B. Comber.

C. Thrixspermum triangulare. Mount Kinabalu. Photo J. B. Comber.

D. Trichoglottis bipenicillata. Mount Kinabalu, Penataran Ridge. Photo A. Lamb.

PLATE 79.

A. Trichoglottis bipenicillata, pale-flowered form. Cult. Bogor Botanic Gardens, Java. Photo J. B. Comber.

B. Trichoglottis bipenicillata, dark-flowered form. Mount Kinabalu. Photo A. Lamb.

PLATE 80.

A. Trichoglottis collenetteae, pale-flowered form. Sabah. Photo A. Lamb.

B. Trichoglottis collenetteae, dark-flowered form. Mount Kinabalu. Photo J. B. Comber.

C. Trichoglottis smithii. Cult. Tenom Orchid Centre. Photo A. Lamb.

D. Trichoglottis smithii. Cult. R. B. G. Kew. Photo R. B. G. Kew.

PLATE 81.

A. **Trichoglottis winkleri** var. **minor.** Mount Kinabalu, Lohan River, cult. Tenom Orchid Centre. Photo J. B. Comber.

B. **Trichotosia brevipedunculata.** Mount Kinabalu, Park Headquarters, *Wood 619.* Photo J. B. Comber.

C. **Trichotosia ferox.** Peninsular Malaysia. Photo J. Dransfield.

PLATE 82.

A. Trichotosia microphylla. Mount Kinabalu, Park Headquarters, *Wood 620.* Photo J. B. Comber.

B. Trichotosia mollicaulis. Sabah. Photo A. Lamb.

PLATE 83.

A. Tropidia saprophytica. Mount Kinabalu, Haye-haye River. Photo A. Lamb.

B. Vanda hastifera var. **gibbsiae.** Sabah. Photo A. Lamb.

PLATE 84.

A. Vanda helvola. Cult. Tenom Orchid Centre. Photo J. B. Comber.

B. Vanilla pilifera. Mount Kinabalu, Hempuen Hill. Photo A. Lamb.

55.120. VRYDAGZYNEA Blume

Fl. Javae ser. 2, 1, Orch.: 71, (1858).

Terrestrial herbs. *Rhizome* decumbent, rooting at nodes. *Stems* weak, fleshy. *Leaves* few, green, sometimes with a median white stripe. *Inflorescence* usually short and densely many-flowered. *Flowers* small, resupinate, not opening widely; *dorsal sepal* and *petals* connivent, forming a hood; *lip* parallel to column, entire, *hypochile* bearing a prominent spur projecting between lateral sepals and containing 2 stalked glands; *column* very short; *stigma* bilobed; *pollinia* 2.

Between 20 and 40 species, according to opinion, distributed from northern India to Taiwan and through Malaysia and Indonesia eastward to New Guinea, Australia and the Pacific islands.

55.120.1. Vrydagzynea albida (Blume) Blume, Fl. Javae, ser. 2, 1, Orch.: 75, t. 19, f. 2: 1–7 (1859).

Etaeria albida Blume, Bijdr., 410 (1825).

Terrestrial. Lowlands, hill forest, lower montane forest. Elevation: 800–1500 m.

General distribution: Sabah, Sarawak, widespread from India, Thailand & Peninsular Malaysia east to the Philippines & probably New Guinea.

Collections. KINATEKI RIVER: 800 m, *Collenette A 115* (BM); KOLOPIS RIVER HEAD: 1400 m, *Carr 3765, SFN 28057* (SING); PORING: *Cribb 89/58* (K); TENOMPOK: 1500 m, *Clemens s.n.* (BM), 1500 m, *29242* (K).

55.120.2. Vrydagzynea argentistriata Carr, Gardens' Bull. 8: 183 (1935). Type: BUNDU TUHAN, 900 m, *Carr 3713, SFN 28051* (holotype SING!; isotype K!).

Terrestrial. Hill forest. Elevation: 900 m. Known only from the type.

Endemic to Mount Kinabalu.

55.120.3. Vrydagzynea bicostata Carr, Gardens' Bull. 8: 185 (1935). Type: KINUNUT VALLEY HEAD, 1200 m, *Carr 3344, SFN 27134* (holotype SING!; isotype K!).

Terrestrial. Hill forest, lower montane forest. Elevation: 1100–1500 m.

General distribution: Sabah.

Additional collections. HAYE-HAYE/TINEKUK RIVERS: 1100 m, *Lamb AL 46/83* (K); MESILAU RIVER: 1500 m, *RSNB 4052* (K).

55.120.4. Vrydagzynea bractescens Ridl., Kew Bull. 1926: 87 (1926).

Terrestrial. Hill forest. Elevation: 1200 m.

General distribution: Sabah, Sarawak, Sumatra.

Collection. LUBANG: 1200 m, *Carr 3766* (K, SING).

55.120.5. Vrydagzynea elata Schltr., Feddes Repert. 9: 430 (1911).

Terrestrial. Hill forest. Elevation: 1200 m.

General distribution: Sabah, Sarawak.

Collection. BUNDU TUHAN: 1200 m, *Carr 3384, SFN 27979* (K, SING).

55.120.6. Vrydagzynea grandis Ames & C. Schweinf., Orch. 6: 16 (1920). Type: KIAU, *Clemens 340* (holotype AMES mf).

Terrestrial. Hill forest, lower montane forest. Elevation: 900–1500 m.

General distribution: Sabah.

Additional collections. HAYE-HAYE/TINEKUK RIVERS: 900–1100 m, *Lamb AL 51/83* (K); KIAU: *Clemens 355* (AMES); LUBANG: 1300 m, *Carr 3195* (BM, K, SING); PENIBUKAN: 1200–1500 m, *Clemens s.n.* (BM, BM); TENOMPOK: 1500 m, *Clemens 28242* (BM), 1500 m, *28344* (BM, K); TINEKUK RIVER: 1100 m, *Clemens 51721* (BM).

55.120.7. Vrydagzynea indet.

Collection. MINITINDUK RIVER: 900 m, *Carr SFN 26740* (SING).

55.121. ZEUXINE Lindl.

Coll. Bot., app., 18, as *Zeuxina* (1826).

Terrestrial herbs. *Rhizome* decumbent, rooting at nodes. *Stems* weak, fleshy. *Leaves* ovate, elliptic or linear, sessile on a broad sheath, or petiolate, sometimes with a coloured median nerve. *Inflorescence* few- to many-flowered, lax or dense. *Flowers* small, resupinate, not opening widely; *dorsal sepal* and *petals* connivent, forming a hood; *lateral sepals* enclosing base of lip; *lip* with a saccate base usually with inflexed margins and containing 2 glands, ± sulcate on lower surface, blade transversely widened and bilobed, small, connected to the saccate base by a short neck or an elongated claw; *column* short, with or without appendages in front; *stigma* bilobed; *pollinia* 2.

Between 40 and 50 species distributed throughout the Old World tropics of Africa, Asia and the Pacific islands. Kores (1989: 32) quotes a figure of 70 species. The circumscription of the genus and the species delimitations are uncertain.

55.121.1. Zeuxine gracilis (Breda) Blume, Fl. Javae, ser. 2, 1, Orch.: 69, t. 18, f. 2 & t. 23D (1858).

Psychechilos gracile Breda, Gen. Sp. Orchid. Asclep., t. 9 (1827).

Terrestrial. Hill forest, sometimes on ultramafic substrate, lower montane forest. Elevation: 800–1900 m.

General distribution: Sabah, Peninsular Malaysia, islands off Sumatra, Java & Krakatau.

Collections. KADAMAIAN RIVER ABOVE MINITINDUK GORGE: 900 m, *Carr 3177, SFN 26756* (K, SING); LOHAN RIVER: 800–1000 m, *Beaman 9346* (K); MESILAU CAVE TRAIL: 1700–1900 m, *Beaman 9147* (K).

55.121.2. Zeuxine papillosa Carr, Gardens' Bull. 8: 189 (1935). Type: KADAMAIAN RIVER NEAR MINITINDUK GORGE, 900 m, *Carr 3159, SFN 26671* (holotype SING!; isotype AMES mf).

Terrestrial. Hill forest by river. Elevation: 900 m. Known only from the type.

Endemic to Mount Kinabalu.

55.121.3. Zeuxine strateumatica (L.) Schltr., Bot. Jahrb. Syst. 45: 394 (1911).

Orchis strateumatica L., Sp. Pl., 943 (1753).

Terrestrial. Hill forest, open grassland. Elevation: 900 m.

General distribution: Sabah, widespread from Afghanistan to Japan, south to New Guinea.

Collections. DALLAS: 900 m, *Carr 3722, SFN 28053* (K, SING), 900 m, *Clemens 27117* (K), 900 m, *27555* (BM, K).

EXCLUDED TAXA

Armodorum sulingii (Blume) Schltr.

Lamb (pers. comm.) comments that some fragments of material at SING, possibly collected by Clemens, were determined as *A. sulingii* by Tan. There is, however, no mention of this by Tan (1975, 1976) in his papers on the taxonomy of the *Arachnis* group, who gives the distribution as Sumatra, Java and Bali.

"Bulbophyllum petreianum"

This name was reported by Burbidge (1880: 264) and repeated by Lamb and Chan (1978: 237). It has not been validly published.

Cheirostylis montana Blume

This rather insignificant forest-floor terrestrial has been found recently in the hills near Tenom. It seems likely that the species may be discovered on Mount Kinabalu in suitable habitats in the hill forest zone.

"Coelogyne racemosa"

This name appears in Burbidge (1880: 264), who states that the species was growing at a ford near Kaung. The name has not been published, and Burbidge probably was making reference to one of the necklace orchids of *Coelogyne* section *Tomentosae.*

Collabium bicameratum (J. J. Sm.) van den Burg, ined.

Collabium is distinguished from *Chrysoglossum* primarily by the immobile lip and column lacking basal fleshy keels. It is currently under revision at Leiden. The species has been found at 1500 m in the Crocker Range. Lamb (pers. comm.) expects that it may occur at about this elevation on Mount Kinabalu.

Dendrobium cinnabarinum Rchb. f. var. **angustitepalum** Carr

This beautiful orange-flowered member of section *Rhopalanthe* is frequently found between 900 and 1400 m in parts of the Crocker Range. So far it has not been observed or collected on Mount Kinabalu. This is peculiar because many sites would appear to offer an ideal habitat.

Erythrorchis altissima (Blume) Blume

This is a spectacular climber reputed to be the largest of all saprophytic orchids. The seeds, unlike the majority of species in the family, are large, most of the size taken up by an irregular, bifurcate wing. The plant has been found in the Crocker Range and on the Bengkoka Peninsula north of Mount Kinabalu. It is unlikely to be overlooked when in flower, and may well turn up on Mount Kinabalu.

Paphiopedilum barbatum (Lindl.) Pfitzer (Syn.: *P. nigritum* (Rchb. f.) Pfitzer)

Paphiopedilum nigritum was described in 1882 and said to have come from Mount Kinabalu. Many early importations were distributed with false provenance to mislead competitors, and it is likely that this taxon originated from elsewhere. *Paphiopedilum barbatum* currently is recognised from Peninsular Malaysia and Penang Island only.

Paphiopedilum lawrenceanum (Rchb. f.) Pfitzer

This is a species from Sarawak that has been reported to occur on the banks of the Tampassuk River on the lower west slopes of Mount Kinabalu. It seems to no longer occur there, and the record remains unsubstantiated.

Spathoglottis confusa J. J. Sm.

This robust, sulphur-yellow flowered species was erroneously attributed to Kinabalu by Lamb and Chan (1978: 249) on the basis of a statement in Burbidge (1880: 107). He described at the foot of Marai Parai "a lovely yellow flowered terrestrial orchid belonging to the genus *Spathoglottis*, but quite distinct from *S. aurea*." *Spathoglottis confusa* was not published by J. J. Smith until 1932. Wc have seen several collections of *S. microchilina* from Marai Parai. Most likely the latter is the species noted by Burbidge. A colour photograph shown on Plate 28 in Luping et al. (1978) probably was taken on either Mt. Alab or Mt. Lotung, where, according to Lamb and Chan, *S. confusa* is abundant.

ACKNOWLEDGMENTS

We are grateful for the opportunities provided by Datuk Lamri Ali, Director of Sabah Parks, to work in Kinabalu Park. Various members of the Park staff, particularly Park Ecologist Jamili Nais, former Park Ecologist Anthea Phillipps, Naturalist Ansow Gunsalam, and Ranger Gabriel Sinit, were very helpful in our visits. Harry Lohok of the Poring Orchid Garden has contributed significantly to the collections of Kinabalu orchids.

A Fulbright Fellowship made it possible for the Beaman team to carry out botanical fieldwork in Sabah in 1983–84. Research on the Kinabalu inventory subsequently has been supported by NSF grants BSR-8507843 and BSR-8822696 to Michigan State University and by the Department of Botany and Plant Pathology, Michigan State University.

We would like to acknowledge the generous financial support provided by Sabah Parks, the TOBU Department Store, Tokyo, Japan, and the American Orchid Society toward publication of the colour plates.

Valuable financial support has also been provided by the Foundation for the Preservation and Study of Wild Orchids, Zürich, Switzerland (Stiftung zum Schutze und zur Erhaltung Wildwachsender Orchideen, Zürich).

Field work of the Beamans was aided by many persons who came to Sabah for their own studies but served effectively as participants in collecting orchids and many other plants. We were fortunate to have been visited by Dr. Kiat W. Tan, Director of the Singapore Botanic Gardens, Dr. John T. Atwood and Mrs. Libby Besse from the Marie Selby Botanical Gardens, Dr. Eric A. Christenson from the University of Connecticut, and Drs. James H. Asher and Stephen C. Bromley from Michigan State University, who stimulated our orchid collecting. Dr. Teofila Beaman was particularly important in helping our guests, spotting orchids that the others of us did not notice, pressing and sorting plants, and helping with the data analysis for preparation of the manuscript.

Prof. Ghazally Ismail, formerly Dean of the Faculty of Science and Natural Resources of the National University of Malaysia, Sabah Campus, facilitated in many ways the field work of the Beamans. Jacinto C. Regalado, Jr. participated extensively in preliminary phases of producing the Enumeration, including computer programming and recording specimen data for the orchids at Singapore. Dr. A. F. Clewell has provided continuing support for the participation of Reed Beaman in the project. Mrs. Sheila Collenette has given many vivid accounts of her orchid-collecting experiences on Kinabalu, contributed several photographs for the colour plates and otherwise significantly facilitated this research. Dr. C. E. Anderson provided many helpful editorial comments on the introductory section of the manuscript.

Mr. Anthony Lamb of the Agricultural Research Station, Tenom, who has a great knowledge of Mount Kinabalu and its orchid flora, has contributed extensively throughout this project. Mr. Chan Chew Lun provided much useful information about orchids while we were in Sabah, and has allowed us to use a number of his fine drawings. Mr. J. J. Vermeulen of the Rijksherbarium, Leiden, provided useful information about the large and difficult genus *Bulbophyllum*. Dr. E. F. de Vogel, also of Leiden, likewise has been most helpful with genera in his area of expertise, particularly those of the subtribe Coelogyninae. Mr. M. A. Clements of the Australian National Botanic Gardens, Canberra, provided helpful information on several nomenclatural and taxonomic problems.

Publication of this treatment has been made possible through the support of the Royal Botanic Gardens, Kew, with the encouragement of the Keeper of the Herbarium, Professor G. Ll. Lucas. The Editor of the Kew Bulletin, Dr. J. M. Lock, has given much useful counsel. We also appreciate the help of Messrs. W. Loader, D. Martindale, K. McPaul and M. Jackson in the Computer Section. Ms C. E. Powell and Dr. R. K. Brummitt provided us access to the list of standardised author abbreviations while it was unpublished. Ms E. A. Dauncey contributed to the description of *Dendrobium hamaticalcar.* Ms E. Catherine illustrated a number of the new taxa and other significant species. Mr. M. Fothergill figured one illustration. Useful technical assistance was provided by Ms Y. Harvey and Ms S. Robbins. The Latin diagnoses have been provided by Mr. A. Radcliffe-Smith. Dr. P. J. Cribb has a first-hand familiarity with the orchids of Mount Kinabalu, and has given much helpful information and editorial counsel in the preparation of the manuscript. We appreciate the use of some of his orchid photographs. Mr. J. B. Comber, Dr. J. Dransfield and Mr. A. Lamb contributed generously from their photographic collections of Kinabalu orchids. Mr. G. Cubitt provided a photograph of *Dendrochilum dewindtianum* and Dr. W. Meijer supplied the photograph of the Pinosuk Plateau in 1964. Dr. W. Rossi provided a photograph of *Nabaluia angustifolia.* Mr. M. J. S. Sands provided photographs of general views of Kinabalu. Remaining photographs were taken by staff of Media Resources, Royal Botanic Gardens, Kew, in particular M. Svanderlik. Mr. R. Vickery of the Natural History Museum, London, facilitated our work in that institution.

We appreciate the opportunity to use the facilities or the loan of specimens from the following herbaria: BM, BO, E, G, K, L, MSC, SAN, SING, SNP (Sabah Parks) and UKMS.

LITERATURE CITED

Ames, O. & Schweinfurth, C. (1920). The Orchids of Mount Kinabalu, British North Borneo. Merrymount Press, Boston.

Anonymous (1989). Orchid Herbarium of Oakes Ames (microfiches 1–307). Chadwyck-Healey. [Accompanied by Index to the Orchid Herbarium of Oakes Ames in the Botanical Museum of Harvard University, by Garay, L. A., Anderson, B. & Robinson, O. (1989). Chadwyck-Healey.]

Atwood, J. T. (1985). Pollination of *Paphiopedilum rothschildianum*: brood-site deception. National Geogr. Research, Spring 1985: 247–254.

Beaman, R. S., Beaman, J. H., Marsh, C. W. & Woods, P. V. (1986). Drought and forest fires in Sabah in 1983. Sabah Soc. J. 8: 10–30.

Beaman, J. H. & Beaman, R. S. (1990). Diversity and distribution patterns in the flora of Mount Kinabalu. *In* Baas, P., Kalkman, K. & Geesink, R., eds. The Plant Diversity of Malesia. Kluwer Academic Publishers, Dordrecht/Boston/London, pp. 147–160.

Beaman, J. H. & Regalado, J. C. Jr. (1989). Development and management of a microcomputer specimen-oriented database for the flora of Mount Kinabalu. Taxon 38: 27–42.

Beaman, J. H., Aman, R. H., Nais, J., Sinit, G. & Biun, A. (In press). Mount Kinabalu place names in Dusun and their meaning. *In* Phillipps, A. & Wong, K. M., eds. Kinabalu: Summit of Borneo, 2nd Ed.

Brummitt, R. K. (1992). Vascular Plant Families and Genera. Royal Botanic Gardens, Kew.

Brummitt, R. K. & Powell, C. E., eds. (1992). Authors of Plant Names. Royal Botanic Gardens, Kew.

Burbidge, F. W. (1880). The Gardens of the Sun. John Murray, London.

Burkill, I. H. (1913). The explosive flowers of *Plocoglottis porphyrophylla*. Gardens' Bull. 1: 190–192.

Carr, C. E. (1935). Two collections of orchids from British North Borneo, Part I. Gardens' Bull. 8: 165–240.

Carter, A. M. (1982). The itinerary of Mary Strong Clemens in Queensland, Australia. Contr. Univ. Michigan Herb. 15: 163–169.

Case, F. W., Jr. (1987). Orchids of the Western Great Lakes Region. Cranbrook Institute of Science Bull. 48, Bloomfield Hills, Michigan.

Christenson, E. A. (1985). Sarcanthine genera 2: *Schoenorchis* Blume. Amer. Orchid Soc. Bull. 54: 850–854.

Christenson, E. A. (1992). An enigmatic blue. Amer. Orchid Soc. Bull. 61: 242–247.

Comber, J. B. (1990). Orchids of Java. Bentham-Moxon Trust/Royal Botanic Gardens, Kew.

Conn, B. J. (1990). Mary Strong Clemens: a botanical collector in New Guinea (1935–1941). *In* Short, P. S., ed. History of systematic botany in Australasia. Australia Systematic Botany Society Inc., pp. 217–229.

Corner, E. J. H. (1964). Royal Society expedition to North Borneo, 1961: Reports. Proc. Linnean Soc. London 175: 9–56, 21 pl.

Cribb, P. (1987). The Genus *Paphiopedilum.* The Royal Botanic Gardens, Kew in association with Collingridge.

Cubitt, G. & Payne, J. (1990). Wild Malaysia. New Holland (Publishers) Ltd.

De Vogel, E. F. (1984). Precursor to a revision of the genera *Entomophobia* (gen. nov.), *Geesinkorchis* (gen. nov.), *Nabaluia* and *Chelonistele* (Orchidaceae–Coelogyninae). Blumea 30: 197–205.

De Vogel, E. F. (1988). Revisions in Coelogyninae (Orchidaceae) III. The genus *Pholidota.* Orch. Monogr. 3: 1–118.

De Vogel, E. F. (1992). Revisions in Coelogyninae (Orchidaceae) IV. *Coelogyne* section *Tomentosae.* Orch. Monogr. 6: 1–42.

Dressler, R. L. (1981). The Orchids: Natural History and Classification. Harvard University Press.

Dressler, R. L. (1990a). The Orchids: Natural History and Classification. Ed. 2. Harvard University Press.

Dressler, R. L. (1990b). The Spiranthoideae: grade or subfamily? Lindleyana 5: 110–116.

Dressler, R. L. (1990c). The major clades of the Orchidaceae-Epidendroideae. Lindleyana 5: 117–125.

Fowlie, J. A. (1984). Malaya revisited. Part XXIX: Rediscovering the habitat of *Paphiopedilum dayanum* on serpentine cliffs on Mount Kinabalu in Eastern Malaysia (formerly north Borneo). Orch. Digest 49: 125–129.

Fowlie, J. A. & Lamb, T. [=A.] (1983). Malaya revisited. Part XXIV: *Paphiopedilum dayanum* and *Paphiopedilum rothschildianum* on serpentine cliffs on Mount Kinabalu, Eastern Malaysia (formerly North Borneo). Orch. Digest 47: 175–182.

Gibbs, L. S. (1914). A contribution to the flora and plant formations of Mount Kinabalu and the highlands of British North Borneo. J. Linn. Soc., Bot. 42: 1–240, 8 pl.

Grell, E., Haas-von Schmude, N. F., Lamb, A. & Bacon, A. (1988). Re-introducing *Paphiopedilum rothschildianum* to Sabah, North Borneo. Amer. Orchid Soc. Bull. 57: 1238–1246.

Holttum, R. E. (1964). Orchids of Malaya. Flora of Malaya 1. 3rd ed. Government Printer, Singapore.

Kitayama, K. (1991). Vegetation of Mount Kinabalu Park, Sabah, Malaysia. Map of physiognomically classified vegetation. East-West Center, Honolulu, Hawaii.

Kores, P. J. (1989). A precursory study of Fijian orchids. Allertonia 5: 1–222.

Lamb, A. & Chan, C. L. (1978). The orchids. *In* Luping, M., Chin, W. & Dingley, E. R., eds. Kinabalu, Summit of Borneo. Sabah Society Monograph.

Lamb, A. & Chan, C. L. (In press). The orchids. *In* Phillipps, A. & Wong, K. M., eds. Kinabalu, Summit of Borneo, 2nd Ed.

Langdon, R. F. (1981). The remarkable Mrs Clemens. *In* Carr, D. J. & Carr, S. G. M., eds. People and Plants in Australia. Academic Press, pp. 374–383.

Lowry, J. B., Lee, D. W. & Stone, B. C. (1973). Effect of drought on Mount Kinabalu. Malayan Nature J. 26: 178–179.

Luping, M., Chin, W. & Dingley, E. R., eds. (1978). Kinabalu: Summit of Borneo. Sabah Society Monograph.

Mackinnon, J. (1975). Borneo. Time-Life Books, Amsterdam.

Merrill, E. D. (1916). On the utility of field labels in herbarium practice. Science n.s. 44: 664–670.

Mueller-Dombois, D. & Ellenberg H. (1974). Aims and Methods of Vegetation Ecology. John Wiley & Sons, New York, etc.

Pain, S. (1989). The case of the stolen slippers. New Scientist 122(1670), 24 June 1989: 48–53.

Parris, B. S., Beaman, R. S. & Beaman, J. H. (1992). The Plants of Mount Kinabalu: 1. Ferns and Fern Allies. Royal Botanic Gardens, Kew.

Pijl, L. van der & Dodson, C. H. (1966). Orchid Flowers: Their Pollination and Evolution. Univ. of Miami Press, Coral Gables.

Rasmussen, F. N. (1985). Orchids. *In* Dahlgren, R. M. T., Clifford, H. T. & Yeo, P. F. The Families of the Monocotyledons. Springer-Verlag, Berlin, etc.

Ridley, H. N. (1888). A revision of the genera *Microstylis* and *Malaxis.* J. Linn. Soc. Bot. 24: 308–351.

Sato, T. (1991). Flowers and Plants of Mt. Kinabalu. Matoba 195, Mizuhashi, Toyama-shi, Japan.

Seidenfaden, G. (1977). Orchid genera in Thailand V. Orchidoideae. Dansk Bot. Arkiv 31(3): 1–149.

Seidenfaden, G. (1982). Orchid genera in Thailand X. *Trichotosia* Bl. and *Eria* Lindl. Opera Bot. 62: 1–157.

Senghas, K. (1973). *In* Brieger F. G., Maatsch, K. & Senghas, K. Rudolph Schlechter, Die Orchideen, 3. Aufl., 4. Lief. Paul Parey, Berlin & Hamburg.

Smith, J. J. (1931). On a collection of Orchidaceae from Central Borneo. Bull. Jard. Bot. Buit., ser. 3, 11: 83–160.

Smith, J. M. B. (1979). Vegetation recovery from drought on Mt. Kinabalu. Malayan Nature J. 32: 341–342.

Stapf, O. (1894). On the flora of Mount Kinabalu, in North Borneo. Trans. Linn. Soc. London, Bot. 4: 69–263, pl. 11–20.

Steenis, C. G. G. J. (1969). Editorial. Flora Males. Bull. 23: 1669–1671.

Steenis-Kruseman, M. J. van (1950). Malaysian Plant Collectors and Collections. Flora Malesiana 1(1): 1–639.

Turner, H. (1992). A revision of the orchid genera *Ania* Lindley, *Hancockia* Rolfe, *Mischobulbum* Schltr. and *Tainia* Blume. Orchid Monogr. 6: 43–100.

Tan, K. W. (1975). Taxonomy of *Arachnis, Armodorum, Esmeralda* & *Dimorphorchis*, Orchidaceae, Part I. Selbyana 1(1): 1–15.

Tan, K. W. (1976). Taxonomy of *Arachnis, Armodorum, Esmeralda* & *Dimorphorchis*, Orchidaceae, Part II. Selbyana 1(4): 365–373.

Turrill, W. B. (1936). J. Clemens. XXXI–Miscell. Notes. Kew Bull. 1936: 287–289.

Vermeulen, J. J. (1991). Orchids of Borneo, 2: *Bulbophyllum*. Royal Botanic Gardens, Kew & Toihaan Publishing Co., Kota Kinabalu.

Whitehead, J. (1893). Exploration of Mount Kina Balu, North Borneo. Gurney and Jackson, London.

Wood, J. J. (1990). The diversity of *Dendrobium* in Borneo. Malayan Orchid Rev. 24: 33–37.

Wood, J. & Bell, S. (1992). 194. *Arachnis longisepala*, Orchidaceae. Kew Magazine 9: 55–59.

INDEX TO NUMBERED COLLECTIONS

Prefixes for collectors' numbers are not included.

(55.62.19); 8667 (55.36.13); 8668 (55.95.6); 8700 (55.17.43); 8701 (55.17.51); 8702 (55.57.2); 8709 (55.10.3); 8745 (55.45.29); 8756 (55.89.2); 8760 (55.20.4); 8761 (55.28.11); 8764 (55.36.13); 8779 (55.57.1); 8782 (55.77.3a); 8796 (55.116.3); 8812a (55.3.2a); 8854 (55.28.6); 8855 (55.83.5); 8856 (55.1.1); 8857 (55.112.4); 8858 (55.66.8); 8981 (55.93.4); 8982 (55.36.37); 8983 (55.94.2); 8984 (55.66.3); 8990 (55.83.2a); 8992 (55.32.2); 8993 (55.28.4); 8994a (55.98.1); 8996 (55.17.43); 8997 (55.28.14); 8998 (55.62.19); 8999 (55.113.7); 9034 (55.95.6); 9035 (55.11.4); 9036a (55.36.7); 9046 (55.89.2); 9047 (55.28.23); 9048a (55.17.13); 9065 (55.66.3); 9066 (55.84.1); 9068 (55.36.58); 9070a (55.105.1); 9077 (55.88.1); 9129 (55.57.2); 9144 (55.17.88); 9146 (55.113.7); 9147 (55.121.1); 9158 (55.88.1); 9189 (55.36.13); 9191 (55.61.1); 9192 (55.87.1); 9253 (55.28.30); 9254 (55.114.7); 9255 (55.48.6); 9256 (55.11.5); 9257 (55.45.43); 9258 (55.18.14); 9259 (55.36.58); 9260 (55.88.3); 9261 (55.67.3); 9262 (55.94.2); 9263 (55.78.9); 9310 (55.11.15); 9340 (55.105.1); 9341 (55.83.5); 9346 (55.121.1); 9350a (55.84.1); 9370 (55.48.4); 9374 (55.45.11); 9385 (55.83.3a); 9387 (55.28.10); 9388 (55.115.10); 9389 (55.28.36); 9390 (55.90.3); 9391 (55.89.2); 9392a (55.45.29); 9393 (55.83.2a); 9455 (55.18.7); 9465 (55.28.10); 9467 (55.28.15); 9482 (55.18.16); 9486 (55.66.9); 9524 (55.38.1); 9525 (55.9.2); 9548 (55.18.4); 9567 (55.43.2); 9568 (55.17.35); 9569 (55.17.31); 9570 (55.32.2); 9571 (55.45.9a); 9572 (55.17.51); 9573 (55.28.15); 9574 (55.36.53); 9575 (55.17.18); 9575a (55.45.29); 9601 (55.28.7); 9602 (55.89.2); 9604 (55.36.27); 9612 (55.28.19b); 9614 (55.66.5); 9615 (55.82.1); 9646 (55.45.19); 9647 (55.93.1); 9648 (55.78.11); 9649 (55.36.59); 9781 (55.28.30); 9782 (55.11.7); 9789 (55.116.3); 9792 (55.10.2); 9812 (55.44.1); 9813 (55.44.1); 9832 (55.45.29); 9886 (55.36.22); 9887 (55.20.2); 9888 (55.37.23); 9889 (55.37.14); 9890 (55.36.3); 9891 (55.28.2); 9892 (55.20.1); 9893 (55.45.41); 9894 (55.45.38b); 9941 (55.28.10); 9948 (55.28.11); 9949 (55.36.27); 9950 (55.95.2); 9951 (55.45.38a); 9952 (55.45.23); 9974 (55.45.18); 10006 (55.114.7); 10007 (55.45.26); 10008 (55.36.15); 10009 (55.27.7); 10010 (55.11.11); 10011 (55.117.1); 10012 (55.95.6); 10013 (55.95.4); 10014 (55.23.2); 10015 (55.108.8); 10321 (55.36.37); 10341 (55.105.1); 10345 (55.36.62); 10346 (55.32.1); 10347 (55.37.15); 10348 (55.42.1); 10349 (55.28.11); 10350 (55.22.1); 10351 (55.115.12); 10352 (55.108.8); 10353 (55.113.9); 10355 (55.22.1); 10356 (55.45.37); 10365 (55.47.2); 10517 (55.17.76); 10518 (55.14.1); 10519 (55.45.13); 10520 (55.45.14); 10521 (55.48.4); 10535 (55.8.2); 10537 (55.35.1); 10551 (55.116.1); 10560 (55.66.14); 10589 (55.97.1); 10596 (55.79.1); 10623 (55.18.11); 10646 (55.57.1); 10659 (55.105.4); 10668 (55.48.4); 10669 (55.17.2); 10670 (55.36.7); 10671 (55.45.28); 10672 (55.45.38a); 10673 (55.19.1); 10674 (55.14.1); 10675 (55.62.19); 10676 (55.79.1); 10677a (55.45.25); 10687 (55.113.7); 10710 (55.94.4); 10711 (55.62.18); 10712 (55.87.1); 10715 (55.95.2); 10718 (55.78.6); 10722 (55.36.6); 10723 (55.43.3); 10754 (55.45.8); 10770 (55.73.2); 10771 (55.45.19); 10772 (55.28.14); 10773 (55.27.8); 10791 (55.45.41); 10800 (55.3.3).

Brentnall 100 (55.89.4); 103 (55.118.2); 107 (55.62.22); 108 (55.17.48); 123 (55.36.21); 124 (55.62.12); 125 (55.36.14); 130 (55.20.5); 133 (55.45.9); 134 (55.17.39); 138 (55.36.7); 146 (55.17.43); 152 (55.36.7); 154 (55.73.1); 156 (55.103.2); 159 (55.6.8); 160 (55.73.1); 165 (55.45.3).

Carr 2665 (55.28.3); 3005 (55.28.6); 3008 (55.28.26); 3009 (55.111.1); 3012 (55.89.4); 3013 (55.11.1); 3015 (55.12.2); 3016 (55.17.49); 3019 (55.36.16); 3020 (55.17.48); 3026 (55.66.2); 3027 (55.11.7); 3030 (55.11.5); 3033 (55.17.31); 3036 (55.83.1); 3037 (55.36.28); 3038 (55.11.4); 3041 (55.62.19); 3043 (55.62.10); 3044 (55.36.37); 3046 (55.11.19); 3047 (55.95.4); 3048 (55.45.11); 3049 (55.17.62); 3056 (55.17.15); 3058 (55.17.18); 3060 (55.17.68); 3061 (55.17.15, 55.17.31); 3063 (55.45.56); 3067 (55.30.2); 3068 (55.36.31); 3069 (55.36.25); 3071 (55.17.35); 3075 (55.28.5b); 3079 (55.115.12); 3084 (55.8.2); 3087 (55.36.27); 3088 (55.36.6); 3089 (55.32.3); 3090 (55.10.2); 3091 (55.28.1b); 3095 (55.95.1); 3096 (55.95.4); 3097 (55.17.54a); 3099 (55.17.52); 3100 (55.70.2); 3103 (55.115.8); 3106 (55.45.28); 3107 (55.103.1); 3110 (55.69.3); 3111 (55.28.33); 3112 (55.42.1); 3114 (55.92.6); 3117 (55.36.60); 3118 (55.78.4); 3119 (55.39.3); 3120 (55.28.19b); 3120A (55.28.19b); 3121 (55.28.22); 3122 (55.28.10); 3123 (55.36.3); 3123A (55.36.3); 3124 (55.43.5); 3124A (55.43.5); 3126 (55.92.7);

3127 (55.48.5); 3128 (55.37.13); 3129 (55.39.1); 3130 (55.59.1); 3130A (55.59.1); 3131 (55.17.31); 3134 (55.37.14); 3135 (55.45.57); 3136 (55.33.4); 3137 (55.61.1); 3138 (55.17.88); 3140 (55.9.2); 3141 (55.17.16); 3144 (55.36.46); 3145 (55.62.7); 3149 (55.28.16); 3150 (55.109.3); 3153 (55.17.6); 3154 (55.18.5); 3155 (55.38.1); 3157 (55.119.2); 3158 (55.109.2); 3159 (55.121.2); 3160 (55.75.3); 3161 (55.11.19); 3162 (55.8.1); 3163 (55.83.3a); 3164 (55.66.13); 3165 (55.17.43); 3166 (55.45.47); 3167 (55.112.3); 3169 (55.67.3); 3172 (55.37.10); 3173 (55.36.11); 3174 (55.13.1); 3176A (55.16.8); 3177 (55.121.1); 3179 (55.53.3); 3186 (55.37.7); 3188 (55.118.1a); 3189 (55.48.1); 3191 (55.17.88); 3192 (55.17.40); 3193 (55.11.19); 3195 (55.120.6); 3196 (55.86.6); 3198 (55.76.1); 3199 (55.34.1); 3199A (55.34.1); 3205 (55.30.3); 3206 (55.60.1); 3208 (55.32.1); 3218 (55.31.1a); 3219 (55.78.4); 3233 (55.37.4); 3236 (55.53.8); 3244 (55.28.25); 3246 (55.39.2); 3254 (55.75.2); 3256 (55.53.2); 3262 (55.28.15); 3264 (55.51.1); 3270 (55.28.2, 55.28.31); 3292 (55.22.4a); 3301 (55.45.57); 3314 (55.24.1); 3315 (55.106.1); 3317 (55.6.13); 3320 (55.86.2); 3324 (55.90.9); 3328 (55.28.11); 3334 (55.64.2); 3336 (55.89.1a); 3343 (55.22.4a); 3344 (55.120.3); 3353 (55.86.1); 3366 (55.28.14); 3368 (55.86.3); 3384 (55.120.5); 3387 (55.77.2); 3389 (55.86.8); 3403A (55.112.6); 3412 (55.37.7); 3417 (55.119.5); 3419 (55.53.3); 3420 (55.71.1); 3422 (55.39.4); 3423 (55.25.1); 3437 (55.53.6); 3474 (55.73.2); 3475 (55.42.1); 3476 (55.37.17); 3476A (55.37.17); 3477 (55.37.33); 3479 (55.28.13); 3488 (55.22.4a); 3489 (55.86.7); 3491 (55.58.1); 3496 (55.28.22); 3499 (55.28.2); 3508 (55.32.2); 3508A (55.32.2); 3509 (55.53.4); 3515 (55.92.1); 3516 (55.28.20); 3518 (55.22.3a); 3521 (55.37.30); 3522 (55.28.18); 3523 (55.92.5); 3523A (55.92.5); 3524 (55.22.2); 3524A (55.22.2); 3528 (55.37.18); 3533 (55.37.11); 3534 (55.37.11); 3535 (55.45.57); 3539 (55.72.1); 3541 (55.37.27); 3545 (55.37.3); 3548 (55.37.27); 3549 (55.92.5); 3549A (55.92.5); 3549B (55.92.5); 3549C (55.92.5); 3550 (55.37.1); 3552 (55.28.27); 3563 (55.89.2); 3564 (55.46.2); 3565 (55.22.4b); 3566 (55.28.2); 3566p.p. (55.28.2, 55.28.23); 3570A (55.90.9); 3572 (55.53.9); 3575 (55.86.4); 3580 (55.22.4a); 3589 (55.45.57); 3597 (55.37.28); 3608 (55.37.22); 3614 (55.58.1); 3620 (55.37.14); 3622 (55.37.21); 3623 (55.37.32); 3638 (55.22.6); 3641 (55.56.1); 3653 (55.37.19); 3654 (55.9.2); 3663 (55.37.26); 3668 (55.37.12); 3669 (55.37.14); 3671 (55.37.9); 3675 (55.37.16); 3678 (55.28.24); 3680 (55.37.16); 3684 (55.37.5); 3692 (55.36.43); 3709 (55.37.14); 3710 (55.37.20); 3712 (55.28.20); 3713 (55.120.2); 3714 (55.28.6); 3715 (55.37.2); 3718 (55.37.12); 3722 (55.121.3); 3736 (55.100.1); 3742 (55.37.21); 3751 (55.37.21); 3752 (55.37.23); 3755 (55.22.1); 3757 (55.53.7); 3758 (55.89.3); 3763 (55.22.4a); 3764 (55.116.3); 3765 (55.120.1); 3766 (55.120.4); 3770 (55.57.1); 3771 (55.46.1); 26261 (55.28.6); 26262 (55.87.6); 26262A (55.87.6); 26263 (55.47.2); 26264 (55.28.26); 26265 (55.111.1); 26266 (55.62.4); 26267 (55.78.3); 26268 (55.11.1); 26276 (55.12.2); 26282 (55.45.57); 26308 (55.66.2); 26309 (55.11.7); 26310 (55.11.5); 26311 (55.66.14); 26312 (55.83.1); 26313 (55.17.31); 26314 (55.11.19); 26314A (55.11.19); 26343 (55.45.57); 26344 (55.66.5); 26344A (55.66.14); 26345 (55.62.10); 26346 (55.11.19); 26347 (55.95.4); 26347A (55.95.4); 26348 (55.45.11); 26349 (55.17.62); 26367 (55.36.37); 26371 (55.78.11); 26372 (55.36.28); 26373 (55.17.15); 26374 (55.17.68); 26378 (55.45.56); 26393 (55.17.31); 26394 (55.6.13); 26395 (55.30.2); 26396 (55.90.9); 26397 (55.105.1); 26409 (55.115.12); 26429 (55.17.35); 26430 (55.36.31); 26431 (55.62.26); 26432 (55.62.26); 26453 (55.28.1b); 26454 (55.28.5b); 26455 (55.36.25); 26456 (55.17.88); 26458 (55.17.88); 26462 (55.66.14); 26466 (55.62.26); 26467 (55.95.1); 26468 (55.95.4); 26469 (55.70.2); 26470 (55.113.12); 26471 (55.45.28); 26472 (55.112.6); 26473 (55.67.3, 55.103.1); 26493 (55.17.15); 26494 (55.115.8); 26495 (55.17.54a); 26496 (55.69.3); 26506 (55.11.4); 26508 (55.69.2); 26510 (55.17.66); 26511 (55.92.6); 26530 (55.28.33); 26532 (55.17.88); 26533 (55.8.2); 26534 (55.87.1); 26535 (55.36.60); 26536 (55.78.4); 26548 (55.36.6); 26549 (55.39.1); 26550 (55.39.3); 26551 (55.17.31); 26552 (55.59.1); 26552A (55.59.1); 26553 (55.43.5); 26553A (55.43.5); 26554 (55.105.3); 26555 (55.36.27); 26556 (55.32.3); 26557 (55.36.3); 26557A (55.36.3); 26560 (55.92.7); 26587 (55.18.17); 26589 (55.45.57); 26590 (55.9.2); 26591 (55.78.11); 26592 (55.36.46); 26593 (55.62.26); 26594 (55.17.88); 26596 (55.33.4); 26597 (55.109.3); 26598 (55.54.1); 26605 (55.17.6); 26606 (55.38.1); 26607 (55.75.3); 26634 (55.8.1); 26635 (55.90.9); 26636 (55.45.57, 55.45.47); 26637 (55.83.3a); 26638 (55.112.3); 26668 (55.37.10); 26669 (55.79.1); 26670 (55.66.13); 26671 (55.121.2); 26692 (55.118.1a); 26693 (55.36.11); 26726 (55.18.5); 26740 (55.120.7); 26742 (55.17.43);

26742a (55.17.43); 26745 (55.37.14); 26756 (55.121.1); 26759 (55.37.7); 26760 (55.118.1a); 26775 (55.48.1); 26776 (55.120.6); 26777 (55.78.11); 26778 (55.34.1); 26778A (55.34.1); 26796 (55.11.19, 55.76.1); 26798 (55.108.5); 26799 (55.33.5); 26800 (55.13.1); 26801 (55.60.1); 26802 (55.108.7); 26828 (55.90.9); 26831 (55.45.57); 26841 (55.30.3); 26842 (55.45.57); 26853 (55.78.11); 26855 (55.36.45); 26856 (55.17.88); 26861 (55.17.64); 26863 (55.17.44); 26864 (55.17.2); 26874 (55.37.4); 26876 (55.28.10); 26877 (55.17.15); 26878 (55.113.12); 26879 (55.62.26); 26892 (55.90.9); 26894 (55.17.54a); 26914 (55.17.88); 26918 (55.18.7); 26918A (55.18.7); 26935 (55.45.23); 26936 (55.115.12); 26937 (55.18.10); 26938 (55.75.2); 26939 (55.53.2); 26940 (55.78.11); 26941 (55.69.3); 26955 (55.43.4); 26969 (55.115.3); 26970 (55.51.1); 26971 (55.69.3); 26973 (55.90.9); 26975 (55.79.2); 26976 (55.17.54b); 26984 (55.17.35); 26985 (55.108.9); 26987 (55.17.11); 26988 (55.28.25); 26994 (55.17.66); 26996 (55.17.88); 26999 (55.45.57); 27003 (55.114.15); 27004 (55.20.7); 27011 (55.45.25); 27012 (55.18.17); 27019 (55.90.9); 27020 (55.17.3, 55.17.16); 27024 (55.36.13); 27025 (55.80.1); 27026 (55.88.1); 27027 (55.18.17); 27028 (55.22.4a); 27029 (55.36.7); 27030 (55.17.18); 27035 (55.115.9); 27038 (55.45.34); 27039 (55.45.13); 27039A (55.45.57); 27039B (55.45.57); 27040 (55.114.3); 27043 (55.45.7); 27049 (55.14.1); 27051 (55.45.57); 27052 (55.36.7); 27053 (55.36.21); 27055 (55.18.14); 27055A (55.18.14); 27055B (55.18.14); 27055C (55.18.14); 27055D (55.18.14); 27059 (55.106.1); 27060 (55.24.1); 27064 (55.45.57); 27064A (55.45.57); 27082 (55.17.31); 27083 (55.10.2); 27092 (55.36.27); 27093 (55.105.4); 27094 (55.45.57); 27106 (55.17.54b, 55.45.57); 27109 (55.17.88); 27110 (55.17.3); 27111 (55.86.3); 27112 (55.28.15); 27130 (55.112.6); 27131 (55.45.33); 27132 (55.32.1); 27133 (55.17.21); 27134 (55.120.3); 27149 (55.86.6); 27150 (55.45.57); 27151 (55.48.2); 27152 (55.17.18); 27153 (55.45.2); 27153A (55.45.2); 27154 (55.3.1); 27168 (55.45.29); 27179 (55.113.12); 27180 (55.22.4a); 27181 (55.113.3); 27189 (55.101.5); 27190 (55.79.1); 27191 (55.3.2a); 27192 (55.90.4); 27192A (55.90.9); 27193 (55.114.15); 27194 (55.17.88); 27195 (55.79.1); 27196 (55.43.3); 27197 (55.96.1); 27198 (55.17.38); 27200 (55.70.1); 27206 (55.90.9); 27207 (55.101.5); 27218 (55.5.1); 27219 (55.113.12); 27220 (55.17.88); 27230 (55.28.36, 55.28.14); 27231 (55.17.8); 27232 (55.45.57); 27234 (55.28.5b); 27235 (55.36.19); 27236 (55.17.35); 27237 (55.17.88); 27243 (55.103.4); 27245 (55.16.1); 27247 (55.17.65); 27248 (55.17.88); 27249 (55.17.42); 27292 (55.17.4); 27306 (55.62.20); 27307 (55.94.1); 27308 (55.86.8); 27310 (55.31.1a); 27312 (55.78.11); 27313 (55.108.4); 27314 (55.67.3); 27315 (55.37.7); 27318 (55.112.2); 27321 (55.114.6); 27333 (55.101.5); 27334 (55.119.5); 27336 (55.36.34); 27337 (55.53.3); 27338 (55.39.4); 27339 (55.25.1); 27343 (55.62.16); 27344 (55.11.10); 27345 (55.115.4); 27346 (55.78.2); 27347 (55.90.9); 27348 (55.17.2); 27349 (55.45.57); 27350 (55.6.11); 27354 (55.17.88); 27358 (55.101.3); 27359 (55.53.8); 27361 (55.28.11); 27362 (55.62.15); 27363 (55.113.12); 27364 (55.17.52); 27365 (55.45.57); 27366 (55.66.7); 27367 (55.113.1); 27368 (55.16.5); 27369 (55.113.12); 27373 (55.113.2); 27374 (55.11.13); 27375 (55.17.49); 27376 (55.17.70); 27388 (55.62.14); 27389 (55.36.41); 27390 (55.11.19); 27391 (55.93.2); 27392 (55.78.11); 27393 (55.112.6); 27393A (55.112.6); 27398 (55.89.1a); 27401 (55.66.4); 27403 (55.17.88); 27404 (55.36.54); 27405 (55.36.2); 27406 (55.45.57); 27406A (55.45.57); 27407 (55.17.37); 27407A (55.45.57, 55.45.25); 27408 (55.108.9); 27409 (55.87.5); 27410 (55.17.35); 27411 (55.67.3); 27412 (55.66.14); 27414 (55.33.8); 27415 (55.45.57); 27416 (55.62.26, 55.62.19); 27417 (55.20.7); 27418 (55.17.76); 27419 (55.108.2); 27420 (55.62.16); 27421 (55.45.57); 27422 (55.27.9); 27424 (55.17.88); 27425 (55.66.10); 27426 (55.6.13); 27427 (55.36.22); 27428 (55.37.13); 27429 (55.17.81); 27430 (55.37.17); 27430A (55.37.17); 27431 (55.37.33); 27432 (55.45.16); 27433 (55.45.57); 27434 (55.62.26); 27441 (55.17.76); 27442 (55.17.78); 27443 (55.22.4a); 27444 (55.45.36); 27445 (55.86.7); 27446 (55.17.71); 27447 (55.85.1); 27448 (55.17.23); 27449 (55.2.1); 27453 (55.17.73); 27455 (55.16.8, 55.86.7); 27455A (55.16.8); 27458 (55.28.22); 27459 (55.62.26); 27460 (55.36.57); 27461 (55.17.12); 27462 (55.45.57); 27464 (55.28.19b); 27465 (55.45.57); 27465A (55.45.57); 27466 (55.28.2); 27467 (55.16.4); 27468 (55.115.7); 27469 (55.17.88); 27470 (55.53.4); 27484 (55.45.22); 27501 (55.28.2, 55.28.31); 27502 (55.17.31); 27503 (55.113.7); 27504 (55.28.20); 27520 (55.17.53); 27520a (55.17.53); 27520b (55.17.53); 27520c (55.17.53); 27521 (55.92.1); 27530 (55.113.12); 27531 (55.37.30); 27532 (55.28.18); 27533 (55.22.2, 55.92.5); 27533A (55.22.2, 55.92.5); 27534 (55.45.16); 27534C (55.45.57); 27548 (55.17.88); 27549 (55.17.3);

27559 (55.17.27); 27560 (55.20.7); 27561 (55.45.57); 27562 (55.37.11); 27563 (55.45.57); 27564 (55.17.88); 27576 (55.17.55); 27577 (55.17.50); 27578 (55.16.6); 27592 (55.52.1, 55.53.6); 27593 (55.90.9); 27595 (55.72.1); 27596 (55.17.19); 27597 (55.37.27); 27598 (55.62.11); 27623 (55.113.12); 27624 (55.37.3); 27634 (55.108.9); 27635 (55.37.27); 27641 (55.92.5); 27641A (55.92.5); 27641B (55.92.5); 27641C (55.92.5); 27642 (55.37.3); 27645 (55.37.1); 27646 (55.108.9); 27661 (55.45.57); 27713 (55.22.3a); 27741 (55.17.9); 27742 (55.115.7); 27743 (55.36.55); 27744 (55.50.1); 27745 (55.18.17); 27746 (55.17.54a); 27747 (55.45.8); 27749 (55.90.9); 27749A (55.90.9); 27757 (55.101.5); 27764 (55.36.23); 27765 (55.45.57); 27766 (55.45.57); 27767 (55.36.51); 27768 (55.17.88); 27769 (55.108.1); 27777 (55.11.19); 27778 (55.101.5); 27779 (55.86.4); 27780 (55.45.57); 27781 (55.89.2); 27782 (55.78.10); 27783 (55.16.3); 27784 (55.113.12); 27784A (55.113.12); 27787 (55.113.12); 27788 (55.113.5); 27790 (55.78.11); 27791 (55.45.57, 55.90.9); 27792 (55.22.4a); 27793 (55.28.27); 27795 (55.36.11); 27796 (55.113.12); 27803 (55.11.19); 27804 (55.101.3); 27805 (55.27.4); 27807 (55.17.85); 27809 (55.45.9a); 27809A (55.45.57); 27810 (55.108.6); 27811 (55.86.2); 27812 (55.101.5); 27813 (55.17.88); 27814 (55.69.3); 27818 (55.36.18); 27819 (55.17.34); 27820 (55.62.26); 27821 (55.17.36); 27822 (55.17.88); 27823 (55.77.2); 27826 (55.17.18); 27828 (55.115.12); 27829 (55.1.1); 27830 (55.17.31); 27831 (55.36.21); 27832 (55.113.12); 27833 (55.17.88); 27833A (55.17.19); 27835 (55.17.83); 27843 (55.17.2); 27849 (55.17.40); 27850 (55.17.88); 27857 (55.18.17); 27858 (55.17.88); 27859 (55.46.2); 27860p.p. (55.28.2, 55.28.23); 27861 (55.58.1); 27863 (55.101.5); 27864 (55.37.18); 27865 (55.52.1); 27866 (55.18.17); 27868 (55.17.88); 27869 (55.48.3); 27871 (55.67.3); 27872 (55.62.8); 27873 (55.62.23); 27881 (55.36.60); 27882 (55.11.3); 27883 (55.17.88); 27884 (55.37.14); 27889 (55.17.88, 55.17.76); 27890 (55.17.88); 27891 (55.37.32); 27892 (55.37.22); 27893 (55.26.1); 27894 (55.36.9); 27897 (55.28.16); 27898 (55.90.9); 27899 (55.17.5); 27901 (55.93.2); 27902 (55.17.88); 27906 (55.17.88); 27907 (55.20.1); 27908 (55.37.21); 27909 (55.17.88); 27910 (55.17.73); 27912 (55.104.6); 27913 (55.20.7); 27914 (55.113.12); 27918 (55.36.33); 27919 (55.17.13); 27919a (55.17.13); 27919b (55.17.13); 27921 (55.17.48); 27924 (55.11.16); 27925 (55.17.16); 27926 (55.17.88); 27931 (55.45.57); 27935 (55.17.55); 27936 (55.17.88); 27937 (55.17.88); 27938 (55.17.60); 27939 (55.17.88); 27943 (55.36.2); 27948 (55.45.57); 27951 (55.61.1); 27953 (55.36.24); 27954 (55.67.3); 27955 (55.36.17a); 27956 (55.17.88); 27957 (55.17.88); 27958 (55.36.56); 27959 (55.20.7); 27960 (55.69.3); 27964 (55.66.14); 27965 (55.28.3); 27967 (55.36.19); 27968 (55.17.84); 27969 (55.104.3); 27970 (55.37.14); 27971 (55.17.18); 27972 (55.115.12); 27975 (55.36.1); 27976 (55.36.53); 27977 (55.53.9); 27978 (55.32.2); 27979 (55.120.5); 27981 (55.101.4); 27982 (55.78.11); 27983 (55.62.26); 27985 (55.113.12); 27986 (55.45.57); 27987 (55.17.88); 27988 (55.101.5); 27989 (55.11.19); 27990 (55.6.2); 27992 (55.78.8); 27997 (55.78.11); 27998 (55.37.19); 27999 (55.17.88); 28000 (55.45.57); 28003 (55.36.62); 28004 (55.37.16); 28005 (55.17.88); 28006 (55.37.16); 28009 (55.113.12); 28019 (55.37.5); 28020 (55.37.26); 28021 (55.36.29); 28022 (55.78.11); 28023 (55.78.11); 28024 (55.36.45, 55.36.17a); 28025 (55.78.11); 28026 (55.103.4); 28027 (55.22.4b); 28028 (55.45.57); 28029 (55.37.12); 28030 (55.113.12); 28032 (55.45.57); 28034 (55.37.14); 28047 (55.37.9); 28049 (55.37.12); 28051 (55.120.2); 28053 (55.121.3); 28054 (55.53.7); 28055 (55.100.1); 28056 (55.56.1); 28057 (55.120.1); 28059 (55.42.1); 36565 (55.37.18); 36566 (55.37.35); 36567 (55.37.21); 36570 (55.37.35).

Chai & Ilias 6007 (55.28.14); 6008 (55.17.43); 6031 (55.17.31).

Chan 20 (55.75.2); 35 (55.36.48); 45 (55.118.2); 49 (55.83.3a); 52/86 (55.114.2); 63/87 (55.17.55); 121/89 (55.62.16).

Chan & Gunsalam 34/87 (55.17.36); 39 (55.53.2); 40/87 (55.22.4b).

Chan & Lamb 53/83 (55.28.1a); 313 (55.62.11).

Chew & Corner 4031 (55.87.5); 4046 (55.17.43); 4052 (55.120.3); 4093 (55.11.4); 4099 (55.115.5); 4100 (55.20.4); 4126 (55.66.14); 4151 (55.28.15); 4176 (55.45.33); 4191

(55.17.72); 4237 (55.28.15); 4258 (55.115.5); 4402 (55.45.28); 4453 (55.28.3, 55.28.36); 4454 (55.45.57); 4463 (55.115.4); 4465 (55.45.57); 4481 (55.62.16); 4483 (55.89.2); 4485 (55.28.22); 4486 (55.62.19); 4527 (55.36.13); 4529 (55.62.14); 4554 (55.45.22); 4556 (55.28.3); 4563 (55.45.29); 4622 (55.45.29); 4624 (55.89.2); 4705 (55.45.28); 4725 (55.18.6); 4775 (55.90.2); 4777 (55.45.23); 4784 (55.17.19); 4787 (55.113.7); 4845 (55.62.26); 4856 (55.17.24); 4857 (55.45.29); 4858 (55.45.19); 4877 (55.89.6); 4941 (55.17.13); 4951 (55.28.10); 4958 (55.83.2a); 4962 (55.36.25); 4972 (55.28.11); 5790 (55.56.3); 7030 (55.87.2); 7062 (55.83.4); 7093 (55.18.14); 7109 (55.45.19); 7113 (55.18.7); 7123 (55.17.84); 7124 (55.59.1).

Chew, Corner & Stainton 179 (55.115.6); 182 (55.37.21); 201 (55.45.38b); 214 (55.9.1); 293 (55.45.13); 733 (55.45.16); 1053 (55.39.3); 1121 (55.45.16); 1299 (55.18.7); 1304 (55.45.28); 1321 (55.83.4); 1323 (55.105.1); 1324 (55.17.43); 1351 (55.10.2); 1670 (55.10.2); 2764 (55.10.2).

Chow & Leopold 74504 (55.103.2); 74511 (55.37.26); 74529 (55.28.16); 74535 (55.17.54a); 74539 (55.112.3); 76430 (55.45.55); 76434 (55.17.31); 76439 (55.62.5).

Christenson 1770 (55.83.5).

Clemens 21 (55.17.57); 24 (55.89.3); 27 (55.78.5); 28 (55.28.33); 29 (55.28.11); 30 (55.28.33); 31 (55.114.3); 32 (55.45.28); 33 (55.105.3); 35 (55.62.23); 35A (55.62.7); 36 (55.17.34); 37 (55.62.7); 38 (55.62.23); 39 (55.33.5); 42 (55.18.14); 43 (55.5.1); 45 (55.86.6); 46 (55.62.7); 48 (55.17.34); 49 (55.62.23); 50 (55.33.4); 51 (55.33.4); 53 (55.18.14); 55 (55.54.1); 56 (55.17.45); 58 (55.104.3); 60 (55.114.8); 61 (55.11.4); 63 (55.11.7); 66 (55.115.9); 67 (55.103.3); 68 (55.48.4); 68A (55.43.3); 69 (55.5.1); 70 (55.20.5); 71 (55.28.25); 74 (55.33.5); 75 (55.66.13); 76 (55.45.28); 77 (55.45.38b, 55.62.7); 78 (55.28.13); 79 (55.28.13); 80 (55.22.1); 81 (55.5.1); 82 (55.33.5); 83 (55.89.3); 84 (55.112.5); 86 (55.66.10); 87 (55.67.1); 88 (55.112.3); 90 (55.62.23); 92 (55.89.3); 93 (55.20.3, 55.109.1); 94 (55.17.21); 96 (55.53.6); 97 (55.17.57); 98 (55.18.10); 99 (55.17.57); 100 (55.83.1); 101 (55.77.2); 102 (55.78.1); 102A (55.78.4); 103 (55.62.16); 104 (55.78.5); 104A (55.78.4); 104B (55.78.8); 105 (55.28.11); 106 (55.17.66); 107 (55.89.1a); 109 (55.112.5); 111 (55.67.1, 55.73.2); 112 (55.45.16); 113 (55.17.19, 55.104.3); 113A (55.17.13); 114 (55.37.23); 115 (55.17.24, 55.37.30); 116 (55.37.24); 116A (55.43.3); 117 (55.62.19); 118 (55.36.47); 119 (55.62.7); 122 (55.28.2); 123 (55.31.1a); 124 (55.11.15); 126 (55.18.10); 127 (55.89.3); 127A (55.89.1a); 128 (55.36.19); 129 (55.11.15); 130 (55.87.6); 131 (55.28.33); 132 (55.18.14); 133 (55.17.3, 55.67.1); 134 (55.66.13); 137 (55.11.5); 138 (55.11.15); 139 (55.62.7); 140 (55.106.1); 141 (55.62.7); 142 (55.28.33); 146 (55.37.14); 147 (55.45.9a); 148 (55.119.1); 149 (55.47.2); 150 (55.105.3); 151 (55.83.1); 154 (55.17.57); 155 (55.18.10); 156 (55.66.13); 160 (55.115.9); 161 (55.28.11); 163 (55.67.1); 164 (55.8.2); 166 (55.103.3); 168 (55.17.72); 172 (55.17.57); 173 (55.28.13); 174 (55.28.26); 175 (55.28.13, 55.62.23); 176 (55.36.19, 55.67.1); 177 (55.20.5); 178 (55.37.14); 179 (55.37.19); 180 (55.62.14); 181 (55.45.7); 183 (55.5.1); 184 (55.87.6); 186 (55.13.1); 188 (55.36.11); 189 (55.20.3); 190 (55.106.1); 193 (55.86.5); 195 (55.17.13); 197 (55.43.2); 199 (55.28.18); 200 (55.28.22, 55.31.1a); 200A (55.28.9); 201 (55.113.7); 202 (55.37.11); 204 (55.28.20); 205 (55.37.21); 206 (55.22.2); 207 (55.28.9); 208 (55.11.7); 209 (55.37.17); 210 (55.73.2); 217 (55.28.2); 219 (55.62.16); 220 (55.62.11); 221 (55.92.3); 222 (55.11.7); 222A (55.36.47); 223 (55.22.2); 224 (55.37.30); 224A (55.37.23); 225 (55.95.6); 225A (55.95.4); 226 (55.36.37); 227 (55.28.1a); 228 (55.105.3); 229 (55.28.11); 230 (55.11.4); 231 (55.22.1); 233 (55.22.1); 236 (55.28.11); 237 (55.11.7); 237A (55.11.4); 238 (55.113.3); 239 (55.22.4a); 240 (55.17.86); 241 (55.6.5); 242 (55.37.23); 244 (55.39.3); 245 (55.43.3); 246 (55.103.3); 247 (55.37.20); 248 (55.37.13); 249 (55.43.2); 250 (55.45.25); 251 (55.28.9); 252 (55.9.2); 253 (55.95.4); 254 (55.17.26); 255 (55.115.5); 257 (55.36.53); 258 (55.66.4); 260 (55.20.2); 261 (55.92.5); 263 (55.8.2); 263A (55.75.1); 265 (55.13.1); 266 (55.39.3); 266A (55.39.1); 267 (55.26.1); 268 (55.73.1); 270 (55.37.5); 272 (55.39.1); 273 (55.89.5); 274 (55.83.3a); 275 (55.78.1); 275A (55.78.8); 276 (55.6.10); 277 (55.13.1); 277A (55.87.6); 278 (55.37.29); 279

(55.42.1); 280 (55.37.23); 281 (55.45.29); 282 (55.11.12); 283 (55.45.11); 284 (55.36.31); 285 (55.37.31); 286 (55.11.9); 287 (55.36.27); 289 (55.37.14); 290 (55.115.10); 291 (55.115.5); 293 (55.28.2); 294 (55.28.33); 295 (55.95.6); 299 (55.28.33); 300 (55.28.33, 55.36.31); 301 (55.18.14); 302 (55.18.10); 303 (55.115.9); 304 (55.36.19); 305 (55.17.31); 306 (55.62.7); 307 (55.5.1); 308 (55.119.1); 311 (55.118.1a); 312 (55.62.23); 314 (55.45.7); 315 (55.28.33, 55.114.3); 316 (55.17.72); 317 (55.17.15); 318 (55.37.19); 320 (55.20.5); 321 (55.83.1); 322 (55.53.6); 323 (55.9.2); 324 (55.62.15); 325 (55.17.34); 326 (55.17.13); 327 (55.17.72); 328 (55.78.4); 329 (55.78.4); 330 (55.67.1); 331 (55.89.1a); 332 (55.37.25); 333 (55.11.13); 334 (55.36.28); 335 (55.48.1); 337 (55.62.23); 338 (55.6.10); 340 (55.120.6); 342 (55.69.2); 343 (55.18.14); 344 (55.105.4); 345 (55.47.2); 346 (55.87.6); 348 (55.11.4); 349 (55.83.1); 350 (55.62.20); 351 (55.11.12); 352 (55.17.34); 353 (55.67.1); 354 (55.17.72); 355 (55.120.6); 356 (55.2.1); 357 (55.28.33); 358 (55.11.15); 360 (55.11.5); 361 (55.11.7, 55.37.14); 363 (55.18.14); 366 (55.36.37); 367 (55.43.2); 368 (55.26.1); 369 (55.28.11); 370 (55.39.3); 370A (55.43.2); 371 (55.47.2); 372 (55.43.3); 373 (55.115.5); 375 (55.22.4a); 376 (55.45.29); 377 (55.37.25); 378 (55.87.6); 380 (55.78.1); 381 (55.18.10); 383 (55.37.2); 384 (55.17.18); 385 (55.37.21); 386 (55.37.23); 387 (55.11.10); 389 (55.16.4); 390 (55.89.2); 395 (55.73.2); 396 (55.37.12); 397 (55.86.5); 398 (55.59.1); 399 (55.32.3); 399A (55.106.1); 400 (55.8.2); 401 (55.53.6); 402 (55.18.10); 403 (55.36.27); 404 (55.11.9); 405 (55.95.6); 408 (55.17.57); 2929? (55.11.18); 3827? (55.36.27); 26019 (55.12.2); 26020 (55.2.1); 26021 (55.112.2); 26022 (55.89.6); 26044 (55.28.6); 26050 (55.17.57); 26122 (55.18.7); 26123 (55.62.14); 26124 (55.33.5); 26125 (55.28.16); 26126 (55.114.3); 26127 (55.28.14); 26128 (55.45.24); 26130 (55.66.8); 26143 (55.45.57); 26233 (55.45.48); 26300 (55.119.2); 26301 (55.89.4); 26302 (55.11.1); 26303 (55.66.7); 26304 (55.3.1); 26305 (55.115.2); 26306 (55.78.4); 26307 (55.118.1a); 26308 (55.53.7); 26309 (55.112.5); 26312 (55.11.15); 26313 (55.45.28); 26314 (55.86.4); 26411 (55.66.4); 26412 (55.45.10); 26413 (55.45.15); 26414 (55.48.7); 26457 (55.45.53); 26458 (55.45.39); 26481 (55.66.7); 26482 (55.86.4); 26502 (55.45.34); 26514 (55.45.44); 26515 (55.36.18); 26516 (55.45.31); 26518 (55.28.11); 26519 (55.93.3); 26520 (55.11.3); 26521 (55.28.6); 26601 (55.115.2); 26602 (55.66.8); 26603 (55.45.44); 26604 (55.115.9); 26605 (55.11.15); 26645 (55.6.7); 26690 (55.66.8); 26690a (55.53.8); 26691 (55.33.5); 26692 (55.66.7); 26722 (55.66.7); 26723 (55.45.28); 26724 (55.18.5); 26725 (55.119.2); 26726 (55.36.11); 26739 (55.1.1); 26766 (55.45.28); 26767 (55.105.3); 26768 (55.87.6); 26769 (55.11.1); 26770 (55.86.6); 26771 (55.33.5); 26772 (55.11.15); 26773 (55.89.4); 26774 (55.6.12); 26775 (55.115.5); 26776 (55.5.1); 26777 (55.6.9); 26778 (55.78.5); 26779 (55.45.56); 26780 (55.5.1); 26781 (55.89.4); 26782 (55.45.28); 26783 (55.78.5); 26784 (55.37.14); 26784A (55.37.12); 26786 (55.17.65); 26791 (55.5.1); 26800 (55.31.1a); 26813 (55.33.9); 26831 (55.62.7); 26832 (55.53.7); 26833 (55.45.56); 26835 (55.89.3); 26836 (55.105.4); 26837 (55.115.5); 26838 (55.45.57, 55.62.7); 26841 (55.37.14); 26842 (55.45.9a, 55.62.23); 26843 (55.17.45); 26844 (55.95.6); 26849 (55.16.5); 26866 (55.18.5); 26874 (55.19.1, 55.62.6, 55.62.20, 55.62.23); 26883 (55.77.2); 26894 (55.62.7); 26895 (55.62.23); 26896 (55.53.7); 26896A (55.53.7); 26899 (55.37.7); 26911 (55.53.8); 26912 (55.66.3); 26913 (55.17.65); 26930 (55.37.14); 26952 (55.28.1a); 26997 (55.53.7); 26998 (55.36.38); 26999 (55.62.7); 27000 (55.62.4); 27002 (55.28.33); 27003 (55.78.5); 27004 (55.28.1a); 27005 (55.66.7); 27006 (55.78.4); 27023 (55.62.23); 27024 (55.34.1); 27117 (55.106.1, 55.121.3); 27138 (55.11.4); 27140 (55.37.11); 27141 (55.37.30); 27142 (55.37.17); 27143 (55.37.21); 27145 (55.37.11); 27146 (55.37.21); 27147 (55.37.23); 27148 (55.37.11); 27149 (55.37.17); 27150 (55.17.19); 27151 (55.17.88); 27152 (55.22.4b); 27153 (55.22.2); 27154 (55.28.18, 55.28.33); 27155 (55.39.3); 27159 (55.73.1); 27160 (55.45.9a); 27162 (55.17.57); 27163 (55.114.6); 27165 (55.45.16); 27166 (55.28.14); 27167 (55.28.1a); 27169 (55.22.4b); 27170 (55.18.10, 55.45.16); 27171 (55.73.1); 27172 (55.28.18); 27173 (55.113.7); 27174 (55.22.1); 27175 (55.45.22); 27176 (55.115.6); 27177 (55.16.4); 27178 (55.62.25); 27180 (55.115.10); 27181 (55.22.1); 27184 (55.73.1); 27185 (55.28.20); 27187 (55.115.5); 27191 (55.28.31); 27207 (55.62.23); 27223 (55.11.18); 27235 (55.43.4); 27236 (55.28.1a, 55.78.7); 27239 (55.89.1a); 27240 (55.36.13); 27241 (55.11.10); 27245 (55.62.12); 27246 (55.45.28); 27247 (55.62.25); 27256 (55.37.12); 27257 (55.11.1, 55.11.15); 27260 (55.62.16); 27264 (55.17.57); 27277 (55.17.64); 27279 (55.11.15); 27323 (55.62.23); 27324 (55.6.12, 55.11.4); 27359

(55.45.57); 27360 (55.78.5); 27361 (55.17.26); 27362 (55.22.1); 27363 (55.95.6); 27364 (55.17.31); 27365 (55.22.4b); 27366 (55.62.7); 27367 (55.45.57); 27368 (55.43.2); 27397 (55.11.5); 27398 (55.89.1a); 27417 (55.114.8); 27418 (55.28.33); 27419 (55.60.1); 27445 (55.13.1, 55.16.5); 27456 (55.11.1, 55.11.15); 27481 (55.6.7); 27491 (55.17.4); 27515 (55.45.57); 27516 (55.45.35); 27530 (55.17.57); 27531 (55.3.2a); 27555 (55.121.3); 27567 (55.9.2); 27615 (55.39.4); 27616 (55.28.26); 27617 (55.105.4); 27618 (55.83.3a); 27629 (55.47.2); 27630 (55.36.11); 27631 (55.89.4); 27634 (55.90.5); 27642 (55.62.19); 27659 (55.17.43, 55.22.1); 27659A (55.22.1); 27685 (55.90.5); 27689 (55.46.2, 55.53.7); 27695 (55.53.3); 27696 (55.77.2); 27697 (55.11.1); 27725 (55.61.1); 27726 (55.112.2); 27727 (55.62.23); 27728 (55.39.4); 27745 (55.6.7); 27752 (55.45.41); 27759 (55.28.3); 27767 (55.78.5); 27768 (55.36.13); 27805 (55.89.2); 27806 (55.75.2); 27807 (55.17.62); 27856 (55.17.62); 27857 (55.36.27); 27858 (55.45.41); 27859 (55.17.88); 27860 (55.37.21); 27861 (55.45.16); 27862 (55.28.18); 27863 (55.17.19); 27864 (55.37.27); 27866 (55.37.17); 27867 (55.37.17); 27868 (55.20.1); 27869 (55.113.7); 27870 (55.37.30); 27871 (55.17.13); 27872 (55.28.18); 27918 (55.20.4); 27978 (55.113.4); 27979 (55.18.7, 55.18.14); 27980 (55.66.8); 27985 (55.13.1); 28036 (55.18.10); 28039 (55.36.21); 28040 (55.113.1); 28051 (55.45.11); 28086 (55.105.3); 28134 (55.36.45); 28135 (55.104.4); 28136 (55.3.1); 28137 (55.114.3); 28141 (55.103.2); 28143 (55.62.4); 28144 (55.106.1); 28171 (55.36.28); 28198 (55.17.31); 28204 (55.36.13); 28224 (55.3.1); 28242 (55.120.6); 28243 (55.11.18); 28244 (55.18.17); 28258 (55.106.1); 28276 (55.104.4); 28291 (55.105.3); 28294 (55.28.14); 28316 (55.28.14); 28323 (55.56.2); 28324 (55.18.7); 28344 (55.120.6); 28356 (55.113.3); 28396 (55.36.19); 28414 (55.78.4); 28417 (55.28.33); 28418 (55.113.1); 28419 (55.45.11); 28420 (55.46.2); 28454 (55.28.14); 28462 (55.113.1); 28482 (55.28.1a); 28516 (55.11.10); 28553 (55.17.51); 28554 (55.103.1); 28582 (55.36.27); 28606 (55.17.31); 28608 (55.43.4); 28614 (55.45.11); 28617 (55.88.1); 28636 (55.36.13); 28665 (55.115.2); 28686 (55.62.6); 28687 (55.45.28); 28715 (55.115.3); 28716 (55.83.3a); 28736 (55.11.7); 28779 (55.115.10); 28780 (55.78.10); 28817 (55.11.15, 55.45.40); 28818 (55.16.6); 28821 (55.36.27); 28829 (55.47.3); 28830 (55.87.3a); 28831 (55.45.3); 28859 (55.45.28); 28865 (55.36.5); 28897 (55.90.6); 28899 (55.17.50); 28949 (55.37.4); 28950 (55.11.5); 28951 (55.28.14); 28952 (55.17.13); 28992 (55.62.19); 28994 (55.20.1); 29000 (55.17.54a, 55.17.66); 29008 (55.45.34); 29009 (55.118.2); 29098 (55.45.24); 29101 (55.36.13); 29119 (55.75.2); 29120 (55.37.30); 29121 (55.3.2a); 29122 (55.36.57); 29123 (55.39.3); 29124 (55.90.6); 29125 (55.17.19); 29126 (55.28.31); 29127 (55.17.53); 29128 (55.37.17); 29129 (55.115.10); 29214 (55.43.2); 29217 (55.72.1); 29219 (55.62.11); 29233 (55.17.53); 29234 (55.90.6); 29235 (55.37.14); 29236 (55.45.48, 55.66.14); 29242 (55.120.1); 29252 (55.103.2); 29253 (55.18.7); 29254 (55.45.16); 29255 (55.17.31); 29284 (55.45.41); 29285 (55.62.19); 29286 (55.115.6); 29287 (55.66.14); 29292 (55.45.23); 29293 (55.28.18); 29295 (55.37.4); 29296 (55.28.15); 29297 (55.11.18); 29302 (55.88.3); 29303 (55.17.43); 29304 (55.48.9); 29317 (55.36.19); 29356 (55.45.18, 55.45.29); 29361 (55.37.4); 29369 (55.5.1); 29376 (55.108.3); 29407 (55.28.33); 29412 (55.37.4); 29440 (55.17.63, 55.17.66); 29453 (55.33.5); 29455 (55.6.8); 29461 (55.115.2); 29464 (55.90.3); 29478 (55.28.10); 29580 (55.99.1, 55.104.4); 29590 (55.36.12); 29631 (55.36.12); 29633 (55.17.51); 29634 (55.65.1, 55.113.3); 29635 (55.45.23, 55.62.12); 29637 (55.87.6); 29638 (55.28.10); 29639 (55.34.1); 29674 (55.17.15, 55.17.77); 29675 (55.45.23); 29676 (55.45.34); 29677 (55.62.14); 29678 (55.45.28); 29679 (55.62.6); 29707 (55.27.9); 29708 (55.66.4); 29749 (55.17.37); 29750 (55.17.50); 29754 (55.43.4); 29761 (55.101.4); 29782 (55.47.3, 55.92.5); 29789 (55.48.4); 29791 (55.17.76); 29792 (55.113.3); 29816 (55.3.1, 55.86.3); 29819 (55.22.4a); 29820 (55.17.42); 29822 (55.45.23); 29823 (55.14.1); 29824 (55.22.4a); 29835 (55.45.38a); 29840 (55.36.2); 29841 (55.43.4, 55.48.4); 29843 (55.11.10, 55.45.28); 29844 (55.45.29); 29845 (55.50.1); 29846 (55.17.76); 29849 (55.45.11, 55.45.32); 29850 (55.27.6); 29877 (55.58.1); 29898 (55.61.1); 29899 (55.45.57); 29901 (55.36.12); 29902 (55.3.1); 29903 (55.45.28); 29904 (55.14.1); 29944 (55.17.76); 29945 (55.45.52); 29953 (55.115.6); 29954 (55.11.2); 29955 (55.28.20); 29992 (55.89.2); 29993 (55.53.2); 29994 (55.22.4a); 29996 (55.45.54); 30015 (55.28.19b); 30016 (55.17.53); 30021 (55.105.5); 30023 (55.28.20); 30087 (55.17.64); 30088 (55.28.33); 30097 (55.105.3); 30101 (55.37.17); 30102 (55.37.27); 30103 (55.37.17); 30105 (55.17.13); 30106 (55.17.51); 30109 (55.17.4); 30112 (55.17.31); 30114

(55.17.31); 30116 (55.17.78); 30117 (55.17.40); 30118 (55.17.44); 30119 (55.18.7); 30120 (55.90.4); 30122 (55.45.34); 30123 (55.115.7); 30125 (55.7.1); 30128 (55.66.14); 30132 (55.78.10); 30133 (55.78.10); 30137 (55.90.6); 30138 (55.78.4); 30139 (55.90.3); 30140 (55.90.4, 55.90.5); 30141 (55.62.12); 30142 (55.37.30); 30143 (55.11.4); 30144 (55.6.11); 30146 (55.17.35); 30147 (55.37.17); 30148 (55.13.1); 30149 (55.13.1); 30150 (55.19.1); 30151 (55.69.2); 30152 (55.45.23); 30153 (55.45.6); 30155 (55.45.22); 30156 (55.45.28, 55.90.6); 30157 (55.28.15); 30158 (55.28.36); 30158A (55.28.15); 30160 (55.104.4); 30161 (55.36.2); 30162 (55.99.1); 30165 (55.36.13); 30166 (55.48.4); 30167 (55.48.4); 30168 (55.11.7); 30169 (55.11.7); 30170 (55.11.10); 30170A (55.11.10); 30171 (55.11.1); 30468 (55.62.6); 30469 (55.115.7); 30470 (55.45.44); 30471 (55.37.16); 30472 (55.36.37); 30473 (55.9.2); 30484 (55.10.3); 30517 (55.8.2); 30522 (55.20.5); 30523 (55.11.4); 30573 (55.37.14); 30574 (55.78.11); 30575 (55.66.5); 30577 (55.66.12); 30578 (55.37.19); 30579 (55.45.38b); 30583 (55.95.1); 30584 (55.11.7, 55.20.2); 30585 (55.36.49, 55.95.1); 30586 (55.36.28, 55.115.7); 30587 (55.36.37); 30588 (55.95.6); 30589 (55.89.2); 30590 (55.22.4a); 30591 (55.28.11); 30592 (55.17.31); 30593 (55.11.15); 30599 (55.62.3); 30600 (55.17.62); 30601 (55.37.25); 30603 (55.20.4); 30604 (55.17.51); 30606 (55.36.49); 30607 (55.42.1); 30608 (55.89.2); 30639 (55.43.3); 30640 (55.37.12); 30641 (55.112.5); 30642 (55.62.4); 30686 (55.36.28); 30696 (55.91.1); 30697 (55.11.7); 30698 (55.62.2, 55.62.10); 30699 (55.45.9a); 30702 (55.37.14); 30704 (55.11.10); 30705 (55.14.1); 30706 (55.28.11); 30706A (55.89.3); 30707 (55.115.9); 30708 (55.11.5); 30775 (55.45.41); 30776 (55.36.6, 55.36.25); 30777 (55.95.6); 30779 (55.90.9); 30780 (55.28.11); 30781 (55.45.44); 30782 (55.62.10); 30783 (55.36.31); 30784 (55.11.19); 30785 (55.83.1); 30786 (55.28.11); 30798 (55.28.13); 30799 (55.37.12); 30800 (55.66.4); 30801 (55.56.1); 30808 (55.92.7); 30809 (55.105.1); 30810 (55.36.27); 30826 (55.37.16); 30827 (55.36.27); 30828 (55.45.23); 30847 (55.89.2); 30848 (55.87.1); 30849 (55.45.11); 30850 (55.45.57); 30851 (55.115.5, 55.115.9); 30852 (55.17.62); 30853 (55.45.25); 30854 (55.17.88); 30855 (55.11.15); 30856 (55.22.4a); 30943 (55.43.2); 30945 (55.20.5); 30946 (55.17.81); 30971 (55.36.25); 30990 (55.114.2); 30991 (55.45.28); 30992 (55.115.12); 30994 (55.28.11); 30995 (55.87.1); 30996 (55.43.4); 30998 (55.36.6); 31000 (55.28.1a); 31002 (55.37.25); 31003 (55.17.35); 31003A (55.17.31); 31004 (55.22.4a); 31005 (55.66.5); 31006 (55.11.15); 31008 (55.95.6); 31010 (55.105.1); 31011 (55.16.3); 31012 (55.66.14); 31019 (55.45.45); 31040 (55.28.31); 31041 (55.62.1); 31042 (55.17.18); 31043 (55.11.4); 31044 (55.62.26); 31045 (55.28.20); 31046 (55.20.2); 31047 (55.28.11); 31048 (55.115.8); 31050 (55.115.7); 31109 (55.18.17); 31110 (55.66.14); 31111 (55.22.4a); 31113 (55.114.15); 31115 (55.29.1); 31116 (55.114.3); 31136 (55.35.2); 31209A (55.89.1a); 31212 (55.36.31); 31252 (55.17.7); 31253 (55.17.35); 31254 (55.17.24); 31255 (55.36.25); 31256 (55.17.19); 31257 (55.22.4a); 31258 (55.45.25); 31337 (55.37.7); 31339 (55.45.31); 31340 (55.66.7); 31341 (55.28.13); 31366 (55.87.1); 31381 (55.115.12); 31382 (55.90.3); 31451 (55.45.57, 55.90.2); 31453 (55.45.57); 31454 (55.45.29); 31455 (55.28.23); 31456 (55.90.8); 31457 (55.37.14); 31510 (55.36.31); 31512 (55.36.13); 31513 (55.66.14); 31514 (55.105.1); 31516 (55.62.10); 31566 (55.20.1); 31567 (55.11.8); 31568 (55.66.5); 31591 (55.45.23); 31625 (55.56.2); 31626 (55.88.1); 31627 (55.28.13); 31650 (55.62.4); 31655 (55.28.13, 55.89.2); 31656 (55.45.19); 31680 (55.45.19); 31681 (55.18.7); 31682 (55.28.20); 31683 (55.37.17); 31731 (55.90.2); 31732 (55.11.4); 31733 (55.45.24); 31735 (55.17.18); 31736 (55.17.13); 31737 (55.17.50); 31741 (55.17.75); 31788 (55.11.10); 31830 (55.37.18); 31831 (55.37.27); 31832 (55.45.23); 31835 (55.37.17); 31836 (55.17.13); 31850 (55.45.11); 31851 (55.17.57); 31886 (55.90.3); 31887 (55.78.7); 31907 (55.28.20); 31930 (55.16.4); 31931 (55.115.12); 31965 (55.18.6); 31966 (55.17.50); 31975 (55.17.7); 32059 (55.17.54a); 32060 (55.112.5); 32061 (55.28.33); 32077 (55.43.3); 32086 (55.115.9); 32087 (55.45.57); 32114 (55.17.62); 32137 (55.2.1); 32138 (55.114.3); 32140 (55.17.46); 32159 (55.56.1); 32170 (55.20.4); 32190 (55.61.1); 32207 (55.17.6, 55.17.24); 32208 (55.43.2); 32219 (55.103.1); 32220 (55.37.16); 32227 (55.105.3); 32228 (55.92.7); 32229 (55.62.1); 32230 (55.28.22); 32242 (55.36.25); 32244 (55.37.13); 32245 (55.36.14); 32248 (55.28.31); 32254 (55.36.28); 32261 (55.36.39); 32271 (55.11.9); 32286 (55.28.15); 32309 (55.45.41); 32310 (55.36.7); 32311 (55.17.50); 32322 (55.37.27); 32324 (55.37.27); 32325 (55.45.41); 32350 (55.62.1); 32351 (55.6.5); 32353 (55.11.9, 55.36.27); 32354 (55.28.20, 55.36.13); 32355 (55.28.11); 32357 (55.66.4); 32358 (55.6.9,

55.36.25); 32359 (55.28.10); 32361 (55.105.1); 32411 (55.11.9); 32432 (55.37.14); 32433 (55.11.4); 32434 (55.45.41); 32435 (55.45.57); 32439 (55.18.17); 32440 (55.45.28); 32442 (55.90.9); 32443 (55.45.57, 55.90.3); 32444 (55.113.8); 32445 (55.17.24); 32449 (55.78.7); 32451 (55.18.10); 32452 (55.115.5); 32453 (55.36.7); 32485 (55.113.8); 32523 (55.11.4); 32524 (55.45.25); 32526 (55.115.4); 32528 (55.87.5, 55.87.6); 32540 (55.30.3); 32547 (55.28.10); 32548 (55.37.13); 32549 (55.45.41); 32550 (55.45.57); 32552 (55.37.14); 32554 (55.17.26); 32555 (55.45.38a); 32556 (55.37.7, 55.45.29); 32557 (55.17.13); 32583 (55.100.1); 32584 (55.78.11); 32585 (55.62.16); 32586 (55.78.4); 32588 (55.17.43); 32608 (55.36.3); 32609 (55.36.13); 32618 (55.45.29); 32666 (55.11.15); 32667 (55.45.38b); 32746 (55.43.2); 32771 (55.11.4); 32772 (55.36.7); 32775 (55.86.7); 32777 (55.45.41); 32778 (55.45.41); 32780 (55.28.3, 55.36.6, 55.115.4); 32781 (55.20.4); 32782 (55.90.6); 32784 (55.45.11); 32785 (55.17.51); 32787 (55.9.2, 55.17.26); 32877 (55.90.9); 32878 (55.17.62); 32879 (55.11.4); 32880 (55.17.41); 32881 (55.78.4); 32887 (55.16.4); 32889 (55.113.7); 32902 (55.45.41); 32903 (55.17.51); 32904 (55.45.5); 32905 (55.11.5); 32906 (55.43.4); 32907 (55.112.5); 32941 (55.28.18); 32975 (55.90.2); 32976 (55.115.12); 32977 (55.11.10); 32978 (55.17.54a); 33003 (55.36.7); 33005 (55.59.1); 33006 (55.90.5); 33026 (55.66.14); 33028 (55.17.88); 33031 (55.62.5); 33036 (55.92.6); 33068 (55.3.2a); 33100 (55.62.16); 33104 (55.11.9); 33106 (55.9.2); 33130 (55.37.33); 33133 (55.43.2); 33134 (55.115.7); 33135 (55.36.39); 33145 (55.113.7); 33153 (55.28.20); 33159 (55.20.2); 33171 (55.45.28); 33172 (55.62.11); 33173 (55.37.17, 55.37.33); 33174 (55.18.6); 33175 (55.17.19); 33176 (55.28.15); 33177 (55.37.30); 33178 (55.37.17); 33179 (55.17.53); 33180 (55.37.17, 55.37.27); 33700 (55.22.4a); 33726 (55.45.28); 33787 (55.45.38b); 33790 (55.62.19); 33791 (55.73.2); 33792 (55.36.57); 33793 (55.115.7); 33862 (55.17.54b); 33894 (55.66.7); 33937 (55.22.4a); 33938 (55.45.38b); 33956 (55.66.14); 33962 (55.45.16); 33966 (55.14.1); 33973 (55.58.1); 33982 (55.17.68); 34040 (55.34.1); 34050 (55.11.7); 34051 (55.36.19); 34052 (55.11.18); 34053 (55.18.7); 34054 (55.90.9); 34058 (55.45.23); 34059 (55.66.12); 34065 (55.17.60); 34097 (55.66.14); 34104 (55.17.62); 34105 (55.115.4); 34106 (55.90.2); 34109 (55.36.13); 34194 (55.17.19); 34238 (55.90.2); 34305 (55.17.13); 34311 (55.114.14); 34314 (55.45.41); 34318 (55.39.1); 34320 (55.45.28); 34321 (55.13.1); 34323 (55.43.1); 34325 (55.78.11); 34327 (55.78.1); 34329 (55.37.16); 34330 (55.45.23); 34331 (55.33.6); 34332 (55.36.37); 34333 (55.28.1b); 34335 (55.17.50); 34336 (55.62.10); 34357 (55.45.41); 34359 (55.28.20); 34362 (55.11.4); 34363 (55.45.13); 34364 (55.17.79); 34365 (55.17.73); 34366 (55.36.13); 34371 (55.17.88); 34374 (55.28.27); 34375 (55.17.13); 34398 (55.90.5); 34399 (55.28.31); 34401 (55.78.11); 34404 (55.66.14); 34411 (55.17.45); 34441 (55.14.1); 34442 (55.17.19); 34443 (55.45.13); 34445 (55.89.6); 34446 (55.66.5); 34481 (55.66.14); 34482 (55.115.10); 34484 (55.113.8); 34486 (55.95.6); 34487 (55.28.31, 55.45.13); 34488 (55.11.15); 35161 (55.83.1); 35162 (55.17.88); 35164 (55.36.19, 55.36.57); 35166 (55.36.13); 35167 (55.95.1, 55.95.3); 35168 (55.11.10); 35169 (55.11.15); 35171 (55.78.4); 35172 (55.78.11); 35173 (55.19.1); 35174 (55.20.1); 35175 (55.32.2); 35176 (55.45.5); 35177 (55.45.24); 35179 (55.45.57); 35180 (55.45.38a); 35181 (55.45.38b); 35182 (55.45.38b); 35183 (55.45.9b); 35184 (55.45.38b); 35185 (55.45.9b); 35186 (55.45.9b); 35187 (55.45.38b); 35188 (55.45.28); 35189 (55.45.29); 35190 (55.45.11); 35191 (55.28.13); 35192 (55.28.2); 35193 (55.28.32); 35194 (55.28.2); 35196 (55.35.2); 35198 (55.103.1); 35199 (55.92.5); 35201 (55.90.9, 55.113.12); 35202 (55.17.13); 35203 (55.17.31); 35204 (55.17.79); 35205 (55.37.25); 40052 (55.115.4); 40053 (55.115.4); 40054 (55.62.10); 40055 (55.17.63); 40056 (55.17.73); 40060 (55.36.37); 40061 (55.45.57); 40064 (55.78.7); 40065 (55.60.1); 40094 (55.28.10); 40111 (55.36.53); 40112 (55.26.1); 40113 (55.33.5); 40116 (55.17.54a); 40126 (55.33.5); 40129 (55.45.57); 40130 (55.17.27); 40132 (55.92.6); 40133 (55.17.24); 40134 (55.37.16); 40135 (55.37.13); 40143 (55.17.31); 40153 (55.83.5); 40195 (55.17.66); 40216 (55.45.31); 40224 (55.28.13); 40226 (55.17.26); 40227 (55.45.41); 40229 (55.28.27); 40230 (55.17.41); 40253 (55.87.1); 40254 (55.28.13); 40255 (55.37.5); 40332 (55.28.13); 40333 (55.9.2); 40334 (55.115.10); 40335 (55.28.11); 40336 (55.78.1); 40337 (55.17.35); 40338 (55.17.66); 40340 (55.17.25); 40341 (55.36.31); 40343 (55.17.31); 40344 (55.26.1); 40345 (55.45.57); 40347 (55.33.5); 40348 (55.17.58); 40349 (55.66.14); 40376 (55.62.14); 40434 (55.37.9); 40444 (55.45.41); 40451 (55.28.6); 40456 (55.33.4); 40458 (55.89.4); 40459 (55.113.12); 40460 (55.87.6); 40464 (55.60.1); 40465 (55.16.2); 40467

(55.37.5); 40502 (55.17.48); 40506 (55.83.1); 40513 (55.8.2, 55.64.2); 40521 (55.17.31); 40522 (55.95.3); 40524 (55.66.14); 40526 (55.17.31); 40527 (55.89.3); 40548 (55.37.29); 40552 (55.115.2); 40558 (55.45.57); 40559 (55.45.31); 40580 (55.31.1a); 40590 (55.67.1); 40602 (55.17.19, 55.17.73); 40603 (55.115.5); 40604 (55.115.4); 40612 (55.18.10); 40630 (55.28.33); 40631 (55.17.73); 40632 (55.6.3); 40633 (55.37.7); 40634 (55.26.1); 40635 (55.17.56); 40657 (55.115.9); 40659 (55.43.3); 40661 (55.115.7); 40662 (55.89.3); 40663 (55.48.4); 40678 (55.17.73); 40693 (55.36.31); 40702 (55.36.53); 40703 (55.45.57); 40786 (55.89.3); 40787 (55.45.57); 40819 (55.89.3); 40820 (55.28.32); 40821 (55.17.46); 40822 (55.78.1); 40824 (55.66.12); 40825 (55.45.50); 40827 (55.115.7); 40828 (55.18.10); 40835 (55.20.1); 40854 (55.28.33); 40854a (55.28.33); 40855 (55.115.7); 40856 (55.45.25); 40858 (55.36.31); 40859 (55.62.16); 40860 (55.45.41); 40861 (55.78.1); 40863 (55.62.14); 40901 (55.28.36); 40906 (55.11.10); 40907 (55.45.57); 40908 (55.62.12); 40909 (55.66.12); 40910 (55.28.31); 40911 (55.45.9a); 40912 (55.45.38b); 40913 (55.45.49); 40915 (55.39.3); 40916 (55.115.5); 40917 (55.45.11); 40918 (55.45.28); 40919 (55.6.8); 40922 (55.78.5); 40923 (55.67.3); 40924 (55.78.1); 40925 (55.37.14); 40926 (55.66.14); 40927 (55.62.16); 40928 (55.62.1); 40931 (55.37.9); 40955 (55.94.3); 40956 (55.6.9); 40957 (55.90.6); 40958 (55.11.10); 48115 (55.58.1); 50057 (55.17.72); 50060 (55.20.4); 50063 (55.17.31); 50064 (55.67.3); 50098 (55.5.1); 50099 (55.62.12); 50100 (55.89.1a); 50104 (55.45.57); 50105 (55.37.29); 50106 (55.19.1, 55.69.2); 50107 (55.17.62); 50109 (55.18.10); 50112 (55.45.57); 50113 (55.28.11); 50141 (55.17.72); 50142 (55.31.1a); 50143 (55.45.9b); 50145 (55.36.31); 50146 (55.17.15); 50147 (55.78.4); 50151 (55.115.8); 50161 (55.45.57); 50162 (55.37.19); 50163 (55.78.5); 50165 (55.45.25); 50166 (55.45.57); 50168 (55.45.57); 50171 (55.28.33); 50172 (55.17.45); 50173 (55.115.5); 50174 (55.6.9); 50176 (55.36.21); 50177 (55.37.14); 50180 (55.90.9); 50181 (55.113.4); 50183 (55.62.26); 50184 (55.37.14); 50199 (55.45.57); 50201 (55.6.9); 50203 (55.43.3); 50206 (55.17.56); 50207 (55.115.9); 50209 (55.45.41); 50230 (55.17.31); 50231 (55.20.4); 50234 (55.37.9); 50235 (55.37.9, 55.37.12, 55.37.21); 50236 (55.37.32); 50237 (55.36.13); 50238 (55.36.28); 50240 (55.78.1, 55.91.1); 50241 (55.90.1); 50244 (55.22.4a); 50246 (55.37.12); 50247 (55.17.13); 50248 (55.89.3); 50251 (55.113.12); 50255 (55.62.2); 50256 (55.37.12); 50257 (55.22.6); 50258 (55.37.9); 50278 (55.37.16); 50285 (55.115.5); 50286 (55.22.1); 50290 (55.89.6); 50291 (55.17.28); 50292 (55.43.2); 50322 (55.37.16); 50323 (55.20.4, 55.95.1); 50324 (55.43.4); 50326 (55.45.38b); 50327 (55.115.7); 50328 (55.115.9); 50340 (55.103.3); 50341 (55.17.51); 50342 (55.45.38b); 50343 (55.36.37); 50344 (55.17.57); 50345 (55.115.4); 50346 (55.89.1a); 50347 (55.37.29); 50349 (55.89.1a); 50350 (55.18.10); 50352 (55.11.4); 50354 (55.78.7); 50355 (55.45.38b); 50357 (55.67.1); 50358 (55.78.11); 50359 (55.11.10); 50360 (55.37.32); 50361 (55.45.57); 50362 (55.28.27); 50363 (55.89.1a); 50364 (55.89.6); 50365 (55.17.56, 55.45.25); 50367 (55.11.10); 50369 (55.17.56); 50370 (55.37.35); 50371 (55.45.57); 50373 (55.115.12); 50375 (55.36.49); 50376 (55.36.12); 50377 (55.91.1); 50378 (55.78.5); 50380 (55.62.4); 50381 (55.45.3); 50382 (55.95.6); 50384 (55.43.3); 50391 (55.45.37); 50393 (55.43.3); 50394 (55.17.31); 50395 (55.115.3); 50396 (55.89.1a); 50397 (55.45.25); 50398 (55.6.4); 50408 (55.28.1a); 50409 (55.43.3); 50410 (55.45.38b); 50411 (55.28.5b); 50412 (55.14.1); 50413 (55.45.57); 50414 (55.22.1); 50415 (55.36.7); 50416 (55.28.3); 50417 (55.115.4); 50418 (55.45.25, 55.45.39); 50421 (55.39.3); 50422 (55.17.73); 50430 (55.89.1a); 50431 (55.115.1); 50442 (55.36.31); 50447 (55.67.3); 50448 (55.11.15); 50450 (55.89.6); 50481 (55.36.12); 50485 (55.17.19); 50487 (55.45.38a); 50490 (55.16.3); 50494 (55.36.7); 50496 (55.115.7); 50511 (55.115.9); 50512 (55.17.56); 50536 (55.45.41); 50540 (55.11.15); 50541 (55.16.6); 50542 (55.32.1); 50556 (55.115.9); 50646 (55.45.41); 50647 (55.45.16); 50648 (55.45.9b, 55.45.16); 50649 (55.22.3a); 50650 (55.37.18); 50652 (55.37.2); 50653 (55.37.30); 50654 (55.37.21); 50655 (55.28.20); 50656 (55.62.25); 50657 (55.113.7); 50658 (55.22.2); 50659 (55.43.2); 50660 (55.45.57); 50661 (55.37.17); 50662 (55.28.18); 50663 (55.113.7); 50664 (55.28.22); 50669 (55.37.3); 50680 (55.28.20); 50699 (55.28.3); 50700 (55.36.28, 55.89.1a); 50703 (55.22.1); 50707 (55.17.35); 50737 (55.28.1a); 50738 (55.36.33); 50740 (55.22.1); 50767 (55.28.22); 50768 (55.37.17); 50770 (55.45.29); 50774 (55.37.11, 55.37.16); 50775 (55.17.19); 50777 (55.37.2); 50801 (55.37.17); 50819 (55.36.3); 50821 (55.106.1); 50903 (55.17.53); 50904 (55.62.11); 50953 (55.28.27); 50954 (55.17.88); 50962 (55.17.57); 51009 (55.113.7); 51012 (55.37.30); 51023

(55.91.1); 51057 (55.104.2); 51079 (55.37.27); 51080 (55.37.25); 51127 (55.62.26); 51144 (55.89.2); 51148 (55.66.8); 51149 (55.90.7); 51192 (55.22.2); 51193 (55.113.7); 51194 (55.92.5); 51302 (55.53.1); 51317 (55.62.26); 51320 (55.73.1); 51424 (55.17.54a); 51425 (55.34.1); 51427 (55.32.1); 51434 (55.11.10); 51462 (55.62.12); 51464 (55.45.57); 51468 (55.20.1); 51469 (55.62.25); 51470 (55.56.2); 51471 (55.18.7); 51472 (55.113.3); 51514 (55.17.57); 51515 (55.62.11); 51534 (55.37.17); 51576 (55.10.2); 51581 (55.87.6); 51588 (55.6.11); 51599 (55.11.10); 51625 (55.109.4); 51626 (55.37.23); 51627 (55.45.9b); 51628 (55.60.1); 51711 (55.51.1); 51712 (55.11.15); 51713 (55.17.25); 51714 (55.37.5); 51715 (55.95.1); 51716 (55.87.1); 51718 (55.105.1); 51719 (55.66.5); 51721 (55.120.6); 51722 (55.62.23); 51723 (55.30.3); 51724 (55.89.3); 51725 (55.114.15); 51726 (55.62.12); 51727 (55.62.25); 51729 (55.17.56); 51730 (55.62.15); 51731 (55.17.52); 51732 (55.115.7); 51733 (55.53.6); 51734 (55.36.3); 51736 (55.28.15); 51741 (55.62.15); 51743 (55.78.5).

Clements 3293 (55.115.12); 3317 (55.36.4); 3350 (55.67.3); 3354 (55.84.1); 3359 (55.101.1); 3367 (55.101.2); 3372 (55.88.3); 3375A (55.66.9); 3385 (55.101.1); 3392 (55.104.1); 3393 (55.27.1); 3397 (55.36.33); 3398A (55.87.4); 3402 (55.36.20).

Cockburn 70120 (55.103.2).

Collenette 1 (55.17.48); 2 (55.36.24, 55.105.1); 3 (55.36.25, 55.45.34); 4 (55.36.37); 5 (55.32.3); 9 (55.35.2); 17 (55.43.2); 18 (55.39.1); 19 (55.28.35); 22A (55.28.27); 22B (55.28.27); 23 (55.28.27); 24 (55.105.3); 25 (55.45.41); 30 (55.92.7); 31 (55.37.13); 33 (55.2.2, 55.33.6); 34 (55.6.5); 35 (55.28.22); 38 (55.28.12); 40 (55.36.53); 41 (55.37.16); 42 (55.28.1b); 43 (55.28.22); 45 (55.28.22); 46 (55.33.6); 47 (55.33.6); 56 (55.28.20); 57 (55.11.9); 62 (55.45.9a); 65 (55.103.1); 67 (55.45.29); 69 (55.36.18); 70 (55.36.17a); 72 (55.28.22); 73 (55.28.22); 74 (55.36.53); 78 (55.17.81); 81 (55.36.25); 82 (55.36.53); 88 (55.113.8); 94 (55.36.21); 102 (55.62.1); 104 (55.37.5); 105 (55.28.2); 106 (55.28.10); 107 (55.62.3); 113 (55.66.12); 115 (55.120.1); 120 (55.66.11); 126 (55.62.23); 127 (55.36.52); 128 (55.28.6); 130 (55.53.7); 133 (55.17.6, 55.17.24); 134 (55.18.2); 135 (55.26.1); 538 (55.17.43); 539 (55.87.5); 540 (55.83.2a); 553 (55.17.53); 584 (55.18.7); 585 (55.83.3a); 586 (55.17.37); 587 (55.17.76); 591 (55.28.33); 592 (55.28.10); 596 (55.18.14, 55.28.10); 597 (55.26.1, 55.114.13); 616 (55.37.27); 623 (55.72.1); 629 (55.28.23); 630 (55.62.9); 635 (55.22.4b); 637 (55.9.2); 640 (55.57.2); 643 (55.119.7); 752 (55.32.1); 753 (55.17.12); 756 (55.26.1); 821 (55.30.3); 825 (55.83.4); 888 (55.113.7); 892 (55.36.53); 897 (55.90.5); 910 (55.17.40); 916 (55.66.5); 917 (55.17.39); 918 (55.17.54a); 1002 (55.17.26); 1010 (55.95.3); 1011 (55.28.35); 1013 (55.18.15); 1020 (55.26.1); 1021 (55.115.3); 1032 (55.28.13); 1039 (55.66.3); 1041 (55.11.18); 1042 (55.37.19); 1044 (55.9.2); 1045 (55.66.14); 1046 (55.30.3); 1047 (55.30.1); 1049 (55.62.5); 1054 (55.45.25); 1055 (55.32.1); 2295 (55.114.2); 2315 (55.17.35); 2361 (55.17.5); 2372 (55.36.27); 2376 (55.17.31); 3214 (55.95.6); 21503 (55.92.4); 21534 (55.37.21); 21535 (55.37.18); 21536 (55.45.41); 21567 (55.62.26); 21574 (55.62.16); 21642 (55.45.29).

Cribb 26 (55.114.7); 27 (55.35.1); 28 (55.66.8); 29 (55.66.9); 30 (55.62.16); 31 (55.36.19); 32 (55.30.3); 33 (55.73.2); 34 (55.18.5); 35 (55.45.28); 36 (55.86.7); 38 (55.17.88); 39 (55.36.7); 50 (55.27.5); 58 (55.120.1); 59 (55.36.36); 60 (55.114.10a); 61 (55.110.1).

Darnton 118 (55.47.3); 129 (55.62.23); 176 (55.111.1); 194 (55.45.3); 195 (55.11.18); 198 (55.20.2); 199 (55.99.1); 200 (55.62.6); 201 (55.90.5); 202 (55.90.4); 203 (55.78.4); 207 (55.113.3); 208 (55.17.54b); 210 (55.17.35); 211 (55.17.31); 212 (55.104.4); 213 (55.90.6); 217 (55.115.8); 223 (55.36.2); 224 (55.113.11); 225 (55.45.57); 226 (55.114.3); 227 (55.45.57); 228 (55.28.33); 268 (55.62.7); 291 (55.11.13); 302 (55.11.1); 303 (55.62.7); 306 (55.111.1); 307 (55.111.1); 308 (55.36.34); 309 (55.33.2a); 310 (55.33.4); 314 (55.33.2a); 316 (55.11.15); 318 (55.62.4); 320 (55.28.26); 352 (55.6.10); 354 (55.6.2); 355 (55.78.3); 356 (55.113.9); 358 (55.39.4); 359 (55.94.1); 372 (55.3.2a); 379 (55.89.4); 385 (55.62.14); 386 (55.27.4); 393 (55.48.8); 394 (55.17.48); 396 (55.113.1); 399 (55.45.41); 403 (55.113.12); 449 (55.45.13); 452 (55.8.2); 482 (55.91.1); 486 (55.32.1); 512 (55.32.2); 513 (55.62.1); 514 (55.17.45); 519

(55.115.8); 520 (55.19.1); 523 (55.90.9); 525 (55.113.10); 536 (55.17.13); 537 (55.62.6); 540 (55.48.4); 541 (55.17.35); 556 (55.62.19); 565 (55.45.28); 572 (55.36.27); 573 (55.59.1); 575 (55.95.6); 596 (55.11.5); 598 (55.87.6); 599 (55.105.3); 602 (55.28.1a).

Dransfield 5558 (55.30.3); 5709 (55.30.3).

Fowlie & Ross 83P912 (55.40.1).

Fuchs 21032 (55.22.4a); 21054 (55.115.12); 21058 (55.45.51); 21072 (55.22.2).

Fuchs & Collenette 21380 (55.83.4); 21381 (55.73.1); 21404 (55.37.21); 21407 (55.62.25); 21498 (55.62.19).

Gibbs 3955 (55.45.4); 3957 (55.105.3); 3958 (55.7.1); 3959 (55.90.4); 3960 (55.45.15); 3970 (55.118.1a); 3971 (55.28.28); 3993 (55.114.3); 3997 (55.53.3); 4003 (55.53.3); 4009 (55.45.9a); 4012 (55.36.31); 4014 (55.62.9); 4016 (55.115.11); 4034 (55.6.5); 4058 (55.11.4); 4059 (55.17.35); 4060 (55.95.5); 4065 (55.66.5); 4076 (55.105.1); 4085 (55.37.14); 4087 (55.37.14); 4090 (55.45.29); 4095 (55.22.3a); 4108 (55.18.7); 4111 (55.67.2); 4117 (55.115.10); 4123 (55.90.2); 4181 (55.37.30); 4215 (55.28.7); 4227 (55.45.22); 4250 (55.37.17); 4252 (55.17.19); 4258 (55.92.4); 4260 (55.22.2); 4261 (55.28.18); 4268 (55.45.16); 4279 (55.115.6); 4301 (55.11.17); 4302 (55.45.22); 4318 (55.80.1); 4319 (55.47.2); 4339 (55.28.6); 4347 (55.83.1).

Gunsalam 1 (55.33.6); 2 (55.113.7); 3 (55.37.18); 4 (55.17.12); 5 (55.17.3); 6 (55.36.13); 7 (55.28.1a); 8 (55.36.27); 9 (55.17.62); 10 (55.37.20); 11 (55.17.53).

Haviland 1097 (55.37.30); 1098 (55.28.18); 1099 (55.17.53); 1100 (55.17.19); 1143 (55.17.54a); 1157 (55.45.16); 1158 (55.92.7); 1164 (55.17.12); 1165 (55.75.2); 1250 (55.45.29); 1251 (55.39.3); 1252 (55.17.31); 1253 (55.43.2); 1291 (55.105.1); 1299 (55.36.31); 1302 (55.11.4); 1381 (55.37.7).

Jukian & Lamb 33 (55.118.2).

Jukian & Lamb in Lamb 4/82 (55.81.1a).

Jumaat 3397 (55.37.21).

Kanis & Meijer 51454 (55.28.13).

Kanis & Sinanggul 51488 (55.45.23); 51496 (55.115.3).

Kidman Cox 902 (55.101.4); 918 (55.17.15); 2502 (55.36.21); 2516 (55.45.28); 2517 (55.14.1); 2518 (55.17.47); 2525 (55.45.13); 2535 (55.95.6).

Lajangah 36159 (55.115.7).

Lamb 1/82 (55.106.1); 2/82 (55.95.6); 3/82 (55.63.1); 4 (55.83.1); 6 (55.39.3); 12/82 (55.17.18); 13/82 (55.47.3); 15/82 (55.20.4); 16/82 (55.45.3); 17/82 (55.36.44); 19/82 (55.36.13); 20/82 (55.43.3); 21/82 (55.17.33); 22/82 (55.62.2); 24/82 (55.86.7); 25/82 (55.87.4); 27/82 (55.45.43); 28/82 (55.55.1); 31/82 (55.103.1); 39/83 (55.87.1); 40/83 (55.43.2); 43/83 (55.33.6); 44/83 (55.18.15); 45/83 (55.105.1); 46/83 (55.120.3); 49/83 (55.92.5); 50/83 (55.56.2); 51/83 (55.120.6); 52/83 (55.36.6); 56/83 (55.17.63); 57 (55.28.17); 58/83 (55.19.1); 75/83 (55.27.1); 155/83 (55.77.2); 159/83 (55.112.5); 164/83 (55.87.4); 185/84 (55.114.1); 275/84 (55.7.1); 303/85 (55.67.1); 304/85 (55.62.24); 305/85 (55.66.14); 308/85 (55.52.1); 310/85 (55.47.1); 311/85 (55.36.36); 312/85 (55.28.17); 361/85 (55.117.1); 386/85 (55.104.5); 390/85 (55.23.2); 433/85 (55.104.1); 442/85

(55.94.1); 443/85 (55.117.1); 500/85 (55.36.58); 503/85 (55.17.47); 504/85 (55.11.6); 551/86 (55.87.2); 577/86 (55.17.39); 585/86 (55.17.3); 586/86 (55.92.4); 587/86 (55.18.13); 588/86 (55.39.1); 592/86 (55.17.67); 651/86 (55.98.1); 665/86 (55.17.58); 715/86 (55.17.29); 726/86 (55.17.2); 733/87 (55.36.53); 734/87 (55.22.2); 737/87 (55.17.81); 738/87 (55.17.57); 742/87 (55.17.59); 746/87 (55.56.2); 785/87 (55.17.17); 828/87 (55.17.2); 829/87 (55.17.40); 838/87 (55.23.1); 846/87 (55.17.69); 851/87 (55.116.2); 864/87 (55.114.9); 868/87 (55.67.3); 877/87 (55.33.7a); 880/87 (55.17.22); 881/87 (55.17.10); 972/88 (55.36.26); 1144/89 (55.3.1); 1164/89 (55.98.1); 1272/90 (55.17.80); 1289/90 (55.17.87); 1365/91 (55.37.6); 1402/92 (55.36.40); 1409/92 (55.11.8); 1467/92 (55.45.20); 1516/92 (55.19.1); 3020 (55.45.56); 3023 (55.36.29); 3156 (55.8.2); 3158 (55.41.1); 3160 (55.45.26); 3164 (55.36.36); 3175 (55.43.2); 3176 (55.20.1); 87134 (55.87.5); 88566 (55.17.12); 88569 (55.53.6); 89640 (55.102.1); 89671 (55.36.39); 89677 (55.26.1); 89678 (55.26.1); 89685 (55.12.1); 89687 (55.36.50); 89693 (55.17.17); 91503 (55.84.1); 91507 (55.33.7a); 91509 (55.18.5); 91511 (55.88.2); 91515 (55.112.1); 91516 (55.27.3); 91528 (55.18.13); 91534 (55.73.2); 91536 (55.57.2); 91537 (55.66.14); 91538 (55.66.14); 91557 (55.66.9); 91573 (55.36.55); 91579 (55.17.19); 91580 (55.17.42); 91581 (55.18.10); 91587 (55.53.8); 92260 (55.114.2); 92328 (55.77.2); 92333 (55.93.4); 92339 (55.47.2); 93351 (55.62.13); 93357 (55.33.3); 93360 (55.36.32); 93361 (55.28.29); 93367 (55.105.2); 93368 (55.36.25); 93369 (55.62.3); 93370 (55.105.2); 93371 (55.105.1); 93375 (55.114.11); 93376 (55.41.2); 93453 (55.27.2); 93469 (55.66.6); 93471 (55.17.66); 93472 (55.36.55).

Lamb & Chan 6/82 (55.75.2).

Lamb & Phillipps 161/83 (55.36.30).

Lamb & Surat 1297/91 (55.17.86); 1302/91 (55.17.30); 1365/91 (55.37.35).

Lamb & Surat in Lamb 1320/91 (55.87.3a).

Lohok in Lamb 355/85 (55.77.1).

Lohok 4 (55.45.21); 5 (55.23.2); 8 (55.117.1); 9 (55.27.9); 10 (55.93.3); 12 (55.68.1); 13 (55.101.4); 15 (55.36.8); 17 (55.36.50); 18 (55.63.1); 19 (55.28.31); 20 (55.101.4); 21 (55.63.2); 22 (55.27.1); 23 (55.45.20); 24 (55.53.8); 25 (55.36.7); 26 (55.17.48); 27 (55.18.1); 28 (55.16.7).

Madani 76464 (55.45.28); 76482 (55.43.4); 89375 (55.77.3a); 89454 (55.102.1); 89459 (55.104.5); 89490 (55.78.5); 89511 (55.119.6); 89513 (55.45.30); 89541 (55.62.3); 89543 (55.28.33); 89544 (55.95.7).

Meijer 20316 (55.45.9a); 20323 (55.45.56); 20326 (55.37.14); 20367 (55.37.11); 20417 (55.45.38b); 22073 (55.28.1b); 24131 (55.37.17); 28560 (55.37.30); 29165 (55.37.17); 29293 (55.45.16); 48111 (55.37.26).

Mikil 33935 (55.11.10); 34516 (55.17.57); 38455 (55.115.7); 46555 (55.28.20); 46571 (55.17.19).

Molesworth-Allen 3313 (55.17.12).

Moulton 103 (55.73.2).

Native collector 68 (55.37.11); 99 (55.37.11).

Patrick & Thomas 248 (55.51.1).

Phillipps 2123 (55.82.1); 2403 (55.17.64); 2467 (55.58.1).

Phillipps & Lamb 3019 (55.116.4).

Poore 87 (55.17.43); 301 (55.45.38b); 424 (55.28.10).

Price 122 (55.28.34); 125 (55.45.29); 145 (55.37.22); 151 (55.17.63); 217 (55.37.4); 220 (55.113.2); 223 (55.17.12); 224 (55.45.56); 226 (55.45.50); 228 (55.17.25).

Puasa 154 (55.45.39); 1540 (55.13.1); 1552 (55.17.31).

Richards 111 (55.28.18).

RSNB 179–2764 under Chew, Corner & Stainton; RSNB 4031–5790, 7030–7124 under Chew & Corner; RSNB 6007–6031 under Chai & Ilias.

Sadau 49690 (55.26.1).

Saikeh 82763 (55.105.6); 82800 (55.17.84).

Sands 3880 (55.37.27); 3881 (55.43.2); 3882 (55.28.14); 3904 (55.17.43); 3976 (55.36.12); 3989 (55.39.3); 4011 (55.66.12).

Sato 162 (55.37.27); 699 (55.37.17); 759 (55.37.3); 762 (55.37.3); 1214 (55.37.34); 1223 (55.37.17); 1558 (55.37.25); 2100 (55.37.21); 2158 (55.37.21).

Sato et al. 970 (55.37.17); 972 (55.37.17); 974 (55.37.17); 1046 (55.37.17); 1047 (55.37.17); 1080 (55.37.17); 1081 (55.37.17); 1082 (55.37.17); 1375 (55.37.3); 1436 (55.37.21).

Shea & Aban 76860 (55.28.6); 76910 (55.11.16); 77158 (55.11.13).

Sidek 50 (55.43.2).

Sidek bin Kiah 38 (55.37.30).

Sinanggul 38313 (55.115.7); 38318 (55.22.3a).

Sinclair et al. 9009 (55.95.6); 9011 (55.17.15); 9027 (55.75.2); 9057 (55.17.12); 9172 (55.17.53); 9180 (55.37.17); 9181 (55.37.27); 9195 (55.28.18); 9197 (55.28.20); 9226 (55.45.22).

Smith & Everard 142 (55.28.15); 150 (55.17.53); 155 (55.37.2).

Sutton 1 (55.73.1); 2 (55.45.38b); 3 (55.73.1); 4 (55.36.25); 5 (55.43.2); 6 (55.28.10); 7 (55.43.2); 8 (55.36.45); 9 (55.43.2); 10 (55.28.20); 11 (55.11.10); 12 (55.22.3a); 13 (55.28.20); 15 (55.69.1); 17 (55.36.53); 20 (55.36.3).

Tan & Phillipps 579 (55.105.3).

Tiong 88658 (55.36.62).

Tiong & Dewol 85738 (55.115.6).

Topping 372 (55.17.53); 1485 (55.95.6); 1523 (55.62.7).

Vermeulen 468 (55.17.54b); 471 (55.17.18); 472 (55.17.19); 475 (55.17.62); 477 (55.17.37); 479 (55.17.2); 480 (55.17.35); 481 (55.17.79); 482 (55.17.39); 483 (55.17.69); 484 (55.17.31); 485 (55.17.54b); 487 (55.17.5); 489 (55.17.10); 492 (55.48.4); 497 (55.17.44); 499 (55.17.64); 513 (55.17.21); 548 (55.45.25); 701 (55.17.83); 1297 (55.17.1); 1308 (55.17.61).

Vermeulen & Chan 390 (55.17.42); 391 (55.45.38b); 408 (55.45.25); 409 (55.6.8); 411 (55.45.46).

Vermeulen & Duistermaat 538 (55.17.24); 539 (55.17.14); 541 (55.17.73); 542 (55.17.50); 543 (55.17.53); 546 (55.17.18); 547 (55.17.36); 662 (55.17.54b).

Vermeulen & Lamb 350 (55.17.53); 353 (55.17.46); 357 (55.17.40); 359 (55.17.52); 365 (55.17.1).

Wood 602 (55.33.7a); 603 (55.66.9); 604 (55.11.11); 605 (55.37.30); 607 (55.28.18); 608 (55.37.17); 609 (55.28.20); 610 (55.115.9); 611 (55.113.7); 612 (55.28.22); 613 (55.17.76); 614 (55.22.3a); 616 (55.30.3); 617 (55.17.62); 618 (55.37.19); 619 (55.115.5); 620 (55.115.8); 621 (55.37.14); 622 (55.9.1); 623 (55.28.24); 824 (55.28.7); 825 (55.66.5); 826 (55.45.28); 827 (55.11.8); 831 (55.36.27); 833 (55.28.20); 834 (55.45.41); 837 (55.36.36); 838 (55.11.5); 839 (55.32.1); 840 (55.62.18); 842 (55.58.1); 844 (55.28.21); 845 (55.32.2); 846 (55.66.12); 847 (55.28.31); 849 (55.92.7); 850 (55.28.15); 851 (55.36.3); 852 (55.28.10).

Wood & Charrington 16370 (55.77.4).

GLOSSARY

ABAXIAL: the side away from the stem, normally the lower surface.

ABSCISSION: the process by which leaves and other organs fall from the plant after formation of a special abscission layer composed of easily broken cells.

ACAULESCENT: becoming stemless.

ACAULOUS: stemless or seemingly so.

ACTINOMORPHIC: having flowers of a regular or star pattern, capable of bisection in two or more planes into similar halves.

ACULEATE: prickle-shaped.

ACUMINATE: having a gradually tapering point.

ACUTE: distinctly and sharply pointed, but not drawn out.

ADAXIAL: the side toward the stem, normally the upper surface.

ADNATE: with one organ united with another.

ADPRESSED: lying flat for the whole length of the organ; appressed.

ALATE: winged.

AMPLEXICAUL: clasping the stem.

ANTHER/ANTHER-CAP: that part of the stamen in which the pollen is produced.

ANTHESIS: that period between the opening of the bud and the withering of the stigma or stamens.

ANTRORSE: turned backwards, directed upward.

APHYLLOUS: without leaves.

APICAL (inflorescence, leaves)**:** borne at the top of the stem or pseudobulb.

APICULATE: furnished with a short and sharp, but not stiff, point.

APPRESSED: lying flat for the whole length of the organ.

APPROXIMATE: drawn close together, but not united.

ARCUATE: curved like a bow.

ARISTATE: awned.

ARTICULATE: jointed.

AURICLE: a small lobe or ear, applied to the lip in e.g. *Malaxis*; also a small lateral outgrowth on the anther of the tribe Orchideae.

AUTOGAMY: self-pollination without the aid of insects or other agents.

AUTOPHYTE: any green plant capable of manufacturing its own food.

AXIL: the point at the angle between a leaf and a stem.

AXILLARY (inflorescence)**:** borne in the axil.

BASAL (inflorescence)**:** at the base of an organ or part such as the stem or pseudobulb.

BASIONYM: the original Latin base name given to an organism.

BICUSPIDATE: having 2 sharp points.

BIFID: divided into 2 shallow segments, usually at the apex.

BIFURCATE: forked.

BIPARTITE: divided nearly to the base into two portions.

BRACT: a frequently leaf-like organ (often very reduced or absent) bearing a flower, inflorescence or partial inflorescence in its axil, sometimes brightly coloured, e.g. *Eria ornata*.

BURSICLE: thc pouch-like expansion of the stigma into which the caudicle of the pollinarium is inserted, found in some Orchidaceae.

CADUCOUS: falling off early, e.g. bracts of certain *Coelogyne* and *Pholidota*.

CAESPITOSE: tufted.

CALCEIFORM or **CALCEOLATE:** slipper-shaped.

CALLUS: a thickened area on the lip.

CALYCULUS: a small cup or circle of bract-like structures outside of the sepals, e.g. in *Lecanorchis*.

CAMPANULATE: bell-shaped.

CANALICULATE: channelled, with a longitudinal groove.

CAPITATE: like a pin-head.

CARINATE: keeled.

CATAPHYLL: the early leaf-forms of a plant or shoot, such as cotyledons, bud-scales, rhizome-scales, etc.

CAUDATE: tailed.

CAUDICLE: the lower stalk-like part of a pollinium, attaching the pollen masses to the stipites or to the sticky disc (viscidium).

CAULESCENT: becoming stalked, where the stalk is clearly apparent.

CAULINE: borne on the stem.

CHLOROPHYLL: the green colouring matter of plants.

CILIATE: having fine hairs at the margin.

CLAVATE: club-shaped, thickened towards the apex.

CLAW: the conspicuously narrowed and attenuate base of an organ; in the orchids applied to the lip.

CLEISTOGAMOUS: flowers which automatically self-pollinate without opening; a form of autogamy.

CLINANDRIUM: the anther-bed, that part of the column in which the anther is concealed.

COCHLEATE: shaped like a snail shell.

COLUMN: an advanced composite structure derived by the fusion of the upper part of the female reproductive organ (pistil) and the lower part of the male reproductive organ (stamen).

COLUMN-FOOT: a ventral extension of the base of the column which has the lip attached at its tip.

COLUMN-WING: a wing- or arm-like appendage of the column, usually lateral; also called a stelid.

COMPLANATE: flattened, compressed.

CONDUPLICATE: folded face to face.

CONGENERIC: a genus identical with another genus.

CONNATE: united, congenitally or subsequently.

CONNECTIVE: the sterile portion of the anther between the two anther cells.

CONNIVENT: coming into contact or converging.

CONSPECIFIC: a species identical with another species.

CONVOLUTE: rolled.

CORDATE: heart-shaped.

CORIACEOUS: leathery.

COROLLA: the inner whorl of the flower.

CORYMB: flat-topped inflorescence.

CRENATE: scalloped, toothed with crenatures.

CRENULATE: crenate, but the teeth small.

CRISTATE: crested.

CUCULLATE: hooded, hood-shaped.

CUNEATE or **CUNEIFORM:** wedge-shaped.

CUSPIDATE: tipped with a sharp, rigid point.

CYMBIFORM: boat-shaped.

DECURRENT: running down, as when leaves are produced beyond their insertion, and thus run down the stem.

DEFLEXED: bent outwards.

DENTATE: toothed.

DENTICULATE: minutely toothed.

DETERMINATE: a habit of growth in which each unit has a limited growth; see sympodial.

DIAPHANOUS: permitting the light to penetrate.

DICOTYLEDON: plants characterised by having net-veined leaves and two seed-leaves, as opposed to monocotyledon.

DISC: the face of any organ, used with reference to the lip in the orchids and sometimes for the removable part of the rostellum projection (as in viscid disc).

DISTAL: away from the base, towards the apex.

DISTICHOUS: leaves, bracts or flowers borne in spikelets alternating in two opposite ranks.

DIVARICATE: extremely divergent.

DORSAL: refers to the upper side of the flower, in orchids, technically the abaxial side of the flower, because of resupination.

DORSIFIXED: fixed on the back or by the back.

DORSIVENTRAL: an organ which has dorsal and ventral surfaces, as a non-terete leaf.

ELLIPSOID: spindle-shaped; narrow and tapering to rounded ends, three-dimensional.

ELLIPTIC: spindle-shaped, two-dimensional.

EMARGINATE: notched, usually at the apex.

ENDEMIC: confined to a given region.

ENSIFORM: sword-shaped.

ENTIRE: simple and with smooth margins.

EPHEMERAL: short-lived, e.g. flowers of *Flickingeria.*

EPICHILE: the terminal part of the lip when it is distant from the basal portion.

EPIDENDROID: any orchid of the subfamily Epidendroideae.

EPIGAEOUS: growing upon the ground.

EPIPHYTE: a plant growing on another plant, but not parasitic.

EQUITANT: said of conduplicate or laterally flattened leaves or bracts which overlap each other in two ranks.

EROSE: bitten or gnawed.

EXALATE: without wings.

EXTINCTORIFORM: shaped like an old fashioned candle-snuffer.

FALCATE: sickle-shaped.

FARINOSE: mealy.

FASCICULATE: clustered or bundled.

FILAMENT: the slender, sterile portion of the stamen which bears the anther; part of the column in most orchids.

FILIFORM: thread-like.

FIMBRIATE: fringed.

FLABELLATE: fan-shaped.

FLEXUOSE: bent alternately in opposite directions.

FOVEA: a small depression or pit, e.g. in lip of *Malaxis.*

FRACTIFLEX: zig-zag.

FURFURACEOUS: scurfy, having soft scales.

FUSIFORM: spindle-shaped.

GALEATE: helmet-shaped.

GENICULATE: abruptly bent like a knee-joint.

GENUS: a taxonomic category above the rank of species.

GIBBOUS: pouched, more convex in one place than another.

GLABROUS: hairless.

GLAND: a secreting structure or a similar protuberance which may not secrete; alternative name for the viscidium.

GLAUCOUS: covered with a bluish grey or sea-green bloom.

GLOBOSE: spherical.

GRANULAR: refers to soft or mealy pollen.

GYNOSTEMIUM: alternative term for the column.

HAMATE: hooked at the tip.

HASTATE: halbert-shaped, i.e. arrow-shaped with the basal lobes turned outward.

HERBARIUM: a botanical museum; a collection of pressed and dried plant specimens arranged in taxonomic order.

HETERANTHOUS: an apical inflorescence produced on a separate shoot which does not develop to produce a pseudobulb and leaves.

HIPPOCREPIFORM: horseshoe-shaped.

HIRSUTE: hairy.

HISPID: bristly.

HYALINE: colourless or translucent.

HYPHAE: individual filaments of a fungal body.

HYPOCHILE: the basal portion of the lip.

HYSTERANTHOUS: an apical inflorescence produced after the pseudobulbs and leaves have developed.

IMBRICATE: overlapping.

INCUMBENT: lying on or against.

INDUMENTUM: any covering, e.g. hairiness.

INFERIOR: inserted below, e.g. the ovary in the orchids.

INFLORESCENCE: the disposition of the flowers on the floral axis or the flowers, bracts and floral axis in toto.

INTERNODE: the section of a stem between two nodes.

ISTHMUS: a narrow portion of a lip or petal.

LABELLUM: a lip, the usually enlarged, often highly modified abaxial petal of the orchid flower.

LACERATE: torn, or irregularly cleft.

LACINIATE: deeply slashed into narrow divisions.

LAMELLA: a membrane or septum.

LAMINA: a blade, the expanded portion of a leaf.

LANCEOLATE: narrow, tapering to each end, lance-shaped.

LATERAL: borne on or near the side.

LAXLY: loose, distant.

LIGULATE: strap- or tongue-shaped.

LINEAR: at least 12 times longer than broad, with the sides parallel.

LIP: alternative term for the labellum.

LITHOPHYTE: growing upon stones or rocks.

LOBULE: a small lobe.

LUNATE: half-moon-shaped.

MEDIAN: on the midline of a bilaterally symmetrical organ.

MENTUM: a chin-like projection formed by the sepals and extended column-foot.

MERISTEM: tissue which retains the capacity for further growth; in monocotyledons, normally in buds and at growing points.

MESOCHILE: the middle portion of a complex lip.

MONANDROUS: having a single anther, the case in all the orchid subfamilies except Apostasioideae and Cypripedioideae.

MONOCOTYLEDON: plants usually characterised by parallel venation and a single seed leaf (absent in orchids).

MONOPODIAL: growth which continues from a terminal bud from season to season.

MUCRO: a sharp terminal point.

MURICATE: rough, with short and hard tubercular excrescences.

MYCORRHIZA: certain fungi having a symbiotic relationship with plants and found in the roots of those plants.

NAVICULAR: boat-shaped.

NECTARY: a nectar-producing structure or gland.

NEOTTIOID: any terrestrial orchid with an erect anther and soft pollinia; any member of the tribe Neottieae.

NERVOSE: prominently nerved.

NODE: that part of a stem which normally bears a leaf or a whorl of leaves.

NON-RESUPINATE: a flower that is not turned upside down.

OBCORDATE: inversely heart-shaped.

OBCUNEATE: inversely wedge-shaped.

OBLANCEOLATE: tapering toward the base more than toward the apex.

OBLONG: much longer than broad, with nearly parallel sides.

OBOVATE: inversely ovate.

OBSOLETE: wanting or rudimentary.

OBTRULLIFORM: inversely trowel-shaped.

OBTUSE: blunt or rounded at the apex.

ONTOGENY: the development of an individual in its various stages.

OPERCULUM: a lid, alternative term for anther-cap.

ORBICULAR: of a flat body with a circular outline.

OVARY: the part of the flower that develops into the fruit.

OVATE: egg-shaped, broader at the base, two-dimensional.

OVOID: egg-shaped, three-dimensional.

OVULE: embryonic seed.

PALEA: flat or terete, leaf-like, moveable appendages attached by a filiform base, found on the sepals and petals in many species of *Bulbophyllum* section *Cirrhopetalum*.

PANDURATE: fiddle-shaped.

PANICLE: a much-branched inflorescence.

PAPILLAE: soft superficial glands or protuberances.

PAPILLOSE: covered with papillae.

PECTINATE: combed.

PEDICEL: the stalk of a single flower in an inflorescence.

PEDUNCLE: the stalk bearing an inflorescence or solitary flower.

PELORIC: an abnormality in which the lip is similar to the petals in form, or vice versa.

PENDENT: hanging.

PERIANTH: the outer, sterile whorls of a flower, often differentiated into calyx and corolla.

PETAL: a white or coloured flower part borne within the sepals; in orchids two of the three inner perianth parts, the third being called the lip.

PETIOLE: leaf stalk.

PILOSE: softly hairy.

PISTIL: the female, or seed-bearing, element in a flower, i.e. the ovary, style and stigma.

PLACENTA: that portion of the ovary that bears the ovules.

PLICATE: folded into many pleats; appearing corrugated.

POLLINARIUM: the male reproductive system of an orchid.

POLLINIA: pollen-masses.

PORATE: set with a pore or pores.

PORRECT: directed outward and forward.

PRAEMORSE: bitten off at the apex.

PROSTRATE: lying flat.

PROTERANTHOUS: of an apical inflorescence produced before the pseudobulbs and leaves on the same shoot.

PROTOCORM: the embryo before primary differentiation is complete.

PROXIMAL: the part nearest the axis.

PSEUDOBULB: a swollen aerial stem.

PSEUDOCOPULATION: a special type of mimicry in which flowers resemble female insects and are pollinated by the males when these attempt to copulate with the flowers.

PSEUDOPOLLEN: mealy, farinose, pollen-like deposit, e.g. on the lip in species of *Eria* section *Mycaranthes,* and *Polystachya.*

PUBESCENT: hairy.

PUSTULATE: having blisters.

PYRIFORM: pear-shaped.

QUADRATE: four-cornered, square.

QUAQUAVERSAL: directed or bending in every direction.

RACEME: a single, elongate, indeterminate inflorescence with pedicellate flowers.

RACHIS: the axis of an inflorescence or compound leaf.

RAMENTACEOUS: possessing thin chaffy scales.

RAPHIDE: needle-like crystal, usually of calcium oxalate, forming bundles in many orchid cells.

RECUMBENT: bent back until the apex is below the base.

REFLEXED: abruptly bent or turned downward or backward.

RENIFORM: kidney-shaped.

RESUPINATE: upside down, or apparently so; applied to the lip, which is usually positioned at the bottom of the flower due to the ovary twisting through 180°.

RETICULATE: netted.

RETRORSE: directed backwards or downwards.

RETUSE: shallowly notched at a rounded apex.

REVOLUTE: rolled back from the margin.

RHIZOME: horizontal, underground or surface-resting stem bearing scale leaves and adventitious roots; in sympodial orchids composed of the bases of successive shoots.

ROOT-STEM TUBEROID: a storage organ, primarily a swollen root, but with a bud and some stem structure at the base; it may push down into the soil placing the plant lower in the soil; also called a sinker or tuber.

ROSETTE: a densely clustered spiral of leaves, usually borne near the ground.

ROSTELLUM: the often beak-like sterile third stigma lying between the functional stigmas and stamen.

ROSTRATE: beaked.

RUGOSE: wrinkled.

RUGULOSE: somewhat wrinkled.

SACCATE: with a conspicuous hollow swelling.

SAGITTATE: arrowhead-shaped.

SAPROPHYTE: a plant which obtains its food materials by absorption of complex organic chemicals from the soil; often without chlorophyll, e.g. *Gastrodia.*

SCANDENT: climbing.

SCAPE: a leafless peduncle arising directly from a rosette of basal leaves; the flowerless lower part of an inflorescence.

SCARIOUS: thin, dry and membranous.

SECTILE: the condition where soft, granular pollinia are subdivided into small packets, usually connected by elastic threads.

SECUND: directed to one side only.

SEPAL: an outer, usually green, perianth segment of a flower; in orchids they are often brightly coloured.

SEPTUM: a partition, e.g. in the lip spur or sac in many Aeridinae.

SERRATE: with sharp, regular teeth, like a saw.

SERRULATE: minutely toothed.

SESSILE: stalkless.

SETA: a bristle.

SIGMOID: S-shaped.

SPATHULATE: oblong and attenuated at the base, like a spatula.

SPIRANTHOID: like *Spiranthes*, any member of the subtribe Spiranthinae.

SPUR: a long or short, usually nectar-containing, tubular projection of a perianth segment, commonly the lip.

STAMEN: the male, or pollen-bearing, element in a flower, made up of filament and anther.

STAMINODE: a sterile stamen, often modified in shape and size.

STELIDIA: column teeth or arms.

STELLATE: star-shaped.

STIGMA: the receptive part or parts of the female sex organs, on which the pollen germinates.

STIGMATOPHORE: an elongated process on either side of the base of the column, often joined to the base of the lip, e.g. in *Habenaria.*

STIPES (plural stipites): the strap of stalk-like columnar tissue which connects the pollinia to the viscidium.

STYLE: the slender part of the pistil which connects the ovary with the stigma; forms a part of the column in orchids.

SUBTEND: to extend under; subtending leaf or bract; the leaf or bract in whose axil a bud or inflorescence is formed.

SUBTERMINAL: just below the apex.

SUBULATE: awl-shaped.

SULCATE: grooved or furrowed.

SUPERPOSED: one above the other.

SYMBIOSIS: an ecological relationship between two different organisms in which both obtain mutual benefit.

SYMPODIAL: growth in which each new shoot is determinate and terminates in a potential inflorescence or solitary flower.

SYMPODIAL MODULE: a sympodial branching system.

SYNANTHOUS: when pseudobulb, leaf and apical inflorescence are produced together.

SYNSEPALUM: two sepals united together, e.g. in *Paphiopedilum.*

TAXON: a taxonomic group of any rank.

TAXONOMY: the science of classification and naming.

TEPAL: a division of the perianth, i.e. sepal or petal.

TERATOLOGICAL: malformed, abnormal.

TERETE: circular in transverse section, cylindric and usually tapering.

TERMINAL: the end or apex.

TERRESTRIAL: ground dwelling.

TESSELLATE: chequered.

THECAE: the hollow spaces in the anther (operculum) into which the pollinia are located.

TOMENTOSE: densely covered with matted woolly or short hairs.

TRIDENTATE: three-toothed.

TRULLATE (trulliform): trowel-shaped.

TRUNCATE: ending abruptly, as though broken off.

TUBER: a thickened and short subterranean branch, beset with buds or 'eyes'. See also root-stem tuberoid.

TYPE: the representative of a group, genus or species, the type specimen being the original specimen from which a description was made and a scientific name applied.

UNCINATE: hook-shaped.

UNGUICULATE: contracted at the base into a claw.

UNINODAL: of one node.

URCEOLATE: urn- or pitcher-shaped.

VANDOID: like *Vanda*, any member of the tribe Vandeae.

VELAMEN: one or more layers of spongy dead cells on the outside of a root, related to the epidermis in origin, which may act as protective insulation.

VENTRAL: on the lower side.

VENTRICOSE: swollen or inflated on one side.

VERRUCOSE: warty.

VERSATILE: hinged and moveable, e.g. the lip in most *Bulbophyllum*.

VISCIDIUM (plural: viscidia): a sticky, disc-like part of the rostellum joined to the pollinium, enabling it to adhere to an insect's body during cross-pollination.

WHORL: a circle of three or more leaves or leaf-like organs attached at the same level on a stem.

ZYGOMORPHIC: bilaterally symmetrical.

APPENDIX

After the manuscript was in page proof we received further information from Mr. Anthony Lamb which results in the addition of 16 species. Another species, *Octarrhena condensata,* is added as a result of redetermination of a specimen in SING. Mr. Lamb has also provided 59 more specimen records that add new locality data for 49 taxa in 28 genera. Some of these specimens also provide documentation for seven taxa included as sight records in the Enumeration (see p. 7, paragraph 1). The collections for which Lamb is the collector are to be deposited in K but were not there when the manuscript was completed.

Data presented in the Introduction (except for the total number of taxa indicated in the last paragraph on page 4 and Table 1) have not been revised to account for these new records. We have learned that specimens cited in the Enumeration with Lohok as the collector were actually made by Dr. P. Cribb. The specimens cited below, however, are actual Lohok collections, numbered differently from those attributed to him in the Enumeration. All of the Lohok specimens seen are spirit collections from plants growing in the Poring Orchid Garden.

ADDITIONAL TAXA

55.12. ARACHNIS

Arachnis breviscapa (J. J. Sm.) J. J. Sm., Natuurk. Tijdschr. Ned.-Indië 72: 74 (1912). Plate 9B.

Arachnanthe breviscapa J. J. Sm., Bull. Dép. Agric. Indes Néerl. 22: 48 (1909).

Epiphyte. Hill forest on ultramafic substrate.

General distribution: Sabah, Sarawak. Elevation: 500 m.

Collection. LOHAN RIVER: 500 m, *Lamb SAN 91518.*

55.17. BULBOPHYLLUM

Bulbophyllum auratum (Lindl.) Rchb. f., Walp. Ann. Bot. Syst. 6: 261 (1861).

Cirrhopetalum auratum Lindl., Bot. Reg. 26, misc. 50, 107 (1840).

Epiphyte. Lowlands.

General distribution: Sabah, Brunei, Kalimantan, Peninsular Malaysia, Thailand, Sumatra.

Collection. MARAK PARAK: *Lohok 91/0345* (SNP).

Bulbophyllum botryophorum Ridl., J. Linn. Soc. Bot. 32: 275 (1896).

Epiphyte. Hill forest on ultramafic substrate. Elevation: 700 m.

General distribution: Sabah, Peninsular Malaysia, Singapore.

Collection. HEMPUEN HILL: 700 m, *Lamb & Surat AL 1290/90.*

Bulbophyllum aff. **hymenanthum** Hook. f., Fl. Brit. Ind. 5: 767 (1890).

Epiphyte. Lower montane forest. Elevation: 1600 m.

Collection. PARK HEADQUARTERS: 1600 m, *Lamb AL 1236/90.*

55.26. CLEISOCENTRON

Cleisocentron sp. 1. Plate 21C.

Epiphyte. Lower montane forest on ultramafic substrate.

This species is represented by a sketch by Lamb (*AL 1211/90)* and a colour slide. The sketch was made from a plant cultivated in the Mountain Garden at Park Headquarters and the slide was taken on Marai Parai Spur. No preserved material has been available for examination.

55.40. DIMORPHORCHIS

Dimorphorchis lowii (Lindl.) Rolfe, Orchid Rev. 27: 149 (1919).

Vanda lowii Lindl., Gard. Chron. 1: 239 (1847).

Epiphyte. Lowlands.

General distribution: Sabah, Kalimantan, Sarawak.

A plant from the Poring area is in cultivation at the Poring Orchid Garden, but was not in flower and a specimen of it had not been prepared at the time the manuscript was completed.

55.45. ERIA

Eria latibracteata Ridl., J. Linn. Soc. Bot. 32: 293 (1896).

Epiphyte. Hill forest on ultramafic substrate. Elevation: 500 m.

General distribution: Sabah, Peninsular Malaysia.

Collection. LOHAN RIVER: 500 m, *Shim & Lamb SAN 93352.*

55.48. FLICKINGERIA

Flickingeria luxurians (J. J. Sm.) A. D. Hawkes, Orchid Weekly 2 (46): 457 (1961).

Dendrobium luxurians J. J. Sm., Bull. Jard. Bot. Buit., ser. 3, 3: 288, t. 4, f. 2 (1921).

Epiphyte. Hill forest on ultramafic substrate. Elevation: 600 m.

General distribution: Sabah, Kalimantan, Sarawak, Peninsular Malaysia, Java, Philippines.

Collection. HEMPUEN HILL: 600 m, *Lamb & Surat AL 1306/91.*

55.79. OCTARRHENA

Octarrhena condensata (Ridl.) Holttum, Gardens' Bull. 11: 285 (1947).

Oberonia condensata Ridl., J. Linn. Soc. Bot. 38: 322 (1908).

Epiphyte. Hill forest. Elevation: 1100 m.

General distribution: Sabah, Kalimantan, Sarawak, Peninsular Malaysia, Sumatra, Java, New Guinea, Solomon Islands, Vanuatu.

Collection. MAHANDEI RIVER: 1100 m, *Carr SFN 26509* (SING).

55.88. PHALAENOPSIS

Phalaenopsis fuscata Rchb. f., Gard. Chron., ser. 2, 2: 6 (1874). Plate 71A.

Epiphyte. Hill forest. Elevation: 600 m.

General distribution: Sabah, Kalimantan, Peninsular Malaysia, Philippines.

Collection. LOHAN/MAMUT COPPER MINE: 600 m, *Shim & Lamb s.n.*

55.97. POMATOCALPA

Pomatocalpa kunstleri (Hook. f.) J. J. Sm., Natuurk. Tijdschr. Ned.-Indië 72: 104 (1912).

Cleisostoma kunstleri Hook. f., Icones Pl. 24, t. 2335 (1894).

Epiphyte. Lowlands.

General distribution: Sabah, Brunei, Kalimantan, Sarawak, Peninsular Malaysia, Thailand, Sumatra, Java, Mentawai, Philippines.

Collection. MARAK PARAK: *Lohok 91/0341* (SNP).

Pomatocalpa spicata Breda, Gen. Sp. Orchid. Asclep. t. 15 (1827).

Epiphyte. Lowlands.

General distribution: Sabah, Kalimantan, Sarawak, widespread from India, Andaman Islands to Burma and Indochina east to the Philippines, north to China (Hainan).

Collection. PORING: *Lohok 92/0198* (SNP).

55.101. PTEROCERAS

Pteroceras aff. **cladostachyum** (Hook. f.) H. A. Pedersen, ined.

Sarcochilus cladostachyum Hook. f., Fl. Brit. Ind. 6: 35 (1890).

Epiphyte or lithophyte. Lowlands.

Collection. MARAK PARAK: *Lohok 91/0340* (SNP).

55.108. TAENIOPHYLLUM

Taeniophyllum aff. **micranthum** Carr, Gardens' Bull. 7: 75, pl. 11B (1932).

Epiphyte. Hill forest on ultramafic substrate. Elevation: 600 m.

Collection. HEMPUEN HILL: 600 m, *Lamb AL 507/85*.

55.112. THELASIS

Thelasis pygmaea (Griff.) Blume, Fl. Javae, ser. 2, 1, Orch.: 22 (1858).

Euproboscis pygmaea Griff., Calcutta J. Nat. Hist. 5: 371, t. 26 (1845).

Epiphyte. Lowlands.

General distribution: Sabah, Kalimantan, Sarawak, widespread in S and SE Asia east to New Guinea and the Solomon Islands.

Collection. PORING: *Lohok 91/0244* (SNP).

55.113. THRIXSPERMUM

Thrixspermum acuminatissimum (Blume) Rchb. f., Xenia Orch. 2: 121 (1867).

Dendrocolla acuminatissima Blume, Bijdr., 288 (1825).

Epiphyte. Lowlands?

General distribution: Sabah, Sarawak, Peninsular Malaysia, Singapore, Thailand, Cambodia, Sumatra, Java, Mentawai, Philippines.

Collection. MARAK PARAK?: *Lohok 91/0354* (SNP).

55.114. TRICHOGLOTTIS

Trichoglottis maculata (J. J. Sm.) J. J. Sm., Bull. Jard. Bot. Buit., ser. 2, 26: 106 (1918).

Trichoglottis lanceolaria Blume var. *maculata* J. J. Sm., Orch. Java, 619 (1905).

Epiphyte. Hill forest on ultramafic substrate.

General distribution: Sabah, Peninsular Malaysia, Sumatra, Java.

Collection. HEMPUEN HILL: *Lohok 87/0141* (SNP).

ADDITIONAL LOCALITY RECORDS

55.2. ACANTHEPHIPPIUM

55.2.1. Acanthephippium javanicum Blume

Collection. PORING: *Lohok 87/0404* (SNP).

55.3. ACRIOPSIS

55.3.1. Acriopsis indica Wight

Collection. PINOSUK PLATEAU: 1800 m, *Lamb AL 1399/91.*

55.4. ADENONCOS

55.4.1. Adenoncos sp. 1

Collections. HEMPUEN HILL: 600 m, *Lamb & Surat AL 1242/90*; PORING: *Lohok 92/0014* (SNP).

55.5. AERIDES

55.5.1. Aerides odorata Lour.

Collection. PORING: *Lohok 87/0031* (SNP).

55.9. APHYLLORCHIS

55.9.2. Aphyllorchis pallida Blume

Collection. PARK HEADQUARTERS: 1500 m, *Lamb SAN 87133.*

55.11. APPENDICULA

55.11.4. **Appendicula congesta** Ridl. in Stapf

Collection. PENATARAN RIVER: 600 m, *Lamb SAN 93374.*

55.12. ARACHNIS

55.12.3. **Arachnis longisepala** (J. J. Wood) Shim & A. Lamb

Collection. HEMPUEN HILL: 900 m, *Lamb SAN 93373.*

55.17. BULBOPHYLLUM

55.17.8. **Bulbophyllum biflorum** Teijsm. & Binn.

Collections. LOHAN RIVER: 600 m, *Lamb SAN 91522*; PORING: *Lohok 91/0097* (SNP).

55.17.11. **Bulbophyllum carinilabium** J. J. Verm.

Collection. PINOSUK PLATEAU: 1500–1800 m, *Lamb AL 770/87.*

55.17.31. **Bulbophyllum flavescens** (Blume) Lindl.

Collection. PARK HEADQUARTERS: 1600 m, *Lamb AL 831/87.*

55.17.34. **Bulbophyllum gibbosum** (Blume) Lindl.

Collection. KUNDASANG: 1200 m, *Lamb & Surat AL 1210/90.*

55.17.61. **Bulbophyllum planibulbe** (Ridl.) Ridl.

Collection. HEMPUEN HILL: 700 m, *Lamb & Surat AL 1300/91.*

55.17.65. **Bulbophyllum purpurascens** Teijsm. & Binn.

Collections. HEMPUEN HILL: 600 m, *Lamb & Surat AL 1240/90, AL 1289/90.*

55.17.68. **Bulbophyllum salaccense** Rchb. f.

Collection. EAST MESILAU RIVER: 1800 m, *Lamb AL 578/86.*

55.17.73. Bulbophyllum sopoetanense Schltr.

Collection. PARK HEADQUARTERS: 1600 m, *Lamb AL 584/86.*

55.17.81. Bulbophyllum trifolium Ridl.

Collection. HEMPUEN HILL: 600 m, *Lamb & Surat AL 1273/90.*

55.21. CHAMAEANTHUS

55.21.1. Chamaeanthus brachystachys Schltr. in J. J. Sm.

Collections. HEMPUEN HILL: 600 m, *Lamb AL 407/85*, 900 m, *AL 1239/90.*

55.24. CHRYSOGLOSSUM

55.24.1. Chrysoglossum reticulatum Carr

Collection. KIAU VIEW TRAIL: 1700 m, *Lamb 87136.*

55.27. CLEISOSTOMA

55.27.2. Cleisostoma discolor Lindl.

Collection. HEMPUEN HILL: *Lohok 91/0540* (SNP).

55.27.5. Cleisostoma striatum (Rchb. f.) Garay

Collection. PORING: *Lohok 91/0209* (SNP).

55.28. COELOGYNE

55.28.6. Coelogyne dayana Rchb. f.

Collection. PARK HEADQUARTERS: 1500 m, *Lamb SAN 88561.*

55.28.8. Coelogyne foerstermannii Rchb. f.

Collection. LOHAN RIVER: 500 m, *Chan & Tan AL 162/83.*

55.28.15. Coelogyne moultonii J. J. Sm.

Collection. PARK HEADQUARTERS: 1500 m, *Lamb SAN 87491.*

55.28.22. Coelogyne radioferens Ames & C. Schweinf.

Collection. PINOSUK PLATEAU: 1800 m, *Lamb AL 1398/91.*

55.32. CRYPTOSTYLIS

55.32.2. Cryptostylis arachnites (Blume) Hassk.

Collection. PARK HEADQUARTERS: 1600 m, *Lamb SAN s.n.*

55.33. CYMBIDIUM

55.33.1. Cymbidium atropurpureum (Lindl.) Rolfe

Collection. PORING: *Lohok 88/0075* (SNP).

55.36. DENDROBIUM

55.36.5. Dendrobium anosmum Lindl.

Collections. LOHAN RIVER: 500 m, *Lamb SAN 93356, Lohok 87/0075* (SNP).

55.36.32. Dendrobium pachyanthum Schltr.

Collection. HEMPUEN HILL: 800–900 m, *Lamb AL 65/83.*

55.36.35. Dendrobium panduriferum Hook. f.

Collections. LOHAN RIVER: 500 m, *Lamb AL 94/83, AL 1109/89.*

55.36.42. Dendrobium sanguinolentum Lindl.

Collection. HEMPUEN HILL: 600 m, *Lamb SAN 93365.*

55.36.43. Dendrobium secundum (Blume) Lindl.

Collection. HEMPUEN HILL: *Lohok 91/0251* (SNP).

55.44. EPIPOGIUM

55.44.1. **Epipogium roseum** (D. Don) Lindl.

Collection. LOHAN RIVER: 800 m, *Lamb SAN 91512*; PORING: *Lohok 92/0318* (SNP).

55.45. ERIA

55.45.19. **Eria javanica** (Sw.) Blume

Collection. PARK HEADQUARTERS: 1500 m, *Lamb SAN 88559.*

55.45.34. **Eria ornata** (Blume) Lindl.

Collections. LOHAN RIVER: 500 m, *Lamb SAN 91520*; PORING: *Lohok 91/0241* (SNP).

55.53. GOODYERA

55.53.8. **Goodyera ustulata** Carr

Collection. PORING: *Lohok 92/0320* (SNP).

55.73. NABALUIA

55.73.2. **Nabaluia clemensii** Ames

Collection. PARK HEADQUARTERS: 1500 m, *Lamb? s.n.*

55.75. NEPHELAPHYLLUM

55.75.2. **Nephelaphyllum pulchrum** Blume

Collection. PARK HEADQUARTERS: 1800 m, *Lamb SAN 88554.*

55.83. PAPHIOPEDILUM

55.83.1. **Paphiopedilum dayanum** (Lindl.) Stein.

Collection. PENATARAN RIVER: 1100 m, *Lamb SAN 93358.*

55.83.2. Paphiopedilum hookerae (Rchb. f.) Stein

a. var. **volonteanum** (Sander ex Rolfe) Kerch.

Collection. HEMPUEN HILL: 800 m, *Lamb SAN 93353.*

55.83.4. Paphiopedilum lowii (Lindl.) Stein

Collection. RANAU: 500 m, *Lamb SAN 93358.*

55.88. PHALAENOPSIS

55.88.1. Phalaenopsis amabilis (L.) Blume

Collection. PORING: *Lohok 91/0332* (SNP).

55.88.3. Phalaenopsis modesta J. J. Sm.

Collection. SAYAP: *Lohok 92/0295* (SNP).

55.91. PILOPHYLLUM

55.91.1. Pilophyllum villosum (Blume) Schltr.

Collection. SAYAP: *Lohok 92/0319* (SNP).

55.101. PTEROCERAS

55.101.1. Pteroceras fragrans (Ridl.) Garay

Collection. PORING: *Lohok 91/0505* (SNP).

55.102. RENANTHERA

55.102.2. Renanthera matutina (Blume) Lindl.

Collection. PORING, *Lohok 87/0254* (SNP).

55.108. TAENIOPHYLLUM

55.108.7. **Taeniophyllum stella** Carr

Collection. LOHAN RIVER: 600 m, *Lamb AL 424/85.*

55.113. THRIXSPERMUM

55.113.6. **Thrixspermum pensile** Schltr.

Collection. LOHAN RIVER: 600 m, *Lamb AL 1238/90.*

55.114. TRICHOGLOTTIS

55.114.2. **Trichoglottis collenetteae** J. J. Wood, C. L. Chan & A. Lamb, sp. nov.

Collection. PENATARAN RIVER: 600 m, *Lamb SAN 93375*; LANGANAN FALLS: *Lohok 89/0004* (SNP).

55.114.10. **Trichoglottis winkleri** J. J. Sm.

a. var. **minor** J. J. Sm.

Collections. HEMPUEN HILL: 600 m, *Lamb & Surat AL 1241/90; Lohok 87/0139* (SNP).

INDEX TO SCIENTIFIC NAMES

Accepted names are in roman type. Synonyms are in *italics*. Taxa not recorded from Mount Kinabalu but mentioned in the text are preceded by a short line. Numbers in **bold** refer to figures. Plates are designated "pl."